CHALLENGES AND INNOVATIONS IN GEOTECHNICS

PROCEEDINGS OF THE 8TH ASIAN YOUNG GEOTECHNICAL ENGINEERS CONFERENCE, ASTANA, KAZAKHSTAN, AUGUST 5–7, 2016

Challenges and Innovations in Geotechnics

Editor

Askar Zhussupbekov
Kazakhstan Geotechnical Society, Astana, Kazakhstan

CRC Press
Taylor & Francis Group
Boca Raton London New York Leiden

CRC Press is an imprint of the
Taylor & Francis Group, an **informa** business
A BALKEMA BOOK

Cover Illustration
Bayterek Tower in Astana, symbol of Kazakhstan
Copyright: Tropinina Olga
Courtesy of www.shutterstock.com

CRC Press/Balkema is an imprint of the Taylor & Francis Group, an informa business

Typeset by MPS Limited, Chennai, India
Printed and bound in Great Britain by CPI Group (UK) Ltd, Croydon, CR0 4YY

Published by: CRC Press/Balkema
P.O. Box 11320, 2301 EH Leiden, The Netherlands
e-mail: Pub.NL@taylorandfrancis.com
www.crcpress.com – www.taylorandfrancis.com

ISBN: 978-1-138-03007-7 (Hbk)
ISBN: 978-1-4987-9974-4 (eBook PDF)

Challenges and Innovations in Geotechnics – Zhussupbekov (Ed.)
© 2016 Taylor & Francis Group, London, ISBN 978-1-138-03007-7

Table of contents

Ground improvement (TS3)

Earthquake and environment (TS4)

Numerical and analytical modeling (TS5)

Advanced soil mechanics (TS6)

Challenges and Innovations in Geotechnics – Zhussupbekov (Ed.)
© 2016 Taylor & Francis Group, London, ISBN 978-1-138-03007-7

Preface

The 8th Asian Young Geotechnical Engineers Conference (8AYGEC) is organized by the Kazakhstan Geotechnical Society (KGS) under the joint auspices of the International Society for Soil Mechanics and Geotechnical Engineering (ISSMGE) and Architecture and Construction Faculty, the Gumilyov Eurasian National University and School of Engineering, Nazarbayev University, Kazakhstan. This conference is the eighth in the series following the first in Bangkok (1991), the second in Singapore (1994), the third in Singapore (1997), the fourth in Seoul (2001), the fifth in Taipei (2004), the sixth in Bangalore (2008) and the seventh in Tokushima (2012). This conference aims at providing interactive possibility for the young geotechnical engineers from Asian regions of ISSMGE.

All the Asian member societies of ISSMGE were invited to nominate two each or more geotechnical engineers and/or researchers under the age of 35 years. The nominees were expected to submit written papers and to present as oral speakers their research works at the conference. About 50 engineers and researchers were nominated by 14 societies in Asia, 3 societies in Europe and one society from United States of America. By this time, the nominees were sent from geotechnical societies of Belorussia, China, India, Germany, Japan, Kazakhstan, Kyrgyzstan, Nepal, Pakistan, Russia, Saudi Arabia, Singapore, South Korea, Syria, Taiwan, Tajikistan, USA and Uzbekistan. The accepted papers cover the areas of Laboratory and Field Testing, Foundation and Underground Structure, Ground Improvement, Earthquake and Environment, Numerical and Analytical Modeling, Advanced Soil Mechanics and Historical Sites.

The general theme of the 8AYGEC has been set as Challenges and Innovations in Geotechnics, which is related to TC 305 of ISSMGE "Geotechnical infrastructure for megacities and new capitals".

On behalf of the Organizing Committee I would like to thank the Scientific Committee and the International Advisory Committee for their great contributions in making this conference a success. We would particularly like to thank Professor Ikuo Towhata, the Vice-President for Asia; Mr. Shigeo Katsu, the President of Nazarbayev University and Professor Asemgul Moldazhanova, Vice Rector of Eurasian National University, for their efforts to planning and organizing this conference. We also sincerely acknowledge the financial sponsorship received from various government institutions, private companies and individuals. Our deep appreciations are also due to all the national societies that have nominated their delegates, to the nominees who have submitted their papers and are going to participate in the conference, to the reviewers who have strongly reviewed and rated the submitted papers, to the senior participants who are going to chair the technical sessions and to the authors of invited lectures for their special contributions to the conference. Finally, we would like to express our special thanks to all members of the local organizing committee and the assistants for their preparations of this conference as well.

Prof. Askar Zhussupbekov,
Chairman of Organizing Committee of the 8AYGEC,
President of Kazakhstan Geotechnical Society

Challenges and Innovations in Geotechnics – Zhussupbekov (Ed.)
© 2016 Taylor & Francis Group, London, ISBN 978-1-138-03007-7

Conference organizers

PUBLISHED BY

The Kazakhstan Geotechnical Society
The International Society for Soil Mechanics and Geotechnical Engineering

LOCAL ORGANIZING COMMITTEE

Prof. Askar Zhussupbekov, Eurasian National University
Prof. Jong Kim, Nazarbayev University
Prof. Anatoli Vakhguelt, Nazarbayev University
Prof. Seriktay Baimukhanov, Eurasian National University
Prof. Talgat Baytassov, Eurasian National University
Mrs. Natalia Zhumagazina, KGS, Ltd
Ms. Ainur Sabirova, KGS, Ltd
Dr. Assem Issina, Eurasian National University
Dr. Asel Tulebekova, Eurasian National University
Dr. Minh Nguyen, Nazarbayev University
Ms. Bibigul Abdrakhmanova, Geotechnical Institute
Dr. Dichuan Zhang, Nazarbayev University
Dr. Sudheesh Thiyya Kkandi, Nazarbayev University
Ms. Aigerim Baidullayeva, Nazarbayev University

SCIENTIFIC COMMITTEE

Mr. Shigeo Katsu, President of Nazarbayev University
Prof. Yerlan Sydykov, Rector of Eurasian National University
Prof. Asemgul Moldazhanova, Vice Rector for Educational work, Eurasian National University
Prof. Victor Popov, General Director of Karaganda GIIZ and K, Ltd
Dr. Gennady Sultanov, Senior Engineer of "PI, Bazis A"
Prof. Bagdat Teltayev, President of KazDorNII, Ltd.
Prof. Adil Zhakulin, Karaganda State Technical University
Prof. Vitaly Khomyakov, Kazakhstan Leading Academy of Construction and Architecture
Dr. Rauan Lukpanov, Eurasian National University
Dr. Zhanbolat Shakhmov, Eurasian National University

INTERNATIONAL ADVISORY COMMITTEE

Prof. Ikuo Towhata, Vice President for Asia, ISSMGE
Prof. Roger Frank, President of ISSMGE
Prof. Neil Taylor, Secretary General of ISSMGE
Prof. Askar Zhussupbekov, President of Kazakhstan Geotechnical Society
Prof. Yongkyu Choi, President of Korean Geotechnical Society
Prof. Chang-Yu Ou, President of Chinese Taipei Geotechnical Society
Dr. Ng Tiong Guan, President of Geotechnical Society of Singapore
Dr. Suttisak Soralump, President of Thai Geotechnical Society
Prof. T.G. Sitharam, Indian Institute of Science
Prof. Uzuoka Ryosuke, Tokushima University

Challenges and Innovations in Geotechnics – Zhussupbekov (Ed.)
© 2016 Taylor & Francis Group, London, ISBN 978-1-138-03007-7

Supporting organizations

SIMSG ISSMGE

ISSMGE TC305
Technical Committee on Geotechnical Infrastructure for Megacities and New Capitals

KGS

NAZARBAYEV UNIVERSITY

20 ANNIVERSARY
L.N. GUMILYOV EURASIAN NATIONAL UNIVERSITY

Special Lecturers

Challenges and Innovations in Geotechnics – Zhussupbekov (Ed.)
© 2016 Taylor & Francis Group, London, ISBN 978-1-138-03007-7

Future challenges for geotechnical engineers (a more contractor orientated perspective)

W. Sondermann
German Geotechnical Society (DGGT), Essen, Germany
Board Keller Group plc, London, UK

C. Kummerer
EMEA division, Keller Holding GmbH, Offenbach, Germany

ABSTRACT: Current market conditions creating new future challenges for the construction industry in combination with the decreasing demand for construction at least in well-developed regions, it becomes more and more important to gain competitive advantages through innovative solutions in order to deal with the continuing pressure on costs, schedule and quality also. For geotechnical engineers the main future challenges are to be familiar with theoretical as well as practical geotechnical applications to identify the optimal combination of technologies including the best design approach to be competitive for a specific project.

1 CURRENT MARKET CONDITIONS AND TRENDS

The ongoing financial crisis in the western world and the efforts not only to consolidate national budgets, but also to rescue certain banks, have not left the building industry unaffected either. Some drastic cutbacks to public budgets, along with the increasingly restrictive granting of credit for the financing of construction projects, have created enormous over-capacities in the construction sector, with purely negative impacts. As a consequence of these negative developments, many companies are experiencing the pressure to restructure more quickly and to reduce costs, but at the same time to keep the consequences of all activities, both externally and internally, within limits acceptable to employees, owners and finance institutions and last but not least costumers. The construction industry is not only undergoing substantial changes because of these basic alterations in the global conditions as during the coming years and decades additional transformations will influence the development:

- the **acceleration in globalisation** must be reckoned, with the economic output of developing countries growing about 3 times faster than those of the industrial nations,
- the **mega centres** of the world will continue to attract more workers and infrastructure problems will accompany the corresponding development.
- **Energy requirements** will increase dramatically and require a more efficient and stable infrastructure.
- The **acquisition of capital** will increasingly move towards private financing, because of the public deficits, which can then act as a driver for further growth.

These future changes in the basic conditions will also force the construction industry to change in order to remain competitive. Attempts to work under these conditions both competitively and cost-efficiently are limited and, in many cases, result in a pure competition on costs. Apart from the cost factor, competitiveness is in addition determined by the following parameters:

- **Costs:** determined by effectiveness and better efficiency as well as industrial safety, technology and competency
- **Quality:** experience and authority of the employees, training and further training as well as management abilities
- **Time:** influenced by productivity and technology
- **Innovation:** in the form of product and/or process innovations

A business can cope with the transformations and changes in progress only by improving its own competitiveness through innovative approaches. The influencing factor of **innovation** offers the largest possibilities for increasing competitiveness and attaining competition advantage through new, innovative ideas.

2 FUTURE CHALLENGES IN CONSTRUCTION

The environment described above clearly shows the always and ever changing nature of tasks, particularly in the geotechnical sector. The significant challenges in the geotechnical world can be summarized as:

a. Construction projects are becoming more and more complex:

The increasing complexity of projects will lead to significantly more attention being paid to experience and competences in such tasks as planning, designing, cost estimation and execution. Only by integrating all these competences into the construction process complexity will be manageable.

b. **Economic optimization requirements will promote global co-operation and increasing global competition:**

In the future, due to more and more global competition from low cost regions, optimizations for economic improvements will create substantial cost pressure and will force cooperation. These developments will create increasing international competition and with increasing internationalization of the market conditions intensified contractual conditions and clauses. To cope with this challenge, extended knowledge of contract and risk management will be essential, alongside the increasing requirement for technical capabilities.

c. **Increasing demand for automation and the automatic monitoring of building processes:**

With increasing complexity and increasing pressure on efficiency, automation and the automatic monitoring of building processes will become increasingly important to maintain the required standard of quality. A significant task for the future will be to minimize risks of faults, to bridge as many gaps in experience as necessary and to avoid negative consequences. But automation will create new requirements, on device manufacturers and suppliers in particular, to directly and promptly optimize productivity and quality using improved, process-oriented monitoring in the construction process.

d. **Clients and consultants will expect a more efficient documentation of the processes and quality:**

Increasing requirements for documentation of construction processes will lead inevitably to improvements in existing quality and risk management systems, to be developed in parallel to the requirements to automation. An even closer cooperation between users and manufacturers of construction equipment will in most cases be adapted as counteraction to these future tasks for process cycle documentation by using measuring and sensor technology.

e. **Environmentally-friendly technologies are becoming more attractive:**

The development and advancement of resource-saving technologies and building processes will become a significant challenge for the entire construction industry, but particularly for geo-technologies. Alongside the reduction in the consumption of materials and of emissions (noise, dust, vibration…), of CO_2 emissions in parallel to the reduction of energy consumption and also the improvement of the logistics process will be at the forefront.

Considering all these future challenges as general threats to the different construction processes, innovative solutions as combination of alternative products, alternative designs, alternative materials and alternative execution procedures could show a way out of this problematic issue.

3 INNOVATION AND BUILDING PRACTICE

Innovations, whether as improvements to products or procedures, are subject naturally likewise to the technical codes and standards which generally and particularly must and should be fulfilled. On top an innovation usually involves an increase in complexity and thus to higher risk levels with warranty entitlements increasing at the same time. In addition, innovations require more investments in crosschecks and quality assessment, in addition to ever faster implementation cycles (time to market). If the competitiveness is to be maintained over a longer period, constant innovation and the improvement of existing products are the basic requirement. In addition, each innovation must to all above deliver an increase in value for the customer in terms of costs, quality and/or time. Only if an innovation fulfils this complementary condition of creating an added value for the customer, it can also sustain in the competition.

In consideration of these economic basic parameters, the construction process usually has also to manage the following technical requirements and framework conditions as a challenge:

- restricted construction areas
- close to some neighbouring buildings which must be integrated
- short building phases with partially overlapping process cycles
- restricted and limited infrastructure and accessibility of the construction area
- high requirements for minimum deformations and settlement/subsidence
- high requirements for minimised emissions (noise, dust, smell, visual, vibrations…)
- high requirements for resource-protecting building methods
- high requirements for work and health protection

This combination of economic and technical requests inevitably require an modification of the current technologies and processes in the construction process as demand led innovation. Contrary to these demand led innovations, through customer/market requirements, new technologies are also brought to the market via innovations promoted by suppliers (supply pushed).

The following projects show how innovative solutions can positively influence competitiveness in terms of costs, quality and execution time.

4 INNOVATIVE SOLUTIONS FOR LATERAL SUPPORT SYSTEM

With the increasing demand in reuse of inner city areas and the urgent need for deeper foundation pits in constricted areas, adjacent structures close to and high ground water level, the well-known established technologies for lateral support especially with small to mid-sized dimensions are often too time consuming and occasionally not economic to install. The alternatives for these specific cases could be the soil mixing technology to create lateral support systems in reinforcing the deep mixing elements additionally to transfer the loads from earth and water pressure. The wet soil mixing system in mixing existing subsoil with hardening additives can be used for such kind of application by embedded reinforcement elements into the soil mixing columns.

The advantages in executing suitable projects using these advanced innovative solutions:

- minimizing material consumption by using existing subsoil as aggregate and reduced steel quantities, no ready mix concrete delivery, limited transportation on and off site
- minimizing mobilisation efforts and space requirements by utilizing limited and adequate plant and tools, no cutting material as waste
- minimizing execution time by using effective and tailor-made tools (productivity)
- potentially can be used in all types of soil as on shore and near shore solution
- minimum vertical or lateral stresses to adjacent structures with no vibrations
- quality (strength, water tightness) of treated material can be adjusted to needs with minimum spoil.

The selected projects for deep excavations should show in the first case how innovative solutions can be used to increase value for clients and tenderers, whereby the second example shows the utilization of an innovation to the benefit of the project and the client.

4.1 *Deep mixing technology as an innovative solution*

4.1.1 *The deep mixing process*

The mechanical mixing process of stabilising soil is using rotating mixing tools and bonding agent introduced into the mechanically structured soil. The bonding agents used are mostly cement or mixtures of cement with fly ash or gypsum. The amount of bonding agent added depends on the type of soil and the required strength (Fig. 1).

Figure 1. Deep mixing process, mixing tool.

4.1.2 *The Royal Castle, Warsaw, Poland*

The Royal Castle Modernisation Project required the construction of an underground double-floor machine hall under the courtyard of the 18th century palace. The total project assumed a 10m deep excavation below the existing courtyard surface. The design and execution were highly restricted by historic structures in the subsoil which resulted in a complex layout of the excavation.

The historical Palace is founded at the Vistula river slope with the upper layer of subsoil built of 2 to 8m thick uncontrolled deposits underlain by fine sands and Pliocene clay. The groundwater level was measured at depths of 2 to 6m below ground level with the hydraulic gradient linked to the Vistula River running at toe of the slope.

The client's design as conventional secant bored piles wall and tie back anchors was bearing a lot of execution risks and disadvantages (such as open borehole stability, required space to adjacent buildings, installation of reinforced cage, discharge material costs and time for execution). To overcome these challenges in complexity and competitive pressure, a much more environmental friendly solution by using existing soil as construction material was designed. The advanced technology in combination with a new design approach as alternative solution was offered to the client and finally awarded.

The temporary construction of the excavation pit down to 10 m depth (Fig. 2) up to 8 m below the existing groundwater level was designed as a structural retaining wall of DSM columns dia. 700 mm in spacing of 550 mm constructed with overlapping Deep mixing columns to provide water tightness and steel H-beams installed into every second column to create the lateral support system and to provide the required overall wall strength. Soldier piles (every second column) reinforced with steel profile were embedded c.a. 4.50 m below the final excavation level. To achieve the required strength in the mixed columns the mixing time was elongated with full restroking of the mixing tool required, to ensure proper conditions for the installation of soldier elements immediately after mixing.

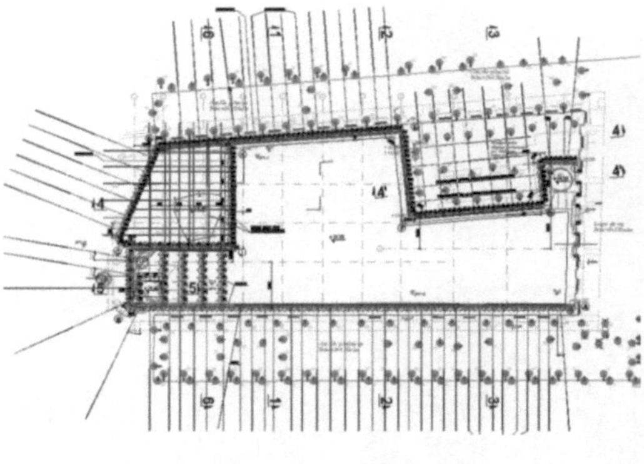

Figure 2. Excavation layout, lateral support system and tie back anchoring scheme.

Figure 3. Single auger deep soil mixing with pre-stressed soil anchors.

During excavation the wall was anchored with a system of two and three rows of pre-stressed anchors (17 to 18 m) in one part of the foundation pit, in order to avoid drilling through the historic structures and monuments, a steel strut system was installed as alternative (Fig. 3).

Due to the complex nature of the total project the structural analysis were conducted with the FEM. Series of variant analysis were conducted using the simple Mohr-Coulomb (M-C) and advance isotropic hardening model (H-S).

To document the quality of execution an automatic measurement device system was implemented. The measurements proved that the excavation support system succeed in limiting soil movements and protecting surrounding buildings. Lateral displacement of the palace's wall did not exceed 4 mm.

With this much more economical approach using a more environmental friendly technology (limited discharge, soil as construction material, limited amount of cement added to the ground) and documenting all the process and impacts on the adjacent structure a highly complex project was managed professionally. The project demonstrates that with advanced approaches options are already given to pass the future challenges.

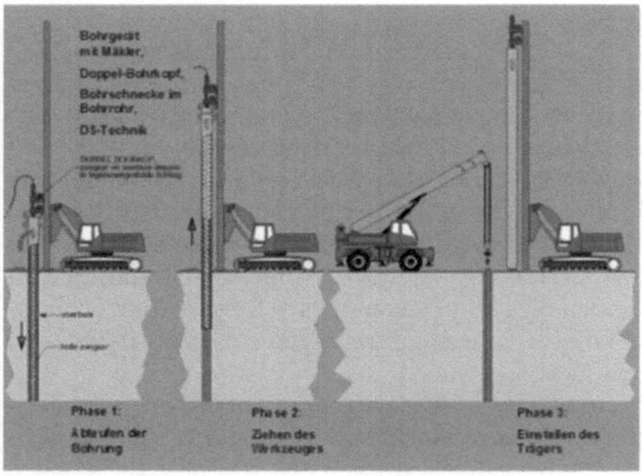

Figure 4. The Tubular Soil Mixing system and process.

4.2 *Cased deep mixing supported by jet grouting The TSM process*

With increase in depth of lateral support systems verticality and overlapping of single columns are becoming the key issues in execution to secure water tightness of the vertical walls. In addition, for some applications where subsoil conditions are not ideal for wet deep mixing technologies the **T**ubular **S**oil **M**ixing system (TSM) as a combination of mechanical mixing with integrated support of hydraulic cutting by jet grouting can provide solutions to overcome these challenges (Fig. 4).

The patented Tubular Soil Mixing process (TSM) can be used to produce a water tide lateral foundation pit support in an inner city environment (Topolnicki, 2002). This process represents a combination of the following two well-known soil stabilization processes:

- the Deep Soil Mixing process (DSM), in which the in-situ soil is loosened mechanically and mixed with a hardening suspension (EN 14579) and
- the jet grouting process in which the in-situ soil is loosened using hydraulic energy and mixed with the grout material (EN 12716).

The boring and mixing tool consists of a auger, which is equipped with removable teeth and suspension nozzles at the tip. Using these nozzles – similar to the jet grouting process – a bonding agent is added to the ground. The auger is guided along its entire length in a counter-rotating casing. The tool is driven by a dual rotary head which is guided on a guide unit. The combination of hydraulic and mechanical mixing ensures a very homogeneous mixing of in-situ soils and produces a high and homogenous firmness in the cubature and/or columns, which exceeds the stability of conventional DSM or jet grouting elements.

The advantage in using this technology for lateral support systems is that the casing secures even in rapidly changing subsoil conditions the designed diameter of the column. The jet grouting assistance is securing the mixing quality to create the homogenous lateral support system.

Figure 5. Boring/mixing tool for the TSM process. Finished TSM wall with inserted H beams.

The combined use of auger and casing allows obstacles like masonry or unreinforced concrete to be penetrated and intersecting columns to be produced. Furthermore, the casing gives a high accuracy in verticality and an absolutely uniform diameter of the columns over the entire casing depth. Immediately after penetrating and mixing from bottom up, a steel girder can be inserted into the fresh material of the column.

The structural system of a shoring wall created using the TSM process is demonstrated in figure 5. Because – unlike the MiP process – when using the TSM process the jet grout technology is used to support the mixing process, the general structural design requirements for the jet grouting elements can be used for the analysis of the cubature manufactured in the TSM process.

Regarding waterproofing and appearance, a temporary pit system made with intersecting TSM columns is comparable to a secant bored pile wall. The points of intersection produced by a superstructure, like evenness, tolerance, mounting possibilities for the insulation and requirements of the formwork also correspond to those of a bored pile wall.

With this innovative technology even complex deep lateral support systems can be managed (flexibility, adaptability) and the whole process is automatically recorded and documented.

5 CONCLUSIONS

In the future it will become even more important for the industry to be able to gain competitive advantage through innovative solutions to deal with the continuing pressure on costs, schedule and even quality.

Innovative solutions in use of new or improved technologies as well as the changing and adjusting of existing technologies to suit the specific requirements of the project can contribute to generating an increase in value for the owner. These approaches can also play their part in meeting the future challenges to reduce complexity, to increase quality through automation and documentation as well as to strengthen the global competitiveness.

For geotechnical engineers the main future challenges are to be familiar with theoretical as well as practical applications of advanced technologies to identify for a specific project the optimal combination of technologies including the best design approach to stay competitive.

To cope with this more general and conceptual formulation of the task the below summarized capabilities and competences of geotechnical engineers are required:

- deep knowledge of technologies and processes including the limits of application of the different technics
- excellent judgement on subsoil characteristics and stress strain relationships of soil materials
- advanced design capabilities including judgement regarding application of design model and approach
- understanding the execution process and the impact of the process on adjacent structures and the environment
- fair judgement on technical requirements versus products involved including alternatives

In addition to these technical competences or expertise soft skills and competences will become more important in a more and more global world with different cultures, diverse approaches and mind-sets cooperating for the success of a project.

To cope with these requirements in addition to the technical skills geotechnical engineers should show good capabilities in:

- Planning and organizing
- Flexibility and adaptability
- Analysing and structuring
- Accuracy and reliability
- Team work and relationship with colleagues
- Decision making

Those geotechnical engineers developing a sound and perfect combination of the technical expertise and the soft competences will be fit for future challenges. As a geotechnical engineer is should be the goal not only to sharpen the geotechnical knowledge but also acquire and shape interpersonal skills.

REFERENCES

DIN EN 12716:2001-12
"Execution of special geotechnical works – jet grouting"

DIN EN 14579:2005-07
"Execution of special geotechnical works – deep-seated soil stabilization"

Kirsch, K., Sondermann, W. (2003)
Ground Improvement, Geotechnical Engineering Handbook, Vol. 2: Procedures, Smoltczyk, U. (editor), Ernst & Sohn, 2003

Raju, V. R. et.al. (2003)
Ground treatment using dry deep mixing for railway embankment in Malaysia, *Proceedings of the 2nd International conference on advances in soft soil Engineering and Technology, July 2–4, 2003, Putrajaya, Malaysia*

Sondermann, W. (2012)
Latest developments in design and execution of deep excavation support systems, *Proceedings Deep Foundation Technologies for Infrastructure Development in India, Sept. 17–20, 2012, Chennai, India*

Topolnicki, M. (2013)
In-situ soil mixing, in *Ground improvement, 3. Edition, Edited by K. Kirsch, A. Bell, CRC press Taylor & Francis group, p. 329–434*

Wehr, W., Topolnicki, M., Sondermann, W. (2013)
Design Risks of Ground Improvement Methods including Rigid Inclusions, *Proceedings 18th Southeast Asian Geotechnical Conference, Singapore May 29–30, 2013*

Challenges and Innovations in Geotechnics – Zhussupbekov (Ed.)
© 2016 Taylor & Francis Group, London, ISBN 978-1-138-03007-7

Reconnaissance in and findings from the 2016 Kumamoto earthquake in Japan

I. Towhata
ISSMGE, Japan

ABSTRACT: The 2016 Kumamoto earthquake sequence in Japan occurred in late April, 2016, in the western island of Kyushu, Japan, and registered M = 7.3 in the biggest event and 6.5 in the second biggest event. These two events were associated with many smaller events and the damage was made worse. Typical damage occurred in traditional houses and volcanic slopes which claimed in total 49 victims. One of the reasons for the substantial damage was the effects of repeated shakings that reduced the seismic resistance of structures with time. Another issue of importance is the engineering mitigation of surface fault ruptures. Because the rainy season is starting in May soon after the earthquake, further disasters in the deteriorated mountain slopes are worried.

1 GENERAL INFORMATION

The earthquake sequence in Kumamoto started on April 14th with an event of M_{JMA} (earthquake magnitude defined by the Meteorological Agency of Japan) = 6.5, after which many events, including the biggest one of M_{JMA} = 7.3, followed (Figure 1). 46 victims were claimed by collapse of houses as well as failures of volcanic slopes. Because the seismic effect accumulated during the duration of the earthquake sequence, damage was made worse with time, several new problems were detected and lessons were learnt.

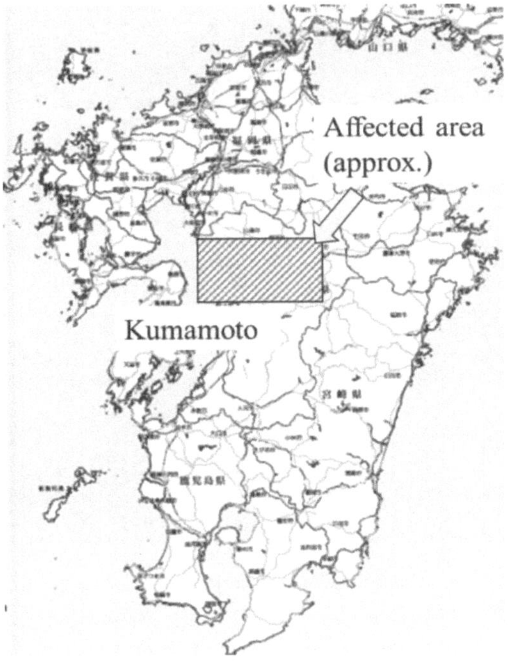

Figure 1. Kumamoto area on Kyushu Island.

2 HOUSE DAMAGE

Many private houses were destroyed by the earthquake sequence; Figure 2. Because the damage extent was so severe in Mashiki Township near Kumamoto, the seismic intensity scale of 7, which is the highest scale of intensity, was declared there for two earthquake events; the one at 21:26 on April 14 (M_{JMA} = 6.5) and the other at 1:25 on April 16 (M_{JMA} = 7.3). Noteworthy is that it was not uncommon that relatively new houses that satisfied the recent seismic design codes were destroyed together with older houses. It is supposed that many of those damaged houses were affected to a certain extent by the first big event, their earthquake resistance was reduced, and thereafter the second event destroyed them. The effects of repeated strong seismic effect is out of scope of the current design principle.

Many houses looked nearly intact after the first event. So, residents came back home. The biggest event of M_{JMA} = 7.3 occurred during the midnight of the next day and the already-deteriorated houses collapsed. Afterwards, many aftershocks continued and the scared tens of thousands of people could not come back home for days and even weeks. Hence, they stayed in shelters for many days and the public sector of the region had to take care of them.

Figure 3 indicates the surface of road pavement in the most damaged area. The destroyed situation of the pavement may suggest that the ground condition was disturbed by shaking as well.

Figure 2. Destroyed private house (Mashiki Town).

Figure 3. Destroyed pavement of a local road (Mashiki Town).

Figure 4. Slope failure on the northern bank of Aso Bridge site.

3 EARTHQUAKE-INDUCED SLOPE FAILURE

The affected area belongs to a very active volcanic area where the active Aso Volcano has a big caldera of 25*18 km^2 in size. Hence, the ground surface is mostly covered by volcanic ash, pumice and other volcanic materials. The induced landslides destroyed human communities in the affected region.

Figure 4 shows the biggest slope failure (500,000 m^3 of soil failed) that happened in the somma (caldera's outer mountain) of Mt. Aso. This movement of soil destroyed the big Aso Bridge of 205.9 m in length. The

Figure 5. Slope failure at Ohkirihata site.

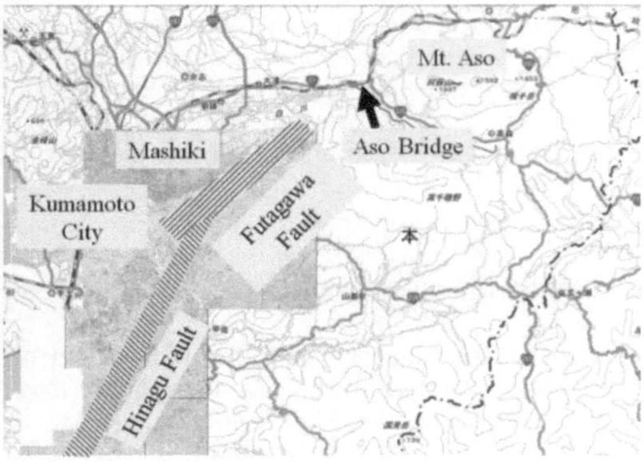

Figure 6. Location of local fault systems (drawn after Active Fault Map in Urban Area by Geospatial Information Authority of Japan).

loss of this important bridge made difficult the rescue and emergency activities. It is further important that there remains cracks near the top of the slope, implying the risk of more slope failures during the coming rainy season. Many more slope failures in the affected region (Figure 5) are potentially unstable during the coming rainy season.

4 SURFACE RUPTURE DUE TO FAULT

The tectonic setting in the northern part of Kyushu Island (Figure 1) undergoes compression in the east-west direction and extension in the north-south direction. More locally, the Kumamoto area has two important fault systems, which are called Futagawa and Hinagu Faults (Figure 6). They are faults of right-lateral strike-slip type.

Many surface ruptures have been found and reported in the Futagawa Fault area (see Figure 7 for example). Figure 8 illustrates an unfortunate example in which a house rests immediately on the rupture. Although it is recently possible to point out the exact location of active faults, there is no measure, except relocation, to mitigate the future risk if houses and communities exist on the rupture. It is possible to trace

Figure 7. Surface rupture caused by underlying fault movement (in Mashiki).

Figure 8. House that was destroyed by underlying right-lateral fault.

the surface evidence of fault action to the east beyond the previously-known range of the Futagawa Fault.

5 LESSONS

After the first strong event on April 14, everybody decided to start rescue and restoration works. It was felt that the damage extent was not so severe although the seismic intensity scale reached 7 in Mashiki, similar to Kobe in 1995. Unfortunately, a bigger shaking took place 28 hours later and many remaining houses collapsed. The current state-of-art seismology cannot forecast the second bigger earthquake but it is understood after the present disaster that such an adverse situation does occur. The existing building design code does not consider such a situation. Note, however, that further reinforcement of buildings against such a difficult situation would increase the construction cost. The acceptance of the increased cost is a matter to be discussed.

It is a pity that little attention is paid by public to soil investigation prior to design and construction of superstructures. In the case of earthquakes, importance of good earthquake-resistant design of houses is understood by public to a certain extent, although not fully accepted. Noteworthy is that the failure of foundation can destroy well-designed superstructures. Figure 9 illustrates failure of a retaining wall that led to significant deformation of the foundation soil. It is very important that owners should recognize the potential seismic hazard under houses and, if necessary, carry out necessary safety measures. The same point can be made of liquefaction problem under houses as well. Fault rupture has not been concerned by the engineering community in the previous time. This is because human activities has been high in alluvial planes where faults in the earth crust are covered by thick soft soil deposits that prevented the fault rupture from propagating to the surface. This fortunate situation is changing in the very recent times because of the expansion of urban areas to the surrounding hilly area wherein the soil coverage is thin and the fault action can easily produce ruptures at the surface. It is necessary to recognize the risk of not only the earthquake-prone construction of fills in the hilly area but also the effect of fault rupture on overlying structures. Although the author does not insist on the urgent development of anti-fault technology of construction, it is necessary to bear in mind the protection of existing structures from fault hazard. The identification of the location of surface rupture is easier nowadays because of the intense study of historical and pre-historical faults. Note further that the probability of the occurrence of surface rupture immediately under one's house is very small.

Figure 9. Failure of retaining wall in a residential area in Mashiki.

The earthquake-induced landslide is a difficult issue of disaster mitigation. First, the onset of earthquakes cannot be predicted and, therefore, that of the landslide either. Hence, emergency warning is not possible. Second, relocation to safer places is difficult to local communities because people need to live in the present place for daily life. Although not very efficient, the only practical way is to develop hazard mappings by which the residents can be aware of the geo-hazard. It is expected thereby that their houses will move to safer places after generations.

Reliable hazard mapping is important in transportation routes as well. As was shown in Figure 4, collapse of a nearby slope can easily hinder the function of important transportation routes and the emergency actions and rescues are prevented. Due to lack of field investigations, the current state of seismic hazard mapping on a national or regional scale uses only the topographical data (elevation, slope gradient etc.) without being able to consider the material properties such as shear strength and rock quality. Local hydrology is certainly out of scope. To improve the situation, it is recommendable to promote soil/rock investigations in slopes at least along important traffic routes so that more reliable hazard maps may be developed.

6 CONCLUSIONS

The 2016 Kumamoto earthquake sequence in Japan occurred in the second half of April and still aftershocks are going on when this draft paper is written. Hence, the content of this paper is tentative. Because of the sequential nature of the seismic events and the volcanic geology of the affected area, several important lessons have been learnt as shown in what follows.

1. Two strong shakings occurred within two days. The first one reduced the seismic resistance of houses and possibly slopes to a certain extent and the second event destroyed them. Such a sequential nature is not considered in existing design codes and regulations.
2. Fault action resulted in ruptures at the surface. Because of the urban expansion in the recent times, existing communities are now prone to those ruptures. A question is raised to what extent a mitigation measure should be developed against such a rare phenomenon.
3. More subsurface investigation is desired under houses and slopes along important traffic routes. Due to lack of concern and shortage of budget, knowledge on risk is not sufficient and unexpected damage occurs. Noteworthy is that a house of good earthquake-resistant design is easily destroyed by the failure of foundation soil.

ACKNOWLEDGMENT

The present reconnaissance study of the author was made together with people of Chuo Kaihatsu Corporation. His special thank is addressed to Dr. H. Yamaguchi of this company. Opinions by Dr. T. Kiyota of University of Tokyo were very valuable in interpretation of the field situation and hence deserve appreciation. The findings by the reconnaissance team of the Japanese Geotechnical Society, which is headed by Prof. Kitazono of Kumamoto University and Prof. Yasufuku of Kyushu University, have been valuable as well. Further supports are expected from those people to develop broader and clearer outputs from the current damage investigation.

Keynote Lecturers

Challenges and Innovations in Geotechnics – Zhussupbekov (Ed.)
© 2016 Taylor & Francis Group, London, ISBN 978-1-138-03007-7

Application of geosynthetics in civil engineers projects

Eun Chul Shin, Min Kyo Jung & Jeong Ku Kang
Department of Civil and Environmental Engineering, Incheon National University, Republic of Korea

ABSTRACT: Geosynthetics are widely used for construction of structures in civil and environmental engineering, and agriculture fields. In the past 30 years, intensive research and development of polymer materials provided to expand the lifetime of geosynthetics and to cope with the adverse environmental conditions. The first part of this paper presents the type of geosynthetics and their functions in practical use. The second part of this paper several geosynthetics applications in Korea are described for the construction of retaining wall, soft ground improvement work, road and railways, coastal protection, and sanitary landfill.

1 INTRODUCTION

The geosynthetics have been widely used since 1960s as a new material for many applications in the geotechnical and geoenvironmental engineerings, and also coastal engineering as well as in transportation engineering. The geosynthetic materials are manufactured in the well quality controlled factory, and it can easy to handle for installation and transportation. The cost of geosynthetics are relatively competitive comparing with conventional materials such as concrete and steel.

With the development of polymer material, geosynthetics are getting popular and provided to expand the lifetime of geosynthetics and to cope with the adverse environmental conditions. In the sense of engineers practical use of geosynthetics, engineers must understand the mechanical function of geosynthetics and appropriate application of geosynthetic materials (Jones, 1996).

Therefore, the frist part of this paper introduces the types of geosynthetic such as geotextiles, geogrids, geonets, geomembraine, geocomposite, geosynthetic clay liner, geopipe, geocell, and geofoam. The functions of each geosynthetics are described with the application of practical use in the civil engineering projects (Shin and Oh, 2007).

In the later part of this paper, several case histories of geosynthetics applications are reported in the order of most usage of geosynthetics, that is, the construction of reinforced retaining walls, soft ground improvement work in the land reclamation projects, road and railways construction work for the subgrade reinforcement as well as construction of overpass way, and application of geotextile tube for shore protection in the coastal regions.

2 TYPES OF GEOSYNTHETICS AND THEIR FUNCTIONS

The geosynthetics are manufactured to use in the various civil engineering projects such as road and railways, retaining wall, dam, land reclamation, sanitary landfill, and embankment as well as foundation.

The geosynthetics are specially manufactured to play a role of key function for the specially designed and constructed structures (Koerner, 2005). The single sheet types of geosynthetics are geotextiles (Figure 1a), geogrids (Figure 1b), geonets (Figure 1c), and geomembrane (Figure 1d). The combined sheet types of geosynthetics are geocomposite (Figure 1e), geoshnthetic clay liner, GCL (Figure 1f), which are made of geotextile with bentonite clay material in between top and bottom geotextile layer or bonded to a geomembrane sheet. The volumetric types of geosynthetics are geopipe (Figure 1g), geocells (Figure 1h), and geofoam or EPS (Figure 1i).

The major functions of geosynthetics are separation, filtration and drainage, and reinforcement. Many cases in the field, one single geosynthetic can be simultaneously used for two functions.

Geotextile are two types, one is named as woven geotextile which is manufactured by continuous knitted fabrics (yarn) sheet. The other geotextile is named as nonwoven geotextile which is manufactured by continuous stitch-bonded fiber (yarns). Geotextile are used in practice for the separation and reinforcement between subgrade and aggregate layer in the road and railway construction as shown in Figure 2(a). It can also use for the purpose of filtration and drainage. These two phenomenon are shown in Figure 2(b, c, d). In the manufacturing of geotextile tube, these two functions are very important parameters in the design.

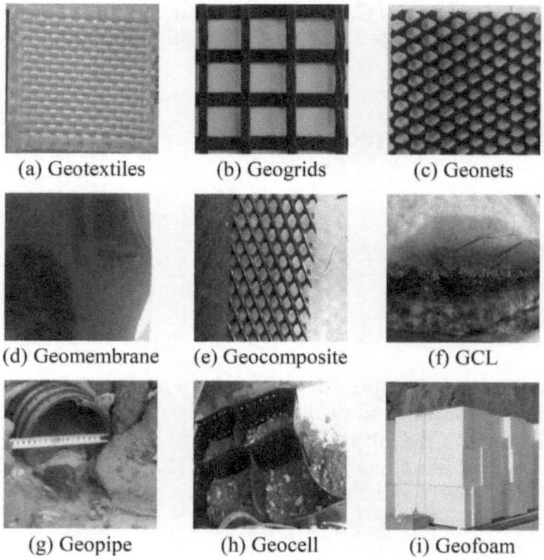

Figure 1. View of various types of geosynthetics.

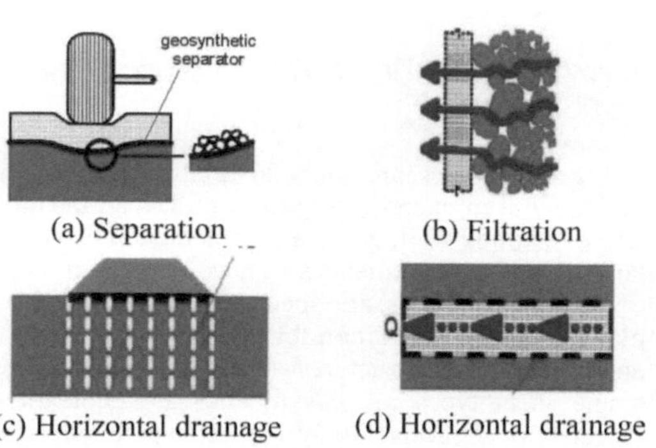

Figure 2. Three phenomena of geosynthetics as separation, filtration and drainage function.

Geogrids are used in practice for the reinforcement of unpaved and paved roads, and the construction of reinforced retaining walls as shown in Figure 3. Geogrids are three types, uniaxial geogrid and biaxial geogrid, and woven bonded. Geogrids are manufactured by polymeric mesh type element with openings through the longitudinal stretching with punching polymer sheet or longitudinal and transverse stretching with punching polymer sheet. The other type of geogrid is made of polyester by means of mechanical bond between longitudinal ribs and transverse ribs together which creates an opening, called apertures. Normally aggregates are interlocked in the apertures of geogrid which induces the reinforcement of soil layers (Tsukada et al., 1993; Sukla, 2002).

The woven geotextile is usually placed in the construction of embankment over very soft foundation soils (Jewell, 1988; Rowe and Li, 1999) to reduce the vertical displacement and prevent the sliding failure of the embankment as shown in Figure 4.

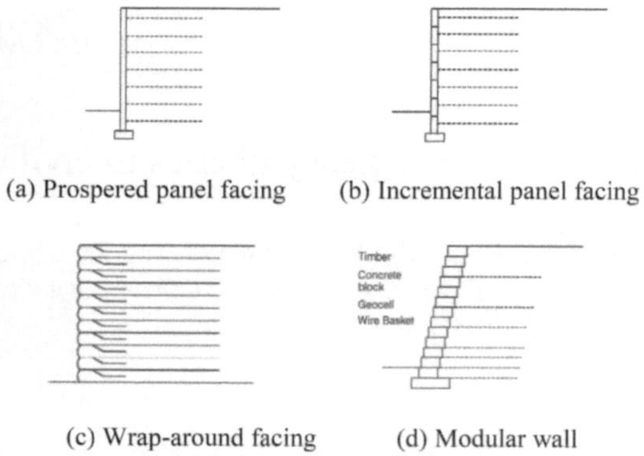

Figure 3. Various types of geosynthetics reinforced walls (Shin and Oh, 2007).

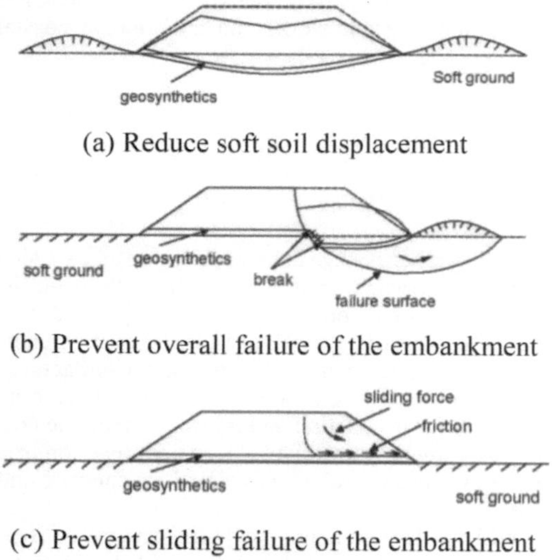

Figure 4. Geosynthetics reinforcement under embankment (Shin and Oh, 2007).

Geonets are extruded polymer meshes with diamond shape apertures. This product is normally placed in the middle of two non-woven geotextiles to make the geocomposite product. The geocomposites are practically used as a drainage in the soft soil and it is also used in the leachate collection system along the side slope of sanitary landfill site.

Geomembrane is made by using the polymetric material as a continuous sheet liner. The function of geomembrane is to prevent the migration of water as a moisture barrier system. It can use in the dam and mostly use in the sanitary landfill system to prevent the leakage of leachate and gas from the waste.

Geosynthetic clay liners are typically used as a liquid barrier system in the canal in the sandy ground or it is commonly used in the side slope of sanitary landfill where the clay liner is not available.

Geopipes are used for collection of leachate or gas in the sanitary landfill system or used in the soft ground to collect the water as a drainage system. It does have perforated holes around polymeric pipes.

Geocell are three-dimensional structures manufactured by using the strip of polymeric belt type sheet. The shape of cells is diamond shape and it can give high reinforcement effect when it fills with aggregate material. Geocell is effectively used in the reinforcement railway ballast, basal reinforcement of highway embankment, slope of embankment or erosion control of shore along coastal lines.

Geofoam is a block type or slab type 3-D system which made of foam from polystyrene. Sometimes, it is called expanded polystyrene (EPS). The weight of geofoam is quite light, so it can replace the heavy weight filling material in the backfill zone of a quay wall or behind retaining wall to reduce the active earth pressure. Geofoam is also preferable used in the backfill area of bridge abutment for the pavement road or railway to minimize the lateral pressure transmitted to the abutment of bridge.

3 CASE HISTORY OF GEOSYNTHETIC APPLICATIONS IN KOREA

3.1 *Introduction*

There are several geosynthetics applications for the construction of retaining wall, soft ground improvement, sanitary landfill, road and railways construction work for the subgrade reinforcement as well as construction of overpass way, application of geotextile tube for shore protection in the coastal regions.

3.2 *Construction of retaining wall*

The reinforcing materials for the reinforced wall are mainly stiff geogrid (PP) and flexible geogrid (PET), geosynthetic strip type, galvanized steel strip, and soil nailing. The wire mesh type reinforced steel bar and steel bar with a cross-reinforcing angle are also used for the reinforced wall.

The most popular basal reinforcing materials for the embankment or land reclamation area are woven and non-woven geotextiles which are also used as a separation purpose between original ground and backfilling soil. However, stiff geogrids are mostly utilized to reinforce the foundation soil for the oil tank storage facilities, airfield, road, railways, and retaining wall structures.

3.2.1 *Flat panel type faced geosynthetic walls*

The wall facing of reinforced earth wall are generally categorized into two types, either hard concrete facing or geosynthetic soft facing. The most of reinforced earth wall are currently utilizing a hard concrete discrete panels, either flat panel type like using in Reinforced Earth Wall (REW) or modular block type. These are usually made of precast concrete with overlapping joints with a compressive filler and are made of various sizes and shapes to fit the design requirement.

The reinforcing material use a belt type continuous strip which made of polyester (PET) with polyethylene (PE) coating. The galvanized metal strip type reinforced earth techniques are mostly used in the highway cut to secure the earth slope against sliding failure. It is shown the construction of an approach embankment for overpass bridge in Figure 5.

Figure 5. Panel type reinforce approach embankment for an overpass bridge of a highway in Korea.

Figure 6. Panel type reinforce approach embankment with galvanized metal strip.

However, the quality is a little bit low as comparing with REW's in terms of mobilizing friction angle and coating along the strip. The galvanized type was used in this project as shown in Figure 6.

3.2.2 *Modular block type faced geosynthetic walls*

The modular block faced geosynthetic reinforced segmental retaining walls are extensively used in Korea. The reinforcing materials for modular block faced reinforced walls are mostly polyester geogrid and polypropylene geogrid. Two type of geogrid are used in practice, uniaxial geogrid and biaxial geogrid, depending on the application purpose. Fasteners and connections are necessary in reinforced modular block wall where geogrids are connected to the facing block. The approach embankment of flyovers as modular block was shown in Figure 7.

3.3 *Soft ground improvement*

3.3.1 *Yeocheon industrial complex*

Yeocheon industrial complex extension project was completed in 2002 after 6 years from the beginning time. Total area of this project was 8,485,000 m^2. Because the filled soil was silty clay in a slurry state, it

Figure 7. Modular block faced walls in an approach of the embankment of a bridge flyover.

Figure 8. Horizontal vacuum consolidation technology applied in the Yeocheon industrial complex extension project.

was necessary to obtain the trafficability of subsurface soil prior to the soil improvement work.

The depth of soil improved in this area is about 5.0 m with the natural water content of 80%. A small barge-ship was used for applying the horizontal vacuum consolidation method as shown in Figure 8.

It was used a pulley having a diameter of 2 m at the top, and 4.5 m long mendrel attached at the bottom of the ship which was used to install the plastic board drain in the soil layer.

As the plastic board drain was fixed at the required interval vertically at the one end of the dike, the barge ship dragged the PBD to the other end of the dike and connected to the drainage head pipe. Backfilled soil surface was covered and sealed by the flexible membrane prior to the operating of the vacuum pump for dewatering.

The vacuum pump was operated until the water content of soil reaches about 70%. The soil with this water content can withstanding the trafficability of deep soil improvement equipment with the help of geotextile and sand layers.

3.3.2 *Busan land reclamation project*

New Busan Port is being under construction by connecting the several islands by the construction of earth dikes from 1997. The purpose of the land reclamation is to build port and harbor facilities and industrial complex.

Figure 9. Ground improvement using horizontal and vertical drains in the Busan industrial complex project.

The soil improvement project in Busan area was the land reclamation project for the industrial complex. Packed drain method was used to improve the soft marine clay. The diameter of packed drain was 12 cm, and 4 drains are simultaneously constructed with the casing pipes. Sand piles are encapsulated by geotextiles and hence give an additional confining pressure to the sand piles.

The construction process of this method is similar to the construction method of sand drains and also it can be constructed up to the maximum depth of 45 m. The location of the land reclamation works in Busan is shown in Figure 9.

3.3.3 *Incheon International Airport project*

The site was formed by the construction of 17.3 km long dikes between Yongjong Island and Yongyu Island in the West Sea of Korea.

The subsoil condition of the site consisted of clay, silt, sand, and formed in the fourth alluvial period. The thickness of soft soil ranged from 3 m to 11 m with an average thickness of 5 m. The field pilot test results indicated that plastic board drain method gives a similar settlement level while obtaining of good quality sand for SD or SCP was not easy and expensive. These similar values of settlement for various soil improvement methods are somewhat influenced due to the predominant of sand seams in the soil layers.

Plastic board drain was selected and used the rest of the site. Consolidation settlement was achieved about 95% in 6 months after installation of PVD with the preloading height of 5.0 m. The permissible residual settlements of various areas were estimated for the period of 30 years. The settlement gauges are placed in the various facilities to measure the future prospective settlement. The settlement gauge at the bottom of runway was recorded as less than 1mm settlement so far since 2000.

It is clear that the secondary consolidation settlement of clayey soil is not occurred. The geogrid installation for different settlement in IIA was shown in Figure 10.

(a) Incheon airport (b) Geogrid installation

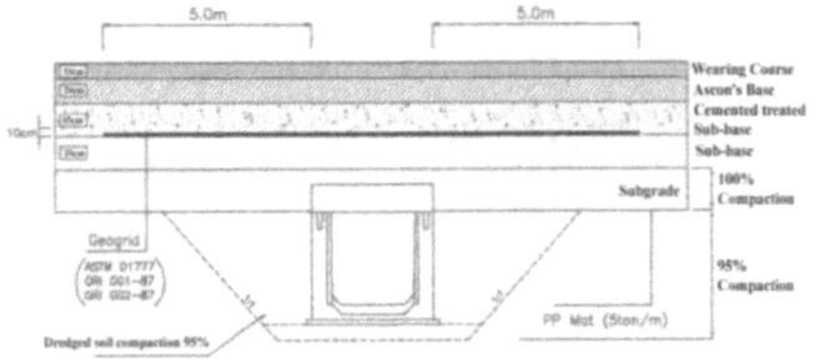

(c) Geogrid for different settlement on box culvert

Figure 10. Geogrid installation for different settlement on box culvert in Incheon International Airport.

Figure 11. Landfill construction work of the waste lime landfill project in Incheon, Korea.

3.4 *Sanitary landfill*

The rate of recycling of waste lime which is produced from the chemical plant and the fertilizer factory as a by-product is excessively low level, so it has been accumulated around site of a plant or a factory during the past 30 years in Incheon. The total amount of waste lime is 8,011,411 m^3. Landfill construction of waste lime landfill project was presented in Figure 11 (Shin, et al., 2010).

From the observation of the first stage landfill construction work, the geosynthetics used in the waste lime landfill project gives an enhancement of slope stability and stability of landfill itself.

A bottom and top liner systems with utilizing various geosynthetics are adopted for design and construction of waste lime landfill. Geosynthetics are used in various applications in waste water facilities. These are extensively used in the design of both base and cover liner systems of landfill facilities.

3.5 *Geotextile tube*

3.5.1 *Shore protection structure*

The geotextile tube technology is mainly used dike and breakwater construction for flood and water control, but they are also used to prevent beach erosion and, for shore protection (Shin and Kang, 2010).

It was reported that dikes can be constructed up to 2.0 m tall to provide flood protection (Perry 1993, Bruin and Loos 1995). Environmental dredging and backfill technology using geotextile tubes was reported by Fowler et al. (1995, 2002) and Mori et al. (2002). The major design considerations are related to the integrity of the units during release and impact, and the accuracy of placement and the stability under current and wave attack (US Army, 1994; Pilarczyk, 1998, 2000).

The size of a single geotextile tube installed in Young-Jin on east coast of Korea was 50.0 m long and 1.8 m height. After the completion of the installation, the variations of shoreline and water depth near shore were monitored over a 12 month period. The variation of shoreline is shown in Figure 12.

(a) Before construction (b) After 12months

Figure 12. Comparison of shoreline variation with elapsed time.

3.5.2 *Temporary staked geotextile structure*

Incheon Grand Bridge linking the Incheon International Airport on Yongjong Island with Songdo City in the Incheon Free Economic Zone (IFEZ). It is 15.7 m wide and has a maximum span length of 145 m. The stacked geotextile tube technology was implemented at the approach bridge of Songdo New City side (Shin, et al., 2012).

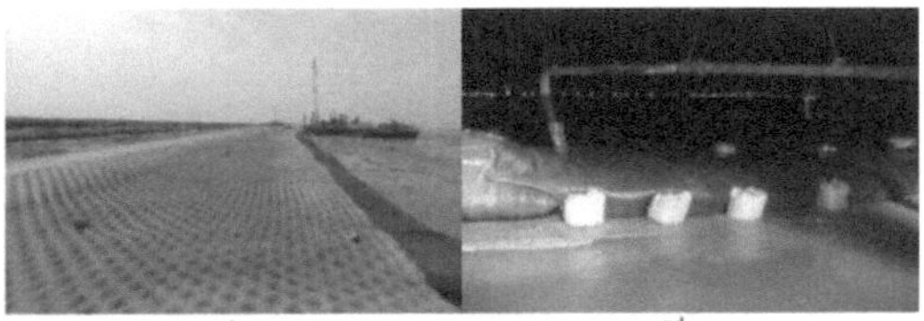

(a) 1st layer (b) 2nd layer

(c) 3rd layer (d) final condition

Figure 13. Temporary stacked geotextile tube in the Incheon Grand Bridge (Total height of 3rd geotextile is 6.2 m).

Three types of geotextile tube were proposed for desired crest height (Figure 13), and the stability analysis against external force was conducted to the each cross section of stacked geotextile tube. The considered external forces are wave force, tidal force and lateral earth pressure by reclamation.

When the temporary dike structure with filling material is constructed, it should be not only considered stability as a structure, but also considered reduction of height caused by settlement such as filling sand with low pressure, scouring of the ground with velocity of tidal level, losing of fine material by flow of water from the geotextile tubes (Hori, 1920; Goda, 1985; Homa and Horikawa, 1965).

4 CONCLUSIONS

Geosynthetics are commonly used for construction of civil engineering structures such as road and railways, retraining wall, soft ground improvement work, coastal protection, and sanitary landfill system since 1960s.

The magnitude of geosynthetics used has been steadily increased and innovative applications of geosynthetics were developed with consideration of construction cost as well as maintenance cost.

This paper has briefly reviewed the various types of geosynthetics and their basic functions in the civil engineering practice. The case histories of construction projects for reinforced retaining wall, over-pass way for road and railways, soft ground improvement work in sea port, airport runway reinforcement work, and sanitary landfill system were presented with the effective use of marginal site due to the dwindling of available land for construction of social-infrastructures.

The contents of this paper would provide a good knowledge to the young geotechnical engineers for better understanding of geosysthetics and its applications in civil engineering projects.

REFERENCES

Fourie, A.B., Johns, D. and Jones, C.J.F.P. 2002 In-situ dewatering of mine tailings using electrokinetic geosynthetics *Proc 6th Int Conf. on Geosynthetics* Nice.

Fowler, J., Toups, D., and Gilbert, P. 1995. Geotextile Contained Contaminated Dredged Material. Marina Del Ray, Los Angeles and Port of Oakland, California, *Fourteenth World Dredging Congress*, Amsterdam, The Netherlands.

Goda, Y. 1985. *Random Seas and Design of Maritime Structures*. University of Tokyo Press, Tokyo, Japan.

Hiroi, 1920. Evaluation of wave pressure. *Journal of JSCE*, Vol. 6, No. 2, pp. 435–449.

Honma, M., Horikawa, K. 1965. Experimental study on total wave force against sea wall. *Coastal Engineering in Japan*, Vol. 8, pp. 119–129.

Jewell, R.A. 1988. The mechanics of reinforced embankments on soft soils. *Geotextiles and Geomembrane*, 7: pp. 237–273.

Jones, C.J.F.P. 1996 *Earth Reinforcement and Soil Structures*. Thomas Telford Publishing, London. p. 379.

Koerner, R.M., 2005. *Design with Geosynthetics*, Fifth Edition, Prentice Hall, New Jersey.

Mori, H., Miki, H., and Tsuneoka, N. 2002. The use of geo-tube method to retard the migration of contaminants in dredged soil, *Proceedings of the Seventh International Conference on Geosynthetics*, 7 ICG-NICE, France, Vol. 3, pp. 1017–1020.

Perry, E.B., 1993. *Innovative Methods for Levee Rehabilitation*. Technical Report, REMR-GT-26, U.S. Army Corps of Engineers, Waterway Experiment Station, Vicksverg, MS.

Pilarczyk, K.W., 1998. Stability criteria for geosystems – an overview – *6th International Conference on Geosynthetics*. Atlanta, USA, Vol. 2, pp. 1165–1172.

Pilarczyk, K.W., 2000. *Geosynthetics and Geosystems in Hydraulic and Coastal Engineering*. A.A. Balkema Publisher, Rotterdam.

Rowe, R.K. and Li, A.L. 1999. Reinforced embankments over soft foundations under undrained and partially drained conditions. *Geotextiles and Geomembranes*, 17: pp. 129–146.

Shin, E. C., Kang, J. K., Kim, K. H. 2010. Design and construction of waste lime landfill construction. *Proceedings of 9th International Conference on Geosynthetics*, São Paulo, Brazil, pp. 1017–1020.

Shin, E. C., Kang, J. K., Kim, S. H. 2012. Construction technology of environmental sustainable shore and harbor structure using stacked geotextile tube3. *Proceedings of Indo-Korean Workshop on Geotechnology for Urban Development (IK-GUD)*, Delhi.

Shin, E. C., Oh, Y. I. 2007. Recent innovative application of geosynthetics in geotechnical engineering. *Proceedings of the 13th Asian Regional Conference on Soil Mechanics and Geotechnical Engineering*. Vol. 1, Post Conference Volume, Allied Publishers Pvt. Ltd., Kolkata, pp. 207–221.

Shukla, S. K. 2002. *Geosynthetics and Their Applications*, Thomas Telford Ltd., London.

Tatsuoka, F. 2013. Keynote lecture: Recent geosynthetic-reinforced soil structures for railways in Japan, *Proceedings of Geotechnics for Sustainable Development-Geotec Hanoi 2013*, Phung (edt), Construction Publisher, pp. 1–26.

Terzaghi, K. 1936. Stress distribution in dry and in saturated sand above a yielding trap door. *Proceedings of 1st Conference on Soil Mechanics and Foundation Engineering*, Harvard University, Cambridge, pp. 307–311.

Tsukada, Y., Isoda, T., and Yamanouchi, T. 1993. Geogrid subgrade reinforcement and deep foundation improvement. *Proceedings of Geosynthetics Case Histories*, International Society for Soil Mechanics and Foundation Engineering, Committee TC9, Yono City, Japan, pp. 158–159.

U.S. Army Corps Engineers 1994, *Shore Protection Manual*, 4th edition. Vol. 2, U.S. Army Corps Engineers, Waterway Experimentation Station, Vicksverg, MS.

Challenges and Innovations in Geotechnics – Zhussupbekov (Ed.)
© 2016 Taylor & Francis Group, London, ISBN 978-1-138-03007-7

Problems of Syrian historical monuments, destroyed by military action: Diagnostics, strengthening and reconstruction

Talal Awwad
Department of Geotechnical Engineering, Damascus University, Syria

ABSTRACT: The entire civilized world is watching the state of the unique monuments of the East, including Syria, where military operations have not completed yet. Some antiquities got destruction that immediately require a detailed investigation; and, as a minimum, to prevent the collapse of the structural elements.

Needless to say that the world practice of restoration damages from the massive bombing and shelling, was most fully represented by publications in period of the Second World War (Russia, Japan and Poland). By using of these proceedings, we can systemize the hazard degree of an object at the time of its possible restoration, and evaluate the damage depending on integrated parameters.

Unfortunately, today it is actual to review this practice in accordance with modern technologies of engineering restoration of damaged monuments and reconstruction destroyed historic buildings. Without this it is impossible to defeat the Vandals of XXI century.

1 INTRODUCTION

In five years of Syrian war, a lot of people have been killed. But another cost has been the loss of irreplaceable historical monuments, the heritage of all humanity. When we lose these monuments, we kill our human history. Actually, there are many important archaeological sites that belong to ancient successive civilizations in Syria. These heritage monuments include temples, theaters, baths, churches, monasteries, and Islamic monuments from the dawn of Islam to the present day. In Syria were the first cities in history, such as the ancient kingdom of Ebla, from 2500 to 2400 BC, the Kingdom of Mary about 2900 BC and Palmyra, the world heritage site and ancient city in the Syrian Desert, which flourished in the second half of the first century BC.

Some of these structures that stood there for thousands years have been forever destroyed and some of them got destruction that immediately require a detailed investigation; and, as a minimum, to prevent the collapse of the structural elements.

2 SYRIA'S ANCIENT LANDMARKS CRUMBLE UNDER ONSLAUGHT OF WARS.

Speaking about problems of Syrian historical monuments, destroyed by military action, two examples immediately spring to mind. The first is the Ancient City of Aleppo, one of six places in Syria classified by UNESCO as a world heritage site and used to be one of the country's main attractions. The other is Palmyra, the world heritage site and ancient city in the Syrian Desert, which flourished in the second half of the first century BC.

Figure 1. Kingdom of Mary. Parts of the walls. It is next to Tall Ḩariri and is located in Deir ez-Zor, Syria.

2.1 *Ancient City of Aleppo*

Located at the crossroads of several trade routes from the 2nd millennium B.C., Aleppo was ruled successively by the Hittites, Assyrians, Arabs, Mongols, Mamelukes and Ottomans. The 13th-century citadel, 12th-century Great Mosque and various 17th-century madrasas, palaces, caravanserais and hammams all form part of the city's cohesive, unique urban fabric, now threatened by war.

The Church of Saint Simeon Stylites is a historical building located about 30 km northwest of Aleppo.

Figure 2. Aleppo's citadel – the Unesco World Heritage site has since suffered damage that will only be open for proper assessment when the war is over.

Figure 3. The loss of the minaret on Umayyad Mosque. The 11th-century mosque.

It is the oldest surviving Byzantine church, dating back to the 5th century. Built on the site of the pillar of St. Simeon Stylites, a famed hermit monk, the Church and surrounding village was designated by United Nations Educational, Scientific and Cultural Organization (UNESCO) as part of the "Ancient Villages of Northern Syria", a World Heritage Site [1].

Figure 4. Historical market in Ancient City of Aleppo.

Figure 5. South facade of Church of Saint Simeon.

2.2 *Palmyra*

The Syrian Desert city known as the Venice of the Sands has suffered another act of vandalism.

Figure 6. Destruction of Temple of Baalshamin.

The Temple of Baal Shamin, the ancient stone ruin located in Palmyra, was targeted and destroyed in the summer of 2015.

The composite of satellite imagery analyses by the United Nations Institute for Training and Research shows that sometime between Aug. 27, 2015, and Aug. 31, 2015, the main building of Temple of Bel and its prized rows of columns, one of the best-preserved Roman-era sites in the Syrian city of Palmyra, were destroyed with explosives. In addition to being the centerpiece of Palmyra's famed ruins, the Temple of Bel was also considered one of the most important heritage relics in the world.

The Monumental Arch, also called the Arch of Triumph or the Arch of Septimius Severus, was a Roman ornamental archway in Palmyra. It was built in the 3rd century. The arch once sat atop the famed colonnaded streets of the ancient city, which linked the Roman Empire to Persia and the East.

The ruins of the Arch of Triumph later became one of the main attractions of Palmyra, until it was destroyed in October 2015. Most of its stonework still survives and there are plans to rebuild it using anastylosis.

3 DIAGNOSTICS, STRENGTHENING AND RECONSTRUCTION.

Most of the historic buildings and monuments in Syria are lightly-loaded structures and consist mainly of domes, stone walls, stone arches, stone cupolas, and stone columns based on strip stone foundations that seated directly on the soil.

Figure 7. Remains of the Temple of Bel after being blown up.

Figure 8. The Arch of Triumph 2,000-year-old structure in the ancient city of Palmyra, had been destroyed.

When the foundation of historical building is destroyed by using explosives, we must carefully examine and measure the volume of the destruction of all the fissure damages, deformations and distortions of masonry and foundations, as well as the state of the soil under the foundation and within the crater from the explosion impact. The soil structure at the

Figure 9. Foundations of Al-Azem Palace.

site of the explosion will be disturbed and, depending on the actions of the shell or bomb, the soil receives the local compaction or loosening. The last reduces bearing capacity of soil.

Bulging of soil from under the base or from the sides is sometimes observed. The presence of open deep crater in the long term may also affect the soil condition (Figure 10).

The destruction of deep strip foundations within the explosion of bombs around the building, usually spreads on insignificant of their length, since the effect of the explosion, and as a result, the air wave is mainly directed to the side part at an angle to the horizontal (Figure 11). If the upper part of the building, on the site of the destruction of strip foundation, preserved, the neighboring part temporarily takes an extra load, in consequence of load redistribution. Therefore, sometimes it will have an increased settlement, in comparison with other parts of the building.

Foundations under pillars and columns are usually wider than strip foundations under the walls. The explosion near this foundation is often destroyed not only its masonry, but also the soil of the base (Figure 12). In such cases it is necessary to establish whether there was not a valuable local settlement and degrees of misalignment.

Restoring the foundations and strengthening the building can be done by increasing the width of the foundation, replacing the masonry that has low strength, underpinning, deepening of the foundation with the pressure transmission to more strength layer of soil, setting the building on piles, foundation supporting by micro piles, using concert or metal beams under the foundation and soil improvement under foundations by mortar injection and chemical grouting by using new technologies.

The historical character of the heritages and monuments should be kept, and any kind or method of restoration and maintenance must achieve construction safety, keep the nature of the building and not to change its historical characters.

Figure 10. Destruction of foundation base of historical buildings by using explosives.

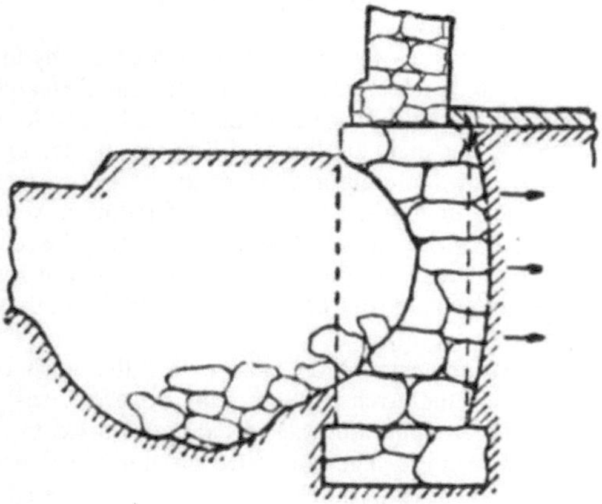

Figure 11. Destruction of deep strip foundations within the explosion.

4 CONCLUSION

Because of the differences of the structures and the architectural materials, it is nearly impossible to make a general rehabilitation rule for all buildings, so we should deal with each case separately.

Supporting of foundations should include:

- Three-dimensional measurements (by using modern scanners).
- Full-scale architectural-Restoration study of the monument.
- Geological and geotechnical surveys.

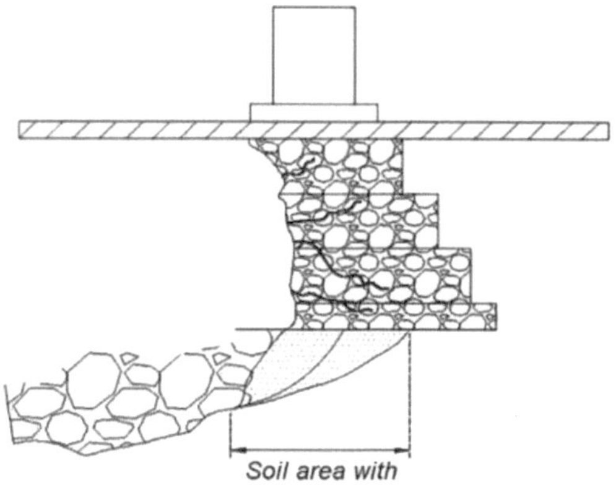

Figure 12. Destruction of Foundations under columns within the explosion.

- Technical inspection (including by using geophysical methods to detect latent defects). Particular attention must be given to examination of unstable objects that allows us to quickly assess the sustainability and to make the necessary calculations and design for the maintenance of fragile objects.
- Design (in accordance with international regulations) and preparation of technical specifications for the reconstruction of destroyed World Heritage sites.
- Constant consultant support of the restoration work on the project, providing interactive design, continuous adjustments to reflect new information. This will speed up recovery times and dramatically reduce the time required, while ensuring the most correct decision-making.

REFERENCES

AAAS, (2016). Ancient History, Modern Destruction: Assessing the Current Status of Syria's World Heritage Sites Using High-Resolution Satellite Imagery. http://www.aaas.org/.

ASOR, CHI, Weekly Reports_ www.asor-syrianheritage.org

Awwad, T. (2013). "The Role of Soil Improvement for Conservation of Syrian Historical Monuments". Proceedings of Workshop ATC19, 18th ICSMGE 18th ICSMGE World Congress, Paris, September, 2013. pp. 89–94.

Awwad, T. (2014). "Analytical Soil-Structure Interaction Study Of a Historical Building Located Near A River Bed (Damascus Citadel Tower Near Banias River)", Journal of AL Baath University-Syria-Homs, Vol. (36). No. (1) 2014. pp. 153–168.

Blasi, C., Coisson E. (2008). Analysis and diagnosis of the damaged structures of Damascus. Citadel, Damascus.

Blasi, C. (2006). Italian Guidelines for the evaluation and redution of theseismic risk in cultural heritage, Roma, Gangemi Ed.

Blasi, C., Rossi, P.P. (1988). Indagine sulle strutture murarie del tempiocosiddetto di Romolo nel Foro Romano: prove non distruttive conmartinetti piatti e analisi numeriche. In II International Conference onNon destructive Tests, Microanalytical methodsand environmentalinvestigations for the study and conservationof cultural heritage, Proceedings, Perugia, Italy.

Bessac J.C., Boqvist, M. (2005). Le chantier de construction de la cittadelle deDamas: méthodologie et résultats preliminaries. Arqueologia de la Arquitectura, 4, 237–249.

Boqvist, M., (2002). L'étude technique du bati: l'état des recherches dans lesectuer de la sale à colonnes, in Berthier, S., AlEjji, E. (2002). Cittadellede Damas. Supplement of the Bulletin d'Et-udes Orientales, LII-LIV, 77–97.

Dietz, K. and Schürmann, A. (2006). "Foundation Improvement of historic buildings by micro piles, Museum Island, Berlin and St. Kolumba, Cologne." 7th ISM workshop, Schrobenhausen 2006.

Guidoboni, E., Comastri A., (2005). Catalogue of earthquakes and tsunamis inthe Mediterranean area from the 11th to the 15th century, Rome: National Institute of Geophysics and Vulcaology.

Hanish, H. (1993). Der Nordostabschnitt der Zitadelle von Damaskus. Damaszener Mitteilungen, 7, 233–296.

Heyman, J. (1966). The stone skeleton, International Journal of Solids and Structures, Vol. 2, 249–279.

Heyman, J. (1998). Structural analysis: A historical approach. Cambridge, Cambridge University Press.

Meinecke, M. (1996). Patterns of Stylistic Changes in Islamic Architecture: Local, New York: NY Press.

Mendes, C., Lourenco, P. (2013). Seismic performance of ancient masonry building: A sensitivity analysis.

Monti, N., Rossi, C., Vassena G. (1996). Methodological approach in analysis of static monitoring system data, in Reports on Surveying and Geodesy, Bologna.

Ottoni, F., Coisson, E., Ejji, E., Blasi, C. (2013). Seismic performance of ancient masonry building: A sensitivity analysis.

Reda Sbeinati, M., Darawcheh, R., Mouty, M., (2005) The historical earthquakes of Syria: an analysis of large and moderate earthquakes from 1365 B.C. to 1900 A.D., Annals of Geophysiscs, 48(3), 347–435.

Rossi, P.P. (1982). Analysis of mechanical characteristics of brick masonrytested by means of nondestructive in situ test. ISMS, Bergamo, Italy.

Sauvaget, J. (1930). La citadelle de Damas, Syria, II, 5990 and 216–241.

Wulzinger, K., Watzinger, C., (1924), Damaskus, Die Islamise Stadt, Berl–Leipzig, 166–187.

Wong, P.K. (2001). Ground improvement case studies chemical lime piles and dynamic replacement. Malaysia and Egypt.

Zhussupbekov, A., Zhunisov, T., Issina, A. and Awwad, T. (2013). "Geotechnical and structural investigations of historical monuments of Kazakhstan". Proceedings of Second International Symposium on Geotechnical Engineering for the Preservation of Monuments and Historic Sites, Naples, Italy, 2013. pp. 779–784.

Challenges and Innovations in Geotechnics – Zhussupbekov (Ed.)
© 2016 Taylor & Francis Group, London, ISBN 978-1-138-03007-7

Geotechnical and construction considerations of pile foundations in problematical soils

A.Zh. Zhussupbekov & A.R. Omarov
Department of Civil Engineering, Eurasian National University, Astana, Kazakhstan

ABSTRACT: Just as every civilization in the history has originated from the riverside, so is the city of Astana – new capital of Kazakhstan has been developed around the Ishim River. As a result there are many bridges across the river. Also, high rise buildings, such Palace of Peace, Abu-Dhabi Plaza, New Railway Station, Expo-2017 constructions sites and many other structures are founded in problematical soil ground of Astana. These unique buildings are supported by deep drawing and boring pile foundations. For designing of piling foundations on difficult soils, is important to have the investigations of behaviour of piles by using dynamic, static and O-cell. This paper includes a new results of several piling tests. These investigations of interaction of piles with soil ground are important for understanding the mechanism of working of different piles on soft and hard soils of Astana. Also, this paper introduced the experiences of piling constructions in winter season on freezing ground. The last part of paper includes recommendations and conclusion with proposing of methodologies for obtaining the bearing capacity and settlements of driving and boring piles on problematical soil ground of Astana.

Keywords: Pile, load, O-cell, bearing capacity, settlement, SLT

1 INTRODUCTION

1.1 *Site railway station*

The existing railway station provides services to more than 7,000 passengers per day and the capacity of the New Railway Station in Astana will be up from 12,000 to 35,000 passengers per day. The main railway gates of the capital will provide bandwidth for all its adjacent directions.

This is a multifunctional and promising part of the city that includes a residential area, several business centers, hotels, schools, pre-schools and other social and cultural facilities. Commercial premises will occupy 656,032 m^2, and 2.8 million square meters will be allotted for housing. Business centers will occupy 507,660 m^2, hotels – 304,375 square meters, utilities and auxiliary infrastructure – 126,000 m^2.

The construction is going to be finished before the international exhibition EXPO-2017 to be held in Astana in summer. The railway station will be built using green technologies, for example, it will have energy-saving lifts and escalators, heat exchangers as well as solar batteries and geothermal power generators, which is very relevant given the fact that the theme of the upcoming EXPO-2017 in Astana is Future Energy. This technology will heat the complex in winter and cool it during summer.

The construction is part of the preparations for EXPO-2017. It will be 11 times bigger than the existing railway station. The height of the six-floor complex will be 49.5 m, the width 116 m and the length 630 m.

A metal-covered railway platform with a length of 2,550 meters is being constructed because the trains will arrive directly into the building.

So far, 87 poles and 17,300 tons or 70 percent of the total metal spans have been installed that are also seen in Figure 2.

Field tests carried out in accordance with requirements GOST 5686 – 94, 1994 and definitions of bearing capacity of the piles in accordance with requirements of SNIP RK 5.01-03-2002, 2002 and ASTM D4945-12, 2012.

This paper includes the short summary about dynamic and static tests by driven piles (cross-section 30×30 cm and length of 12 m). The methodologies of definition bearing capacity of the pile by aforementioned methods were also given. As an example for those methods, paper describes the results of the dynamic, traditional static and the new PDA (Pile Dynamic Analyzer) tests of cooperative work of soils

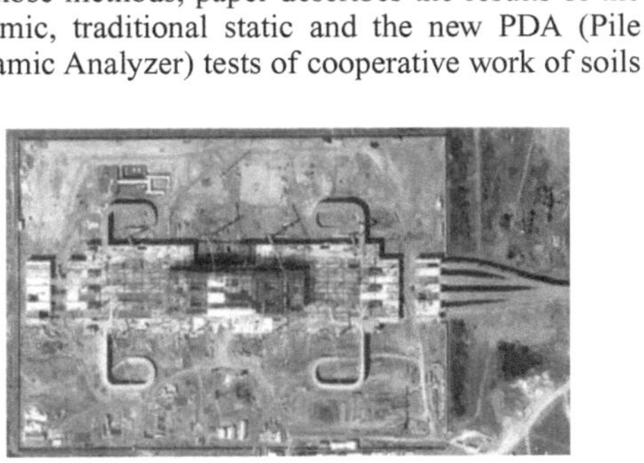

Figure 1. Plan of the new railway stations (Astana, Kazakhstan).

Figure 2. Railway platforms (Astana, Kazakhstan).

Figure 3. Project Expo-2017 (Astana, Kazakhstan).

Figure 4. Kazakhstan platform of Expo-2017.

and piles performed in the construction site of the New Railway Station in Astana.

1.2 *Site Expo-2017*

The symbol of the exhibition EXPO-2017 will be the Kazakhstan platform itself made in the sphere form several floors high with 24000 m^2 in total (see Figure 3 and Figure 4). The symbol of the exhibition is located in the center of the exhibition village. It is surrounded by international, thematic and enterprise platforms.

The results and comparing of soil tests of the piles for following methods: Vertical static test Static Load Test (hereinafter SLT) and the Bi-Directional Static Load Test (hereinafter BDSLT) Experienced bored piles with a length of 31.5 m, diameter 1000 mm. Bi-directional static load tests and Static load tests carried out in accordance with ASTM D1143.

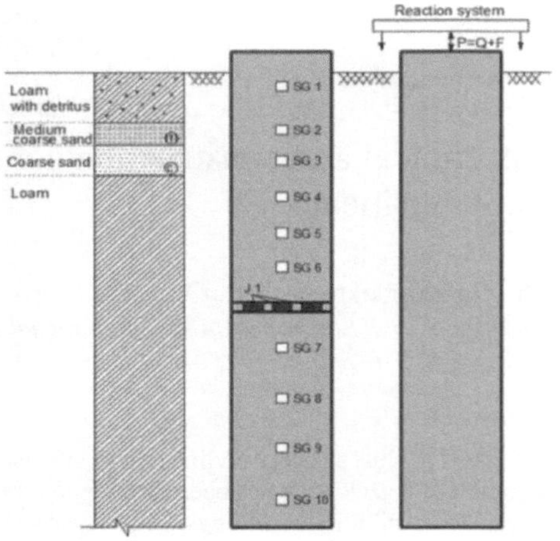

Figure 5. Geotechnical condition of construction site and details of piles (diameter 1000 mm and length of 31.5 m).

Typical geotechnical conditions of the Expo 2017 construction site as mentioned before are represented by the following soils:

Soil 1 – loam with detritus is covered by top soil and fill-up soil in the depth of 0.20-0.40 m. the thickness of layer varies from 3.6 to 8.1 m. On the field description loam is brown, carbonated, and with middle coarse sand band of thickness equal to 5 cm and with 5–10 cm thickness loamy sand layer.

Soil 2 – middle coarse sand is brown, with loamy soil which is equal to 5–10 cm thick and water-saturated. The middle coarse sand is of thickness 0.60–4.50 m.

Soil 3 – Coarse sand thickness is 1.0–3.9 m. Coarse sand is described by semi-gravel, color is brown, and with middle course sand band of thickness equal to 5–10 cm, water-saturated and includes 10 percent gravel fragment.

Soil 4 – loam is deep-brown and mahogany color. And includes debris and detritus up to 10 percent.

Geotechnical condition of construction site and details of piles are shown in Figure 5.

1.3 *Plate Load Test*

Plate Load Test is a field test for determining the ultimate bearing capacity of soil and the likely settlement under a given load.

The Plate Load Test basically consists of loading a steel plate placed at the foundation level and recording the settlements corresponding to each load increment.

The test load is gradually increased till the plate starts to sink at a rapid rate. The total value of load on the plate in such a stage divided by the area of the steel plate gives the value of the ultimate bearing capacity of soil.

The ultimate bearing capacity of soil is divided by suitable factor of safety to arrive at the value of safe bearing capacity of soil.

Field tests carried out in accordance with requirement GOST 20276-99, 1999.

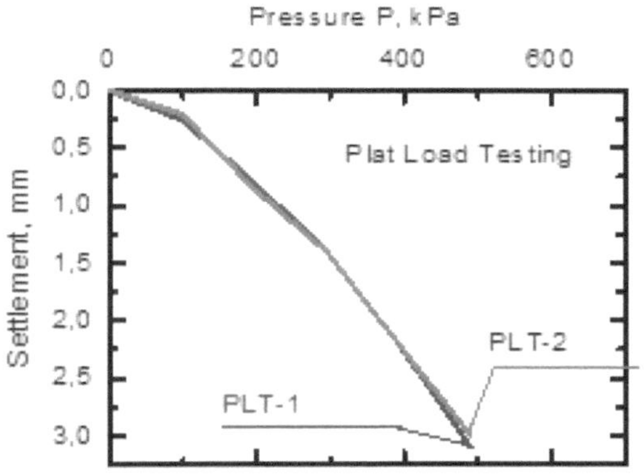

Figure 6. Load-settlement diagram from PLT results.

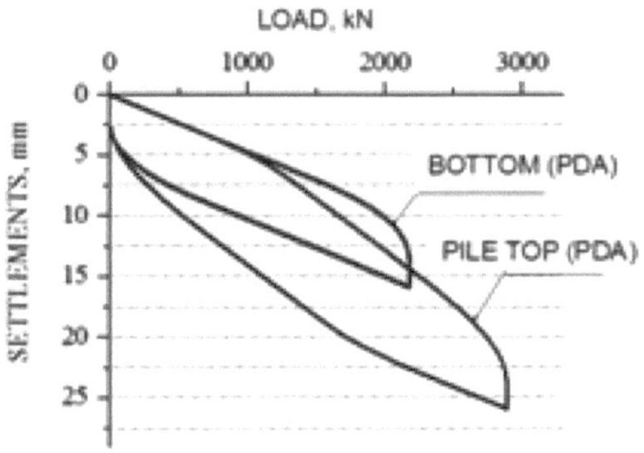

Figure 7. Load-settlement diagrams method PDA.

Totally two tests, one test in Plate Load Test 1 and second test in Plate Load Test 2 at a depth of 12 m had been made. Diameter of the plate D = 276.4 mm and area A = 600 mm^2.

Figure 6 shows the result of the pressure-settlement diagrams of PLT tests of PLT – 1 and PLT – 2.

Modulus deformation of soils in BR04 – 20 MPa and BR17 – 22 MPa, at a depth of 12 m.

2 METHODOLOGIES OF THE PILED FIELD TESTS

2.1 *Dynamic Load Test (DLT)*

All the reinforced concrete piles are having a width b – 0.3 m, cross-section area $A = 0.09\,m^2$, length $L = 12$ m, weight – 2730 kg, modulus of elasticity $E_p = 27500$ MPa, and density $\rho_p = 2500\,kg/m^3$.

In Kazakhstan, DLT is carried out by using different types of pile driving machines and hammers. Before starting the test, pile surface along the whole length had been painted through each 1 meter by marks; last one meter is painted through each 0.1 meters.

For our project, pile driving was performed by using the driving machine "Junttan PM-25" with hydraulic hammer HHK-7A. The weight of the hummer is 7000 kg and the headband weight is 990 kg.

During the pile driving process the number of blows of the each 1 meter of pile penetration into the soil ground and of the last one meter in each 0.1 meter were counted. The falling heights of blowing part of the hammer were recorded at the same time. Pile driving was continuing till the design refusals – 0.5 cm (cm/blow).

The highest average refusal received during the re-driving of piles after their "rest". These values of refusals had been used for determination of bearing capacity of piles. According to SNIP RK 5.01-03-2002, 2002, the rest time for the piles immersed into clayey soils should be 6 days.

Re-driving of test piles was carried out sequentially by three and five hammer blows. The strain gauges with the length of 10 cm were attached on top of the piles before starting re-driving. Strain gauges had been fixed on 60 cm from the pile head.

It is common practice to use the following driving equation to estimate the driving resistance (ultimate pile capacity), F_u, in Kazakhstan (GOST 20176 2009):

$$F_u = \frac{\eta AM}{2}\left[\sqrt{1 + \frac{4Ed(m_1 + \varepsilon^2(m_2 + m_3))}{\eta AS(m_1 + m_2 + m_3)}} - 1\right] \quad (1)$$

where: η – coefficient dependent on the concrete strength of the piles 1500 kPa; A – cross sectional area of pile; M – coefficient (1.0 for hammer impact); Ed – potential energy of hammer – WH-hammer weight; H – fall height of hammer; $h = 0.7$, S-set per blow; ε-coefficient of restitution; m_1 – total mass of hammer device; m_2 – total mass of pile and pile cap, and m_3 – mass placed between pile head and pile cap.

Allowable bearing capacity of the piles with an allowance for safety factor ($FS = 1.4$) equal. to 540 kN.

2.2 *Pile Dynamic Analyzer (PDA)*

Today, in process of adaption into Kazakhstan practice is an alternative load test method which precluded disadvantages of both SLT and DLT – Pile Driving Analysis (further PDA) – allow performing up to 10 piles per day and much cost effective than SLT.

Figure 7 presents the monitoring results of PDA test showing pile dynamic compression and tension stresses, static pile capacity and blow counts versus pile penetration depth.

CAPWAP analysis results that include plots of measured pile head data obtained under the hammer blows from the end of driving and associated simulated pile head and toe static load-movement relationships are presented in the Figure 7.

Allowable bearing capacity of the piles with an allowance for safety factor ($FS = 1.4$) equal. to 714 kN.

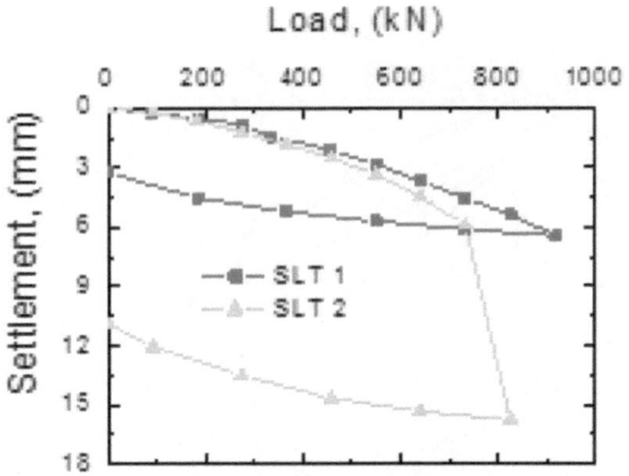

Figure 8. Comparisons of SLT load-settlement diagrams.

2.3 *Results of Static Load Tests (driving pile testing)*

Static Load Test (further SLT) one of the more reliable field tests in analyzing pile bearing capacity. SLT should be carried out for driving piles after the "rest".

SLTs were carried out for two piles on the construction site. The measured relationships between the pile head load, L, and the head displacement, S, of the test piles are shown in Figure 8.

It is seen from Figure 8 that the load-displacement curves of piles No. 3 and No. 24 are almost identical, having an ultimate shaft capacity of 920 kN (No. 3) and 825 kN (No. 24) respectively.

2.4 *O-Cell test or static testing of subsoil by the piles with bidirectional load*

The method suggested by J. Osterberg allows determining the calculated subsoil resistance under the lower end of the pile and on its lateral surface at the same time. The specific thing of the O-Cell test is that the load is applied not on the pile head, but on the pile body where the adjustable jack is set. It works in two directions.

The test pile was a 1000 mm diameter bored pile. The hydraulic jack assembly comprising of three (3) 500-tonne capacity bi-directional hydraulic jacks, was installed at 16.80 m (330.60 m RL) below the Cut off Level.

The hydraulic jack assembly and steel cages were joined and lowered into the bored hole. The pile was concreted according to the contractor's method statement.

There was pair each of tell-tale rod installed at the top and the bottom of the hydraulic cell assembly. Their movements were measured against a reference frame constructed by the contractor.

Ten levels of vibrating wire-type strain gauges (Geokon- 4911 Sister bar type), comprising four units at each level, were installed in the test pile to measure strain at nominated locations. The strain gauges were mounted at designated Level 1 to Level 10 as shown in

Level Reference	Level description	Depth, m (RL)	Depth, m
COL	Cut of Level	347.40	0.000
SG 1	Strain Gauges	346.90	0.500
SG 2	Strain Gauges	344.20	3.200
SG 3	Strain Gauges	341.50	5.900
SG 4	Strain Gauges	338.80	3.200
SG 5	Strain Gauges	336.10	11.300
SG 6	Strain Gauges	333.60	13.800
J 1	Hydraulic jack assembly (3 x 500tonnes)	330.60	16.800
SG 7	Strain Gauges	327.60	19.800
SG 8	Strain Gauges	323.90	23.500
SG 9	Strain Gauges	320.20	27.200
SG 10	Strain Gauges	317.40	31.000
PT	Pile Toe	316.90	31.500

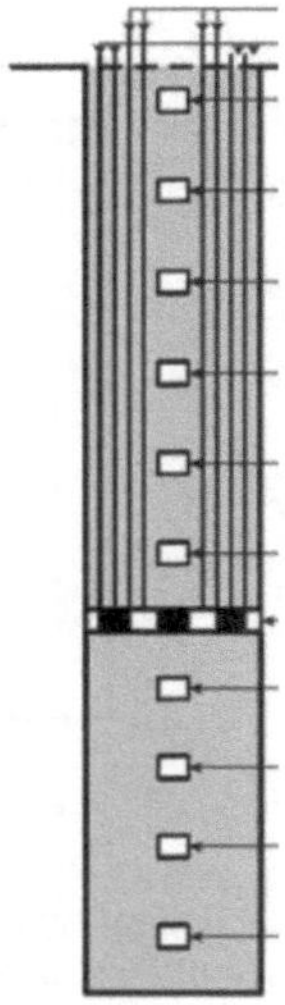

Figure 9. The scheme O-cell test.

Figure 10. Test piles static loading method O-cell.

the pile layout provided. Displacement gauges which are used during the O-cell load test are shown in Figure 9. The location and placement of the instrumentation are shown in Figure 9. The test pile details are summarized in Figure 10.

A millimeter scale was fixed to the reference frame and direct readings from a dumpy level to this scale were observed to check that there were no errors in the displacement transducer readings. The displacement, load and strain data were automatically recorded at 1-minute intervals.

Before the tests, 10 strain-measuring transducers connected to a data detector (data-logger) were installed in the body of the experimental pile. Unlike a traditional static testing, O-Cell allows to obtain two dependences "load-subsidence": one curve characterizes the resistance of the pile under the bottom end, the second one – on its lateral surface. Therefore, using these two curves we can obtain an equivalent curve "load-subsidence", which is analogous to the curve SLT.

A high-pressure bourdon gauge and calibrated electronic pressure transducer are used to monitor the hydraulic cell pressure, from which the loading is derived by applying the calibration factor of the hydraulic cell to the pressure. Electronic pressure

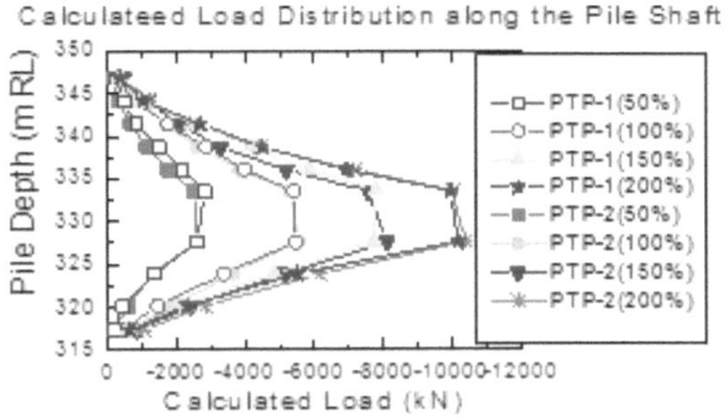

Figure 11. Total calculated load of piles PTP-1 and PTP-2.

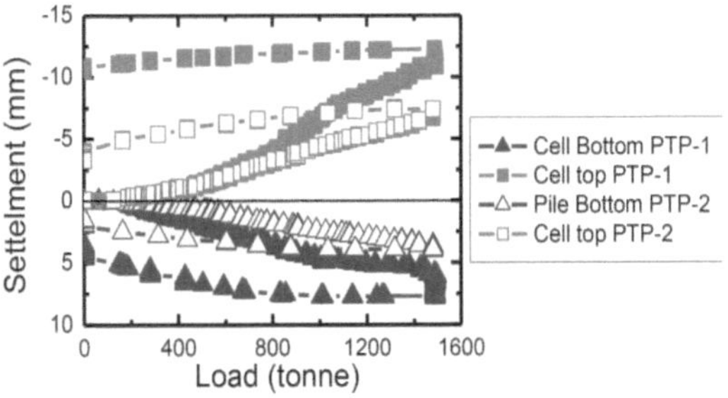

Figure 12. Results of BDSLT piles PTP-1 and PTP-2.

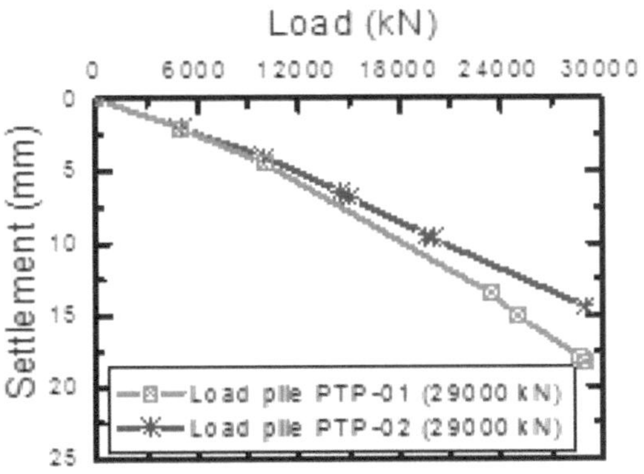

Figure 13. The results of the test O-Cell.

Figure 14. Static tests of piles with the SLT method.

transducer which used for O-cell load test is shown in Figure 10.

Figure 11 shows the results of strain-measuring transducers. Figure 9 presents the load distribution along the length of the piles. The graph shows that even at maximum load, the pile is kept by lateral resistance of the subsoil. Only a small part of the load accrues to the pile edge. The indicators of lateral resistance of the pile on the depth are presented in Figure 11.

Average unit side shear between any two strain gauge levels of the pile was calculated by the change in load divided by the circumferential area between the two strain gauge levels. The maximum mobilized unit side shears between the adjacent strain gauge levels were computed.

Pile of PTP-1 the pile section above the jack, the shear distribution indicated an increase in unit skin friction from 76 kN/m^2 to 481 kN/m^2 at 200% of the working load. For the pile section below the jack, the shear distribution indicated an increase in unit skin friction from 190 kN/m^2 to 458 kN/m^2 at 200% of the working load.

Pile of PTP-2 the pile section above the jack, the shear distribution indicated an increase in unit skin friction from 83 kN/m^2 to 477 kN/m^2 at 200% of the working load. For the pile section below the jack, the shear distribution indicated an increase in unit skin friction from 207 kN/m^2 to 437 kN/m^2 at 200% of the working load.

The O-Cell test results are presented in Figure 10. At the maximum test load of 100% (14500 kN), the maximum displacements of the piles are PTP-1 – 7,30 mm and PTP-2 – 6,50 mm, and at the maximum workload of 200% (29000 kN), displacements of the piles are PTP-1 – 18,35 mm, and PTP-2 – 14,40 mm (see Figure 12).

Figure 13 shows the comparison of the results of piles test by O-Cell method (the equivalent curve).

2.5 *Static load test carried out in accordance with ASTM D1143*

Vertical static tests of piles by SLT method is one of the most reliable field test methods of soils for analysis of the bearing capacity of piles (see Figure 14).

The increment of the load was 25%. The sequence of application of the force is as follows: 25, 50, 75, 100, 50, 0, 25, 50, 75, 100, 125, 150, 175, 200, 150, 100, 50, and 0% of the project (6000 kN). In the first cycle of experimental pile up to 100% of the project pressurization, in the second cycle load up to 200%. The exposure time of the intermediate stages of loading amounted to 30 min, unloading – 20 min. Exposure time of peak load amounted to 120 min and 240 min respectively.

Experienced bored piles with a length of 31.5 m, diameter 1000 mm. Load test on pile amounted to 600 m and 1200 tons. Tests were carried out after reaching more than 80% the strength of the concrete piles.

Figure 15 presents the results of the SLT method. The first test cycle up to 6000 kN load represented by the red line, with full draught amounted to 2.09 mm.

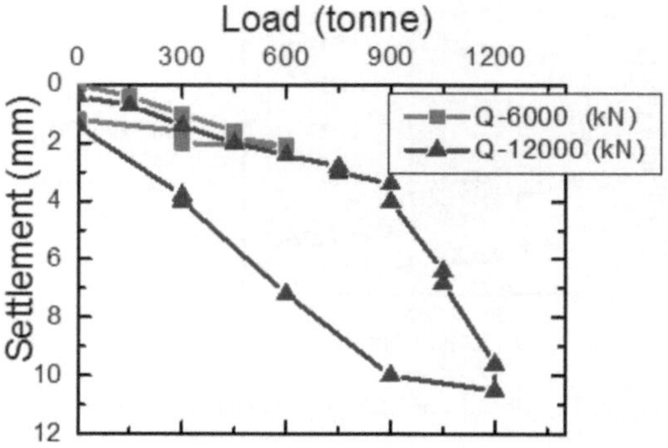

Figure 15. Schedule of load-settlement with the SLT method of pile No. 166 (load 6000 kN and 12000 kN).

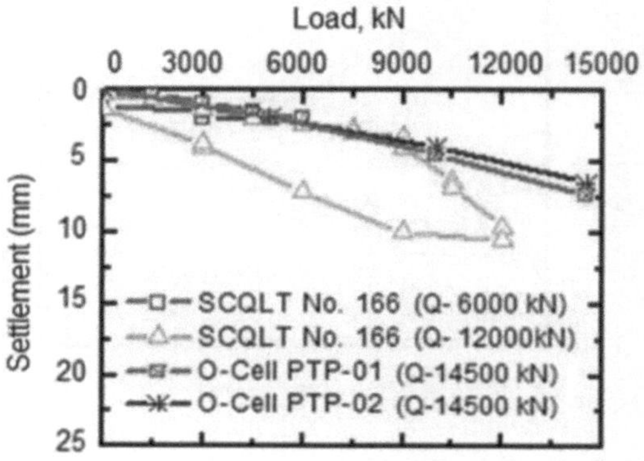

Figure 16. Comparison between the results of O-cell load test and static load test.

The second cycle of up to 12000 kN load – blue line, full of sediment was 10.51 mm.

Figure 16 shows the comparison of the results of pile test method O-cell (equivalent curve) with the results SLT.

3 CONCLUSION

Existing pile foundation standards practiced in Kazakhstan are out-of-date and are in urgent need for modernization. This paper presented very short descriptions of coming changes to the concept of Kazakhstan pile foundation design.

According to the results of DLT of driven piles (30×30 cm and length of 12 m) the bearing capacity of the piles amounted to 540 kN. The bearing capacity of driven piles according to the results of SLT1 and SLT2 amounted to 767 kN and 687.5 kN, respectively.

According to the results of PDA bearing capacity of the piles are equal to 714.3 kN. PDA and DLT are more economical issues than SLT. PDA dynamic test showed more coincide with SLT result than traditional DLT test.

Field O-cell load test was carried out and results of field test were analyzed for the first time in Astana, Kazakhstan. The method of O-cell has been studied and advantages of using O-cell load test for large diameter pile foundations are described.

The tests results are as follows: At the maximum work loading of 200% (29000 kN), displacements of the pile are PTP-1 – 18.35 mm, PTP-2 – 14.40 mm.

Following results are obtained from the mentioned static load test (top download):

- the maximum settlement up to 12000 kN is 10.51 mm;
- the residual settlement after unloading to zero is 0.32 mm.

The O-cell testing method provides some important advantages. There is no structural loading system at the ground surface. Load can be applied at or very close to the pile tip for measurement of base resistance. In conventional top load testing, most or all of the shaft resistance must be mobilized before there is significant load transfer to the tip.

The development of bidirectional load on high bearing capacity of the piles gives engineers a new powerful tool for assessing the interaction of the piles with the subsoil. Finally this method is saves funds and time, because then is no necessity to use anchoring system.

In spite of benefits of O-cell tests there are some limitations. The applied method is mainly used for bored piles. The principle disadvantage is: adjustable jack and transducer for measuring displacements have to be pre-installed before testing and after testing they remain in the pile.

Along with the disadvantages, the major advantage of the O-cell is that it allows determining both pile tip and shaft resistance, which is has special value for the analysis and evaluation of bearing capacity of piles of a large diameter.

REFERENCES

ASTM D4945-12. 2012. *Standard Test Method for High-Strain Dynamic Testing of Deep Foundations.* ASTM International, West Conshohocken, PA.

ASTM D1143. "Standard Test Methods for Deep Foundations Under Static Axial Compressive Load".

GOST 5686-94. 1994. *Methods for field testing by piles.* Gersevanov Research Institute of Bases and Underground Structures (NIIOSP), Moscow.

GOST 20276-99. 1999. *Field methods for determining the strength and strain characteristics.* Gersevanov Research Institute of Bases and *Underground* Structures (NIIOSP), Moscow.

Guoliang Dai, Weiming Gong. (2012). "Application of bi-directional static loading test to deep foundations". J. of Rock Mechanics and Geotechnical Engineering, 4(3): 269–275.

SNIP RK 5.01-03-2002. 2002. *Pile foundations.* KAZGOR *Design* Academy, Almaty.

Challenges and Innovations in Geotechnics – Zhussupbekov (Ed.)
© 2016 Taylor & Francis Group, London, ISBN 978-1-138-03007-7

On interim results of sand production research in weak sandstone formations in Kazakhstan

J.R. Kim & M.H. Nguyen
Nazarbayev University, Astana, Kazakhstan

ABSTRACT: Sand production is one of the main issues in the development of heavy oil reservoirs in Kazakhstan, such as Karazhanbas, Kenkiyak, North Buzachi oilfields. Both sand production onset and volumetric sand production prediction have been objectives of many studies, however, main attention has been directed to the effect of flow rate and pressure, and also mechanical strength of the samples. However, little attention has been made to mircoscopic properties of sand particles and cement. Currently, the research project titled 'Micromechanics of perforation penetration and sand production in weak sandstone formations in Kazakhstan' has been carried out at Nazarbayev University. This research is focusing on the impact of microscopic parameters of sand particles and cementation between them and the purpose of this paper is to present the interim results of this project.

1 INTRODUCTION

1.1 *Background*

Almost 70% of the world's oil and gas reserves are produced from weakly consolidated or non-consolidated reservoirs (Bianco and Halleck, 2001). Main problem associated with such unconsolidated reservoirs is sand production. Sand may erode both downhole equipment and surface facilities. Uncontrolled catastrophic amount of sand may lead to wellbore instability and collapse. In addition, environmental issues arise when disposing the produced dirty sand. Many different sand control technologies have been used, including flow rate restriction, selective completion, chemical consolidation, slotted liners or screens, and gravel packing. Most of these technologies are costly and those which are less expensive result in a significant loss in productivity. In addition, environmental issues arise while disposing the produced dirty sand. It is well reported that sand control management cost the oil industry billions of dollars annually.

Currently, the research project titled 'Micromechanics of perforation penetration and sand production in weak sandstone formations in Kazakhstan' has been carried out by the NU 'Petroleum Engineering and Geomechanics Research Group. The objective of this research project is to study the mechanics of sand production, to make a research of impact parameters that can effect on sanding process in weak sandstone formations in Kazakhstan using laboratory experiments of the perforation and sand production process, which are combined with modeling works using a microscopic coupled modeling technique of Discrete-Element-Method (DEM) and Computational-Fluid-Dynamics (CFD) and a FEM modeling of boundary value problems. This research consists of three main stages: 1) Laboratory tests of the perforation process is conducted on artificial cemented sand samples to study the relation between penetration resistance, penetration depth and material properties. DEM simulations is conducted to understand the effect of the historical stress condition on the perforation penetration for horizontal wellbores, and to investigate the perforation damage to the cemented sand material. 2) Sand production investigations are conducted on samples prepared in stage 1 using laboratory tests and numerical modeling. Fluid flows are generated from the sample boundaries towards the perforation tunnel(s) and the amount of sand production are monitored in real time; furthermore, a mini hydro-cyclone is designed to find an effective method to separate the eroded sand particles from the outflow. 3) Laboratory tests measuring the constitutive behavior of cemented sand are conducted and combined with DEM simulations. They are used to calibrate the input parameters for the FEM simulation of a boundary value problem of a horizontal well to study the effect of perforation direction on sand production.

The objective of this paper is to introduce the research activity of Petroleum Engineering and Geomechanics Research Group at NU and to present the interim results of on-going project.

2 LITERATURE REVIEW

There are two main mechanisms involved in sand production process: firstly, it is a rock matrix failure due to stresses exceeding its strength; and secondly, detachment of the sand grains and transportation of

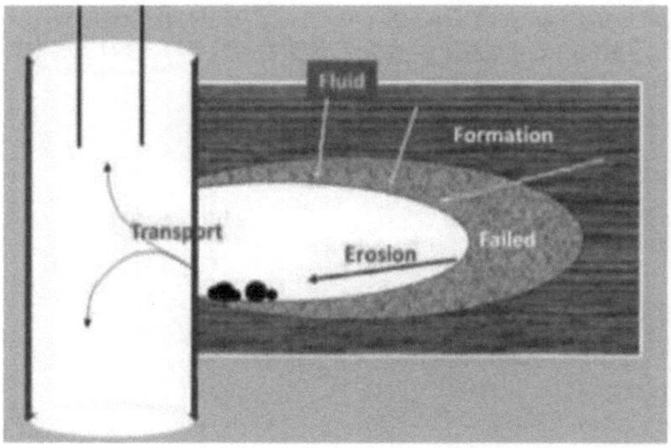

Figure 1. Plastic zone around the perforation cavity (Kim et al, 2012).

them by fluid flow (erosion). Intact formation is in its hydro-mechanical equilibrium until formation is drilled and perforated. Stress around the perforation tunnel is redistributed. Rock matrix failure occurs once this stress (shear or tensile) becomes larger than rock strength. However it doesn't always lead to sand production. Disaggregated particles are still held by capillary and cohesion forces. Once fluid flow energy exceeds these forces, sand particles are detached from the rock matrix, and sand production is initiated. This zone of failed material is called plastic zone, and volume of sand production depends on the size of this plastic zone (Figure 1).

Three main types of sand production have been identified from the field observations and laboratory studies: Transient sand production; Continuous sand production; and Catastrophic sand production.

Just after the perforation burst of sand production is observed due to perforation debris coming out with reservoir fluid. If the conditions are favorable, sanding arch is formed (Figure 2) and sand production ceases. This is called *transient sand production*. This phenomenon can also be observed after fluid rate or drawdown changes, when the sand arch is destroyed under new flow rate and drawdown values, producing considerable amount of sand, followed by forming a new sand arch. At some critical fluid rate or drawdown sand arch never forms, and sand is produced continuously at constant rate. This is called *continuous sand production*. At higher drawdown pressures the plastic zone enlarges, and at some destructive drawdown pressure *catastrophic sand production* occurs, which leads to borehole collapse and well shut-in.

Parameters influencing sand production can be grouped as rock and fluid properties, completion design and production conditions (Cheung, 2010). There is a strong relationship between rock strength and the geometry of the wellbore breakout (Rahmati et al, 2013). The post-failure mechanical properties of rock coupled with fluid flow have a significant impact on the volume of produced sand (Kim et al, 2012). It is observed both from the field and experimental data, that sand production increases with water

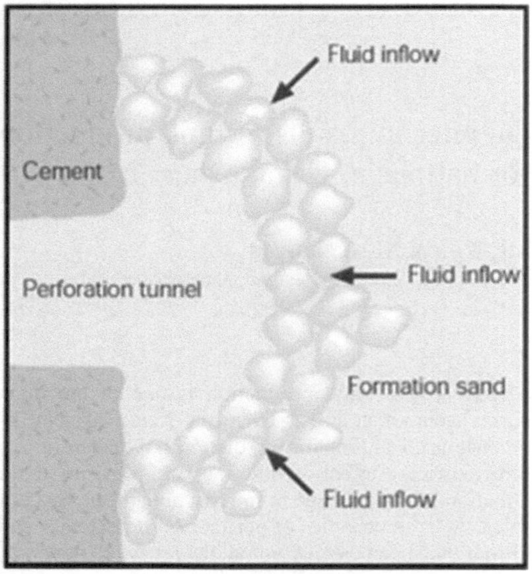

Figure 2. Sanding arch around the entrance to a perforation cavity (Carlson et al, 1992).

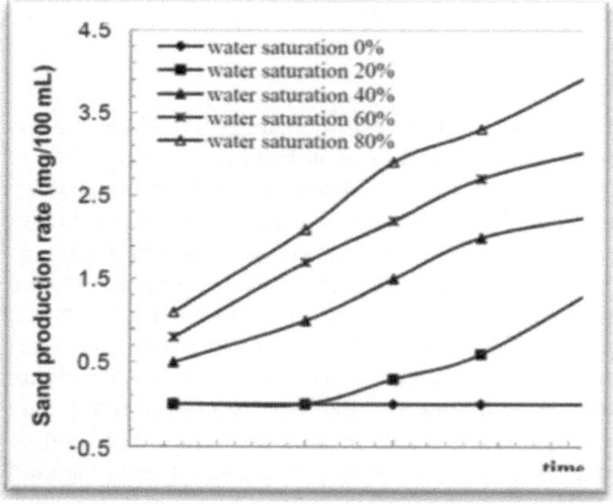

Figure 3. Sand production rate at different water saturation levels (Luo et al, 2012).

breakthrough. It is explained by rock strength reduction and capillary force decrease. Slight increase in water saturation leads to the most decrease in rock strength (Han et al, 2005). Young's modulus of saturated rock is about 20–30% lower than dry rock (Kim et al, 2012). Properties of the cement and injected water are also important for sand production (Luo et al, 2012). Swelling clay minerals can block the pores and reduce the permeability, which leads to higher drawdowns.

It is well reported in the literature that high stresses cause high sand production (Nouri et al, 2006). The effect of flow rate on sand production greatly depends on the phase of the flowing fluid (Bianco and Halleck, 2001). According to Papamichos et al, (2010)

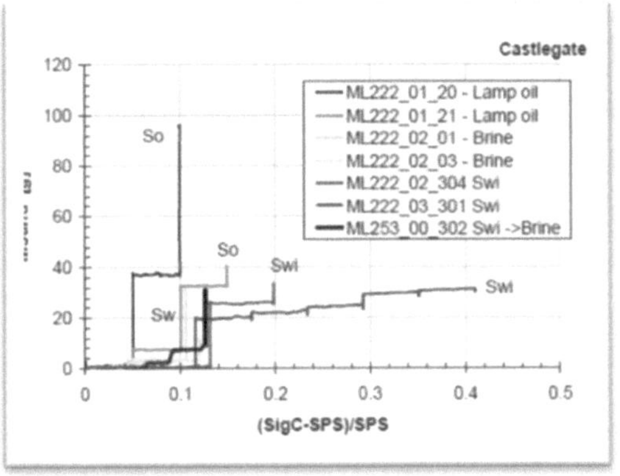

Figure 4. Sand production for single phase and multiphase flows (Papamichos, 2010).

and Bianco and Halleck (2001) volumetric sand production is less for multiphase flow than that of one phase flow.

Fattahpour et al, (2012), concluded that microproperties of sand grains and their relation with cement bonds have great impact on post-failure behaviour of the sandstone, thus on sand production volume.

Various analytical models have been developed based on different constitutive laws. The principle is to derive mathematical formulations for the perforation stability and failure mechanisms. Analytical models are fast and easy to use; however they are mainly based on simplifying assumptions and therefore the estimation of sand production for most cases is not accurate. Laboratory experiments are very useful to study the mechanisms of sand production, though its results are strongly influenced by boundary conditions and it is not always straight forward to apply them to the field-scale problems. Thick-walled cylinder tests are the typical sand production experiments, where hollow cylinder is compressed until sand is produced. The stress at which the initial failure of the cylinder material occurred is considered to be the stress at the onset of sand production.

Numerical models can be grouped under continuum and discontinuum methods. In the continuum method, rock is treated as a continuous material and constitutive laws are employed when describing its behavior. From the literature, the models based on critical state theory are the most accurate and allow capturing shear, tensile and compressional failures (Rahmati et al, 2013; Ranjith et al, 2014). The main disadvantage of continuum models is that they cannot describe local discontinuities as particle detachment. For this case discontinuum methods can be used. The 3D Discrete Element Modeling (DEM) can capture the motion of individual grains, and therefore provide information on sanding mechanism at micro scale. Considering each individual particle makes DEM computationally expensive and time-consuming, which prevents using DEM for large-scale problems. For eliminating weaknesses of continuum and discontinuum methods, hybrid FEM-DEM models can be used for sand production problems. Computational Fluid Dynamics (CFD) is used with DEM to describe fluid flow and sand erosion (Zhou et al, 2011).

3 RESEARCH PROCEDURES

3.1 *Preparation of artificial sandstone sample*

Artificial sandstone sample with properties similar to that of real rock will be prepared using the different methods. Ismail et al. (2000) prepared cemented sand samples using different cementing agents and observed that the stress-strain behavior of samples having the same unconfined compression strength was dependent on the preparation method, which was due to the effect of different microstructures. The Calcite-In-situ-Precipitation-System (CIPS) method (Kucharski et al., 1996) was recommended for the preparation of artificial sandstone as it was able to replicate more accurately the natural process of calcite precipitation. A sand production experimental plan using the CIPS method and Portland cement as the cementing agents will be developed. Field conditions (confining pressure, flow rate, etc.) will be taken into account and factors controlling sand production as given in the literature (e.g., Carlson et al., 1992) will be examined.

3.2 *Study of the material properties, microstructure and composition of samples*

Specific gravity, Unconfined Compressive Stress (UCS), porosity, permeability, and stress-strain behaviour of the artificial samples will be investigated. As a result, methodology of preparing artificial sandstones with specified petrophysical properties and mechanical strength will be developed. Microstructural analysis is important for investigation of cement bond properties, stress effect on size and shape of sand particles. Mineralogical composition of the sample greatly influences its properties, therefore has to be properly described and studied. For this purpose, X-ray diffraction analysis and Scanning Electron Microscopy and Energy Dispersive X-ray Spectroscopy (SEM-EDS) analysis on samples will be performed. Other technologies (e.g., 3-D X-ray computed tomography, 3-D CT) will be employed if necessary. Based on literature review, some selected material properties of artificial sandstones will be determined following American Society for Testing and Materials (ASTM) standards. Other literature will be referred to if proper methods are not available in ASTM for any specific properties.

3.3 *Perforating the samples*

Weak sandstones are almost never produced from open hole. The wellbore is cased and perforated before

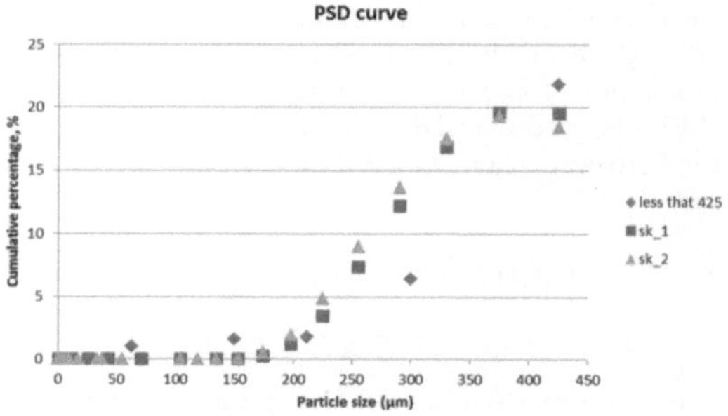

Figure 5. Typical particle size distribution curves for reservoir sand produced from the oilfield near Aktau.

inflow initiation. In a laboratory we have to reproduce perforation process in order to get samples with perforation tunnels, from which sand is transported to the wellbore.

3.4 *Experimental tests of sand production process*

Sand production will be studied using a new high pressure apparatus. The apparatus has loading frame to apply stress and the cell, which has several ports to apply different fluids, like water, crude oil and gas, etc. Produced sand is flowed from top of the sample.

3.5 *Sand production modelling at microscopic and macroscopic levels*

DEM simulation of discrete particles will be run to calculate fluid-particle interaction force for each particle; cementation bonding is modelled as adhesion force between particles. FEM simulation runs will be conducted to model the water saturation effects. Heat transfer in particle-fluid flow system will be modelled by thermal DEM-CFD approach.

4 LABORATORY EXPERIMENTS

Sample preparation method by literature (Shabdirova et al., 2016) was used to prepare artificial sandstone samples with the reservoir sand produced from the oilfield near by Aktau. The main components of this sand are quartz (Silicon oxide, SiO_2) and Kaolin (Aluminum silicate hydroxide, $Al_2Si_2O_5(OH)_4$).

Sieve Analysis and Mastersizer 3000 with dispersion unit Aero S was conducted to identify particle size distribution of field sand. Samples after treatment from oil were prepared and grouped as cohesionless and fine-grained soils. The test sample was obtained by quartering and was oven dried at 105°C to 110°C. Dry sieving method was used for more homogeneous material, while wet sieving was applied for samples to remove silt and clay with a 75 μm sieve. Figure 5 shows the typical particle size distribution curves for field sand.

In order to study the material properties and stress-strain behavior of sample, consolidated undrained test was performed using Digital Tritest 50 apparatus.

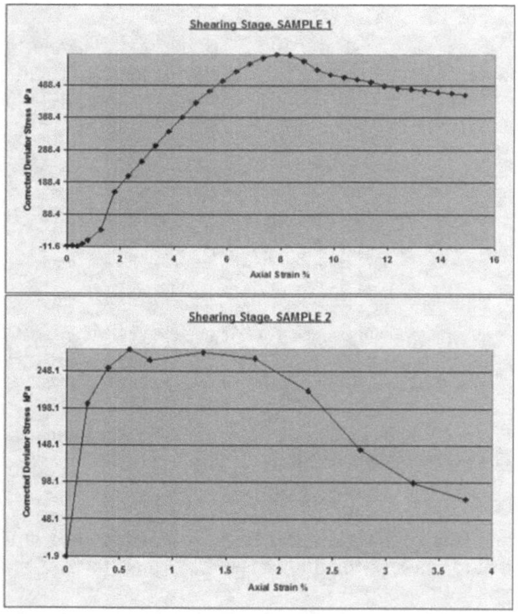

Figure 6. Stress-strain relations during the shearing stage of the CU Test.

Figure 6 shows the shearing stages of CU Tests for Sample 1 and 2.

5 NUMERICAL SIMULATIONS

The pilot modeling of triaxial test and perforation were conducted with FBGRAN code from the Granular Dynamics Research Group of the University of Birmingham, UK. DEM Modeling and simulation procedures were followed by the literature (Thornton, 2000).

5.1 *Triaxial test simulation*

The total numbers of 3620 spheres were randomly generated within a cuboidal periodic cell (4.6 mm) with a solid fraction 0.5 and no interparticle contacts. Five different sizes of spheres were used during the specimen generation:

1. 0.135 mm – 190 particles;
2. 0.200 mm – 1325 particles;
3. 0.265 mm – 990 particles;
4. 0.330 mm – 925 particles;
5. 0.400 mm – 190 particles.

The following mechanical properties were adjusted for the all particles:

1. Young's modulus E = 70 GPa;
2. Poisson ratio $\nu = 0.3$;
3. Solid density $\rho = 2650\ kg/m^3$.

Isotropic compression: Prepared specimen was compressed with an isotropic stress using a strain rate of

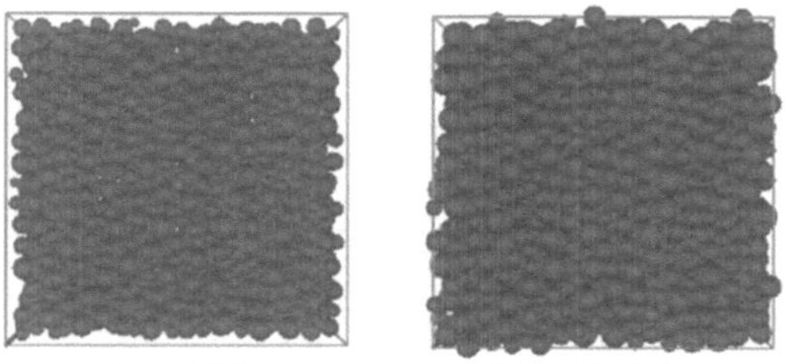

(a) p=0 kPa (b) p=100 kPa

Figure 7. Isotropic compression.

(a) at 650 cycles (b) at 750,000 cycles (c) at 850,000 cycles

Figure 8. Specimen compression during triaxial simulations.

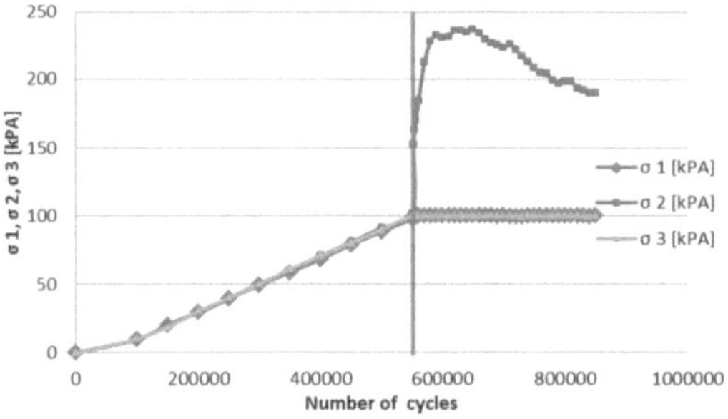

Figure 9. Evolution of stresses with the number of cycles.

10^{-4}/sec until 100 kPa. A servo-control was periodically introduced in order to maintain a desired stress level until the solid fraction and coordination number have attained constant values and a quasi-equilibrium state had been achieved. Figure 7 shows the specimen compression during the isotropic compression when the isotropic pressure p changes from 0 kPa to 100 kPa.

Triaxial (axisymmetric) compression: Isotopically compressed specimens were prepared for the triaxial compression stage. During the axisymmetric compression a strain rate of 10-5/sec was used. The imposed strain-rate tensor was adjusted to minimize the difference between the desired and measured values of the stress tensor. Figure 8 shows the specimen compression during the triaxial compression. Figures 9 to12 show typical results from the triaxial test simulations.

5.2 *Perforation simulation*

The perforation simulation consists of 3 stages: particle generation, deposition of particles and perforation simulation.

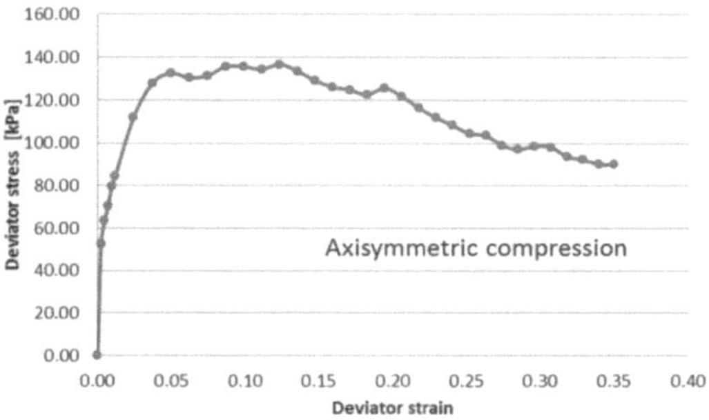

Figure 10. Evolution of deviator stress during axisymmetric compression (constant mean stress at 100 kPa).

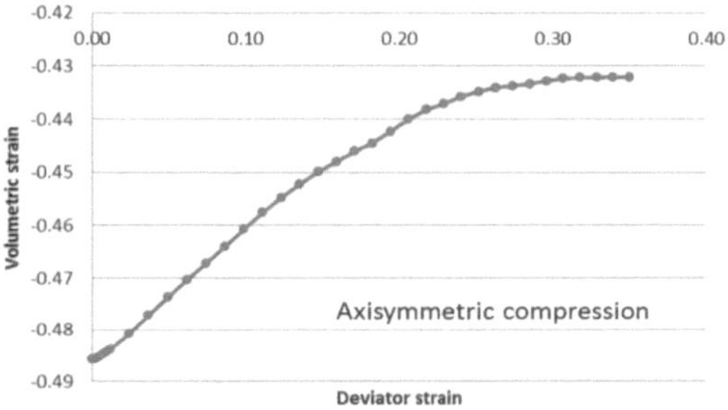

Figure 11. Evolution of volumetric strain during axisymmetric compression (constant mean stress at 100 kPa).

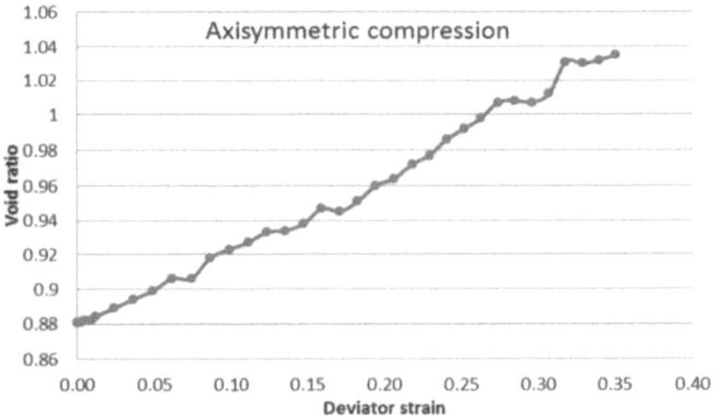

Figure 12. Evolution of void ratio during axisymmetric compression (constant mean stress at 100 kPa).

In the first stage, the preferred number of particles was randomly generated. Mechanical properties for two types of materials (particles and walls) were set. The base wall was plane infinite and other walls were periodic. In the second, stage the particles needed to be pluvial deposited to form the bed before the finite wall (perforator) was introduced. It would continue in a similar way until having a settled bed in an approximately equilibrium state. The porosity, number of contacts, maximum and average ball velocities and maximum Y value (height of bed) were monitored to check the equilibrium state. In the third stage, the small finite wall was created at the top of workspace as a penetrometer. By specifying the velocity in the vertical direction for this small wall and by setting the

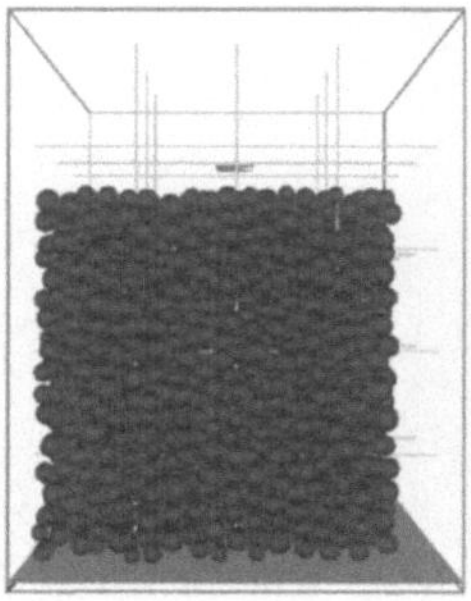

(a) 2D model

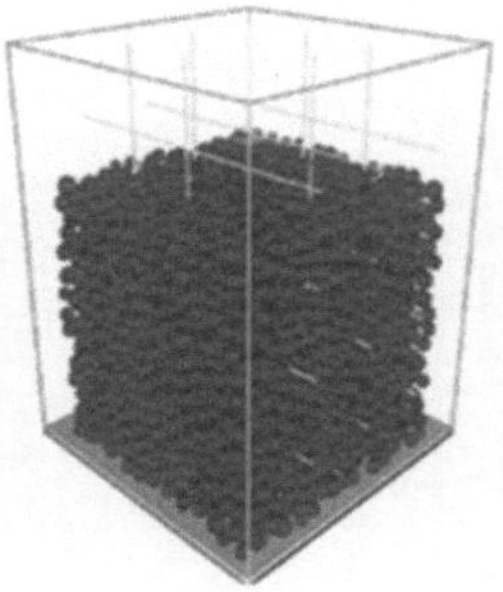

(b) 3D mode

Figure 13. DEM models of perforation.

number cycles this wall will go down and penetrate the top center of the specimen. Figure 13 shows examples of the 2D and 3D models of perforation.

6 CONCLUDING REMARKS

The following results will be expected to achieve from this project:

1) Development of analytical relationships to relate the perforation penetration to perforation energy and material properties; additional factors to be considered are stress history, perforation diameter, perforation spacing and perforation sequence.
2) Quantification of the effect of the perforation damage on the amount of sand production in weak sandstones and to develop criteria for sanding initiation and sanding severity in terms of the fracture damage level in the parent rock, the critical drag force and the depletion level over the timescale of a well-life.
3) Development and calibration of a continuum based modeling method for boundary value problems of sand production which can incorporate the microscopic characteristics of the sanding mechanisms discovered in this study.
4) Development of a strategy for sand production management in Kazakhstan including an efficient method to separate the eroded sands from the production flow based on the research outcomes.

ACKNOWLEDGMENTS

This research was sponsored by Nazarbayev University. The authors thank Misses A. Shabdirova, A. Rakhimzhanova and Z. Bissekenova for their work for experiments and numerical simulations.

REFERENCES

Bianco, L., Halleck, L. (2001), *Mechanisms of arch instability and sand production in two-phase saturated poorly consolidated sandstones* SPE European Formation Damage Conference, The Hague, Netherlands.

Carlson, J., Gurley, D., King, G., Price-Smith, C., Waters, F. (1992), *Sand control: Why and How?* Oilfield review.

Cheung, L.Y.G. (2010), *Micromechanics of Sand Production in Oil Wells*, Ph.D. thesis, Imperial College of London.

Fattahpour, V., Mahdi, M., Mahdi, M. (2012), *An experimental investigation on the effect of grain size on oil-well sand production,* Journal of Petroleum Science and Engineering. Vol. 9 pp. 343–353.

Han, G. and Dusseault, M. B. (2015), *Sand stress analysis around a producing wellbore with a simplified capillarity model*, International Journal of Rock Mechanics and Mining Sciences, Vol. 42, Issues 7–8, pp. 1015–1027.

Ismail, M. A., Joer, H. A., & Randolph, M. F. (2000), *Sample Preparation Technique for Artificially Cemented Soils*, Geotechnical Testing Journal, Vol. 23, No. 2, pp. 171–177.

Kim, A.S., Sharma, M.M. (2012), *A Predictive Model for Sand Production in Realistic Downhole Condition,* 46th US Rock Mechanics / Geomechanics Symposium, Chicago, IL, USA

Kucharski, E., Price, G., Li, H., and Joer H. A. (1996), *Engineering Properties of CIPS Cemented Calcareous sand*, Proceedings of the 30th International Geological Congress, Beijing, China.

Luo, W., Xu, S., Torabi, F. (2012), *Laboratory Study of Sand Production in Unconsolidated Reservoir*, SPE Annual Technical Conference and Exhibition, San Antonio, TX, USA

Nouri, A., Vaziri, H., Kuru, E., & Islam, R. (2006), *A comparison of two sanding criteria in physical and numerical modeling of sand production,* Journal of Petroleum Science and Engineering, Vol. 50, No. 1, pp. 5570.

Papamichos, E., Tronvoll, J., Skjærstein, A., Unander, T. E. (2010), *Hole stability of Red Wildmoor sandstone under anisotropic stresses and sand production criterion,* Journal of Petroleum Science and Engineering, Vol. 72, Issues 1–2, pp. 7892

Rahmati, H., Jafarpour, M., Azadbakht, S., Nouri, A., Vaziri, H., Chan, D., Xiao, Y. (2013), *Review of Sand Production Prediction Models*, Journal of Petroleum Engineering, Vol. 2013

Ranjith, P.G., Perera, M.S.A., Perera, W.K.G., Choi, S.K., Yasar, E. (2014), *Sand Production during the Extrusion of Hydrocarbons from Geological formations: A review*, Journal of Petroleum Science and Engineering, Vol. 124, pp. 72–82

Shabdirova, A.D., Bissekenova Z., Minh, N.H., Kim, J.R. (2016), *Sample preparation method of clay-rich sandstone analogue of sandstone reservoirs in Kazakhstan*, 50th US Rock Mechanics/Geomechanics Symposium, Houston, Texas, USA,

Thornton, C. (2000), *Numerical simulations of deviatoric shear deformation of granular media*, Geotechnique, Vol. 50, No. 1, pp. 43–53

Zhou, Z.Y., Yu, A.B., Choi, S.K. (2011), *Numerical simulation of the liquid-induced erosion in a weakly bonded sand assembly*, Powder Technology Vol. 211, pp. 237–249

Laboratory and field testing (TS1)

Challenges and Innovations in Geotechnics – Zhussupbekov (Ed.)
© 2016 Taylor & Francis Group, London, ISBN 978-1-138-03007-7

Centrifuge applications in micropile foundations

A. Alnuaim
Civil Engineering Department, King Saud University, Riyadh, Saudi Arabia

M.H. El Naggar
Department of Civil and Environmental Engineering, The University of Western Ontario, London, ON, Canada

H. El Naggar
Dalhousie University, Halifax, NS, Canada

ABSTRACT: The geotechnical centrifuge testing is a useful tool to simulate complex geotechnical problems such as soil–structure interaction (SSI), offshore pipelines and dams. In deep foundation engineering, the geotechnical centrifuge testing can offer insight into the behavior of piles and its interaction with soil or with other structures such as a raft (mat foundation) or adjacent structures. This paper will present some application of geotechnical centrifuge testing on the complex interaction of micropiled raft (MPR) for both sand clay soils. Four MPRs centrifuge tests were conducted: three tests in sandy soil and one test in clay soil. In addition, single micropile and isolated raft foundation centrifuge tests were carried out in both sand and clay soils. The results of the centrifuge tests were used to calibrate and verify the non-linear three-dimensional finite element models for both the sand and clay soils. Subsequently, the verified models were employed to conduct a comprehensive parametric study. The parametric study focused on providing additional insights regarding the performance of micropiled raft system that should be helpful for the design engineers. The physical dimensions of MPR structural components and parameters considered in the parametric study are within the range used in the current practice. The results of the centrifuge tests and numerical parametric study were analyzed to establish design guidelines for micropiled rafts. The results of both centrifuge testing and numerical analysis can be used to develop some design guidelines and assess the available analytical solutions and offer some adjustment factors.

Keywords: Centrifuge, micropile, micropiled raft, sand, clay.

1 INTRODUCTION

Performing a full-scale testing program for micropiled raft foundations could be extremely expensive if all important factors that influence their performance are to be investigated. Centrifuge testing offers an acceptable alternative for full-scale testing because it provides approximately the same stress level of the prototype case, which is important to replicate the soil stiffness and strength.

1.1 *Concept of centrifuge modelling*

The concept of this test is to rotate a body of certain mass (m) at a constant radius (r) with a steady speed (υ) about an axis. To maintain the circular path, the body will experience a constant radial centrifugal acceleration equal to $r\omega^2$. The body will be subjected to a radial force equal to $me\omega^2$, where $r\omega^2 = v\hat{}2/r$ ($\omega =$ angular speed). This radial acceleration can be normalized by earth's gravity, g, so the body is subjected to an acceleration equal to ng where n is scale factor, which can be given by:

$$n = (r\omega\hat{}2)/g \tag{1}$$

Under these conditions, the stresses in the centrifuge model will increase to maintain equilibrium and will be similar to the stresses at equivalent points on the prototype. However, the centrifuge model dimensions are scaled down by a linear scale equal to n as will be discussed later. However, the scale is not linear in all situations. (Wood, 2003).

Figure 1 shows a comparison between the soil under 1 g gravity field and the soil in the centrifuge test under ng gravity field. Table 1 shows a comparison between the vertical stress level in 1-g model and centrifuge model to the vertical stress level in the prototype.

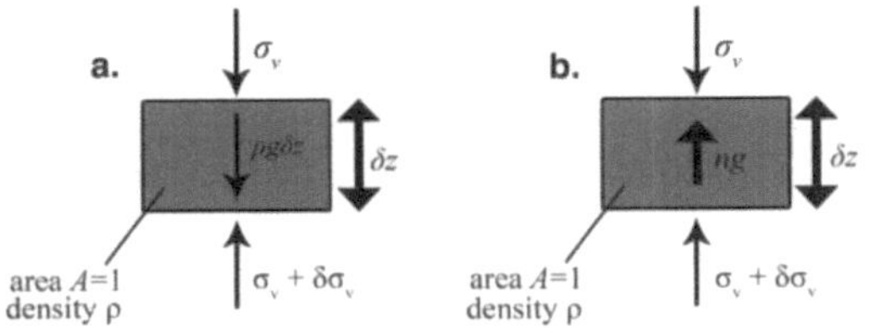

Figure 1. An element of soil: (a) at 1g gravity field and (b) on centrifuge test under ng gravity field (after Wood, 2003).

Table 1. Scaling factors for parameters used to calculate the vertical stress in 1 g testing and centrifuge testing (ng).

	1 g testing		centrifuge testing (ng)	
Quantity	Prototype	Model	Prototype	Model
Depth, z (m)	h	h/n	h	h/n
Density, ρ_s (Kg/m^3)	1	1	1	1
Unit Weight, γ (kN/m^3)	$\rho_s g$	$\rho_s g$	$\rho_s g$	$\rho_s ng$
Stress = z* γ (kPa)	$h\rho_s g$	$h\rho_s g/n$	$h\rho_s g$	$h\rho_s g$

Table 1 shows that the 1-g small-scale model reduces the stresses level at any equivalent point of the prototype, which can lead to some errors, especially if the model is sensitive in respect to the vertical and horizontal stresses in the soil such as the case for pile-soil interaction modelling. On the other hand, the small-scale centrifuge model can replicate the same state of stress as the prototype case.

1.2 *Scaling Laws of centrifuge modelling*

The main advantage of the centrifuge testing is to model the prototype using a model with smaller dimensions while maintaining the soil state of stress unchanged between the prototype and the model. However, the scaling of the prototype has to follow certain scaling laws to simulate the real behavior of the prototype for the problem under consideration. (Garnier et al., 2007) presented a set of scaling laws necessary for accurate modeling of piled raft foundations under *ng* gravity field. Table 2 summarizes the scaling laws applicable to micropiled raft foundations.

1.3 *Micropiles*

A micropile is a small diameter "cast-in-place" pile. Micropiles were initially used to repair historic buildings that were damaged during World War II, which entailed installing micropiles by drilling through the existing foundation and filling the holes with cement grout and a steel bar (Lizzi, 1982). Nowadays, different types of micropiles are used to provide the load carrying capacity of new foundation systems (FHWA, 2005).

Micropiles were introduced into North America in the mid 1980's and their use has been rapidly growing ever since. Their construction methods have been improved and new construction techniques have been developed. The advancements in drilling equipment have resulted in the ability to drill through almost any ground condition to install micropiles at any angle with minimum noise, vibration and disturbance. In addition, the relatively small size of the equipment has allowed the underpinning of existing foundations even in restricted access situations (Bruce and DiMillio, 1995).

Table 2. Scaling laws used in centrifuge testing.

Characteristic	Model/Prototype
Gravity	n
Density	1
Stress	1
Length	$1/n$
Acceleration	n
Dynamic Frequency	n
Velocity	1
Axial Rigidity, EA	$1/n$^2
Flexural Rigidity, EI	$1/n$^4
Force	$1/n$^2
Mass	$1/n$^3
Moment	$1/n$^3

In most cases, the micropiles are constructed by drilling the pile shaft (the bore hole) to the desired depth using an appropriate drilling method depending on the depth and size of the micropiles and the sensitivity of the site to disturbance. During drilling, a casing is installed simultaneously to support the shaft wall. After the drilling is completed, a steel reinforcement, (e.g. steel bars) is placed, occupying approximately 8% of the total volume. Subsequently, grout is placed in the predrilled micropile hole; in most cases the grout is pressurized after removing the temporary casing to increase the diameter of the bonded length and to increase the friction of the micropile shaft with the soil. In order to increase the lateral stiffness and capacity of the micropile, the top portion of the temporary casing can be left in ground and is not fully removed (Shong and Chung, 2003).

Micropiles are used in a variety of applications, including: (1) to serve as a main foundation system for static and dynamic loading; (2) to upgrade an existing foundation system; and (3) to stabilize slopes and reinforce the soil (Bruce and DiMillio, 1995).

The concept of underpinning foundation systems using micropiles is similar to the concept of the piled raft, which is a composite structure with three components: subsoil, raft and piles. These components interact through a complex soil-structure interaction scheme, including the pile-soil interaction, pile-soil-pile interaction, raft-soil interaction, and finally the piles-raft interaction.

This paper presents some of the geotechnical centrifuge applications in micropile foundation systems in sand and clay soils.

2 CENTRIFUGE TESTING PROGRAM

All the centrifuge testing was carried out at C-Core Centrifuge facility located at Memorial University, St. John's, Newfoundland, Canada. For the sand soil, the centrifuge testing program consisted of the following tests: (1) one test on a single micropile; (2) one test on a raft with a thickness equivalent to 0.6 m in prototype scale; and (3) three tests on micropiled groups with different raft thicknesses (i.e. 0.3 m, 0.45 m and 0.6 m in

Table 3. Scaling laws for centrifuge modeling and models and equivalent prototype dimensions.

Description	Scaling law	Prototype	Model
Micropile Diameter	E_pA_p/n^2	150 mm	9.53 mm
Axial Rigidity (EA)		516.8 MN	207 kN
Micropile Length	1/n	10 m	200 mm
Raft Width and Length	1/n	5.25 m	105 mm
Raft Thickness	E_pI_p/n^4	0.6 m	16.4 mm
Flexural Rigidity		698 MN. m^2	0.1 kN.m^2
Raft Thickness	$E_{pp}I_p/n^4$	0.45 m	12.3 mm
Flexural Rigidity		295 MN. m^2	0.05 kN.m^2
Raft Thickness	E_pI_p/n^4	0.3 m	8.2 mm
Flexural Rigidity		87.25 MN. m^2	0.014 kN. m^2

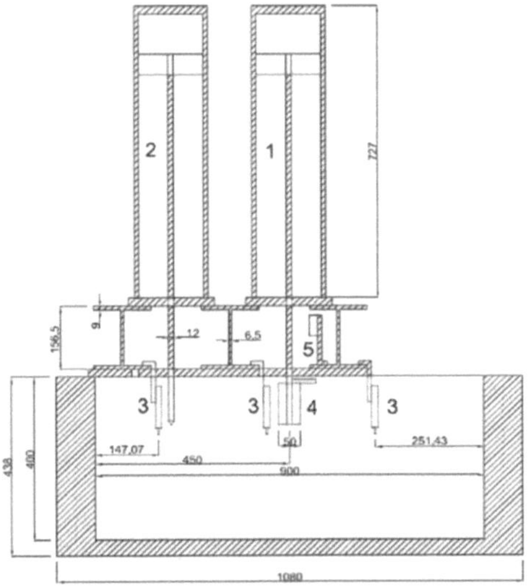

Figure 2. Vertical cross-section of centrifuge package consists of: (1) vertical actuator for applying the load; (2) sand cone for CPT; (3) LVDTs; (4) load cell; and (5) laser (all dimensions in mm).

prototype scale). All tests were performed on dry sand soil with a relative density (D_r) of 70% and under centrifugal acceleration of 50 g (Alnuaim et al., 2015b). In addition, 3 testes were carried out in clay soil as following: (1) one test on micropiled raft; (2) one test on a single micropile; and (3) one test on an isolated raft. The soil bed in all tests was kaolin-silt mixture (K-S), which was consolidated from slurry and had an average undrained shear strength, $s_u = 30$ kPa. All tests were performed under centrifugal acceleration of 50 g (Alnuaim et al., 2015a). Table 3 presents all the model and prototype dimensions for the tests. Figure 2 and 3 shows the setup for centrifuge testing programs in both sand and clay soils.

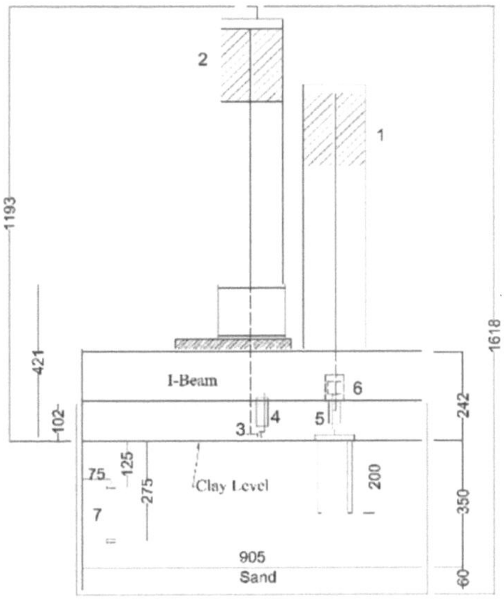

Figure 3. Vertical cross-section of centrifuge package consists of: (1) vertical actuator for applying the load to the model, (2) actuator used for T-bar, (3) T-bar, (4) LVDTs to measure clay settlement, (5) load cell, (6) laser, and (7) PPTs (all dimensions in mm).

3 APPLICATION OF CENTRIFUGE TESTING PROGRAM IN MICROPILE FOUNDATION

3.1 *Estimating micropile capacity*

The geotechnical centrifuge testing has the ability to estimate the capacity of micropile, micropile group or micropiled raft (MPR) in cheaper way than conducting a file testing. Moreover, in the case of micropile group and micropiled raft, it is very difficult to perform a field test to confirm the design capacity for those systems. However, by using model scale in centrifuge which will duplicate the field conditions and therefore the load bearing capacity can be estimated.

Figure 4 and 5 show the load displacement curve for single micropile in sand and clay, respectively. From the load displacement curve the bearing capacity can be estimated and then converted into prototype scale.

3.2 *Benchmark for numerical modeling*

One of the benefit of having a small scale geotechnical centrifuge testing results, is the ability to calibrate and verify the numerical modeling. As mentioned before the model in the centrifuge will be under similar stresses as the prototype which yield accurate results compared to the field tests. This section presents the development and calibration of the FEM that was used to carry out the numerical parametric study for a micropiled raft (MPR) installed in sand and clay. The 3D FEM was established using the computer program PLAXIS 3D (PLAXIS, 2013) considering an appropriate size mesh and a number of elements following a sensitivity study. The results of the centrifuge study of

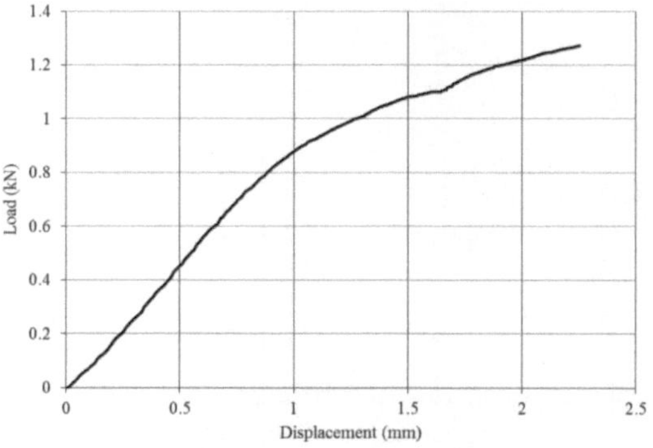

Figure 4. The load-settlement curve for a single micropile in sand (model scale).

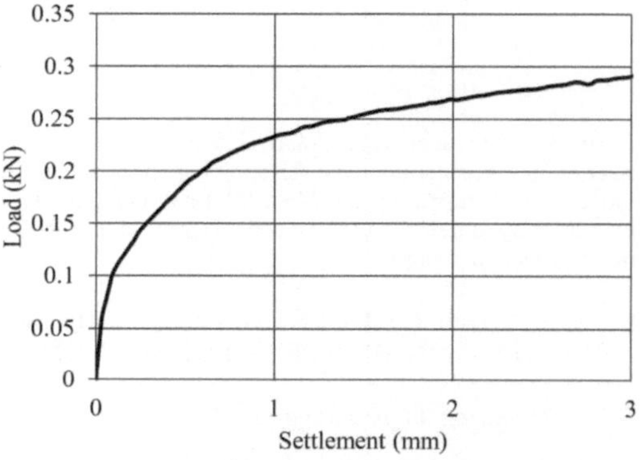

Figure 5. The load-settlement curve for a single micropile in clay (model scale).

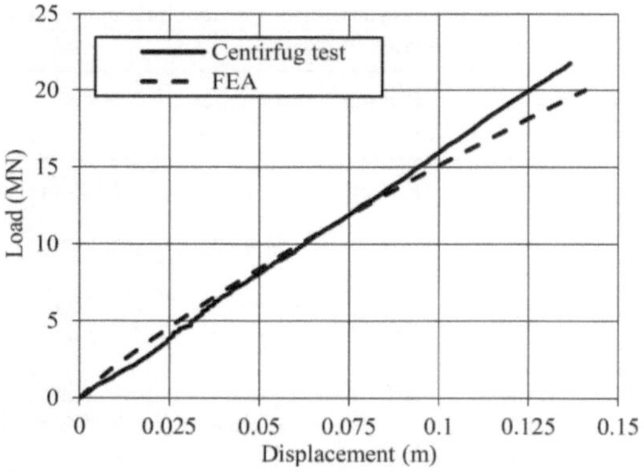

Figure 6. 3D FEA verification result for MPR with 0.6 m raft in sand.

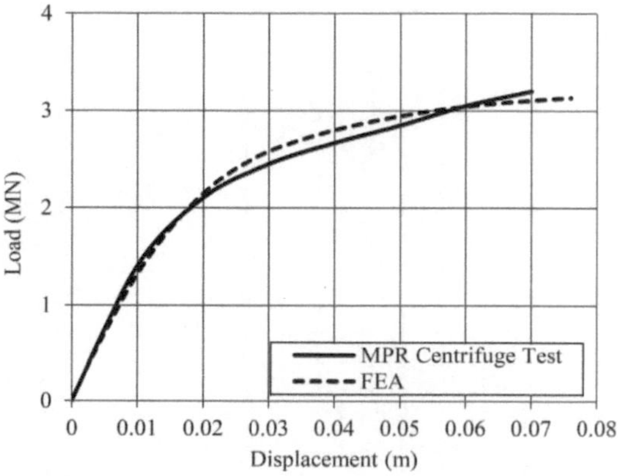

Figure 7. 3D FEA verification result for MPR with 0.6 m raft in clay.

the MPR were then used to calibrate the FEM considering the prototype dimensions of the MPR tested in the centrifuge. The calibrated FEM was then employed to perform the parametric study to evaluate the effect of different factors on the performance of the MPR system. Figures 6 and 7 show the verification results of numerical model to the centrifuge testing results of MPR in both sand and clay. It can be notice the good agreement between the numerical and centrifuge results which indicate that the numerical model is capable to predict the behavior of MPR accurately and can be utilize to understand the effect of different factors on the performance of MPR.

3.3 *Load sharing in MPR*

The concept of underpinning foundation systems using micropiles is similar to the concept of the piled raft, which is a composite structure with three components: subsoil, raft and piles. These components interact through a complex soil-structure interaction scheme, including the pile-soil interaction, pile-soil-pile interaction, raft-soil interaction, and finally the piles-raft interaction. These interactions will affect the load sharing between the micropiles and the raft which is very difficult to estimated analytically and should be evaluated using micropiled raft load test, geotechnical centrifuge or calibrated numerical model.

For the micropiled raft in sand, the axial load carried by the micropiles in a micropiled raft system is affected by the raft flexibility, which is influenced primarily by the raft thickness and the spacing between the micropiles. Figure 8 presents the percentage of load carried by each component of the micropiled raft system with different values of raft thickness at different load levels. At the initial load stage, most of the load was carried by the micropiles; this is believed to be due to the lack of intimate contact between the raft and underlying soil, as the soil tends to settle and compact during the increase of g level. Similar behavior was reported by Horikoshi and Randolph (1996). As the applied load increased, the proportion of the load carried by the micropiles dropped significantly at about 1 kN and continued to decrease gradually after that. At approximately 4 kN, the load transferred by the micropiles reached a plateau and became almost constant. At this load level, the percentage of load carried by the raft was 41%, 49% and 58% of the total

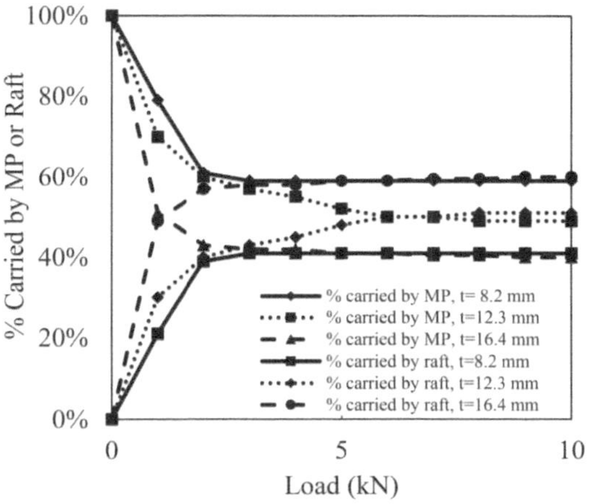

Figure 8. Load sharing between the micropiles and the raft for different raft thicknesses.

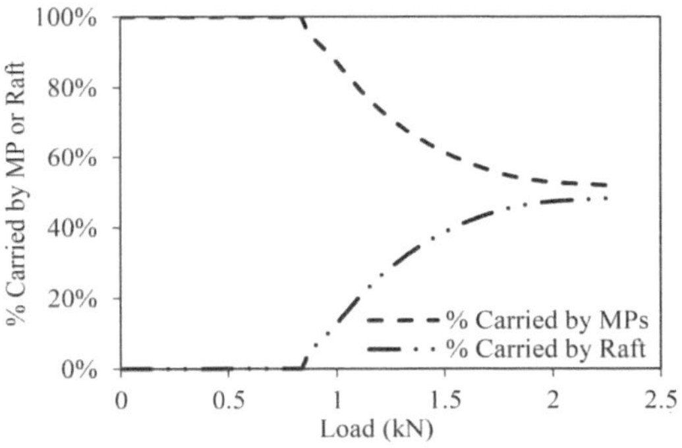

Figure 9. Load sharing between the micropiles and the raft for different raft thicknesses.

applied load for a raft thickness of 8.2 mm, 12.3 mm and 16.4 mm, respectively.

For micropiled in clay, Figure 9 presents the percentage of load carried by the micropiles and the raft at different load levels. At the initial loading stage (less than 0.75 kN), most of the load was carried by the micropiles; this is believed to be due to the lack of intimate contact between the raft and underlying soil, as the clay was consolidated by about 2.25 mm during the in-flight consolidation process. Similar behavior was reported by Horikoshi and Randolph (1996).

4 CONCLUSIONS

The results of the geotechnical centrifuge testing program presented in this paper clearly indicate the benefit of such testing program to understand the complex interactions in micropiled raft (MPR) and the effect of different parameters on the overall performance of MPR. In addition, it offers a great tool to calibrate and verify a numerical model for further and deep investigations.

REFERENCES

Alnuaim, A., El Naggar, M. & El Naggar, H. (2015a) Performance of micropiled raft in clay subjected to vertical concentrated load: centrifuge modeling. *Canadian Geotechnical Journal,* 52, 2017–2029.

Alnuaim, A. M., El Naggar, H. & El Naggar, M. H. (2015b) Performance of micropiled raft in sand subjected to vertical concentrated load: centrifuge modeling. *Canadian Geotechnical Journal,* 52, 33–45.

Bruce, D. A. & Dimillio, A. F. (1995) Introduction to Micropiles: An International Perspective. *Foundation Upgrading and Repair for Infrastructure Improvement, ASCE, Geotechnical Special Publication, No. 50,* 1–26.

FHWA (2005) Micropile Design and Construction Guidelines, Implementation Manual. National Highway Institute.

Garnier, J., Gaudin, C., Springman, S., Culligan, P., Goodings, D., Konig, D., Kutter, B., Phillips, R., Randolph, M. & Thorel, L. (2007) Catalogue of scaling laws and similitude questions in geotechnical centrifuge modelling. *International Journal of Physical Modelling in Geotechnics,* 7, 1.

Horikoshi, K. & Randolph, M. (1996) Centrifuge modelling of piled raft foundations on clay. *Geotechnique,* 46, 741–752.

Lizzi, F. (1982) The static restoration of monuments. *Basic creteria-Case histories-Strengthening of buildings damaged by earthquakes.*

Plaxis, B. V. (2013) *Reference Manual,* Delft, PLAXIS BV.

Shong, I. L. S. & Chung, F. C. (2003) Design & Construction of Micropiles. *Geotechnical Course for Pile Foundation Design and Construction, Ipoh,* 29–30.

Wood, D. M. (2003) *Geotechnical modelling,* CRC Press.

Challenges and Innovations in Geotechnics – Zhussupbekov (Ed.)
 ISBN 978-1-138-03007-7

Analyzing of soil ground to frost heaving of structures

Zhanbolat Shakhmov, Akmaral Tleubayeva, Elmira Smagulova, Laura Utepbergenova, Yessengeldy Togabayev & Daniyar Bazarbayev
L.N. Gumilyov Eurasian National University, Astana, Kazakhstan

ABSTRACT: Construction of structures in seasonal freezing condition of soil ground should be considered by frost susceptibility. Usually we review just vertical and tangent forces to structures of buildings. These forces relate to construct shallow foundations. When construction is gone in big depth it have to understand horizontal forces to underground wall, deep foundations by frost heaving. So understanding and consideration of different frost heaving direction in construction design is very important measure for prediction limiting deformations of structures.

1 INTRODUCTION

The deformation of underground structures of buildings is main geotechnical problem in seasonal freezing ground of Kazakhstan. So according to freezing-thawing of soil ground there are important to predict frost heaving, freezing force. There are tangent frost heaving, vertical frost heaving, horizontal frost heaving which influence underground structures and could deform those structures. Frost heaving of soil directly depend on moisture content of soil, type of soil, chemical composition. Especially soil has high frost susceptibility if pores of soil fulfilled with water.

Hydrodynamical forces of frost heaving tend to move up foundation with relegation soil. If statical relegation links between soil and foundation will be less than weight of foundation and structures, so after that begin to slip a frozen soil layer with lateral surface of foundation. The resistance to those slippage in freezing of soil and foundation define tangent forces of frost heaving. There is a regelation of soil with lateral surface of underground structure during the freezing of soil ground by depth.

2 INFLUENCE OF FROST HEAVING ON RETAINING WALLS

2.1 *Frost heaving principles*

Frost-heaving is probably the most troublesome practical consequence of soil freezing. Earlier researches of frost-heaving like Bescow show that the cause of frost-heaving is the formation of ice-lenses in the soil and that a thin water layer must exist between soil particles and ice so that the ice-lens may be segregated from the soils (Beskow, G 1947). It once was assumed that the frost damage was caused by expansion of water upon freezing, but this 9 percent volume change is not sufficient to explain heaves measured in frictions of a meter. A close examination of the soil involved in frost heave reveal a series of layers of pure ice separating soil layers. The cumulative thickness of the ice layers represents the total amount of frost heave, which can lift and crack building floors and foundations and pavements.

There are three necessary conditions for frost heaving: frost susceptible soil, freezing temperatures, and the presence of water. The frost-susceptibility of a soil depends on the amount of fines passing the no. 200 (0.074 mm) sieve. In particular, Casagrande (1931) found that he could index the frost-susceptibility of the soil as a function of the material finer than 0.02 mm. He found soils having more than 3% of material finer than the 0.02 mm size appear to be highly frost-susceptible. In addition to having a frost-susceptible soil, freezing temperatures and a near-by source of water are also required for frost heave to occur [6].

Freezing frost heaving of soil ground could provide vertical uplifting of foundation one or two stores building. And heterogeneity of soil, difference of freezing depth and pressure under foundation usually show uneven uplifting of foundation causing to collapsing of building. Even placing of the foundation sole lower than freezing depth of the region isn't keeping safety of the building because also important to take into account tangent frost heaving. Soil could freeze with walls of the foundation and uplift it by tangent force of frost heaving. Tangent force of frost heaving can provide rupture of low loading reinforced concrete piles with a wire.

Seasonal freezing of soil also is irregular in road or pavement. Along the pavement of it is due to the heterogeneity of the soil ground of foundations of the composition and moisture content, influence of the communication, culverts, underground crossings. In a transverse direction main reason is accumulating snow at the roadside. Irregular frost heaving leads to

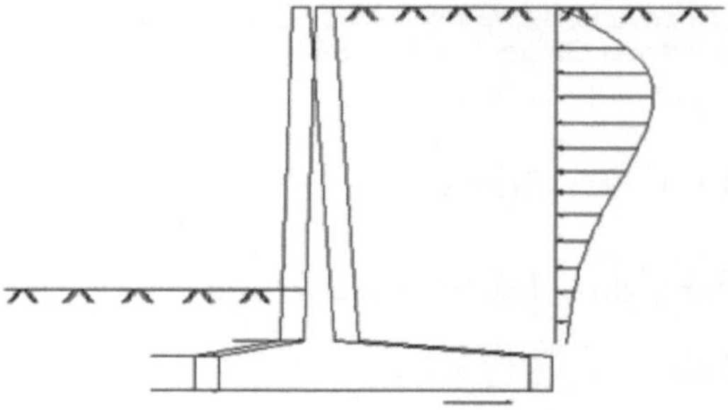

Figure 1. Influence of frost heaving on retaining wall (slippage).

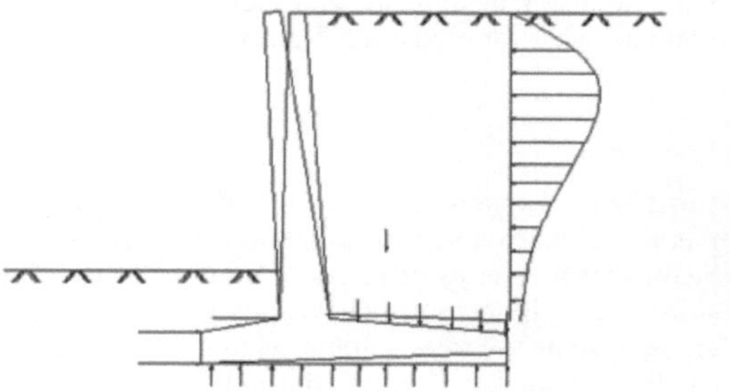

Figure 2. Influence of frost heaving on retaining wall (overturning).

formation cracks on surface of pavement. A great danger represents freeze heaving for vertical structures, interacting with the soil, retaining walls, the walls of the unheated cellars, underground crossings, wells and others.

2.2 *Deformability of retaining wall*

Susceptibility of soil to frost heaving is called heaving and frost heaving soil sometime it's called frost susceptiblity.

Frost heaving has deformability influence to foundations and underground walls. Freezing of soil ground on retaining walls is done vertically and horizontally. So therefore freezing for retaining walls is more complex comparing with typically foundations.

According to pictures above its seen two type deformations of retaining walls by freezing process of soil. There are slippage and overturning which could happen with retaining wall during freezing in the cold.

3 SOIL GROUND PROPERTIES AND FREEZING EXPERIMENTS

It is important to know soil ground properties for defining freezing susceptibility, which is important for design and construction of the buildings. Frost susceptibility could be estimated roughly by freezing soil normative according to soil properties (Zhussupbekov, A. et al. 2011, Zhussupbekov, A. et al. 2012, Zhussupbekov, A. et al. 2013).

Table 1. Properties of the soil specimens.

Soil number Description	I CL	II SP	III SM	IV CL
Specific gravity (G_s, kN/m^3)	2.62	2.63	2.67	2.63
Natural water content (w_n, %)	21	N.P	N.P	27
Particle size passed #200 (%)	52	19.6	1.9	86.1
Liquid limit (LL, %)	27.01	N.P	N.P	42
Plastic limit (PL, %)	17.75	N.P	N.P	25
Maximum dry unit weight (γ_{dmax}, kN/m^3)	1.79	1.92	1.93	1.76
Optimum water content (w_{opt}, %)	15.9	11.5	12.4	18
USCS	CL	SP	SM	CL

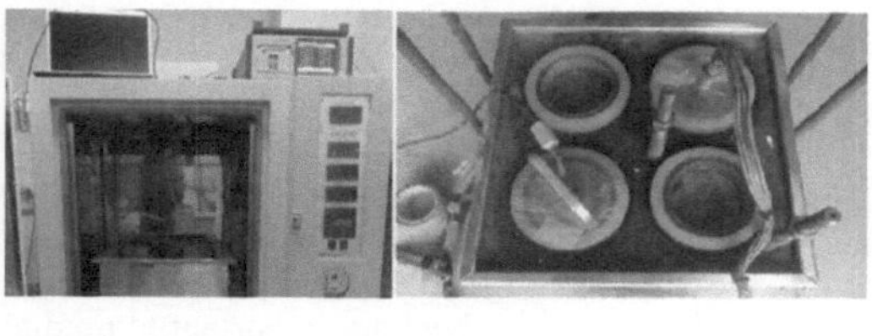

Figure 3. Freezing camera.

The special mold which can eliminate the side friction between soil and wall is used in the laboratory freezing test. Prior to the laboratory freezing test, geotechnical properties of the soils are determined throw the basic tests. The geotechnical properties of soil specimens are presented in Table 1.

The specimen is remolded and prepared in mold which is appropriate for freezing chamber. Only after saturation, they were putted in the freezing apparatus.

General view of freezing chamber filled in with soil and freezing equipment is presented in Figure 3.

The grain size distribution curves of the soils are presented in Figure 2. The first 24 hours is a conditioning period. Both the top and bottom plates are kept the temperature at 3°C. The first freeze starts at the beginning of the second 24-hours period. The temperature of the top plate was lowered and hold it up at −3°C. The temperature of the bottom plate was kept at 3°C for 8 hours.

4 RESULTS AND DISCUSSIONS

The experimental freezing tests were performed for understanding frost susceptibility of the soil and predetermination possible frost action to the structure. The frost heaving results of soil specimen after freezing test are presented in Figure 5. There are some curves

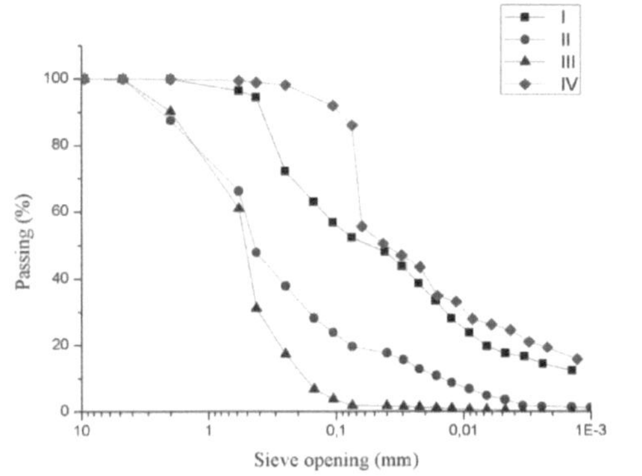

Figure 4. Grain size distribution curves of soil specimens.

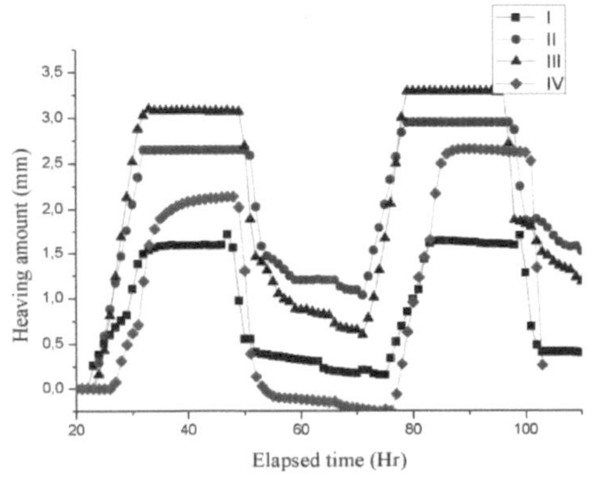

Figure 5. The frost heaving amount of the soil specimens by ASTM.

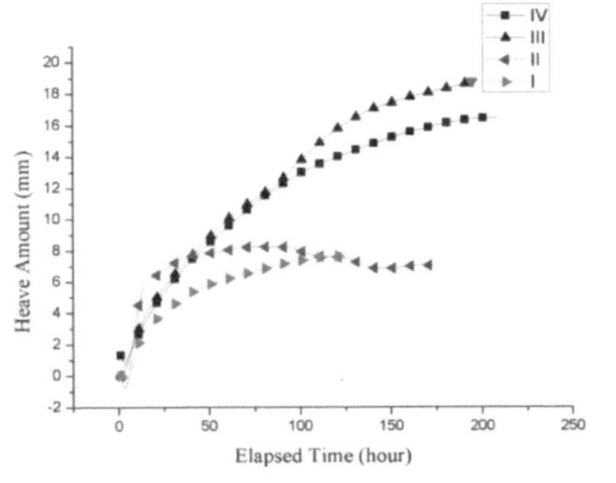

Figure 6. The frost heaving amount of soil specimens by TRRL.

which show frost different frost heaving resulted by freezing test.

Also freezing test according ASTM for comparison with TRRL testing provided by experiments throw freezing chamber. The results shows some differences and pasted below in Figure 6.

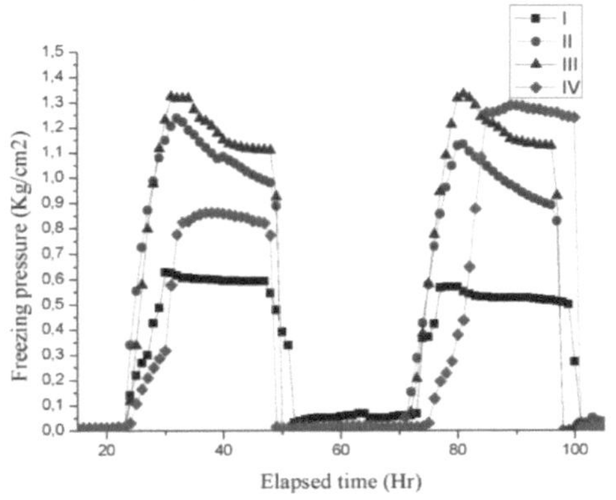

Figure 7. The freezing pressure curves of soil specimens by ASTM.

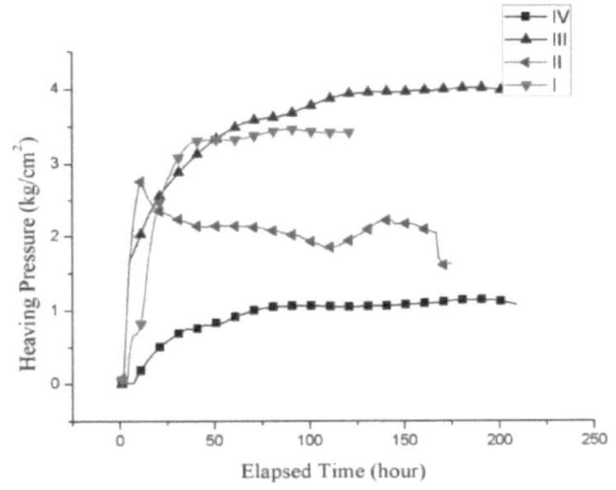

Figure 8. The freezing pressure curves of soil specimens according TRRL.

There also presented results of frost heaving pressure below.

The freezing pressure, heaving amount of soil specimens gives different results which show frost susceptibility of the soil specimens. According to the results TRRL frost susceptibility properties like frost heaving could reach more than 18 mm and heaving pressure 400 kPa. These results could damage roads and another light-weight constructions.

5 CONCLUSIONS

Freezing experiments show high frost susceptibility criteria for silty-clay soil types. Heaving amount for silty-clay soil ranging from 8 mm to approximately 20 mm. While heaving pressure ranging from 1 to 4 kg/cm^2. These results indicate sensibility of silty-clay soils to freezing process. Especially its important to consider frost susceptibility results during design of retaining wall. Also the results of freezing tests show dangerous data for retaining structures in condition of Kazakhstan.

REFERENCES

Beskow G, 1935. Soil Freezing and frost heaving with special application to roads and railroads. *Swed. Geol. Soc., Ser. C.* (translated by J.O.Osterberg, Northwestern Univ., 1947), Vol. 375, No. 3: 67–75.

Zhussupbekov A, Zhunisov T and Shakhmov Z, 2011. Influence of geological conditions on historic monuments in Kazakhstan. *Proceedings of Special Session on Geo-Engineering for Conservation of Cultural Heritage and Historical Sites: The 14th Asian Regional Conference on Soil Mechanics and Geotechnical Engineering,* Hong Kong, China: 111–114.

Zhussupbekov A, Shakhmov Zh, Shin EC and Krasnikov S, 2012. Challenges for transportation geotechnics in extreme climates of Kazakhstan and Korea. *Advences in Transportation Geotechnics II: CRC Press Taylor&Francis Group – 2nd International Conference on Transportation Geotechnics*, Hokkaido, Japan: 655–660.

Zhussupbekov A, Utepov Ye, Shakhmov Zh, Ling H, 2013. Model testing of piles in a centrifuge for prediction of their in-situ performance. *Journal Soil Mechanics and Foundation Engineering*, 50, No. 3: 92–96.

Challenges and Innovations in Geotechnics – Zhussupbekov (Ed.)
© 2016 Taylor & Francis Group, London, ISBN 978-1-138-03007-7

Advantages of the piles testing methods according to the USA standards

A.S. Tulebekova, N.T. Alibekova, I.T. Zhumadilov & G.R. Alipbayeva
Eurasian National University, Astana, Kazakhstan

ABSTRACT: Methodic of testing piles conforming to the requirements of ASTM D1143/D1143M-07 (USA) standard and GOST 5686–94 (Kazakhstan) standard is presented in this paper. The methodic of testing piles conforming to the requirements of these standards have some differences. Discussion of using control equipments, technological features, advantages and disadvantages of aforementioned methodic might be important for understanding the different points. The papers include the results of static tests of piles by using two different standards. These full field scale piling tests were performed in problematical soil ground of construction sites of new capital Astana (Kazakhstan).

1 INTRODUCTION

Development of high-rise buildings and constructions in complex hydro-geological condition of some regions of Kazakhstan requires reliable design of foundation, which leads to improvement of the geotechnical Standards. Nowadays many international projects are realized in Kazakhstan, it demands to use international Standard, moreover, for realization unique project usage of leading foreign high-tech, economic, ecological and energy-efficient technology, including technology for pile installation, equipment for geological investigation is required.

Unfortunately, present Standards are confined to application of modern technology of pile foundation installation, indicating incomplete usage of advanced technology. The results of research will be directed to modernization of the Kazakhstan Standards and oriented to adaptation of advanced geotechnologies.

Pile foundations are commonly used in Kazakhstan especially in such regions as Astana, Karaganda, Atyrau, Aktau, Pavlodar, Kostanay and Kokshetau. Field test are carried out in accordance with requirements of GOST 5686–94 "Soils. Methods of the field tests with piles" [1].

Definitions of bearing capacity of the piles are in accordance with requirements of SNIP RK 5.01-03-2002 "Pile foundations"

2 METHODOLOGIES OF THE FIELD TESTS

2.1 *Dynamic load test regarding GOST 5686–94 (Kazakhstan)*

In Kazakhstan dynamic test is carried out by using many types of pile driving machine. Before pile driving special marks through 1 m are painted on its surface along the whole length of the pile, but on last meter through each 0.1 m (see Figure 1).

In process of the pile driving the numbers of blows are recorded for each meter of pile penetration and last one meter through each 0.1 m. At the same time it is necessary to fix height of the fall of the blowing part of hammer. Test pile driving is terminated till designing refusals (cm/blows) (See Figure 2).

For definition of the bearing capacities of piles it is required to use average refusal which are obtained during redriving of the piles after their "rest". The "rest" time depends on soil condition of site: for clayey soil 6–10 days, for sandy and gravel soils up to 3 days.

Redriving of test piles is carried out by two step-by-step guarantees at three and five blows of the hammer.

Figure 1. Painted test pile.

Figure 2. Dynamic test of pie driving by machine Junttan PM-25.

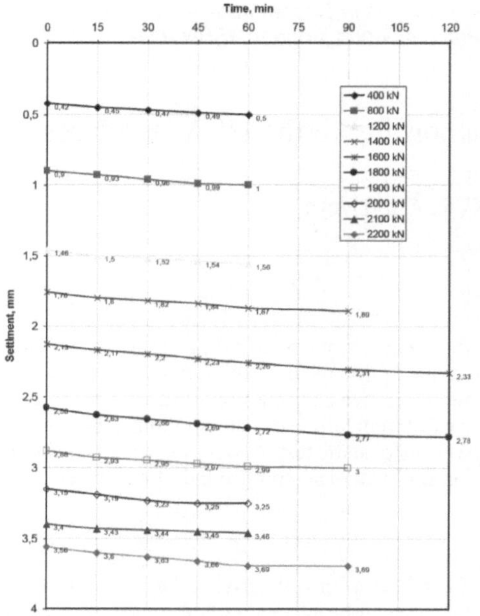

Figure 3. Criteria of conditional stabilization of pile.

Before beginning redriving on test piles measured tape with lengths of 10 cm divided by 1 mm is attached. The control of the test pile penetration is conducted through theodolit.

2.2 *Static load test regarding GOST 5686–94 (Kazakhstan)*

Static load test should be carried out for driving piles after the "rest" and for bored piles after achievements of the concrete strength more than 80%. For static load tests the following equipment is used:

- hydraulic jack SMJ-158A – 200 ton;
- caving in-measurers of the type 6ПАО.

The pressure in the jack was created by the help of manual oil pump station MNSR-400 with power up to 800 kg/cm^2, the moving of steel piles was fixed by caving in-measurers of the type 6-PAO, which were installed on the both sides of unmovable bearings with the benchmark system.

The first record was performed just after putting the loading, then consequently 4 records with an interval of 15 minutes, 2 records with an interval of 30 minutes and further for every hour until the conditional stabilization of pile settlement. For the criterion of conditional stabilization of pie settlement was taken when the speed of settlement of piles on the given stage of loading did not exceed 0.1 mm during the last 1–2 hours of observations (see Figure 3). Reloading (unloading) conducted half stages of the loading. Carrying out of static test is shown in Figure 4. Data of static field tests drawing load – settlement diagrams are shown in Figure 5.

Figure 4. Static load test of bored pile performance.

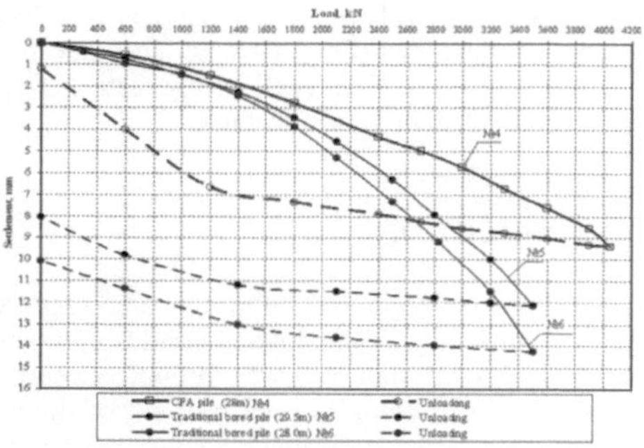

Figure 5. Diagrams of settlements S – loading P.

3 DEFINITIONS OF BEARING CAPACITY R ACCORDING TO GOST 5686–94

Bearing capacity of the piles, F_u, by dynamic load test is defined as:

$$F_u = \frac{\eta AM}{2}\left[\sqrt{1+\frac{4E_d}{\eta AS_a}\cdot\frac{m_1+\varepsilon^2(m_2+m_3)}{m_1+m_2+m_3}}-1\right] \qquad (1)$$

where η = factor, dependent on concrete strength of the piles; A = area of the pile section; M = 1 – factor, dependent on pile driving hammer's impact; E_d = effective energy of blows of the hammer, kNм, calculated

$$E_d = G \cdot H \qquad (2)$$

where G = weight of blowing part of the hammer, kN; H = drop height of blowing part of the hammer.

Static load involving settlement of the tested pile, which equals S and determined according to the below equation shall be taken as the particular value of the pile ultimate resistance to the pressing load and defined as:

$$S = \zeta S_{u,mt} \qquad (3)$$

where ζ = coefficient for conversion factor of the limit value of mean settlement of the building foundation or

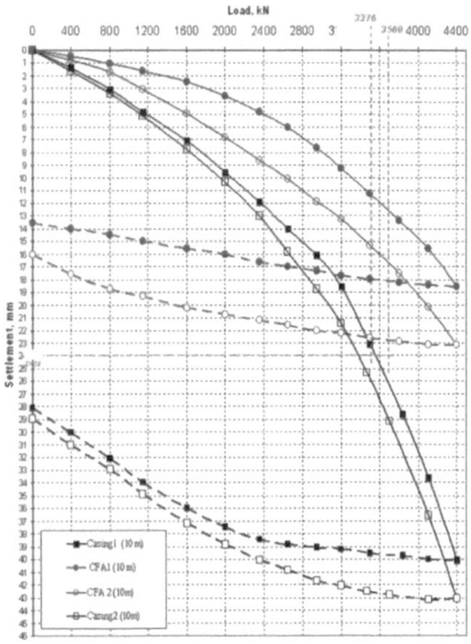

Figure 6. Diagram of dependences of settlements S from the loading P for the definition of bearing capacity of the piles.

Figure 7. The USA Embassy in Astana.

structure $S_{u,mt}$ into pile settlement obtained while static tests at conventional settlement stabilization; according to the requirements $\zeta = 0.2$ shall be taken as the coefficient value;

$S_{u,mt}$ = s maximum permissible value of mean foundation settlement of the designed building or structure as stated either in the project statement or requirements SNIP RK 5.01-01-2002. "Soil grounds of the buildings".

4 DYNAMIC AND STATIC LOAD TEST ON THE CONSTRUCTION SITE IN ASTANA CONFORMING TO THE REQUIREMENTS OF ASTM

Now we analyze dynamic and static load tests of driven steel H-pile on construction site of apartment "the USA Embassy in Astana" which is shown in Figure 7.

Table 1. Geology of the construction site of the USA Embassy in Astana.

Element	E, кg/cm^2	ν	γ, кg/cm^3	c, кg/cm^2	φ
Density Clay loam	210	0.35	0.00198	0.59	23
Semi-solid Clay	51	0.35	0.00189	0.27	19

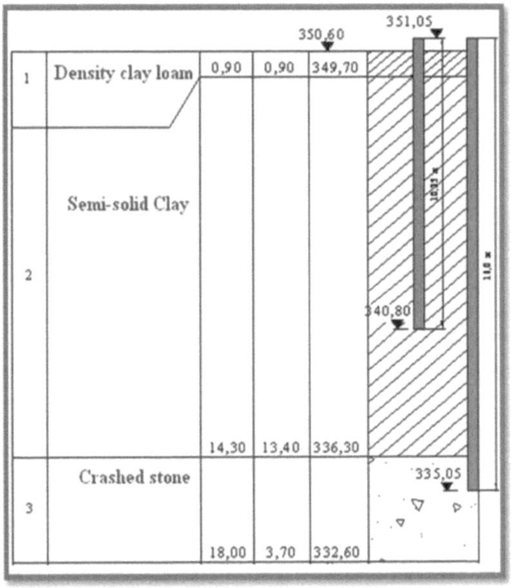

Figure 8. Geology of the construction site of the Embassy of the USA in Astana.

Geology of construction site of the USA Embassy in Astana is shown in Table 1 and Figure 8, 9.

Dynamic test:

Steel H-piles were driven according to design load, preliminary criteria of refusal:

- *on 600 kN working loading, refusal of a pile should be equal to1.25 cm (25 cm/20 blows);*
- *on 400 kN working loading, refusal of a pile should be equal to 1.67 cm (25 cm/15 blows).*

Results of dynamic tests are shown in Table 2 and Figure 10.

Static load test:

The ultimate testing load for steel H-piles doubly exceeds the allowable design load (according to the Standard ASTM D 1143-81): for 600 kN allowable loading design – 1200 kN ultimate testing load, and for 400 kN accordingly 800 kN. Results of dynamic tests are shown in Table 3 and Figure 11.

Figure 9. Construction site of the USA Embassy in Astana.

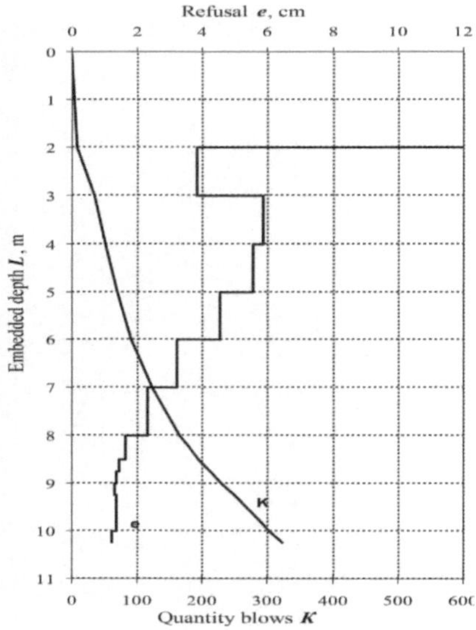

Figure 10. Results of the dynamic load test.

Table 2. Results of the dynamic test.

Number of steel H-piles	Embedded depth in soil, m	Design load, kN	Refusal of pile at driving, cm
LT-1	7.00	600	1.00
LT-2	9.00	600	1.09
LT-3	10.00	600	1.14
LT-7	8.25	600	1.00
LT-8	10.25	600	1.25
LT-9	9.25	600	1.25
LT-4	8.00	400	1.56
LT-5	8.25	400	1.47
LT-6	7.75	400	1.67

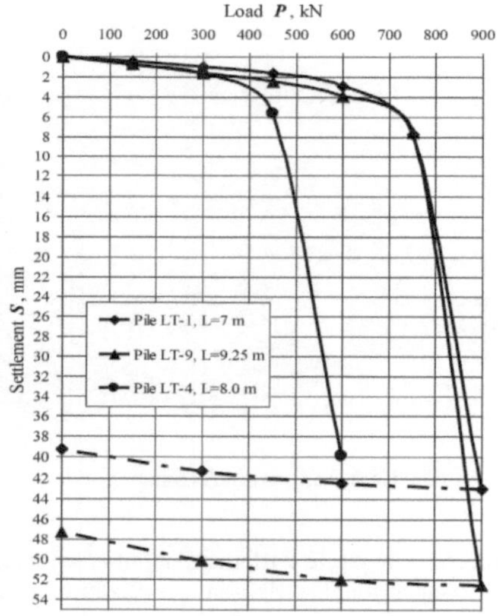

Figure 11. Results of the static load test.

Table 3. Results of the static load test.

Number of pile	Embedded depth, m	Design load, kN	Design load kN	Settlement, mm	Applied load kN
LT-1	7.00	600	1.00	43.03	900
LT-9	9.25	600	1.25	52.55	900
LT-4	8.00	400	1.56	39.88	600

After extension lengths of a steel H-piles by electric welding were driven according to design load and preliminary criteria of refusal:

- for 600 kN, refusal of 0.33 cm (25 cm/75 blows);
- for 400 kN, refusal of 0.83 cm (25 cm/30 blows).
- and static load tests were carried out again.

5 COMPARATIVE ANALYSES

Technical specificity of ASTM D1143/D1143M-07 [2] has some differences with national Kazakhstan standard GOST 5686–94. Some devices and control equipment were not used by requirement of GOST 5686–94 "Soils. Field test". Principal differences between American Standard and Kazakhstan norms are shown in Table 5.

In practice requirements of American Standard showed that results of test are safer [3] because of using many control equipment and devices for determination, measurement reverberated.

Also GOST is regulated only two out of six measurements with reducers provided by ASTM. By requirement of both standards of loading on pile is transferred by jack. But GOST does not consider that each jack needs to be provided by manometer if we use more than one jack. It is important for control of work of jack and to do correct test.

In other case sometimes reducers of axial displacement of pile showed undirected displacement and regarding requirement of GOST is not instructions. In this case regarding ASTM it is necessary to fix reducers of sway. By requirement of both standards loading on pile is transferred by jack.

Transducer of measurement of axial displacement serves for measure the movement of the bottom of the pile on its top. This measurement more exactly is determined by settlement of soil.

For the purposes, ASTM proposes to use a system of measurement of the indicator displacement installed on top of the pile.

Force transducer to measure the lateral resistance pile is used to measure the friction force on his side the entire length of the pile.

Thus, we can determine the distribution of the lateral resistance of piles.

Table 4. Results of dynamic and static tests.

Number of pile	Embedded depth, in soil m	Design load kN	Refusal of pile at driving, cm	Settlement, mm	Applied load kN
LT-1	9.75	600	0.31	4.66	1200
LT-2	15.00	600	0.31	7.34	1200
LT-3	13.00	600	0.32	7.00	1200
LT-7	11.00	600	0.28	4.41	1200
LT-8	16.00	600	0.30	6.27	1200
LT-9	12.75	600	0.27	4.96	1200
LT-4	11.25	400	0.78	3.38	800
LT-5	11.25	400	0.74	2.17	800
LT-6	9.75	400	0.63	3.17	800

Table 5. Principal differences between American Standard and Kazakhstan norms.

	GOST	ASTM	Advantages
Parameter of experimental stand for test			
Distance between testing pile till anchoring pile	$5d<L_1>2.5$ m	$3d<L_1>1.5$ m	–
Distance between testing pile till	$5d<L_2>2.5$ m	$L_2<2$ m	–
Devices and equipment For loading	jack	Jack with Spherical prop	allow to control the operation of jacks and prevent possible non-uniformity of their work, because that would result in failure of the test
Measurement of load on top pile	manometer	manometer	
	–	Dynamometer (more 100 tc) fixed for each jack	
Measurement of load on all length pile	–	Tensometer	
Measurement of axial displacement of top pile	Transducer of axial displacement of cap	Transducer of axial displacement of cap	allow to track the eccentric loads at shifting the focus of the load transfer to the axis of the pile and compensate them by computational methods. This will avoid incorrect test results
	–	Visual control	
	–	Optical instrumental control	
Measurement of sway top pile	–	Transducer of sway top pile	will more accurately determine the draft soil under the edge of the pile, and the presence of voids in the pile to separate the real from the soil sediment compression defective piles
	–	Optical control	

6 CONCLUSION

Since 1994 Kazakhstan Standard has not changed, ASTM standard was updated in 2007, and therefore takes latest developments in technology and technical terms and provides for the use of more modern equipment.

The analysis in the article shows that the tests with ASTM makes them more reliable and gives detailed information about the process of testing and the results.

Actual question today is to update the national standards, harmonize with international standards.

The introduction of innovations described will avoid from making wrong results of test. Along with TC55 work Design Academy "KAZGOR" conducts research on the existing international experience of technical regulations to ensure the safe construction processes and products.

And members of these Committees registered that in practice tester sometimes has situations when it makes wrong design because in normative there is not detailed information.

ASTM showed new improvement in pile tests. Method of pile testing regarding Kazakhstan Standard in comparison with American Standard becomes obvious disadvantage.

Unfortunately, present Kazakhstan Standards are confined to application of modern pile technology and big difference between experimental bearing capacity obtained by static loading test (SLT) and design value obtained by Kazakhstan Standard indicates incomplete usage of modern pile technology. And so research of advanced pile technologies is very important for the feature Kazakhstan geotechnical development.

REFERENCES

ASTM D1143/D1143M-07. Standard. Test Methods for Deep Foundations Under Static Axial Compressive Load., ASTM International.

GOST 5686–1994. Soils. Field test methods of piles. M.:Standards Publishing House.

Smolin, B.S., Zakharov, V.V., Puzanov, V.V. 2010. Features of pile load test by ASTM Proc. intern. symp.Russia.

Zhusupbekov, A.Zh., Hoe Ling, Lukpanov, R., Tulebekova, A.S., Yenkebayev, S., Sultanov, G.A. (2012) Geotechnical issues of piling foundations on problematical soil ground of Kazakhstan. Сборник статей научно-технической конференции ≪Численные методы расчетов в практической геотехнике, 389–397.

Challenges and Innovations in Geotechnics – Zhussupbekov (Ed.)
© 2016 Taylor & Francis Group, London, ISBN 978-1-138-03007-7

Measured temperature and moisture distribution in the subgrade of the "Almaty-Bishkek" highway

A. Nugmanova & B. Teltayev
Kazakhstan Highway Research Institute, Almaty, Kazakhstan

ABSTRACT: Climate conditions have observable effects on temperature and moisture changes in the subgrade of highways. It is widely known that the climate conditions in Kazakhstan are sufficiently different. Particularly, Almaty region's average air temperature during summer is equal to +35°C and it is equal to −30°C during winter. Moreover, zero values of temperature measurements occur frequently. In addition, it is estimated that the highest subgrade temperature of "Almaty-Bishkek" highway obtains a value of +31°C and the lowest value is equal to −4°C. The freezing depth of this particular highway may provide a value of 150 cm. Special sensors are used to determine temperature and moisture change characteristics mentioned.

Keywords: Highway, subgrade, temperature and moisture distribution, specific sensors.

1 INTRODUCTION

Investigation of water and thermal regime for road pavement structures and road subgrade in different climatic, hydrological and geological conditions are essentially important, as the distribution of temperature and moisture in pavement layers and soil subgrade has a strong influence on their deformation and strength characteristics. This is also mentioned in the Published Project Report PPR 184 by Willway, Baldachin, Reeves, Harding, McHale, Nunn (2008), where it said "The vulnerability of a pavement to climate depends on factors such as pavement type and condition, but also on location–specific factors such as geology, traffic flow and proximity to water sources. The climate variables identified as having the most impact on pavements are temperature (extremes and average), precipitation, soil moisture". Experimental method is important in investigation of this problem. Experimental results are unique in obtaining the characteristics and regularities of distribution and heat and moisture displacement in the above structural elements of roads.

The article presents some results of experimental investigation for temperature and moisture distribution in subgrade of the highway "Almaty-Bishkek" within the period from May 25, 2013 to September 3, 2015. "Almaty-Bishkek" is the most important road, which represents the connection between two centers of the highest significance, Almaty – the southern capital of Kazakhstan and Bishkek – the capital of the Kyrgyz Republic.

Actual data for temperature and moisture distribution at the points of subgrade is obtained by using special sensors for temperature and moisture.

2 EXPERIMENTAL ROAD SECTION

Experimental road section, where sensors for temperature and moisture were installed, is close to Almaty city on the highway "Almaty-Bishkek" (58 km + 895 m). This road according to "Construction norms and regularities of the Republic of Kazakhstan. Highways" document falls into the 2-nd technical category, it has 2 lanes and is in operation after reconstruction in 2005.

Pavement structure on this highway consists of the following layers: fine-grained asphalt (5 cm), coarse-grained asphalt (10 cm), over previously used fine-grained asphalt (6.5 cm) and previously used cold mix asphalt (15 cm), sand and gravel mixture (60 cm). Subgrade soil is heavy sandy clay loam.

3 TEMPERATURE AND MOISTURE SENSORS

The temperature and moisture sensors on the above mentioned experimental road section were installed in summer of 2013. Feature of these sensors is that one temperature sensor, operating by the principle of thermal resistance variation, and another moisture sensor, operating by the principle of dielectric permittivity variation, are mounted in a single metal capsule. According to Jacksona, Mansfielda, Saafib, Colmanc, Romined (2008) "Current techniques for detecting soil moisture and temperature such as gamma attenuation, soil heat flux, and GPR are mostly surface measurements and these surfacemeasurements cannot provide profound temperature and moisture profile". However, in our experiment specific sensors combined in a metal capsule allow to obtain data results on temperature and

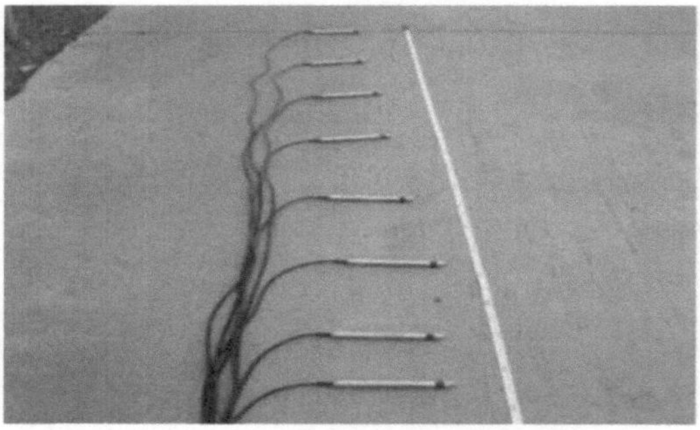

Figure 1. General view of temperature and moisture sensors.

Figure 2. Location of sensors at the surface of the highway "Almaty-Bishkek" (58 km + 895 m).

moisture in particular points of road structure simultaneously, not only just at the surface. More detailed information about these sensors is given by Teltayev (2011, 2013). General view of temperature and moisture sensors and their location at the surface of the highway are shown in Figures 1 and 2. Temperature and moisture are automatically measured in each hour. The data results are recorded in a memory device.

4 TEMPERATURE

Figures 3 and 4 show graphs for comparison of average monthly air temperature and hourly temperatures in subgrade at the depths of 100 cm, 135 cm, 170 cm, 205 cm and 240 cm during the period from May 25, 2013 to September 3, 2015.

It is seen, that in the big depths (h + 100 cm and more) the behavior of average monthly air temperature variation is similar to hourly temperature variation for these depths. It should be mentioned, that extremum graphs of subgrade points delay in time compared with extremum graph of air temperature variation. This phenomenon can be explained by thermal inertia of the materials for pavement layers. Meanwhile, the deeper the point is located, the more the delay is.

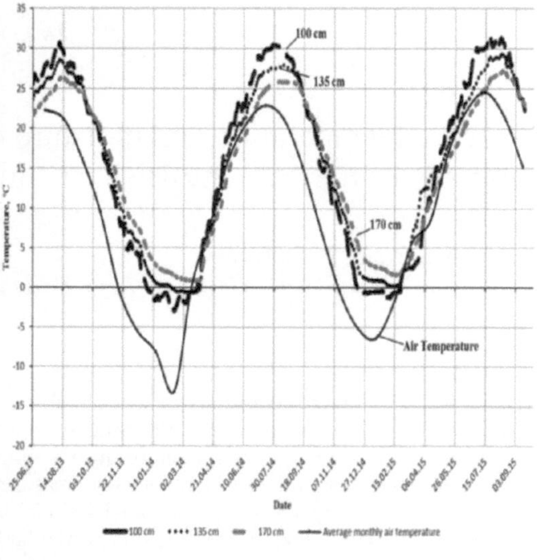

Figure 3. Graphs of temperature variation at subgrade points and average monthly air temperature on experimental road section.

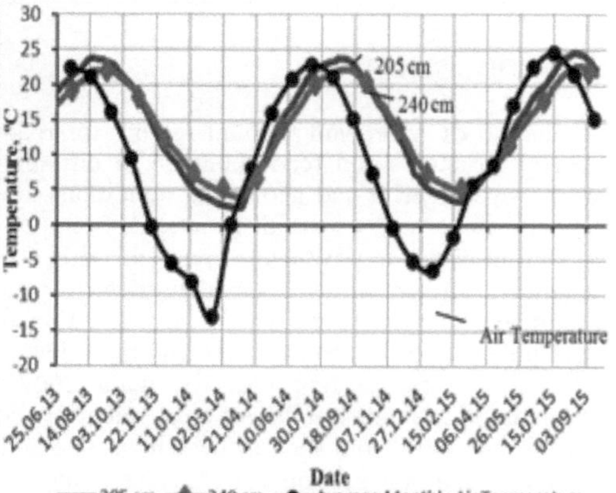

Figure 4. Graphs of temperature variation at subgrade points and average monthly air temperature on experimental road section.

Temperature falls down with depth increase during the warm period, and, on the contrary, temperature rises with depth decrease during the cold period. It seems that temperature is stabilized in some depth and remains constant disregarding season. It can be assumed that from the depth of 100 cm (road subgrade surface) daily fluctuations of air temperature do not impact temperature in subgrade, i.e. temperature variation in subgrade is of seasonal nature. At the beginning of observation period, temperature in subgrade had values +21°C ... +26°C, and at the end of observation period the values were +22°C ... +25°C.

Temperature sensor, located in the lowest part of experimental road section, i.e. in the depth of 240 cm, records that temperature reduction continues till the

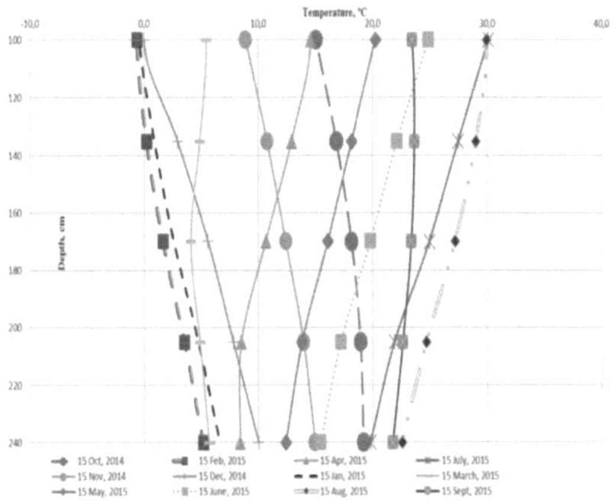

Figure 5. Graphs of temperature distribution along subgrade depth on experimental road section.

end of March in both 2014 and 2015. Meanwhile, for the period from the middle of September 2013 to the end of March 2014 the temperature in the depth of 240 cm varied from +22°C to +4°C, and during the period from the middle of September 2014 to the end of March 2015 temperature varied for similar values.

Figure 5 shows graphs of temperature distribution along highway depth (100–240 cm). As expected, temperature values vary essentially in relation to season of the year. Thus, the maximum temperature in the depth of 100 cm on July 15, 2015 (summer) reaches +33°, on 15 October 2014 (autumn) it is equal to +15°C, on 15 February 2015 (winter) it is approximately equal to 0°C, and on 15 April 2015 (spring) it is equal to +14°C. Temperature at the depth of 240 cm in annual cycle is always positive and varies within a range (+4°C ... +24°C).

The graphs, shown above, describe the behavior of temperature distribution in the depth in relation to season of the year. For example, gradual temperature reduction occurs in autumn close to the surface of subgrade, whereas the reverse process can be observed in spring.

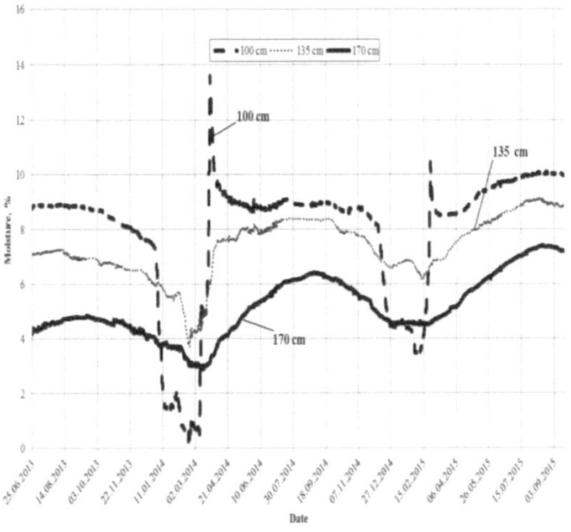

Figure 6. Graphs of moisture variation in points of subgrade for experimental road section.

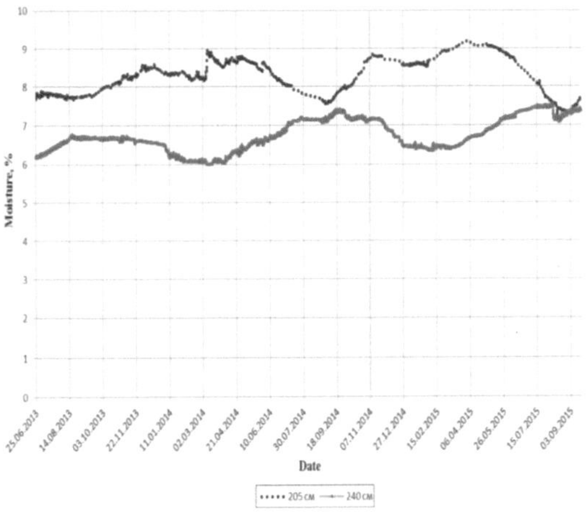

Figure 7. Graphs of moisture variation in points of subgrade for experimental road section.

5 MOISTURE

Graphs of moisture variation at the points of road subgrade are shown in Figures 6 and 7.

These graphs show that moisture variation in the cold period is different in various depths. So, at a depth of 205 cm moisture value is within 7–9%, and in the depth of 240 cm it is 6–7%. At a depth of 100 cm from the moment of temperature transition from positive values to negative ones, moisture value was significantly reduced both in 2014 and 2015. The maximum moisture content in this depth, on March 25, 2014 amounted to 13%, and minimum moisture content was equal to zero on 21 February 2014. Then, at the moment of temperature transition from negative values to positive ones, moisture content increased again. This phenomenon is due to the transition of water into ice at low temperatures. The same phenomenon was also observed at a depth of 135 cm. It is due to the fact that the absolute value of the minimum temperature in this depth less than in the depth of 100 cm both in 2014 and 2015.

Graphs of moisture content distribution in highway depth at different time intervals are shown in Figure 8.

It is seen that moisture content is essentially variable at the upper part of subgrade (to a depth of 170 cm). The values of moisture content vary substantially at the subgrade surface (4% in winter and 10% in summer). At depths of 205 and 240 seasonal moisture content variation does not exceed 1–1.5%.

Therefore, it can be assumed that if ground water level is deep, then the moisture content variation in

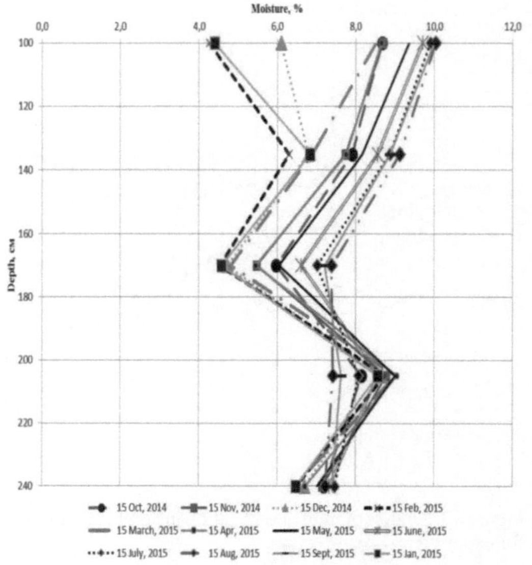

Figure 8. Graphs of moisture distribution on subgrade for experimental road section.

points of subgrade for the highway is impacted by atmospheric precipitation and phase transitions.

6 CONCLUSION

Based on the results obtained from these experiments, the following conclusion can be drawn:

1. The behavior of average monthly air temperature variation complies with hourly temperature variation in big depths.
2. Extremum temperatures at subgrade points in lower layers occur later than in its upper layers. This can be explained by thermal inertia of subgrade material.
3. In spring moisture content increases due to ice melting from upper layers gradually extending into the depth of subgrade.
4. it can be said that if ground water level is deep, then the moisture content variation in points of subgrade for the highway is affected by atmospheric precipitation and phase transitions

REFERENCES

3.03-09-2006* (2014). Construction norms and regularities of the Republic of Kazakhstan. Highways, Astana, Kazakhstan.

Jacksona, T; Mansfielda, K; Saafib, M; Colmanc, T; Romined, P. (2008). Measuring soil temperature and moisture using wireless MEMS sensors. Measurement. Volume 41 (Issue 4), pp. 381–390.

Teltayev, B.B. (2011). Regularities of water and thermal regime of the highway "Astana-Burabay", Bulletin of the Kazakhstan Highway Research Institute, Almaty, Kazakhstan, pp. 36–54.

Teltayev B. Road Soil Basement Temperature and Moisture Variations. Proceedings of the 5th International Geotechnical Symposium "Geotechnical Engineering for Disaster Preventions and Reduction, Environmentally Sustainable Development". Incheon, Korea, 2013, pp. 493–500.

Willway, T; Baldachin, L; Reeves, S; Harding, M; McHale, M; Nunn, M. (2008). The effects of climate change on carriageway and footway pavement maintenance. Published Project Report PPR 184.

Challenges and Innovations in Geotechnics – Zhussupbekov (Ed.)
© 2016 Taylor & Francis Group, London, ISBN 978-1-138-03007-7

Research of the mechanical properties of soil basis by an equivalent material

B. Kaldanova, G. Zhukenova, A.Zh. Zhussupbekov, Y. Muzdybayev, T. Muzdybayeva &
B. Dosmukhambetova
Eurasian National University, Astana, Kazakhstan

ABSTRACT: This report described the research of the mechanical properties of the soil basis by an equivalent material. This material is suitable for producing models of a base so as to quickly attenuate deformations in him the next step after the application of the load. It was applied cohesive soils the Karaganda coal basin and research carried out in a special device, which is designed to conduct experiments on the diagram plane strain.

Keywords: equivalent materials, isotropic compression, shear, triaxial compression, load path's.

1 INTRODUCTION

The continuously growing volume of capital construction in our country, improving its overall technical level of entering the top 50 developed competitive countries are closely related to the development, research and implementation in practice of the construction of new, advanced materials and designs.

The extensive theoretical and experimental research in the field of constructions and buildings held by domestic and foreign scientists, point to a number of features in their work, complicating the purely theoretical approach to the calculation of real structures.

The complexity of these problems, especially in the design, in spite of the high efficiency of modern numerical methods of the theory of elasticity and plasticity, requires experimental studies. Considerable difficulties arise, for example, in establishing the state of the material equations and limit state assessment of structural elements exposed to a complex combination of power and other effects, which lead to the elastic-plastic deformation and accumulation of damage over time.

The purpose of laboratory simulation was to study the nature of the interaction of the processed base and foundation construction, and also features collaboration system structure – the base with expansion joint rolling.

In the study the effect of the base motion to the nature of its interaction with the foundations of applied modeling on equivalent materials [2].

Turning to the specific task of establishing the criteria of similarity and scaling constants (transient factors) in the simulation by equivalent materials, it is necessary to establish, first of all, the basic forces that define the essential features of the process under study. In the first approximation, we can confine ourselves to the two kinds of power, namely, the external forces – gravity and internal forces – the stresses arising in the dirt thicker. If we assume that these two kinds of forces, along with the likeness of the geometrical properties of the system, the initial of her condition and the likeness of the boundary conditions uniquely determine the behavior of our system, on the basis of the general law of Newton, and the method of dimensions, you can get a determining criterion of similarity corresponding to the action of these kinds of forces [1]. Of course, in fact, in the study the physical impact of the role played by other forces, however, their effect can be given to the state of the soil mass reflects its mechanical strength and deformability.

According to the method of equivalent materials, proposed by G.N. Kuznetsov [1], with data on the characteristics of the mechanical properties and the full-scale model of soil N_N and N_M, we can for a given relationship γ_M/γ_N and N_M/N_N and to calculate the scale of the model $1/L = m_1$. For determining the physical and mechanical characteristics to be taken are the characteristics that play in the process leading role.

In the selection of materials – loose soil can be equivalent to a first approximation as the defining characteristics of using a set of coupling values and internal friction angle C; φ. To ensure the destruction process is similar to the conditions necessary to comply with the equation [1.5]:

$$C_m = 1/L \cdot \gamma_M/\gamma_N \cdot C_N \tag{1}$$

$$tg\varphi_m = tg\varphi_N$$

where $1/L = m_1$ – linear scale model;

φ_m, φ_N – the proportion of materials and patterns of nature;

C_m, C_N – grip material models and the real nature of soil.

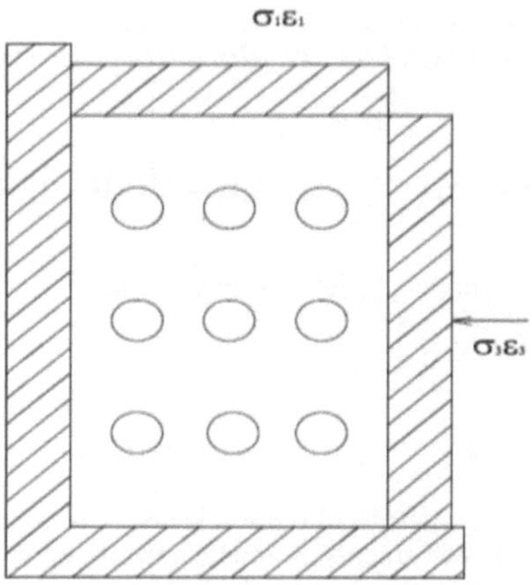

Figure 1. Special installation.

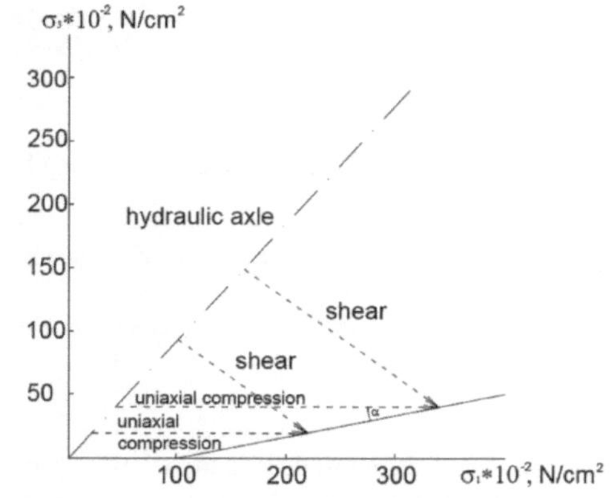

Figure 2. The passport of the equivalent strength of the material on the test results.

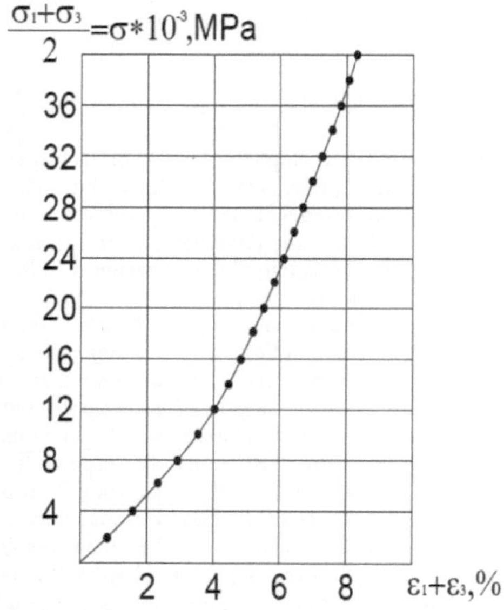

Figure 3. The results of experiments with the equivalent material on isotropic compression with dumping.

In the simulation, cohesive soils (loams) Karaganda's coal basin according to an equivalent material, to establish the scale of the model, first of all, should determine the following physical and mechanical properties of the sand mixture: c, φ, E, γ.

2 THE METHODOLOGY AND RESULTS OF THE STUDY OF THE MECHANICAL PROPERTIES EQUIVALENT MATERIAL

The material of the soil foundation model is taken mixture consisting of 97% of fine quartz sand and 3% by weight spindle oil having a clutch that allows modeling of cohesive soils of Karaganda's coal basin [6]. This material is suitable for the manufacture of the base model, the strain decay rapidly after application it loads the next stage.

In connection with the development of numerical methods for the study of mechanical properties of soils in recent years there have been a lot of work, aimed at the development of simple phenomenological models to describe complex set of elastic-plastic properties of the latter [3]. In practical terms, the most interesting are the study, the result will be a simple expansion model requires experimental determination of the minimum number of characteristics. The overwhelming majority of studies carried out on the devices triaxial (stabilometric) [7].

Research is being conducted on a special installation, diagram is shown in Figure 1.

The sample has a 100 × 100 dimensions in plane strain and length 200 mm.

Voltages are measured dynamometers type brand DOSM-3-01 and deformation indicators dial scale division of 0.01 mm.

The survey taken following loading scheme:

a) isotropic compression – the sample is loaded gradually and simultaneously increases equal to the stress $\sigma_1 = \sigma_3$;
b) uniaxial loading – the sample is loaded to a predetermined level is $\sigma_1 = \sigma_3$ isotropic, then the stress σ_1 is gradually increased;
c) shear – the sample is loaded to a certain $\sigma_1 = \sigma_3$ isotropic, and then σ_1 increases and σ_3 decreases in equal steps.

Experiments (a, b, c) are conducted at an initial value $\sigma_1 = \sigma_3 =$ 0.20; 0.40; 0.80; 0.90; 1.80; 3.20 H/cm^2. For each loading scheme carried out 10 test series. Loading scheme shown in Figure 2.

The test results for isotropic compression are shown in Figure 3. The uniaxial compression (Fig. 4) in the first stage loading is accompanied by a decrease in volume of the sample, and then the plastic flow becomes

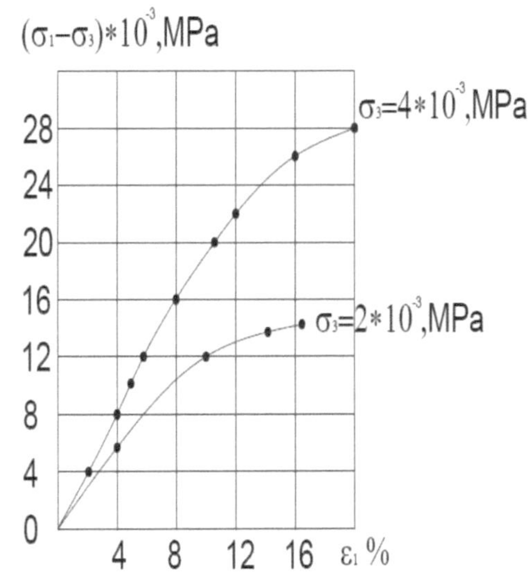

Figure 4. The results of experiments with an equivalent material on uniaxial compression with dumping.

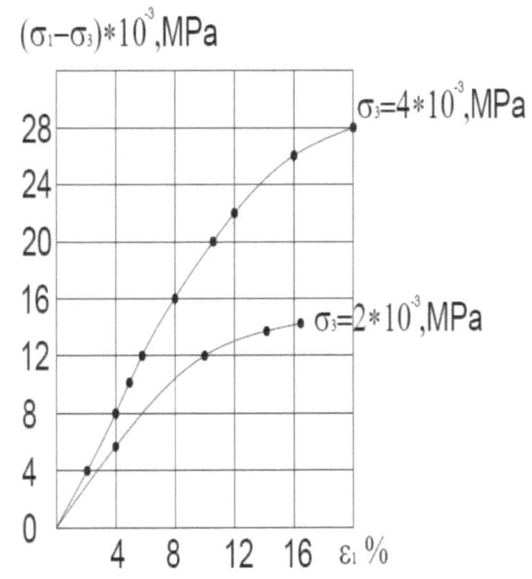

Figure 5. The results of experiments with an equivalent material on shearing with dumping.

Table 1. Physical and mechanical properties of natural soil and equivalent material.

Name of soil pumps and model material	Specific gravity, γ, kN/m^3	Cohesion C, kPa	The angle of internal friction, degree	Deformation module E, MPa	Coefficient of Puasson υ
1	2	3	4	5	6
Loam	2,05	40	22	20	0,3
An equivalent material	1,7	0,90	39	0,27	0,25

coextensive. Out in critical condition with a single-layer compression is not clearly fixed, as this output occurs at a small angle of the meeting $<\alpha$, and deformation, and the change of the sample is accompanied by a seemingly limitless hardening sizes.

Shear test results are shown in Figure 5. The graphs clearly flatten out when reaching the critical surfaces. From the strength of the passport (Figure 2) that the path to the loading plane strain conditions does not affect the strength of the soil. It should be noted that 12-17% of the tests of the plane strain diagram of the internal friction angle lower than with triaxial tests [4].

According to the presented experimental tests obtained the necessary parameters (E, C, φ, γ) to select the model scale.

Physical and mechanical characteristics of the full-scale soil (loam Karaganda) and equivalent material are shown in Table 1.

Substituting the values of cohesion and the share of the Karaganda loam and equivalent material in the formula (1), after some transformations define the scale of the model:

$$m_1 = \frac{C_m}{C_N} \cdot \frac{\gamma_N}{\gamma_m} = \frac{0{,}90}{40} \cdot \frac{2{,}05}{1{,}7} = \frac{1}{40}$$

Consequently, the linear scale and full-scale model of the object (building, foundations, structures), is defined by the strength properties (adhesion) and loam equivalent material and is 1:40.

3 CONCLUSIONS

A result of research of the mechanical properties of the soil basis by an equivalent material revealed that after application of the next load stage is rapid deformation of damping that is due to the fact that the load path does not affect the strength of the soil. Also as a result of the study were obtained the necessary parameters (E, C, φ, γ) to select the scale model and experimentally obtained the ratio of the strength properties (adhesion) and loam equivalent material equal to 1:40.

REFERENCES

Afanasyev G. V. Experimental study of deformation models of square slabs on grade layer, the underlying base to replace a non-compressible. – In the book: "Building constructions", issue 1. – Kiev: Budivelnik, 1975, p. 93–101.

Barden L. and Mockton I. Tests on model pile groups in soft and stiff clay. – Geotechnique, London, vol. 20, N1, 1980, p. 94–96.

Kaldanova B., A. Hasegawa., K. Kaneko. Comparison of mechanical properties of sand by using a triaxial compression device. The Journal of Hachinohe Institute of Technology. Hachinohe, Japan. 2015, No. 34.

Kuznetsov G.N. and et. all Study of rock pressure on models from equivalent materials. – M. Ugletehizdat, 1973, 180p.

Zhusupbekov A. Zh., Cholod A.F., Jousoupbekova V.V. Investigations of soil straining under vertical load by horizontal displacement. Proc. of the Inr. Geol. Congress, Kyoto, Japan, 1992, p.110–111.

Zhusupbekov A. Zh., Baitasov T.M. Influence of horizontal curvature strain on the treated building. – Proceedings of the international symposium "Reconstruction of St. Petersburg-2005", 5–11 October 1992, v.2. – p. 21–28.

Zhussupbekov A.Zh., Tanaka T., and Aldungarova., A.K. "The effect of reinforcement on stability of model of the dam on undermining soil ground". Proceedings of the international symposium "New Innovations and Sustainability". Fukuoka International Congress Center, 9–13 November, 2015, 184.

Challenges and Innovations in Geotechnics – Zhussupbekov (Ed.)
 ISBN 978-1-138-03007-7

Evaluation of wind power unit reliability according to the results of field studies on the example of Ereymentau wind power station

D.K. Orazova, A.Zh. Zhussupbekov, R.E. Lukpanov & S.B. Yenkebayev
Department of Civil Engineering, Eurasian National University, Astana, Kazakhstan

ABSTRACT: This paper includes the analysis of vibrational effect from the tower block to the foundation of the wind generating unit of Ereymentau WPS. The paper shows results of vibromonitoring of wind power tower (WPT). The analysis of vibrating influence from the tower fluctuating load on WPU foundation had been made by the results of in-situ monitoring. The results of research are presented in diagram dependence of vibrating characteristics (frequency, amplitude, acceleration) from wind pressure intensity. There was given an extrapolation of potential efforts arising in the foundation at maximum wind load in that region. Plaxis 2D software program shows the calculation of vibration impacts corresponding to the foundation of wind power unit (WPU) from the tower.

1 INTRODUCTION

Wind power is the most dynamically developing type of renewable energy sources. Having studied the energy potential of wind in Kazakhstan, the Government of the Republic of Kazakhstan together with the UN development Program "Kazakhstan is the initiative of the development of wind power market" has resolved that the Ereymentau district of Akmola region is the most perspective area for the construction of wind power stations [1].

The first steps in the Program realization were taken in Ereymentau district of Akmola region.

Within the context of an upcoming exhibition "EXPO-2017" in Astana it is planned to provide power supply of the exhibition facilities by using the energy of Ereymentau WPS.

Figure 1. Current Ereymentau WPS.

2 CLIMATIC AND HYDROGEOLOGICAL CONDITIONS OF THE SITE

Annual measurements of speed and direction of wind were made at the site within the context of UNDP project on wind power. Measurements were performed in accordance with international standards in the field of measuring wind speed in order to estimate wind potential (IEA/IEC) [2].

The distribution of wind speed and Weibull parameters at the height of 51 m (the axis of the gondola) for the site of Ereymentau WPS are shown in Figure 2.

According to the results of wind pressure measurements the diagram of seasonal distribution of wind speed was made. It demonstrates the changes in the wind flow speed by month in relation to average annual wind speed (Figure 3).

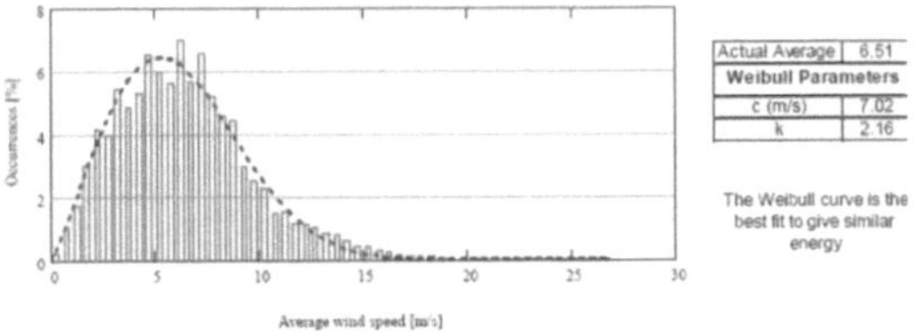

Figure 2. Distribution of wind speed and Weibull parameters at the height of 51 m.

The geological structure of this territory consists of sedimentary and metamorphic rocks of the proterozoic and the paleozoic broken by intrusions in the north–eastern part of the city, and covered by eluvial–deluvial quaternary sediments represented by clays (Figure 4).

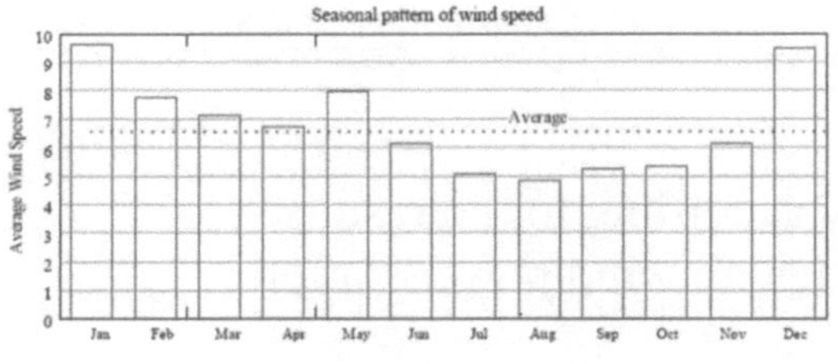

Figure 3. Monthly average wind speed at the height of 50 m.

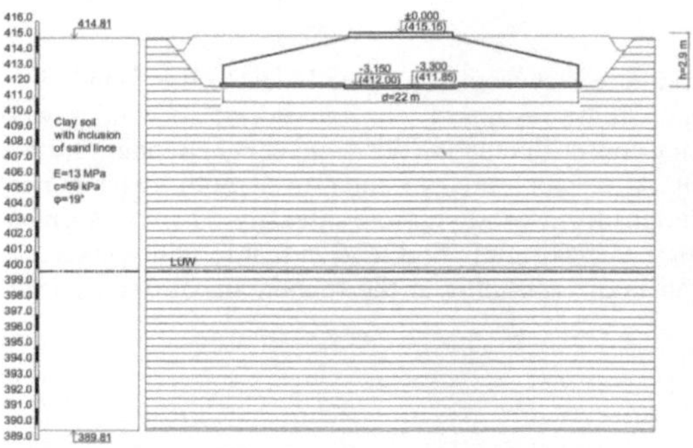

Figure 4. Geological section of the site.

3 THE PROCEDURE OF VARIATIONS AND VIBRATIONS MEASURING WITH THE HELP OF VIBRA PROFOUND INDICATOR

The field studies and measuring vibrational effect on WGU foundation were made with the help of VIBRA Profound indicators. These instruments allow defining the speed, acceleration, vibration frequency and foundation displacement caused by wind pressure. The instrument system meets national and international standards SBR 2002, DIN 4150 and DIN 45669.

The source of vibration is the WPU tower located on a reinforced concrete foundation with a thickness of 2.9 m and a diameter of 22 m. Fluctuations caused by the tower, are passed to the ground base through the foundation.

Vibration loads from the tower to the foundation were measured by sensors with the help of VIBRA Profound device. Movements are recorded on the WPU tower (Figure 5). Intermediary measurements were counted every 10 seconds.

Vibration impact measurements were made at the Ereymentau site, from August, 26 to September, 27 2015. It took a month to make all the measurements. During a month there were impermanent wind speeds. Control points data was obtained according to the diagrams of the device. The relationship between wind pressure and vibration impacts on the WPU foundation from the tower was obtained after data analyzing. The method of extrapolation can allow measuring the impact on the foundation at a maximum wind speed of 25 m/s.

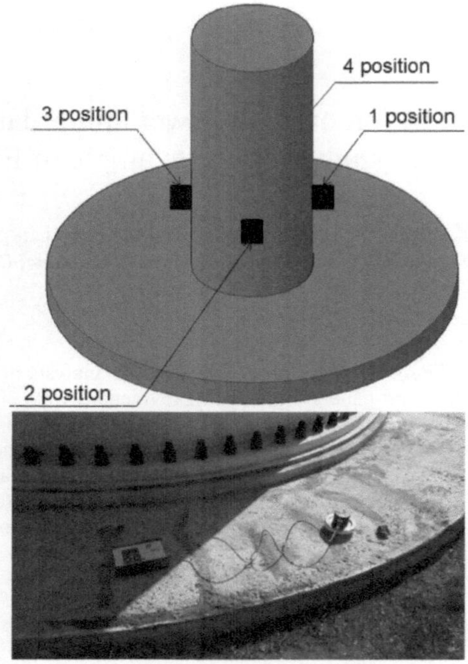

a)

b)

Figure 5. The location and the sensor measurements of vibrations: a) the location of the sensor on the tower wind turbine, b) sensor readings maximum displacements of the tower wind turbine.

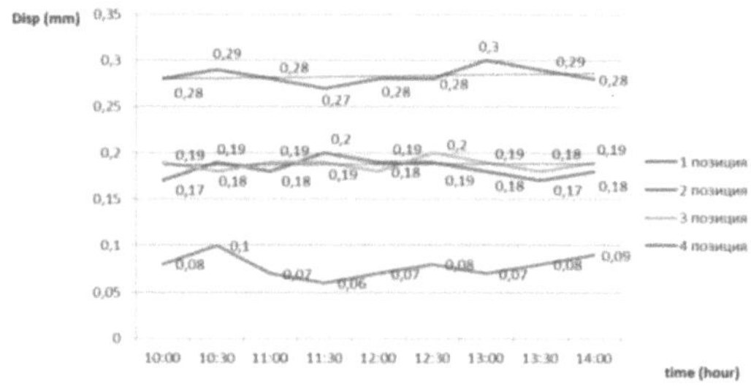

Figure 6. Graph of dependence of displacement and time.

With the help of the instrument we got the diagram of dependence of foundation absolute displacements and the time in this area (Figure 6). The diagram shows the maximum values of displacements for each position. The maximum values of displacements for the 1 position are 0,1 mm, the maximum values of displacements for the 2 and 3 positions are 0,2 mm, and 0,3 mm – for the 4 position, at the maximum wind pressure of 4,75 m/s.

The measurement results of vibration shown in the graphs of dependence of displacements, accelerations, frequencies of vibration, velocity of vibration from wind speed (Figure 7).

4 DYNAMIC ANALYSIS OF WIND TURBINES ON AN ELASTIC-PLASTIC FOUNDATION

PLAXIS 2D allows modeling the interaction between soil and construction. The influence of vibration work on the soil on which it is installed was considered.

The dimensions of the soil mass are determined based on the computation of compressible stratum, represented by a clay layer of 60 m. Soil surface is set at the point $y = 0$. The foundation is represented by clay, which is considered here as an elastic-plastic material. A dataset is formed by selecting the *soil* type and *interface*, according to the information given in Table 1. Dynamic stiffness of soil is significantly higher than its static stiffness, as dynamic loading is usually rapid and it results in ultra-small deformations.

Material data set for footing according to the information given in Table 2.

A distributed load is applied to the foundation, as well as a horizontal force and a concentrated moment for modeling the wind turbine weight, and fluctuations that it causes. The geometric model is shown in Figure 8.

Figure 9 shows the deformed mesh of the foundation.

Figures 10, 11, 12 show the results of the accelerations, the results velocity of vibration values and the diagram of the dependence of vertical displacement on time.

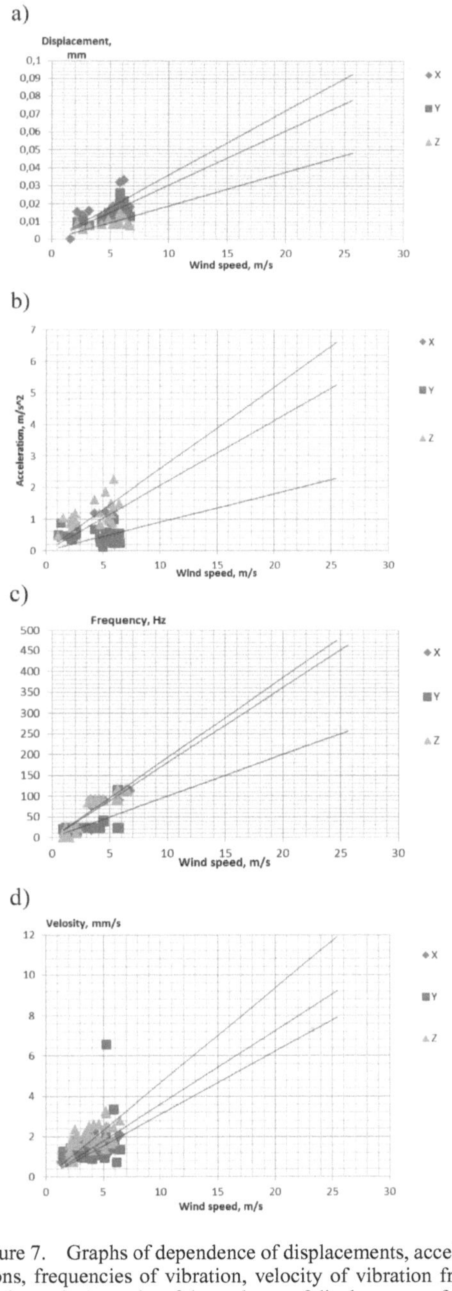

Figure 7. Graphs of dependence of displacements, accelerations, frequencies of vibration, velocity of vibration from wind speed: a) graphs of dependence of displacements from wind speed, b) graphs of dependence frequencies of vibration from wind speed, c) graphs of dependence of accelerations from wind speed, d) graphs of dependence of velocity of vibration from wind speed.

5 CONCLUSION

According to the measurement results the prognosis of vibration parameters of wind turbine foundation from wind pressure was made.

Table 1. Material properties of the subsoil.

Parameter	Name	Value	Unit
General			
Material model	Model	Mohr – Coulomb	–
Type of material behavior	Type	Drained	
Soil unit weight above phreatic level	γ_{unsat}	16	kN/m^3
Soil unit weight below phreatic level	γ_{sat}	18	kN/m^3
Parameters			
Young's modulus (constant)	E′	13*10^4	kN/m^2
Poisson's ratio	ν'	0,38	
Initial			–
K_o determination	–	Automatic	–
Lateral earth pressure coefficient	$K_{o,x}$	0,674	–

Table 2. Material properties of the footing.

Parameter	Name	Value	Unit
General			
Material model	Model	Linear elastic	–
Type of material behavior	Type	Non–porous	–
Soil unit weight above phreatic level	γ_{unsat}	24	kH/m^3
Parameters			
Young's modulus (constant)	E′	30*10^6	kH/m^2
Poisson's ratio	ν'	0,2	
Initial			–
K_o determination	–	Automatic	–
Lateral earth pressure coefficient	$K_{o,x}$	1	–

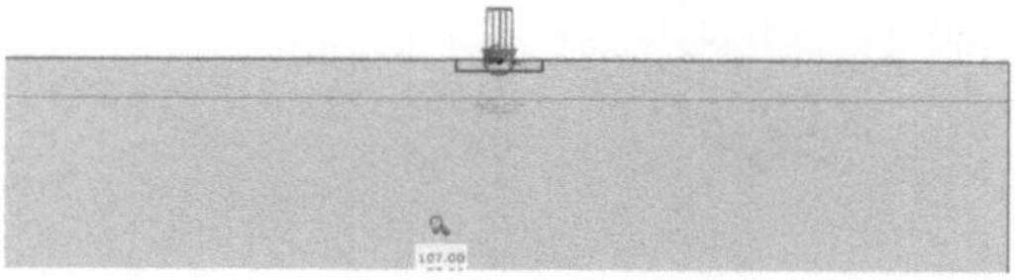

Figure 8. Geometric model of the foundation.

Maximum values of the parameters at the maximum wind pressure in this region are:

1) Maximum displacement on X = 0,09 mm, Y = 0,076 mm, Z = 0,048 mm;
2) Maximum acceleration on X = 5,14 m/s^2, Y = 2,24 m/s^2, Z = 6,48 m/s^2;
3) Maximum frequencies of vibration on X = 450 Hz, Y = 200 Hz, Z = 480 Hz;
4) Maximum velocity of vibration on X = 7,8 mm/s, Y = 9 mm/s, Z = 11,8 mm/s.

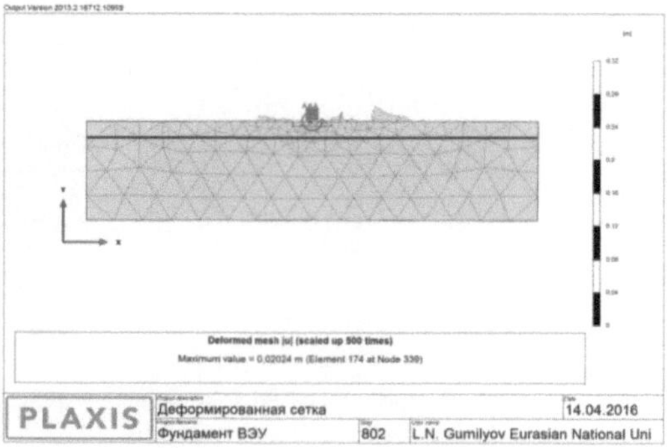

Figure 9. Deformed mesh of the foundation.

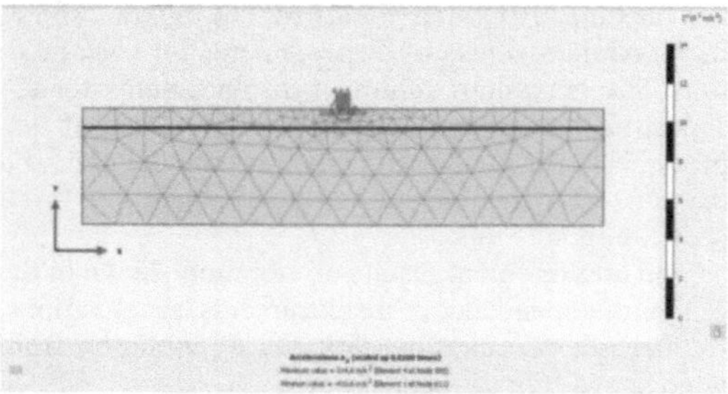

Figure 10. The results of the accelerations.

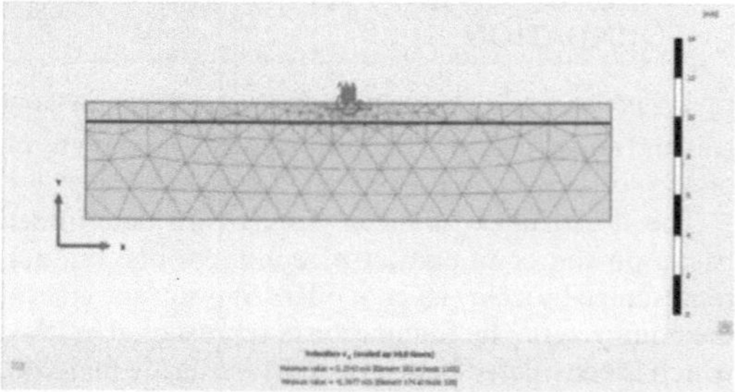

Figure 11. The results velocity of vibration values.

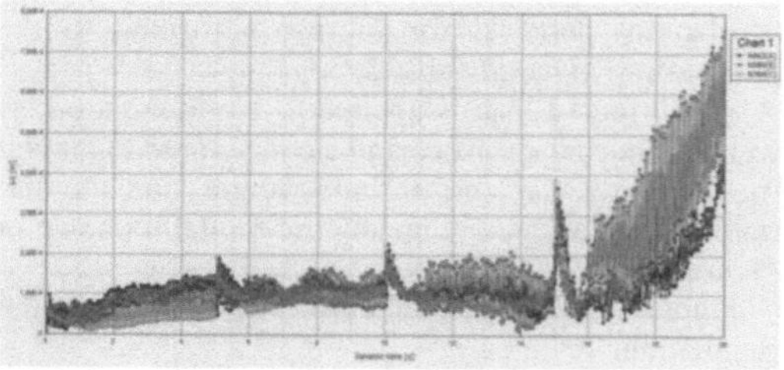

Figure 12. The diagram of the dependence of vertical displacement on time.

The obtained vibration impacts from the tower to the foundation under maximum wind pressure were integrated in Plaxis 2D software program. Having applied static and dynamic loads in the program we got the displacement with a certain frequency and acceleration, and vibrations speed.

The diagram of the dependence of vertical displacement on time shows that these displacements do not exceed a limit value for WPU foundations [3].

REFERENCES

National program of wind energy development until 2015 with the perspective till 2024 (project) Almaty//Astana. – 2007, 3–5.

Project of the Government of Kazakhstan and the Development Program of the United Nations Organization "Kazakhstan – Initiative of the Development of Wind Power Engineering Market", the wind power plant near Ereymentau city, Pre-investing research//Almaty. – 2008, 7–9.

DIN 45669. Human exposure to vibration in buildings, Germany, 1995–2006, p. 8.

Challenges and Innovations in Geotechnics – Zhussupbekov (Ed.)
© 2016 Taylor & Francis Group, London, ISBN 978-1-138-03007-7

Influence of blow energy of the hammer on the bearing capacity of piles during dynamic testing

Rauan E. Lukpanov, Denis V. Tsigulyov, Serik B. Yenkebayev & Dauen T. Askarov
L.N. Gumilyov Eurasian National University, Astana, Kazakhstan

ABSTRACT: The paper presents the results of dynamic testing of piles, and also presents an analysis of estimates of the blow energy of the hammer on bearing capacity of piles during dynamic testing of soils in the same hydro-geological conditions.

Keywords: dynamic test of pile, energy of blow, bearing capacity

1 INTRODUCTION

Nowadays, the use of pile foundations at construction sites in Kazakhstan is the most popular. The usefulness is explained by the need to ensure high bearing capacity of buildings and structure. Qualitative analysis of bearing capacity of the pile is also a challenge for the geotechnical engineer. Dynamic testing is one of the most common methods for determining bearing capacity of the pile in natural conditions. It is an express method. It has several disadvantages. For example, the theme of blow energy of the hammer is still actual. This scientific paper presents the study of blow energy of the hammer influence on bearing capacity of experimental piles.

Figure 1. Dynamic testing on the construction site "Urban Romance".

2 DESCRIPTION OF THE CONSTRUCTION SITE

Dynamic testing of soil with driven piles were conducted on the construction site: Multifunctional complex "Urban Romance" (block number 10), which is located on the right bank of the Ishim River, south of Kaldayakov street in Astana, 3 phase of construction (Figure 1).

3 GEOLOGICAL CONDITIONS

On the basis geological investigation of soil and results of drilled boreholes and laboratory testing of soils it was found out: The geological structure on construction site revealed medium upper quaternary age – modern deposits represented by clay soil (loam), sand of average grain and gravel sand and alluvial formations, represented by loam including siltstones, crushed stone from Lower Carboniferous period, represented by sandstones including siltstones and argillites. All of these deposits are covered by bulk soils represented by fertile layer of soil and construction debris (crushed stone, gravel, crushed brick, fragments concrete structures, reinforcement, etc.). The bearing layer (under the pile tip) is crushed stone ground. According to geological investigation, the crushed stone is greenish-gray and yellowish brown, with loamy filler 20÷30%. Crushed stone is from low to high mechanical strength. It is weathered. Crushed stone ground is placed at a depth of 3,0 m and 9,2 m, with thickness of layer from 0,8 m to 2,80 m [1].

4 DRIVING MACHINE

The driving and redriving of the pile had been made by driving machine SP49, the parameters of which is presented in table 1.

Table 1. Parameters of the pile driving machine.

Parameters	Value
MAXIMUM cross-section of pile	35 × 35 cm
maximum length of the pile	12,0 m
Capacity	12,0 t/c
Mast incline	
– right and left	7 degree
– back and forth	18 degree
Changing the length of the mast	0,4 m
The width of the guide mast	625 mm
Dimensions in the working position:	
– length	4728
– width	5045
– height	18465
Dimensions in the transport position:	
– length	10610
– width	4300
– height	3455
The mass hinged part:	
– (without the basic machine)	8,6 T
– pile driver weight (with a hammer CII-77)	26,8 T

5 DYNAMIC TESTING OF SOILS BY DRIVEN PILES

Before the testing the marks by 100 mm of spacing were put along the whole pile surface, and the last meter was marked by 10 mm of spacing. During the process of pile-driving were counted the number of hammer blows, record the height of fall. At the same time tester should fix the movement of the pile during the pile-driving process visually or by the leveling. Redriving of piles performed after rest (3–6 days depending on the type of soil). For observing the movement of the pile during redriving on the pile surface attached tape measure (scale interval is 1 mm) [2].

During dynamic testing of piles the value of ultimate resistance F_u in accordance with results of pile penetration and refusal determined by the following equation [2,3]:

$$F_u = \frac{\eta A M}{2}\left[\sqrt{1 + \frac{4E_d}{\eta A s_a} \cdot \frac{m_1 + \varepsilon^2(m_2 + m_3)}{m_1 + m_2 + m_3}} - 1\right], \quad [1]$$

where $\eta = 150$ tf/m^2 – coefficient depending on the material of the pile, (in the table 10 [3] 1500kH/м^2); $A = 0,09$ m^2– cross section of pile; $M = 1$ – coefficient for pile-driving by hammer; $E_d = G(H\text{-}h)$, *tf·m* – effective energy of hammer, take on table 12 [2], G – weight of a blowing part of the hammer; H – drop height of blowing part of hammer during the pile-redriving; s_a – the actual residual refusal, equal to the value of pile penetration per one blow, m; m_1 – total mass of hammer device; ε – coefficient of restitution of blow, for driving precast concrete pile by hammer with wood dashpot, $\varepsilon^2 = 0,2$; m_2 – total mass of pile and pile cap; $m_3 = 0$ – mass of driving hood.

Table 2. Private values of refusals and bearing capacity of piles.

Drop weight of hammer, cm		Average value of refusal per blow, (cm)/ bearing capacity of piles (kN)							
		ĕ	2	3	4	5	6	7	8
h=120	3	0.067/147	0.033/210	0.033/210	0.033/210	0.033/210	-	-	-
	5	0.02/273	0.02/273	0.02/273	0.02/273	0.02/273	-	-	-
h=150	3	0.3/82	0.1/147	0.2/102	0.067/181	0.67/181	0.1/147	0.133/126	-
	5	0.26/89	0.12/133	0.28/85	0.140/123	0.18/108	0.06/191	0.22/97	-
h= 180	3	0.4/82	0.2/119	0.4/82	0.533/70	0.4/82	0.167/131	0.433/79	0.40/82
	5	0.5/73	0.26/103	0.48/74	0.54/70	0.34/90	0.38/84	0.62/65	0.40/82

For analyzing the effect of blow energy of hammer was decided to perform dynamic tests of piles with different drop height of hammer: 1.20 m, 1.50 m and 1.80 m, accordingly energy of blow was: 3.00, 3.75 and 4.5 t·m.

Table 2 shows the results of bearing capacity of piles, which were obtained by the equation 1, and by the results of refusal of piles during dynamic tests.

According to the empirical dependence of bearing capacity of pile by the results of dynamic tests – the bearing capacity of pile is directly proportional to energy of hammer and is inversely proportional to value of refusal. That is, the greater the drop height of the hammer the greater bearing capacity. The greater the refusal, the less bearing capacity. Considering that pile driving conditions are the same for all test piles, the equation reduces to the analysis of two unknown: Fu = Ed/Sa[...] (i.e, in our case, all the variables of the equation are same, except for the blow energy and refusal).

For analyzing the effect of blow energy in the equation (1) substitute a fixed value of drop height. Then substitute different values of refusal from 0.1 to 1 cm, we get the curve, characterizing the dependence of the bearing capacity of the different refusal when energy is a fixed.

Figure 2 shows the three curves of the bearing capacity by refusal for different heights (energy): 1.20, 1.50 and 1.80 m.

Figure 3 shows two curves which correspond to the energies of 1.20 and 1.50, and points (the actual

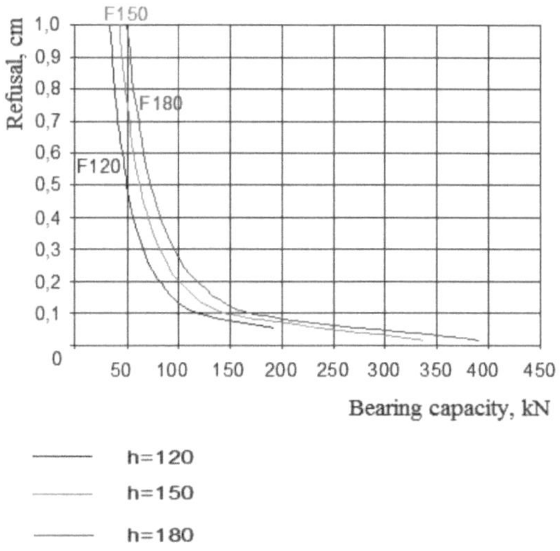

Figure 2. Dynamic testing results.

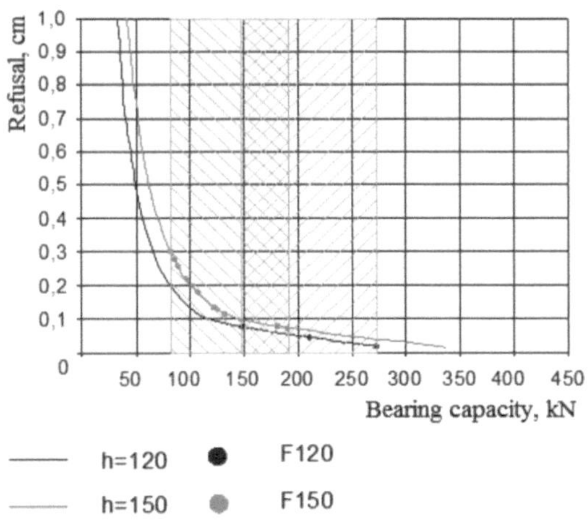

Figure 3. Dynamic testing results.

bearing capacity of piles). These points are based on results of tested piles. In the Figure 4 and 5 – for the curves of 1.50 and 1.80, 1.20 and 1.80 respectively. For the analysis of the empirical equation of reliability in the graphs show the range corresponding to the minimum and the maximum value of the bearing capacity for different blow energy. In case of full convergence (reliability) of the results, all the ranges should be the same. Different ranges of the test shows an error of test or limitation of empirical equation.

Quantitative assessment (in percentage) will be given on the basis proportions of coincidental range area (on the graph) to uncoincidental range area (part of the range lying outside overlay area).

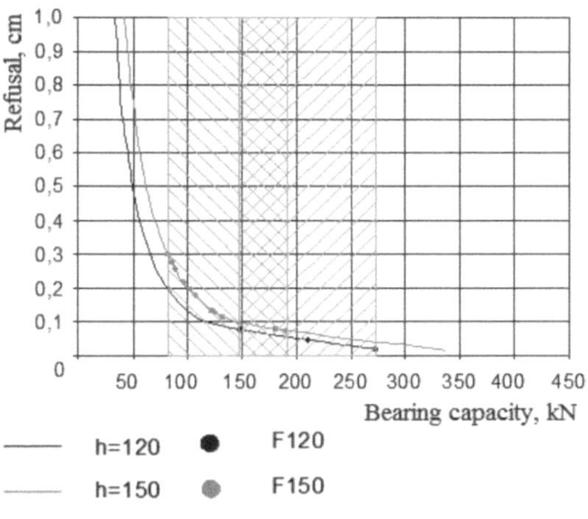

Figure 4. Dynamic testing results.

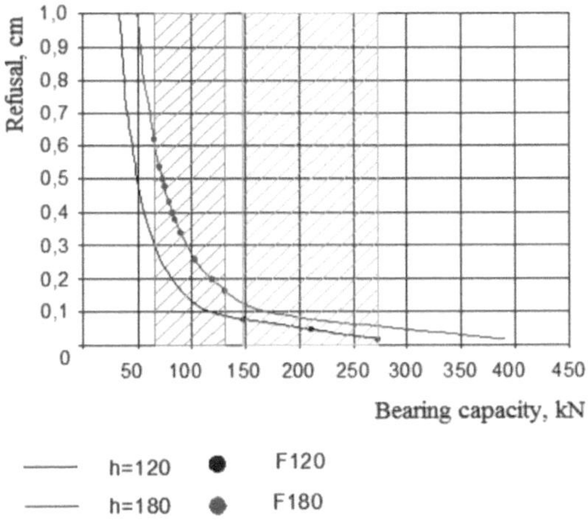

Figure 5. Dynamic testing results.

Also it should be noted that the refusal of pile has not reached the maximum allowable 0.2 cm, during dynamic test by energy of 1.20. Therefore test results by energy 1.20, can not be used for evaluation of the empirical dependence of bearing capacity. According to the graph 4 overlay range as a percentage is $(115–80) \cdot 100/(115–60) = 64\%$ (without bearing capacity which corresponds refusal of less than 0.2 cm). The graph 5 ranges do not overlap, therefore, this once again confirms that the test results at an energy of 120 cm is not valid for analysis.

6 CONCLUSION

According to the results of the comparative analysis, empirical equation to definition of piles bearing capacity by the dynamic testing (in these pile-driving, hydrological and climatic conditions) has an error of 65% (blow energy in the range of 150–180 cm). The comparisons also confirm that this dependence does not apply for refusal of less than 0.2 cm.

In future more complex dynamic tests will be planned for a more precise analysis the effect of blow energy of hammer on bearing capacity of piles.

REFERENCES

Report on the dynamic testing of piles (Arch. N0. 29/2-2014) "Astana Geotechnical Consulting" LLP, 2014: p. 25.

GOST 5686-94. Soils. Field test methods by piles. – M.: Standards Publishing House, 1994: p. 3–5.

SNIP RK 5.01-03-2002. Pile foundations, Almaty: KazGOR, 2003: p. 22–24.

Foundation and underground structure (TS2)

Challenges and Innovations in Geotechnics – Zhussupbekov (Ed.)
 ISBN 978-1-138-03007-7

A case study on zoned excavation of a large foundation pit in proximity to existing tunnels

Q. Li, Z.H. Xu, J. Li & X.J. Wang
Department of Underground Structure and Geotechnical Engineering, East China Architectural Design and Research Institute Co., Ltd., Shanghai, People's Republic of China
Shanghai Engineering Research Center of Safety Control for Facilities Adjacent to Deep Excavations, Shanghai, People's Republic of China

ABSTRACT: As available ground resource in congested urban areas in Shanghai is scarce, a number of excavations have to be carried out in close proximity to existing tunnels. It is of great concern to not only ensure the safety of excavation itself but also to minimize the adverse excavation-induced influences on adjacent tunnels in services. The performance of a large excavation for a basement covering an area of 23,000 m^2 near existing tunnels is presented in this paper. Zoned excavation technique was utilized to construct the foundation pit. The behaviour of the excavation and corresponding tunnel deformations have been investigated based on three-dimensional finite element analyses and extensive field monitoring. Hardening soil model with small-strain stiffness (HSS model) whose parameters were determined and calibrated against comprehensive laboratory tests was applied. The field monitoring results show that taking advantage of the zoned excavation procedure the lateral wall displacements were smaller along the nearby tunnels than those at other sides. The measured maximum horizontal tunnel movement was found to be 7 mm, which was within acceptable limits. Numerical analyses considering the small strain stiffness of soil give rise to fairly reasonable results as compared with the measured results.

Keywords: case study, zoned excavation, tunnel, small strain stiffness

1 INTRODUCTION

As available ground resource in congested urban areas in Shanghai are scarce, excavations become much larger and deeper than before to maximize the land use as much as possible. In addition, more and more excavations have to be carried out in close proximity to existing structures. Therefore, it is of great concern to not only ensure the safety of excavation itself but to mitigate the effect of excavation on adjacent environment. The construction of an excavation would inevitably lead to significant influences on the deformation of adjacent tunnels. Several authors have presented field monitoring data and numerical analyses relating to excavation-induced influences on existing tunnels in Taipei Chang et al., 2001), Singapore Sharma et al., 2001) and Shanghai (Hu et al., 2003).

Zoned excavation method is often applied in large-scale excavations where the construction site can be separated into smaller zones and the excavation process is carried out zone by zone. Once the area reduces, it can speed up the excavation process and construction of struts for each zone to form a complete retaining system early. In Shanghai, excavations in close proximity to existing tunnels are commonly constructed by using the zoned excavation technique (Wang et al., 2015).

In this paper, monitoring data relating to the performance of a large excavation adjacent to existing tunnels and excavation-induced tunnel movements are presented. Moreover, the effect of the excavation made using the zoned construction method on the movements of adjacent tunnels has been investigated by carrying three dimensional finite element analyses. The results from the field monitoring and numerical analyses are compared and discussed.

2 DESCRIPTION OF THE CASE HISTORY

2.1 *The project*

The site was for the development of the Green Valley Project, which was constructed in the location of Expo 2010 Shanghai in Pudong district. The project involved three 8–14 storey high-rise buildings and 2–4 storey podiums. The basement covered an area of 38000 m^2 and the excavation ranges from 11.4~15.1 m in depth. Fig. 1 shows the plan view of the site. It is clearly seen that the west side of the excavation is quite close to the existing Xi Zang road tunnels which are 11.36 m in diameter. The tunnels were constructed by using the shield tunneling method and lined by reinforced concrete segments. The two tunnels are

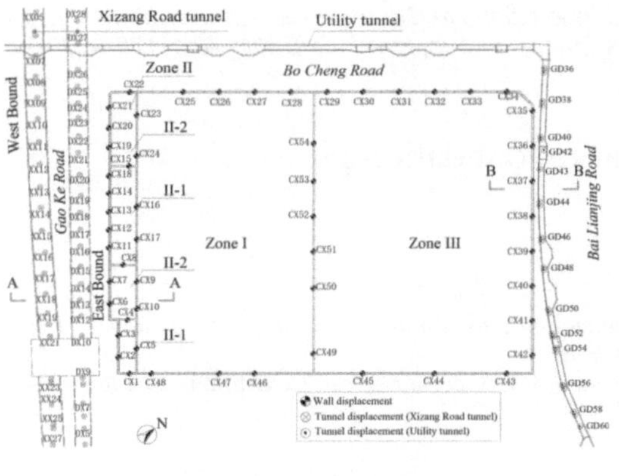

Figure 1. Plan view of the project and layout of instrumentation.

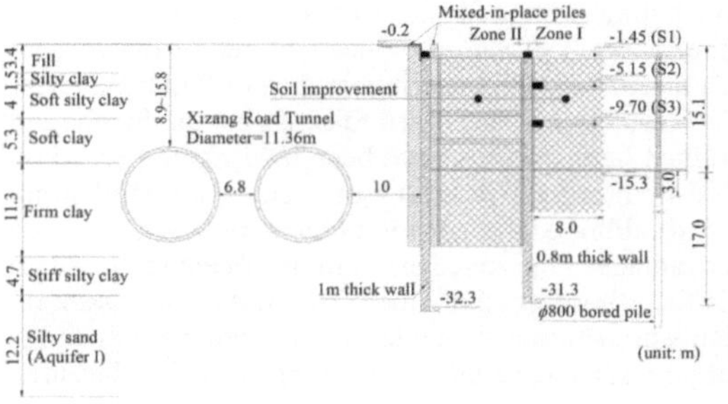

Figure 2. Relationship between the excavation and Xi Zang Road tunnel (along cross section A-A).

6.8 m apart and embedded about 8.9–15.8 m below the ground surface. As shown in Fig. 2, the closest distance between the excavation and the tunnels is about 10 m. To the east and north sides lies a utility tunnel which measures 3.8 m×3.3 m in cross section. The closest distance between the excavation and the utility tunnel is about 5.3 m.

2.2 *Geotechnical conditions*

The top two layers at this site are fill and silty clay followed by a 10 m-thick soft clay layer, which has relatively high water content and void ratio, but low shear strength. Underneath the soft clay layer lie a firm clay layer and a stiff silty clay layer. The silty sand layer (Aquifer I), is located about 30 m below ground. The water table is located at 0.5 m below ground surface.

2.3 *Design considerations*

As the excavation is quite close to the Xi Zang road tunnels at the west and the utility tunnel at the east, the excavation at these two sides were retained by 1000 mm-thick diaphragm walls which were supported by three levels of concrete struts. Two rows of mixed-in-place piles were constructed along the either side of the diaphragm wall to reduce the influence of wall construction on the existing tunnels. Peripheral

Table 1. Construction sequences.

Stage	Activities
1	Construction of diaphragm walls and soil improvement
2~ 6	Excavation and installation of concrete struts of Zone I
7~ 10	Removal of struts and construction of slabs of Zone I
11~ 15	Excavation and installation of struts of Zone II-1
16~ 18	Removal of struts and construction of slabs of Zone II-1
19~ 22	Excavation and installation of struts of Zone II-2
23~ 26	Removal of struts and construction of slabs of Zone II-2
27~ 30	Excavation and installation of concrete struts of Zone III
31~ 34	Removal of struts and construction of slabs of Zone III

8 m-wide soil inside the excavation was improved from the ground surface to 5 m below the formation level.

Considering the excavation area, the excavation depth, and the protection requirements of adjacent tunnels, zoned excavation technique was adopted in this project. The excavation was divided into two relatively large pits (Zone I and III) and a small one (Zone II), as shown in Fig. 1. Zone II was generally 14 m wide and adjacent to the metro tunnels. It was improved by deep soil mixing piles in a full type and supported by one level of reinforced concrete struts and three levels of steel struts. Zone I was constructed first followed by Zone II and Zone III subsequently. Principal construction activities adopted are summarized in Table 1.

3 PROGRAM OF FIELD MONITORING AND NUMERICAL ANALYSES

3.1 *Layout of field monitoring*

To verify the design considerations, and to monitor the performance of the excavation as well as adjacent tunnels, a number of instruments were installed. The layout of the monitoring points is shown in Fig. 1. Lateral wall displacements were measured by using inclinometers. The precise leveling system was adopted to measure the movements of both Xi Zang road tunnel and the utility tunnel.

3.2 *Three dimensional numerical analyses*

Figs 3 (a) and (b) show a 3D view of the finite element mesh generated in the Plaxis software (PLAXIS, 2015). The model consists of subsoil layers, retaining structures, struts, and tunnels. All soil layers were modeled using 10-node elements. The tunnels and retaining walls are modeled by 6-node triangular plate elements (shell), while the struts are simulated by

Table 2. Parameters for the HSS model in numerical analyses.

Soil layer	γ (kN/m^3)	E_{oed}^{ref} (MPa)	E_{50}^{ref} (MPa)	E_{ur}^{ref} (MPa)	G_0^{ref} (MPa)	c' (kPa)	φ' (°)	ψ (°)	$\gamma_{0.7}$	ν_{ur}	p^{ref} (kPa)	m	R_{inter}	R_f
②Silty clay	18.6	4.37	5.24	30.56	122.22	1	25.2	0	2×10^{-4}	0.2	100	0.8	0.65	0.9
③Soft silty clay	17.5	2.82	3.38	19.72	78.88	1	31.4	0	2×10^{-4}	0.2	100	0.8	0.65	0.6
④Soft clay	16.7	2.01	2.41	14.05	56.20	1	24.2	0	2×10^{-4}	0.2	100	0.8	0.65	0.6
⑤Firm clay	17.8	3.36	4.03	23.50	94.00	1	27	0	2×10^{-4}	0.2	100	0.8	0.65	0.9
⑥Stiff silty clay	19.7	7.59	9.10	53.11	212.44	1	30	0	2×10^{-4}	0.2	100	0.8	0.65	0.9
⑦Silty sand	19.1	10.97	10.97	43.88	219.40	1	33	3	2×10^{-4}	0.2	100	0.8	0.7	0.9

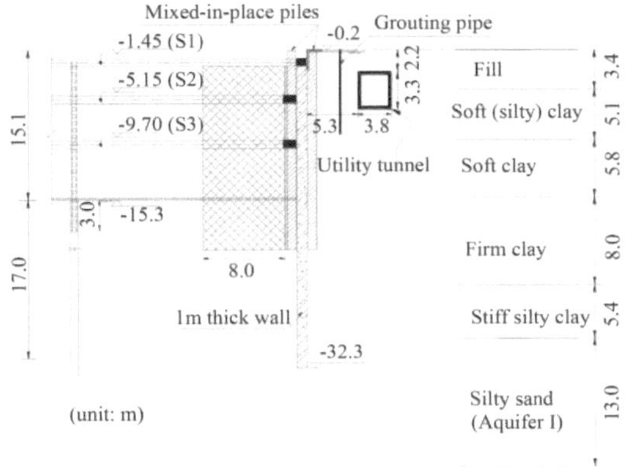

Figure 3. Relationship between the excavation and the utility tunnel (along cross section B-B).

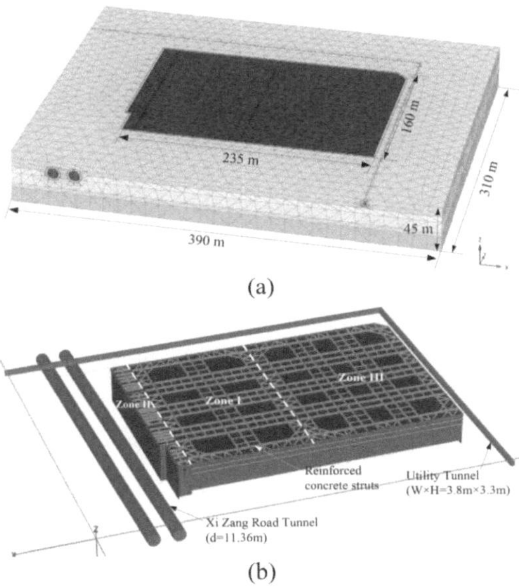

Figure 4. (a) 3D finite element model and mesh generation; (b) model of tunnels and reinforced concrete struts.

beam elements. The length of the mesh in the lateral (x) direction is 390 m. The width (y) and depth (z) are 310 m and 45 m, respectively. The model consists of 86266 elements and a total of 131663 nodes. The movements in all directions and rotations are fixed at the bottom of the mesh. Roller boundaries are adopted for the vertical faces of the mesh.

The hardening soil model with small-strain stiffness (HSS), which is a development of the original hardening soil model (HS) taking the small strain stiffness of soils into account (Benz, 2007), was applied in numerical analyses. In addition to the parameters for the HS model, two parameters, namely the small strain shear modulus G_0 and the reference shear strain $\gamma_{0.7}$ are adopted to capture the soil stiffness at small strain levels. By carrying out extensive triaxial and resonant column tests, all input parameters for the HSS model are obtained and summarized in Table 2.

4 COMPARISON BETWEEN FIELD AND NUMERICAL RESULTS

4.1 *Wall displacement*

Fig. 5 shows the measured lateral wall displacements at typical stages. The wall deformed in a deep-seated profile and the maximum wall displacement occurred near the formation level. The maximum wall deformation δ_{max} fall in the range of 0.13%H~0.47%H (H is excavation depth). It is clearly shown that the lateral deformations of diaphragm walls in Zone II (CX9 and CX20) were much smaller than those at north (CX26) and east side (CX40). The smaller wall displacement lies in (1) the scale of each excavation adjacent to the tunnels is significantly reduced as a result of applying zoned excavation construction; (2) the soil is improved by deep soil mixing piles giving rise to higher stiffness. The noticeable difference demonstrates the feasibility of zoned excavation method in reducing wall displacements for a large excavation. At CX9, more than 70% of the wall displacement took place in the course of the excavation of Zone I. The computed wall displacements based on the 3D analyses using the HSS model are also included in the figure for comparison. All the computed results agree fairly well with the measured data.

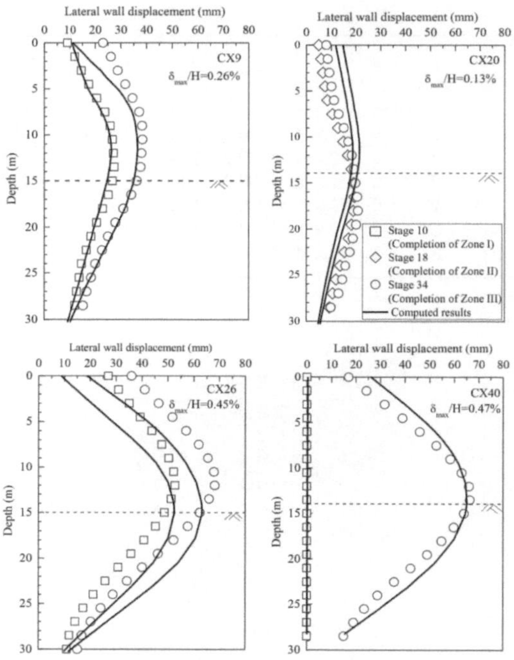

Figure 5. Comparison between measured and computed wall displacements.

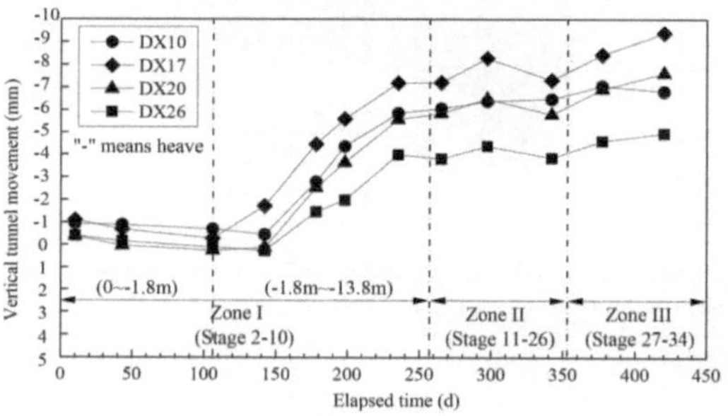

Figure 6. Development of vertical tunnel movement with time (east bound).

4.2 *Tunnel movement*

Fig. 6 shows the development of vertical movements of the east bound tunnel with time. The tunnel continued to heave during the entire course of excavation, except for the time when shallow soil in Zone I was removed. It is postulated that the ground heave due to stress relief was dominant in the period and the soil took the tunnel to move upward together. The maximum tunnel heave was about 9.5 mm which was less than the limit of 20 mm in Shanghai. It is obviously shown in the figure that the tunnel moved upward significantly during the construction of Zone I. Afterwards, the vertical movement increased slightly. A similar case reporting continuous heave of exiting tunnels adjacent to an excavation was presented by Li et al. (2015).

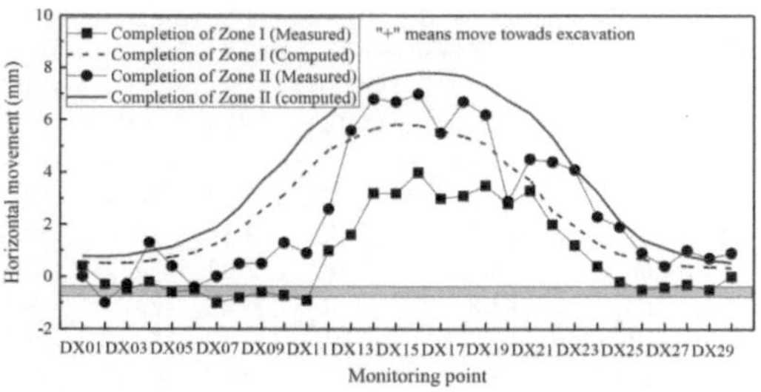

Figure 7. Variation of horizontal tunnel movement with excavation (east bound of Xi Zang Road tunnel).

Fig.7 compares the computed and field monitored horizontal tunnel movements measured at the tunnel invert. The computed and measured results are fairly comparable. Most tunnel linings moved towards the excavation. Three dimensional distributions of horizontal movements can be observed. A maximum horizontal movement of 7 mm occurred at DX16, near the middle of the west side. The structural integrity of the tunnels was maintained well throughout the excavation process. Due to the limited length of the paper, the movements of the utility tunnel are not presented.

5 CONCLUSIONS

A case history of a large braced excavation made using the zoned construction method in proximity to existing tunnels was presented. The effect of the excavation on the adjacent tunnels was investigated using field monitoring data and finite element analyses. Hardening soil model with small-strain stiffness (HSS model) was applied. The field monitoring results show that the lateral wall displacements were smaller along the nearby tunnels than those at other sides by using the zoned excavation technique. The east bound of Xi Zang road tunnel moved 9 mm upwards and 7 mm towards the excavation, less than the regulation of 20 mm in Shanghai. Numerical analyses considering the stiffness of soil at small strain levels give rise to fairly reasonable results as compared with the measured results

ACKNOWLEDGEMENT

The authors would like to acknowledge financial support from research grant 15pj1433500 provided by shanghai pujiang program

REFERENCES

Benz, T. 2007. Small-strain stiffness of soils and its numerical consequence. *PHD thesis, Institute of Geotechnical Engineering, University of Stuttgart, Germany,*

Chang, C. T., Sun, C. W., Duann, S. W. and Hwang, R. N. 2001. Response of a Taipei Rapid Transit System (TRTS) tunnel to adjacent excavation. *Tunnelling and Underground Space Technology,* 16(3), 151–158.

Hu, Z. F., Yue, Z. Q., Zhou, J. and Tham, L. G. 2003. Design and construction of a deep excavation in soft soils adjacent to the Shanghai Metro tunnels. *Canadian Geotechnical Journal,* 40(5), 933–948.

Li, Q., Wang, W. D. and Xu, Z. H. 2015. Design and behaviour of a deep large excavation adjacent to existing buildings and tunnels. *In Proceedings of International Conference on Geotechnical Engineering. Colombo, Sri Lanka,* 455–458.

PLAXIS 2015. PLAXIS 3D aniversary edition-reference manual. Netherlands, 2015.

Sharma, J. S., Hefny, A. M., Zhao, J. and Chan, C. W. 2001. Effect of large excavation on deformation of adjacent MRT tunnels. *Tunnelling and Underground Space Technology,* 16(2), 93–98.

Wang, W. D., Xu, Z. H. and Li, Q. 2015. Design and case histories of large deep excavations in complex urban environment in Shanghai. *ISSMGE Bulletin,* 9(6), 18–39.

Challenges and Innovations in Geotechnics – Zhussupbekov (Ed.)
© 2016 Taylor & Francis Group, London, ISBN 978-1-138-03007-7

Reliability evaluation of the bearing capacity of a nodular cast-in-place concrete pile

K. Watanabe
Technical Research Institute, Obayashi Corporation, Tokyo, Japan

A. Mitsumori
Obayashi Corporation, Tokyo, Japan

H. Nishioka & M. Koda
Structures Technology Division, Railway Technical Research Institute, Tokyo, Japan

ABSTRACT: In recent years, both the height and weight of buildings have increased. This trend is especially noticeable in the central urban areas of Japan. Both tensile and compressive forces occur in foundations such as these piles because of the overturning moment from earthquake and wind loads. To address these situations, new types of foundations need to be developed for high-rise superstructures. One approach is the nodular cast-in-place concrete pile, which has a nodular part at the mid-section of the pile. In situ full-scale load tests were carried out to estimate the bearing capacity of the nodular cast-in-place concrete pile. The bearing capacity was also estimated according to the Design Standards for Railway Structures and Commentary. This paper summarizes the results of standard bearing capacity based on the data from the in situ full-scale load tests and the estimation of ground resistance coefficient for nodular cast-in-place concrete piles.

1 INTRODUCTION

Significant compressive and tensile forces due to both ordinary and earthquake loads act on the foundations of ultrahigh-rise buildings with large aspect ratios, which are increasing in number in the urban areas of Japan. A nodular cast-in-place concrete pile has been developed to handle this issue; this is a cast-in-place concrete pile with an increased diameter for the middle section containing a nodular part, as shown in Figure 1. There have already been a number of applications in the architecture field (Watanabe, K. et al. 2008, Watanabe, K. et al. 2009, Watanabe, K. et al. 2010, Watanabe, K. et al. 2012). In the civil engineering field, when the space above railway tracks is utilized, the compressive force can be large because of the large spans, and the tensile force can be large because of high floor heights. This study evaluated the application of nodular cast-in-place concrete piles in civil engineering, particularly for the railway sector, with regard to their compliance with the Design Standards for Railway Structures (hereafter referred to as Railway Standards). This evaluation considered the variance in the bearing capacities of cast-in-place concrete piles and the different degrees of reaction forces for various displacement levels to reflect the above characteristics for end and nodular part resistances as well as the shaft resistance according to the standard pile end bearing capacity ratio. Establishing an evaluation method will make it possible to apply nodular cast-in-place concrete piles to railway structures. This report outlines the results with regard to the degree of standard bearing capacity based on the load test data. The report also provides the results for the ground resistance coefficient, which is used to calculate the design bearing capacity.

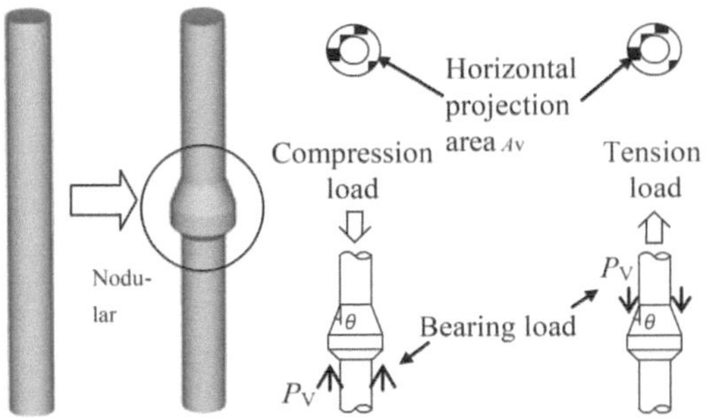

Figure 1. Schematic view of nodular cast-in-place concrete pile.

2 CONSTRUCTION METHOD FOR CAST-IN-PLACE CONCRETE PILES

Figure 2 shows the construction procedure for nodular cast-in-place concrete piles. First, the hole for the shaft section is excavated using the earth drill construction method. The pile tip is then enlarged with further

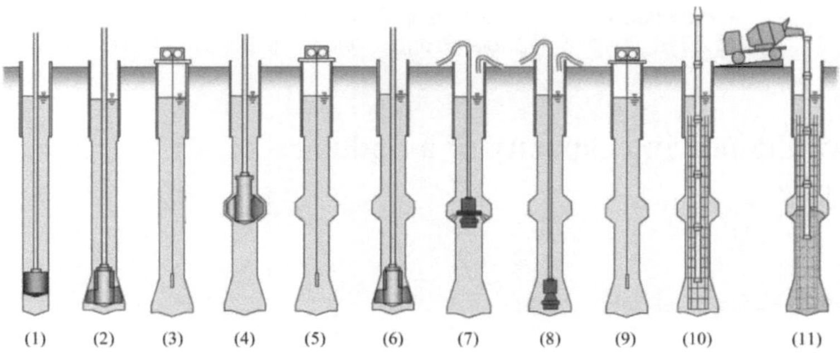

Procedure of construction
(1) Excavation of pile shaft
(2) Excavation of under-reamed part
(3) Measurement of pile shape
(4) Excavation of nodular part
(5) Measurement of pile shape
(6) Slime treatment
(7) Cleaning of nodular part
(8) Slime treatment
(9) Measurement of pile shape
(10) Insertion of steel reinforcement cage
(11) Concreting

Figure 2. Construction procedure of a cast-in-place concrete pile.

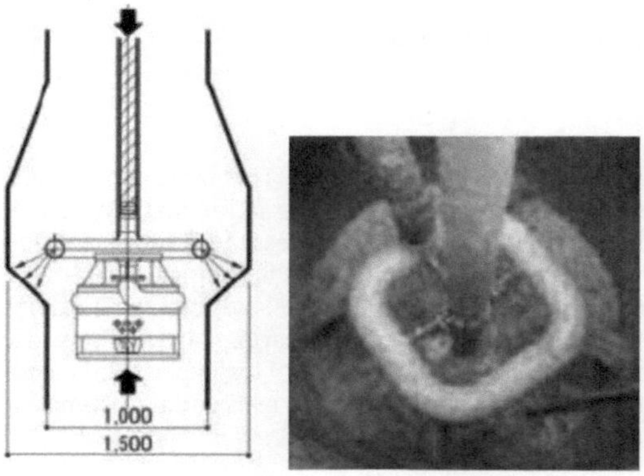

Figure 3. Slime cleaner for the nodular part.

excavation. The nodular part section is then excavated using a special excavation device for the nodular part. Once the excavation is completed, the slime at the nodular part and pile tip sections is treated with slime cleaner. Then, the slurry in the cavities is replaced with a better liquid. Figure 3 shows a special slime cleaner that is used for the slime treatment of the nodular part sections. The steel reinforcement cage is then inserted, and the concrete is poured in a tremie pipe. The shape of the nodular part sections is verified by ultrasound measurements.

3 RESULTS OF THE INVESTIGATION OF THE STANDARD BEARING CAPACITY

The load test data used for the evaluation were taken from two sites. All load tests were performed in compliance with the Method for Load Test of Single Piles (The Japanese Geotechnical Society, 2002). The loading continued until the displacement at the pile tip exceeded 10% of the pile diameter. The displacement at the nodular part was defined as the displacement measured in the middle of the nodular part.

The standard bearing capacity was examined according to the concepts in the Railway Standards (Railway Technical Research Institute, 2012), where the maximum load was defined as the value when the displacement at the pile tip reached 10% of the pile diameter. Figure 4 shows the results of the investigation on the standard bearing capacity. The data were taken from two load tests in which piles were loaded with sufficient displacement (at least 10% of the pile diameter). The results of standard bearing capacity were categorized into the end bearing capacity, bearing pressure at nodular part, and shaft friction. The end bearing capacity was calculated by dividing the axial force reaching the pile tip by the pile tip area. The bearing pressure at the nodular part was calculated by dividing the difference in the axial forces above and below the nodular part by the horizontal projection area of the nodular part. Moreover, the shaft friction was calculated by dividing the difference in axial forces of each section by the peripheral area between the corresponding sections. Eqs. (1), (2-a) and (2-b) were used to estimate the standard bearing capacity of cast-in-place concrete piles according to the Railway Standards.

Standard end bearing capacity:

$$q_{tk} = 60\ N\ (\text{kN/m}^2) \tag{1}$$

(sandy soil: $qtk \leq 3500$, gravel soil: $qtk \leq 7500$)
Standard shaft friction (sandy soil):

$$\gamma_{fk} = 3N\ (\leq 150)\ (\text{kN/m}^2) \tag{2-a}$$

Standard shaft friction (clayey soil):

$$\gamma_{fk} = 6N\ (\leq 150)\ (\text{kN/m}^2) \tag{2-b}$$

where, N is the N value obtained from the standard penetration test.

Figure 4(a) shows the relationship between end bearing capacity, bearing pressure at nodular part and SPT N value. This figure also plots the standard end bearing capacity line for cast-in-place concrete piles as indicated by the Railway Standards. The end bearing capacity and bearing pressure at nodular part derived from the load tests of the nodular cast-in-place concrete piles form a relationship with values generally the same or above the standard end bearing capacities provided by the Railway Standards. Figure 4(b) shows the relationship between the shaft friction and SPT N

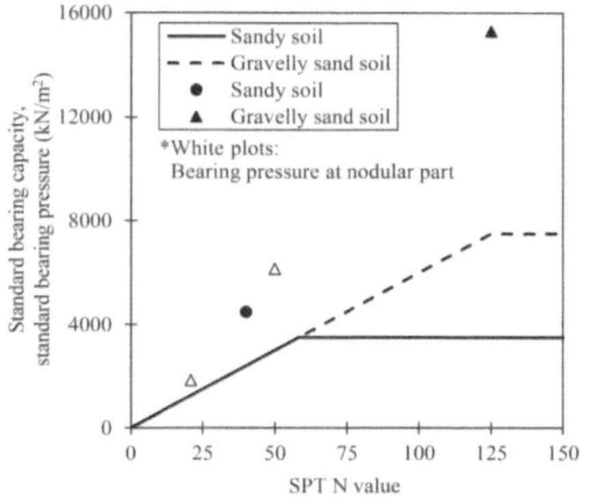

(a) Relationships between end bearing capacity, bearing pressure at nodular part and SPT N value

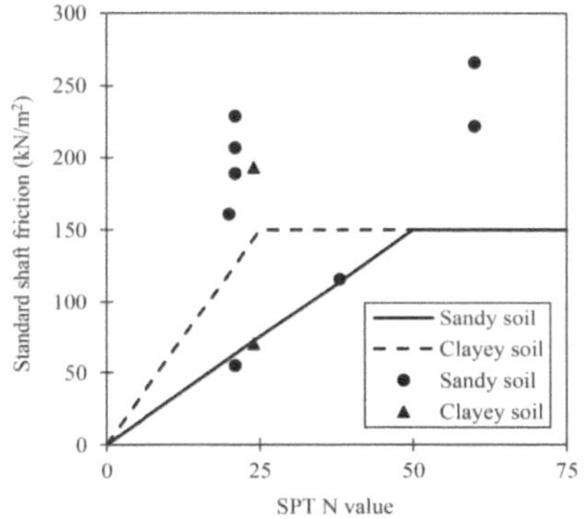

(b) Relationships between shaft friction and SPT N value

Figure 4. Design and tested bearing capacities.

value. The standard shaft friction for cast-in-place concrete piles according to the Railway Standards is also plotted with the shaft friction. The calculated values from the equations provided by the Railway Standards were generally exceeded by the shaft friction. The shaft friction for cohesive soil was significantly lower than the standard shaft friction because the data were derived from the results of the tension test performed after the compression test and thus were subject to the effects of a significant deformation. Figure 5 compares the observed values and the calculated values for the standard bearing capacity. The observed values clearly exceeded the calculated values.

4 CALCULATION OF GROUND RESISTANCE COEFFICIENT

4.1 *Calculation method for ground resistance coefficient*

The design vertical bearing capacity R_{vd} of a pile is calculated by using Eqs. (3) ~ (5) from the Railway Standards.

$$R_{vd} = f_r R_k = f_{rt} R_{tk} + f_{rf} R_{fk} \tag{3}$$

$$R_{tk} = q_{tk} A_t \tag{4}$$

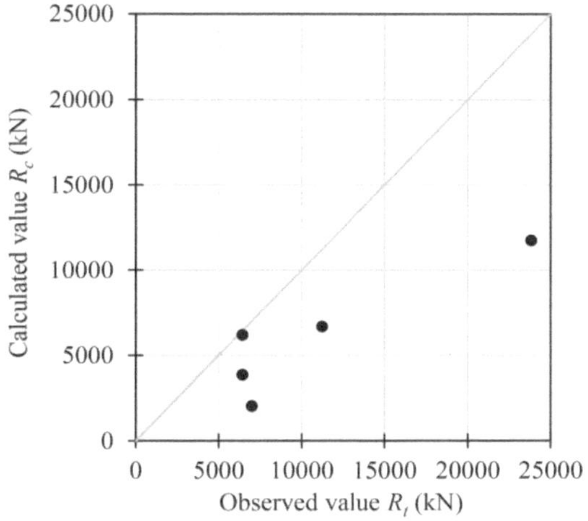

Figure 5. Relationship between calculated and observed values for design bearing capacity.

$$R_{fk} = \gamma_{tk} U \Delta l \tag{5}$$

where f_r: Ground resistance coefficient, R_k: Standard bearing capacity, f_{rt}: Ground resistance coefficient for pile end, R_{tk}: Standard end bearing capacity, f_{rf}: Ground resistance coefficient for pile shaft, R_{fk}: Standard shaft friction, q_{tk}: Standard end bearing capacity, A_t: Pile tip area, γ_{fk}: Standard shaft friction for different soil layers, U: Peripheral length of the pile, Δl: Length of different soil layers.

Beside the standard bearing capacity, the ground resistance coefficient needs to be evaluated in order to calculate the design vertical bearing capacity with Eqs. (3) ~ (5). The ground resistance coefficients are stipulated to ensure that the prescribed reliability is secured for different required performance.

Let the actual measured values of the pile end resistance and shaft resistance when a certain limit state has been reached (i.e., the settlement has reached a certain standard displacement) in a vertical load test of a single pile be R_t^t and R_f^t, respectively. Then, the performance function Z for the action of the pile head load S can be expressed by Eq. (6) using R_t^t, R_f^t, and S as random variables. The subscript t indicates an actual measured value.

$$Z = R_t^t + R_f^t - S \tag{6}$$

The performance function Z generally becomes complex and it becomes difficult to easily derive the limit state exceedance probability. The first-order proximity reliability design method (i.e., first-order reliability method (FORM)) is used to approximate the probability variable with a normal distribution function, and a Taylor expansion is performed at the point where the performance function reaches zero (design point) to suspend the series with a first-order term. This allows linearization to be performed and the limit state exceedance probability to be calculated. If the performance function is expressed as a sum of the linearized normal distribution function when this method is used, then the limit state exceedance probability can be calculated without requiring convergence.

First, the probability variables R_t^t, R_f^t, and S are normalized using the characteristic values for the standard bearing capacities R_{tk} and R_{fk} and the characteristic value S_k of the load acting on the pile head. These are then assumed to be probability variables with mutually independent normal distributions. Then, the performance function Z given in Eq. (6) can be expressed as Eq. (7) by using the probability variables Y_t, Y_f, and Y_S. These have normal distributions for which the average value is 0 and standard deviation is 1.

$$Z = R_{tk}\mu_t + R_{fk}\mu_f - S_k\mu_s + R_{tk}\sigma_t Y_t + R_{fk}\sigma_f Y_f - S_k\sigma_s Y_s \quad (7)$$

where, μ_t, μ_f, μ_s, σ_t, σ_f, σ_s: Average values and standard deviations of the probability variables.

This transformation derives the point located at the shortest distance to the plane expressed by Eq. (7) from the origin located at (Y_t, Y_f, Y_s) as the design point. The safety index β can be derived with Eq. (8) and is defined as the shortest distance from the origin to the design point.

$$\beta = \frac{\mu_Z}{\sigma_Z} = \frac{R_{tk}\mu_t + R_{fk}\mu_f - S_k\mu_s}{\sqrt{(R_{tk}\sigma_t)^2 + (R_{fk}\sigma_f)^2 + (S_k\sigma_s)^2}} \quad (8)$$

If the standard probability variables are mutually independent according to the normal distribution function and if the performance function can be expressed as a linear function of the standard probability variables, then the safety index and limit state exceedance probability $P_f(Z \leq 0)$ have the following relationship:

$$P_f = \Phi(-\beta) \quad (9)$$

where, Φ is a standard normal probability distribution function and expresses how far the average value, μ_z is located from the point at which the limit state is reached ($Z = 0$) as a scale factor with respect to the standard deviation σ_z.

The ground resistance coefficient equivalent to the safety coefficient for the resisting side is derived by following the method described by Nishioka et al. (2008). The performance function in Eq. (6) can be expressed as Eq. (10) through a transformation that uses the ground resistance coefficients f_{rt} and f_{rf}, and activity coefficient γ_s

$$Z^d = f_{rt}R_{tk} + f_{rf}R_{fk} - \gamma_s S_k \quad (10)$$

Because Eq. (10) satisfies $Z^d = 0$ when the target safety index β_a is achieved, the ground resistance coefficient can be expressed by Eqs. (11-a) and (11-b).

$$f_{rt} = \mu_t - \beta_a\alpha_t\sigma_t \quad (11\text{-a})$$

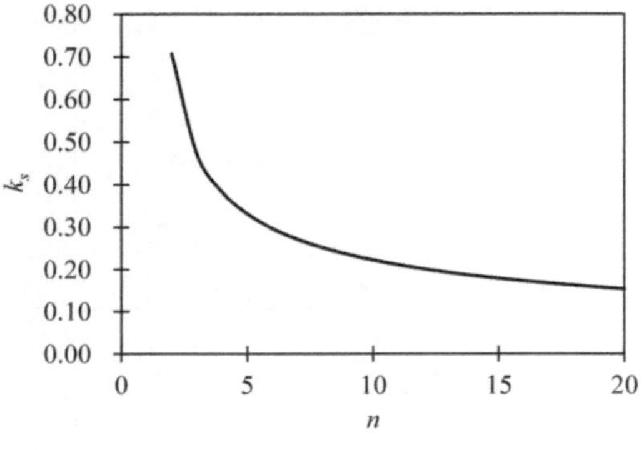

Figure 6. Coefficient k_s for data (confidence level: 0.75).

$$f_{rf} = \mu_f - \beta_a\alpha_f\sigma_f \quad (11\text{-b})$$

where μ_t and μ_f: Average values of the probability variables, σ_t and σ_f: Standard deviations of the probability variables, β_a: Target safety index, α_t and α_f: Sensitivity coefficients.

The target safety index β_a expresses the safety margin for respective limit states. Because the sensitivity coefficients α_t and α_f change in response to the change in the standard pile end bearing capacity ratio p_t, a ground resistance coefficient that reflects whether the designed pile is a friction pile or bearing pile can be calculated. The sensitivity coefficients α_t and α_f and the standard pile end bearing capacity ratio p_t are given by Eqs. (12) ~ (14).

$$\alpha_t = p_t\sigma_t \Big/ \sqrt{p_t^2\sigma_t^2 + (1-p_t)^2\sigma_f^2} \quad (12)$$

$$\alpha_t = (1-p_t)\sigma_f \Big/ \sqrt{p_t^2\sigma_t^2 + (1-p_t)^2\sigma_f^2} \quad (13)$$

$$p_t = R_{tk}/(R_{tk} + R_{fk}) \quad (14)$$

Because the average values μ_t, and μ_f are estimated from a small number of specimens, they are estimated from Eq. (15), which has a lower limit of 0.75 for the confidence level corresponding to the number of data points n.

$$\mu = \mu - k_s\sigma \quad (15)$$

where, μ: Estimated average value, $\bar{\mu}$: Average value for the load test data, σ: Standard deviation of the load test data, k_s: Coefficient that corresponds to the number of data points n, as indicated in Figure 6.

Table 1 lists the required performance and limit state necessary to calculate the ground resistance coefficient. Such values comply with the Railway Standards, and the five performance parameters were statistically analyzed.

4.2 *Calculation results for ground resistance coefficients*

Table 2 lists the average values and coefficients of variation of the statistical analysis. The table also

Table 1. Required performance and limit state.

(a) Excluding seismic motions

Required performance	Performance lists	Standard displacement (mm)	Target safety index β_a Exceedance probability of standard displacement P_f	Limit state
Usability	Long term bearing performance	20(13.3) ※1	β_a=1.6, P_f=5%	Crack by differential settlement
	Short term bearing performance	20	β_a=1.3, P_f=10%	
Safety	Stability	min(50 or 0.05D) ※2	β_a=1.0, P_f=16%	Noticeable displacement at railway

※1: The value in braket considering displacement increase of long term load test is used in statistical work. (Aoki et al., 1985)

※2: D means pile diameter.

(b) Seismic motions

Required performance	Performance lists	Standard displacement (mm)	Target safety index β_a Exceedance probability of standard displacement P_f	Limit state
Restorability and Performance level 1	Residual displacement	50 or 0.05D	β_a=0.85, P_f=20%	Maintenance and repair of railway
Restorability and performance level 2		0.1D	β_a=0.5, P_f=30%	Maintenance and repair of structure

Table 2. Resistance factors for nodular cast-in-place concrete pile.

		Pile shaft				Pile tip or Pile tip + nodular part (Nodular cast-in-place concrete pile)			
	Required performance	Usability	Usability	Safety / Restorability and performance level 1	Restorability and performance level 2	Usability	Usability	Safety / Restorability and performance level 1	Restorability and performance level 2
	Performance lists	Long term bearing performance	Short term bearing performance	Stability Residual displacement	Residual displacement	Long term bearing performance	Short term bearing performance	Stability Residual displacement	Residual displacement
	Standard displacement	Long term20mm	20mm	min (50mm, 5%D)	10%D	Long term20mm	20mm	min (50mm, 5%D)	10%D
Cast-in-place concrete pile	Effective data number n	7	6	6	6	9	8	8	8
	Average value μ	1.16	1.16	1.31	1.37	0.5	0.56	0.85	1.39
	Variation coefficient V	38%	38%	40%	33%	60%	46%	44%	49%
Nodular cast-in-plsce concrete pile	Effective data number n	3 (5)	3 (5)	3 (5)	3 (5)	2	2	2	2
	Average value μ	1.06	1.22	1.62	1.74	0.63	0.73	1.06	1.35
	Variation coefficient V	21%	16%	25%	65%	22%	27%	41%	65%

presents the results of the statistical analysis for the cast-in-place concrete piles compared with the Railway Standards. The table inside the brackets next to the number of effective data points regarding the shaft friction is the total number of data points used for the investigation on the standard bearing capacity, as described in Section 4.1. The number of data points outside the brackets was adopted for the statistical analysis to eliminate data that contained significant effects from the loading history, such as the tension load test implemented after the tension part of the alternating load test or the compression test that caused significant deformation. Table 2 reveals that the cast-in-place concrete piles had a large coefficient of variation for the pile tip in the long term bearing performance with a small standard displacement. On the other hand, the nodular cast-in-place concrete piles tended to have smaller coefficients of variation at the pile end and nodular part in the long term and short term bearing performance with a small standard displacement. This differed from the bearing capacity at the pile end of the cast-in-place concrete piles, which are conventionally considered to have a large variance. This suggests that there is a small variance in the bearing capacity at the nodular part and pile end sections of nodular cast-in-place concrete piles. The slime treatment using a slime cleaner specially made for the nodular part and the replacement of the slurry with a better liquid after the slime treatment may be reasons for this tendency. The pile shaft can be conservatively evaluated by applying the same evaluation as that for the cast-in-place concrete piles because the coefficients of variation related to the pile shaft were also smaller than those of the cast-in-place concrete piles. The target safety index in Table 1 and the statistical values and standard pile end bearing capacity ratios in Table 2 were then used to calculate the ground resistance coefficient using the method described in Section 4.2. It is necessary to consider the uncertainties of the ground survey in actual designs, even for the statistical values with smaller coefficients of variation (less than 30%) in Table 2. For the purpose of comparison, the procedures in the Railway Standards were followed. Thus, even when the coefficients of variation were under 30%, they were still considered to be 30% for the purpose of this report. Figure 7 shows the relationship between the ground resistance coefficient and standard pile end bearing capacity ratio. From the bottom to the top, each line represents the ground resistance coefficients for different required performance: the usability and long term bearing performance, usability and short term bearing performance, safety and non-seismic motions, restorability and performance level 1, and restorability and performance level 2. Figure 7 shows the ground resistance coefficients for cast-in-place concrete piles as described by the Railway Standards with dotted lines for the purpose of comparison. For the required performance, the ground resistance coefficient tended to be small when the standard pile end bearing capacity ratio was large. The ground resistance coefficients were generally larger than those of the cast-in-place concrete piles because the average values were greater and the standard deviations were smaller. For the restorability and performance level 2, the average value of the pile shaft was sufficiently larger than those of the cast-in-place concrete piles. However, because the coefficients of variation were

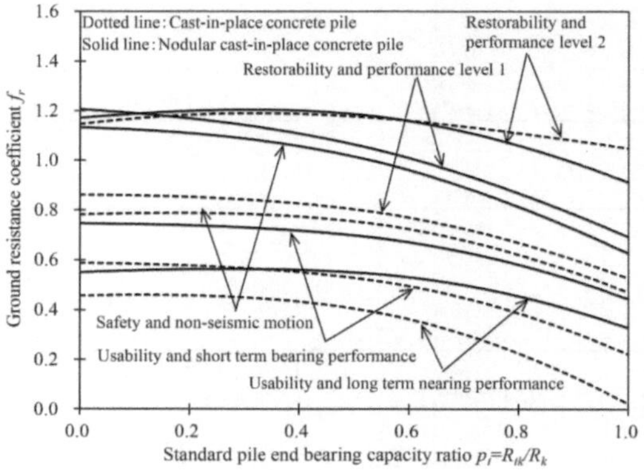

Figure 7. Ground resistance coefficient for a nodular cast-in-place concrete pile.

large, the ground resistance coefficients were about the same in regions where the standard pile end bearing capacity ratio was small. The values were lower than those of the cast-in-place concrete piles in the region where the standard pile end bearing capacity ratios were large because the coefficients of variation were large. This is despite the fact that the average values for the pile end and nodular part were about the same as those of the cast-in-place concrete piles. The ground resistance coefficients of the nodular cast-in-place concrete piles were inferred to be about the same as those of the cast-in-place concrete piles when the slime treatment using a slime cleaner specially made for the nodular part was not performed or when the slurry was not replaced with better liquid after the slime treatment. The ground resistance coefficients were believed to increase with the bearing resistance in the nodular part when the above construction process was performed.

5 CONCLUSIONS

This paper presents the results of standard bearing capacity and the evaluation of the bearing capacity for nodular cast-in-place concrete piles. The results are summarized as follows.

1) The calculation formula for the standard bearing capacity of cast-in-place concrete piles described in the Railway Standards was compared with the data from load tests on nodular cast-in-place concrete piles. The bearing capacity of the nodular cast-in-place concrete piles was found to be comparable to the values from the standard bearing capacity calculation formula for cast-in-place concrete piles described by the Railway Standards.
2) The ground resistance coefficients for nodular cast-in-place concrete piles were calculated in compliance with the Railway Standards. The results showed that they generally exceeded the ground resistance coefficients of the cast-in-place concrete piles.

REFERENCES

Aoki, K. et al. 1985. Coefficient of subgrade reaction used in design of piles. *Structural Design Documents*: No. 83, 20–25. (in Japanese).

Nishioka, H. et al. 2008. Calculation of bearing capacity by pile construction methods using statistical analysis of load test data. *Railway Technical Research Institute Report*: Vol. 22, No. 10, 41–46. (in Japanese).

Railway Technical Research Institute. 2012. Design Standards for Railway Structures and Commentary. (in Japanese).

The Japanese Geotechnical Society. 2002. JGS standard, Method for Axial Load Test of Single Piles. (in Japanese).

Watanabe, K. 2008. Static axial reciprocal and tensile load tests of single cast-in-place concrete nodular piles (Part 1 to 4). *Proceedings of Annual National Conference of Architectural Institute of Japan*:567–576. (in Japanese).

Watanabe, K. et al. 2009. Static axial tensile and compressive load tests of nodular diaphragm wall supporting high-rise tower (Part 1 to 5). *Proceedings of Annual National Conference of Architectural Institute of Japan*: 439–450. (in Japanese).

Watanabe, K. et al. 2010. Static Axial Compressive and Tensile Load Test of Single Cast-in-place Nodular Concrete Pile (Part 1 to 4). *Proceedings of Annual National Conference of Architectural Institute of Japan*: 613–620. (in Japanese).

Watanabe, K. et al. 2012. Static load tests of nodular cast-in-place concrete pile and estimation on resistance of nodular part. *Proceedings of IS-Kanazawa*: 749–756.

Challenges and Innovations in Geotechnics – Zhussupbekov (Ed.)
© 2016 Taylor & Francis Group, London, ISBN 978-1-138-03007-7

Comparative analysis of allowable load carrying capacity of laterally loaded piles

F.N. Sheikh, A.A. Gill & T. Masood
Berkeley Associates (Pvt.) Ltd., Lahore, Pakistan

M. Irfan
University of Engineering & Technology Lahore, Pakistan

ABSTRACT: In developing countries, financial constraints often govern the scope of in-situ geotechnical investigations. The performance of large scale tests, like the lateral pile load tests, is scarce and is limited only to the mega projects. Theoretical approaches, like Brom (1964) and Davisson and Gill (1963) methods, are therefore relied upon to estimate the allowable load carrying capacities of piles in lateral direction. In this paper, the results from seven lateral pile load tests (three tests under fixed head and four under free head conditions) conducted at the 1223 MW combined cycle power plant at Balloki, Pakistan are assessed. Analysis has been conducted to compare the allowable pile load capacities and the corresponding pile head deflections measured through in-situ full scale pile load tests with theoretical and computer aided pile capacity estimates. The analysis results and subsequent discussions signifies the superiority of computer aided pile capacity estimates over theoretical evaluations. The use of computer aided pile capacity estimates is envisaged to assist in enhancing safety as well as economizing the design.

1 INTRODUCTION

The cost of geotechnical investigations is typically around 1% to 3% of the total project cost. In developing countries, where the budgets are generally tight and regulations pertaining to infrastructure development are neither well-documented nor properly implemented, the main focus remains on squeezing the project cost. The scope of geotechnical investigations is thus, often governed by the financial limitations. The performance of large scale tests to ascertain the estimated bearing characteristics of foundation-soil system is even lesser, and limited only to mega projects.

Lateral capacity of piles is important to incorporate the effects of machine loading, earthquake, wind, earth pressure, traffic loading, etc. In Pakistan, the performance of full scale lateral pile load tests, for the determination of lateral load carrying capacity of piles, is limited only to mega projects. For majority of the projects, lateral pile capacity evaluations are done through theoretical approaches like Brom (1964) method or the Elastic method (Davisson and Gill, 1963). The applicability of these theoretical approaches in the local alluvial planes has not been verified. Due to such uncertainties, the designers are inclined to use high safety factors, thereby making the design uneconomical.

In this research, test data from lateral pile load tests conducted at the 1223 MW combined cycle power plant at Balloki, Pakistan has been analyzed. A comparative analysis has been conducted between lateral pile load capacities interpreted from the pile load tests, theoretical approaches and computer-aided pile capacity estimates. The compiled database, and the subsequent analysis would facilitate the designers in selecting the most appropriate methodology for the estimation of lateral pile load capacities.

2 PROJECT BRIEF

Since the last decade or so, Pakistan has been facing acute power shortage. Keeping in view the power crisis in the country and to meet future demand, 1223 MW CCPP Balloki is being constructed in Balloki, District Kasur, Punjab – Pakistan. The project site expands over an area of 360,847 m^2. Geotechnical investigations conducted at the project site comprised drilling of boreholes (642 in number) with the performance of Standard Penetration Test (SPT) at a general depth interval of 1.0 m in each borehole, performance of field permeability tests (16 in number), excavation of test pits (17 in number), in-situ density determination in each test pit, etc. The generalized soil profile, as explored from the geotechnical investigations comprised top layer of soft to very stiff lean clay/silty clay extending down to a general depth of 13.0 m below the average existing ground level (EGL). A medium dense to very dense sandy layer follows, which continues

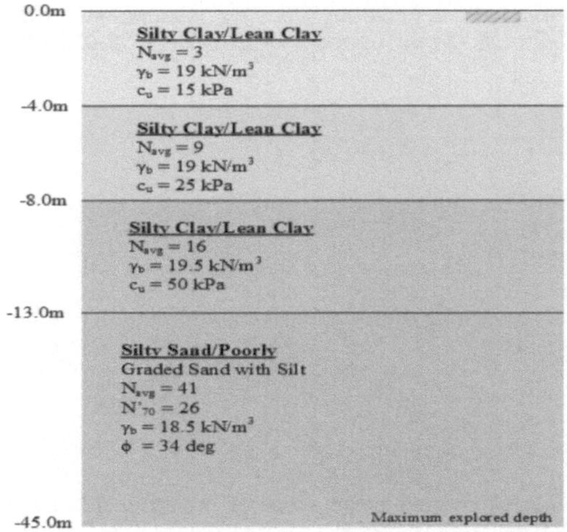

Figure 1. Generalized stratigraphy at the project site.

down to the maximum explored depth of 45.0 m below EGL. A comprehensive laboratory testing program was designed to ascertain the soil stratigraphy and it geotechnical characteristics. The subsurface soil profile considered for the evaluation of pile load carrying capacities is schematically illustrated in Figure 1.

3 RESEARCH METHODOLOGY

Considering the objectives planned for this research, lateral pile load capacities were evaluated by means of theoretical approaches, computer-aided analysis and full-scale pile load tests.

Two theoretical methods of analysis were employed in this research, i.e. Brom (1964) method, and Davisson and Gill (1963) method. The Brom (1964) approach, which is the most commonly used and the widely accepted approach in Pakistan, is based on earth pressure theory. Ultimate lateral pile load capacity for free-end as well as fixed-ended piles can be estimated by this approach. However, the estimation procedure is limited to uniform cohesive and non-cohesive soils only. The prevailing soil stratigraphy has to be converted into an equivalent cohesive/non-cohesive soil layer for the estimation of pile capacity. Such approximation can lead to quite unrealistic pile capacity estimates. The Davisson and Gill (1963) method is based on the elasticity theory. The basic derivation is based on the analogy of elastic settlement caused by a beam resting on foundation soil. The beam is considered to be supported by elastic springs whose stiffness is a function of soil's modulus of subgrade reaction. This method outlines the procedure for the estimation of lateral pile load capacity in cohesive soils only. An equivalent cohesive soil strata is therefore required to be generated which puts a serious question mark on the estimated pile load capacity in stratified soils.

Table 1. Basic information about pile load tests conducted at the project site.

Pile Designation	Test Type	Diameter (mm)	Length (m)	Test Load (kN)
K2′	Free Head	760	30	400
K3′	Free Head	760	20	400
K4′	Free Head	760	20	400
KZ1′	Fixed Head	760	30	550
KZ2′	Fixed Head	760	30	550
KZ3′	Fixed Head	760	20	550
KZ4′	Fixed Head	760	20	550

Computer-aided analysis was conducted by *LPile* software (Ver. 5.0.47) from *Ensoft Inc.* This software performs evaluations for the estimation of lateral pile load capacity and the corresponding deflections by considering the various in-built p-y curve models for each soil layer in the pile embedment.

Seven full-scale pile load tests were performed at the project site. Three of these tests were performed under free-end conditions while four tests were performed under fixed-end conditions. Basic information about each of these test piles is summarized in Table 1. All the tests were carried out in accordance with ASTM D3966-07. Pile head deflection was recorded under the test load, which was applied in increments. The results of pile load test were interpreted using IBC (2012) one inch deflection criteria for the determination of ultimate, and subsequently the allowable lateral load carrying capacity of piles.

A comprehensive comparative analysis was conducted among the results obtained from theoretical approaches, computer-aided estimates and estimates from full-scale pile load tests. The comparison results and the related discussions are presented in the following sections.

4 RESULTS AND DISCUSSIONS

In this research, allowable lateral load carrying capacity of pile was estimated using both classical approaches and *LPile* software. The capacities were estimated for free as well as fixed head conditions. Allowable load carrying capacities were evaluated against tolerable lateral deflection of 12.7 mm. However, load deflection response of laterally loaded pile was also checked using *LPile*. Values of allowable lateral load carrying capacity of piles evaluated through aforementioned approaches are summarized in Table 2.

These estimated values were then compared with the results of full scale pile load tests conducted at the project site. The load test data was analyzed using the IBC (2012) approach to determine ultimate load carrying capacity of test pile under lateral loading. Following the guidelines of IBC (2012), the allowable loads were determined as half of the loads

Table 2. Basic information about pile load tests conducted at the project site.

Evaluation Method	Allowable Lateral Load Carrying Capacity (kN)	
	Free Head	Fixed Head
Brom's Method	60	125
Elastic Method	71	
LPile	88	215

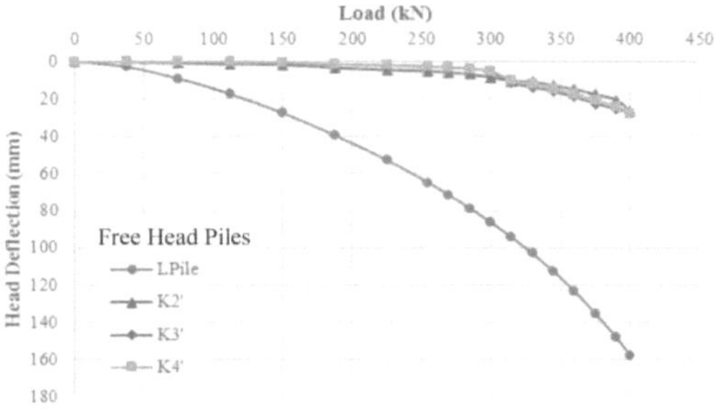

Figure 2. Comparison of load deflection behavior of free head piles from *LPile* and full scale pile load tests.

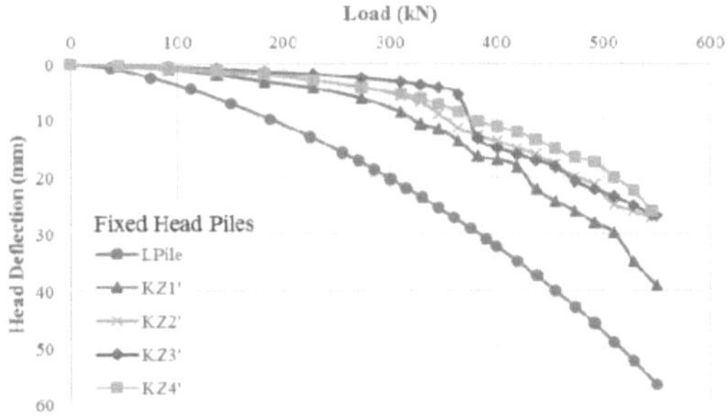

Figure 3. Comparison of load deflection behavior of fixed head piles from *LPile* and full scale pile load tests.

producing gross lateral deflection of 25 mm. Results of lateral pile load test conducted at the project site revealed average lateral load carrying capacities of 199 kN and 233 kN under free and fixed head conditions respectively. The actual pile response from the pile load tests, compared with the simulated response (using *LPile*) for free head and fixed head piles are shown in Figure 2 and Figure 3 respectively. Significantly smaller head deflections were observed for both free head and fixed head piles, compared to the *LPile* simulated response. The overestimation of head deflections may be attributed to conservativeness of soil parameters considered for the *LPile* analysis.

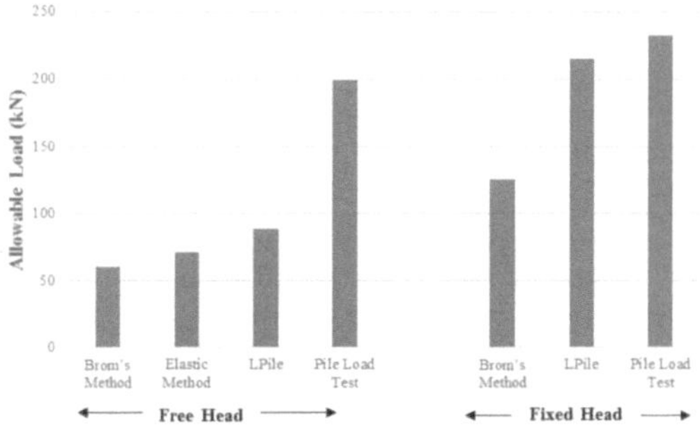

Figure 4. Comparison of average allowable lateral load carrying capacity evaluated from theoretical methods, *LPile* computer analysis, and full scale pile load tests.

Figure 4 depicts a comparison between the lateral pile load capacities estimated by theoretical methods, *LPile* analysis, and average capacity from pile load tests. Through comparison it can be observed that actual load carrying capacities of piles in lateral direction are significantly higher than values estimated through both classical approaches and *LPile*. This can be associated with adopting conservative geotechnical design parameters during evaluation.

Furthermore, estimation using *LPile* software yielded higher lateral capacities under both free and fixed head conditions than the ones estimated through theoretical methods. This is due to the limitation of classical theoretical approaches to consider stratified soil in their analysis. In theoretical approaches the analysis is primarily based upon considering the pile to be embedded only in a single soil layer, i.e. the top stratum. Since the top stratum is generally weak, hence the analysis results from theoretical approaches are on the conservative side, as evident from Figure 2–4. On the contrary in *LPile*, complete soil stratigraphy can be modelled and the corresponding increase in soil strength with depth is taken into account. The estimated load carrying capacities are therefore much closer to reality.

Furthermore, this limitation of theoretical approaches to consider a uniform soil profile along pile embedment can lead to unsafe results in cases where the top stratum is stronger and weak soil underlies it. The estimated load carrying capacity of the piles in such case can be higher than the actual pile load capacity.

Considering the aforementioned analysis and discussions, the use of theoretical pile capacity evaluation methods, lead to highly over-conservative design resulting in an uneconomical solution. In the absence of full scale pile load tests, the use of computer-aided analysis through softwares like *LPile*, is recommended for preliminary estimation of lateral load carrying capacity of piles. These estimated capacities should always be verified through full scale pile load tests. The test number and extents of loading may be selected considering the project's budget.

5 CONCLUSIONS AND RECOMMENDATIONS

Comparison among the theoretically estimated allowable pile load carrying capacity values, *LPile* evaluations and results of pile load tests yields following conclusions;

- Actual load carrying capacities of piles in lateral direction are significantly higher than values estimated theoretically through classical approaches (i.e. Brom (1964) method and Davisson and Gill (1963) method) and *LPile* computer analysis. This can be associated with adaptation of conservative geotechnical design parameters for pile capacity evaluation.
- Compared to the theoretical methods, estimation using *LPile* computer analysis yielded higher lateral capacities under both free and fixed head conditions. This is due to the fact that contribution of each individual soil layer is considered for *LPile* computer analysis while the theoretical methods only consider a single uniform soil layer for the estimation of lateral pile load capacity. Also the uniform soil strata considered for theoretical methods is generally based upon strength parameters equivalent to the top most layer, which is generally weak in nature.
- Theoretical approaches have limitation of using uniform soil stratigraphy and hence the uniform geotechnical design parameters. Thus, they have tendency to yield either conservative or unsafe estimates depending upon the geotechnical design parameters of top soil layer. It is therefore worth-investing in computer analysis softwares like *LPile* which can help in considerably economizing the design.
- For regions like Pakistan where alluvial deposits exist, it is recommended to employ computer analysis softwares like *LPile* for preliminary estimation of lateral load carrying capacity of piles. These estimated capacities should always be verified through full scale pile load tests. The test number and extents of loading may be selected considering the project's budget.

REFERENCES

Brom, B. 1964. Lateral resistance of piles in cohesionless soils. *Journal of the Soil Mechanics and Foundations, ASCE,* 90, 123–156.

Davisson, M. & Gill, H. 1963. Laterally loaded piles in a layered soil system. *Journal of the Soil Mechanics and Foundations Division,* 89, 63–94.

IBC, I. 2012. International Code Council. *International Building Code. International Code Council: Washington DC, United States.*

Challenges and Innovations in Geotechnics – Zhussupbekov (Ed.)
 ISBN 978-1-138-03007-7

Mid-tunnel underground docking of tunnel boring machines in Singapore

G.T. Senthilnath
Geoconsult Asia Singapore, Singapore

ABSTRACT: Due to increased depth of tunneling, the time and cost involved in constructing a deep shaft is very high. In order to minimize the number of shafts and to shorten the construction period, it may be a viable option to drive TBM (Tunnel Boring Machine) from either side (of stations/portals) so they could meet midway. The TBMs could be then docked to establish a safe working chamber for concluding construction works including the TBM dismantling, internal concrete lining etc., without the need for any intermediate TBM retrieval shaft. This paper presents a case study of such TBM mid-tunnel docking performed in Singapore and also discusses details of structural design of temporary ground support system, analysis of settlement, assessment of potential impact on structures above the docking location and excavation sequencing.

Keywords: TBM, Underground docking, Numerical modelling, Jurong formation

1 INTRODUCTION

With the population of Singapore expected to rise to 6.9 million by 2030, underground infrastructure for transportation, utilities and commercial and mixed-use purposes has become a priority in the city state of Singapore (PMO Singapore 2013). With the increasing population and limited available land, underground infrastructures help to free surface space to improve the living condition of the city.

Figure 1 shows various forms of underground space utilization in Singapore. However, over the years, it has been difficult to find surface space even for constructing vertical shafts, which are used for launching and receiving the Tunnel Boring Machines (TBM). In order to address this issue (while maintaining a shorter construction time), a direct underground TBM docking method has been adopted for a project in Singapore. The paper discusses about design and construction details of two 7 m diameter TBMs docked at approximately 65 m below ground surface without any intermediate shaft.

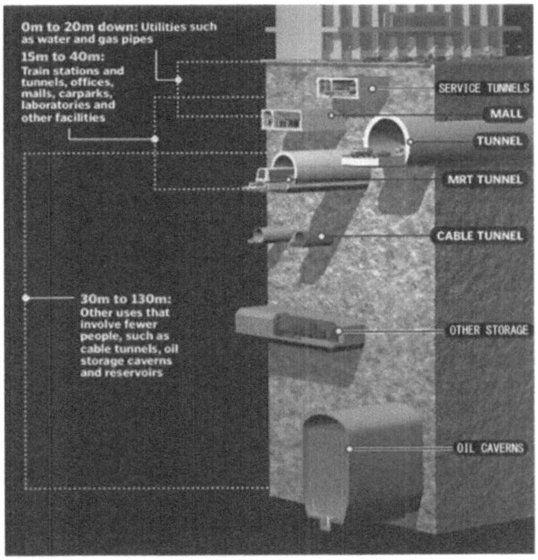

Figure 1. Underground space utilization in Singapore (Woon & Yingxin 2010).

2 PURPOSE OF TBM DOCKING

The purpose of docking is illustrated in Figure 2. Figure 2a shows that two nos. of shaft are required to drive a tunnel from Point A to Point B. If the construction time needs to be optimized, we can engage two TBMs (TBM1 from Shaft A and TBM2 from Shaft B) to meet midway and can be retrieved at Shaft C as illustrated in Figure 2b. However if we want to further optimize the construction cost and surface space utilization, we could avoid Shaft C by performing a mid-tunnel underground docking as illustrated in Figure 2c. The main performance objective during the TBM docking is that the joint section should have sufficient water cut-off performance without any water pressure dropdown and minimum impact to the surface structures above the docking location. This can be achieved in various ways. So far, few schemes of underground TBM docking methods have been tried worldwide. One of them is the use of special kind of

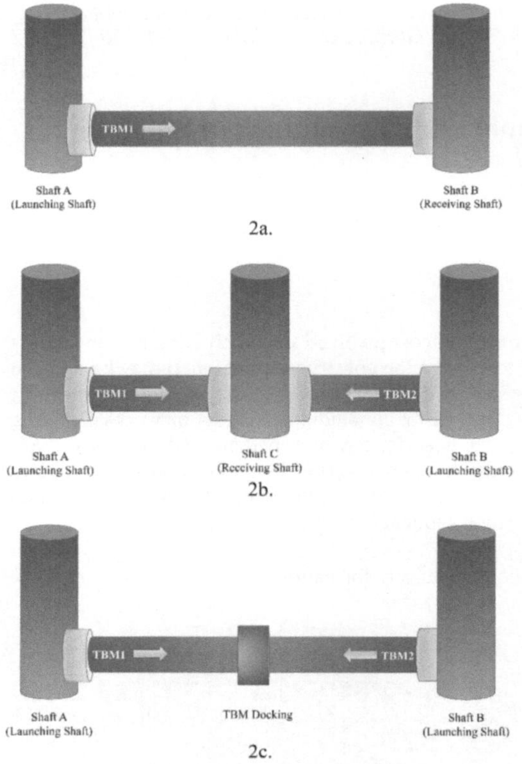

Figure 2. Purpose of TBM docking.

TBM specifically made for such docking scheme during the manufacturing stage itself. In such method, two shield machines, one on the penetrating size and the other on the accepting side move ahead excavating soil towards each other and when they meet face to face, their cutter head diameters are reduced. The steel ring built in the shield machine penetrates into the accepting room of the shield machine on the accepting side so that the two machines dock to each other mechanically. More details about such method has been elaborated by Nakamura et al. (Nakamura 1992). The other method of docking without the need for any specially prepared TBMs, is the overdrive method. In this scheme of docking, the first TBM reaches the docking location and a shotcrete soft eye dome is prepared for the second TBM to breakthrough. This scheme of docking needs detailed assessment and design of temporary structure/soft eye for successful docking.

3 PROPOSED SCHEME OF DOCKING

The docking scheme discussed in this paper is similar to the overdrive docking scheme except that there is no soft eye dome prepared for the breakthrough. The soil trapped in between the TBMs needs to be excavated manually and a temporary support scheme is designed to support the ground between the two TBMs.

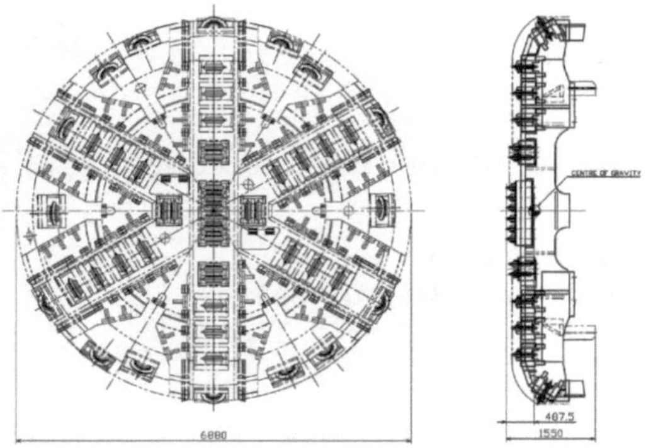

Figure 3. Slurry and EPB cutter head.

The scheme includes installation of temporary support system between the two TBMs. Before the excavation, chemical grouting is performed while maintaining the bulkhead, followed by dismantling of TBMs once the water ingress is found to be less than predetermined allowable water ingress.

3.1 *Description of TBMs*

Both the TBMs used in this docking operation are of Hitachi make with an excavation diameter of 7 m and shield length of 11.5 m. One of the TBM (TBM1) is a hard rock slurry TBM and the other (TBM2) is an earth pressure balance (EPB) TBM. Both the TBMs have a very similar cutter head with same opening ratio. Cutter head of the TBMs are shown in Figure 3. Along the bored tunnel stretch, pre-cast concrete segmental lining is designed to provide the final permanent support. However, at the TBM docking portion (between the two cutter heads which is manually excavated) a cast in-situ lining is used as permanent structure.

3.2 *Ground condition*

The docking location was expected to be in slightly weathered to sound sedimentary Jurong Rock formation (SIV to SIII). Jurong formation at this location is predominantly reddish-brown and purple consisting of interbedded layers of sandstone, siltstone and mudstone. SPT values of SIV are usually from 50 to more than 100 blows/300 mm. The Rock Quality Designation (RQD) for grade SIV cores range between 0% and 85% with the majority of data being less than 50%. This implies that the grade SIV rocks are generally of very poor to poor quality. Moderately weathered to slightly weathered materials (SIII and better) are generally described as 'bedrock' in engineering terms. Generally, SIII rocks exhibit a wide range of fracture conditions from RQD = 0% to 100%. Ground water level is close to surface level. The piezometer water head was typically recorded as being between 0 m to 1.5 m. For the purpose of analysis and design, ground water level was taken at 1m below ground level.

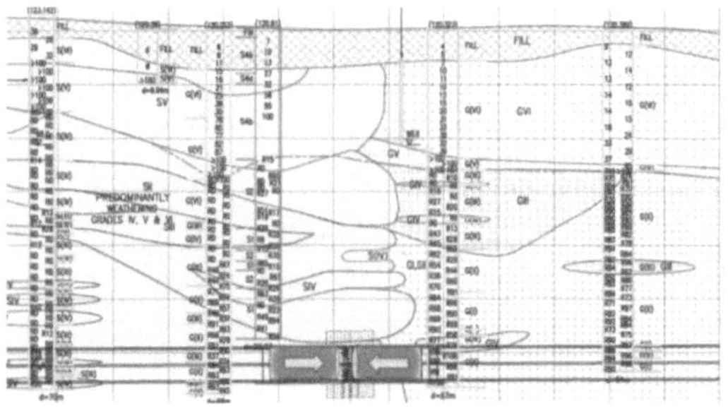

Figure 4. Geological profile at TBM docking.

The proposed docking location is exactly 30 m below a live functioning metro (Mass Rapid Transit) station with many commercial and residential buildings falling inside the influence zone of the TBM docking operation. Geological profile at the docking location is shown in Figure 4.

3.3 *Temporary and permanent structure*

FEM program PLAXIS was adopted for the soil structure interaction between the tunnel and the surface structure. Finite element model is created based on the soil layers derived with reference to the nearest borehole. General sequence for analysis is model initialization followed by dewatering of ground water in the model due to grouting around the docking location followed by excavation between the TBMs and installation of steel supports. Since the TBM shield is utilized during excavation between the cutter heads, a relaxation factor of 70% is considered in 2D model for elastic ground behavior based on the initial stress approximation given in Equation 1 (Sakurai 1978)

$$P(t) = 0.3 + 0.7(1 - e^{-\frac{2d}{r}}) \tag{1}$$

where, $P(t)$ represents the initial stress in the ground which is time dependent indicating the rate of progression of tunnel, d is the distance between the TBM shields and r is the radius of the tunnel opening. This can be implemented in PLAXIS by deactivating the tunnel cluster and applying an ultimate level of unloading ($\sum M_{Stage}$) equal to above value before activating the temporary support.

Both 2D and 3D models of the docking mechanism (shown in Figure 5 & 6) are studied to derive the forces on temporary support. In 2D longitudinal model, the support provided by the TBM shield on either side of the temporary structure is simulated using a vertical support stiffness derived based on Young-Laplace equation for thin shells (Boresi et al. 1993) as given in Equation 2.

$$P = \frac{\sigma_\theta t}{r} \tag{2}$$

where P is the internal pressure (representing vertical support), t is the thickness of the skin shield, σ_θ is

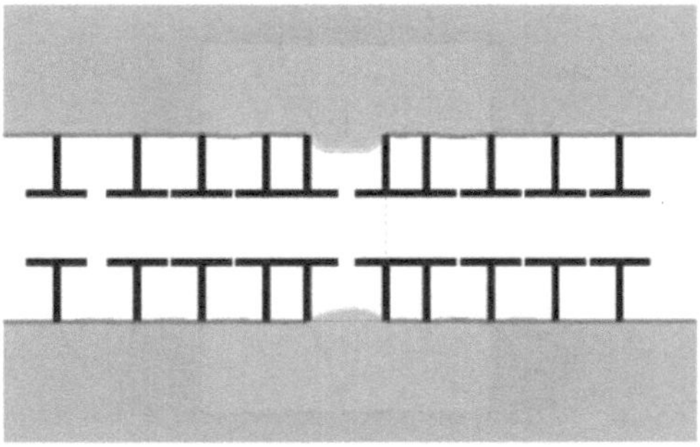

Figure 5. Longitudinal analysis section.

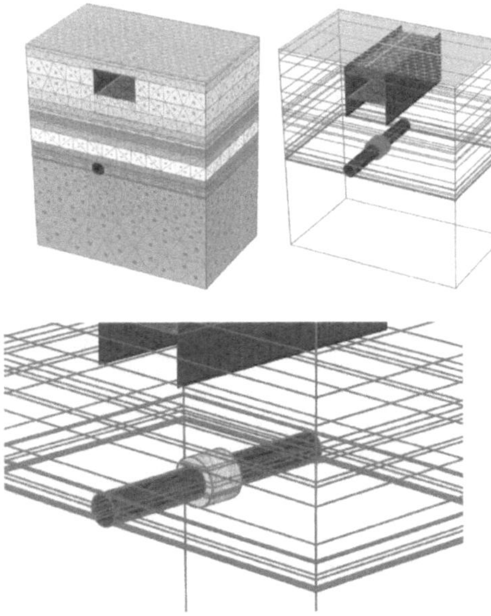

Figure 6. 3D PLAXIS model for analysis.

the hoop stress and r is the mean radius of the TBM skin shield. This relation is used to derive the relation between support stiffness and hoop stiffness.

Based on the results of these analysis, two options of temporary support system were arrived at. First scheme was using sprayed concrete with wire mesh and the other scheme was using continuous steel plates spanning between the TBM shields. Owing to logistical issues and the requirement to work in confined space, the contractor preferred to use the temporary support scheme using steel supports. In order to install the steel plate spanning between the TBM shield skins, the cutter head was dismantled in stages (as shown in Figure 7) to excavate the soil and weld the steel support between the TBMs. The permanent cast in-situ lining is designed by ignoring the presence of temporary support systems and TBM skin shield for the load combinations likely to occur during the lifetime of the tunnel for a minimum fire rating of 4 hours.

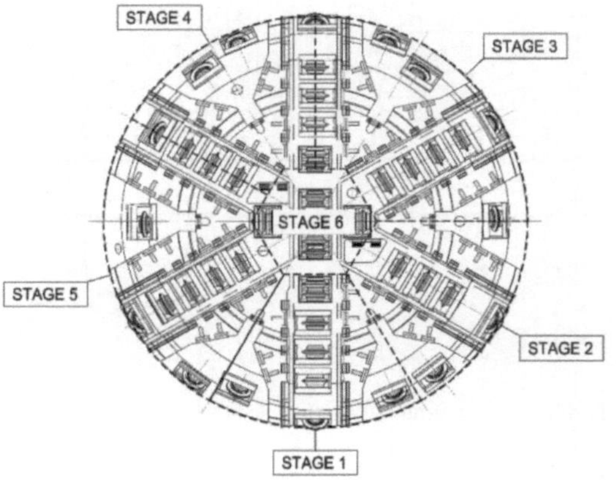

Figure 7. Dismantling stages of the cutter head.

3.4 *Wedge stability and face stability*

During the docking operation, the rock between the two TBM cutter heads was exposed for a short period of time (before temporary support installation). Under the influence of gravity, roof and wall wedges may fail either by falling, sliding or by rotating out of their sockets. Rock wedge stability program (Unwedge by Rocscience) was used for wedge analysis for potential failure mechanism (Kavvadas 2005) (by assuming no in-situ stress) with a global factor of safety of 1.5. Wedge analysis was based on the face mapping from first TBM's cutter head intervention at the docking location and based on known joint orientations. It was concluded that no additional support using rock bolts would be required.

Face stability was not a concern in this scheme of docking as the TBMs maintained full bulkhead till the grouting was complete and by the time the bulkhead was dismantled, the TBMs were facing each other at less than 1m clear distance.

3.5 *Settlement and impact assessment*

The influence of the docking operation on surface structures was studied based on Volume loss assumption and based on water inflow assessment. In order to assess the settlements during tunnel drive, a volume loss of 1% was considered. However to study the impact due to TBM docking operation, a higher volume loss of 3% was considered to account for stress relaxation in the ground between the cutter heads. Based on the settlement trough, it is estimated that maximum settlement because of 3% volume loss at the surface level is less than 5 mm.

3.6 *Contingency measures and instrumentation*

Since the docking operation was planned in weathered to sound rock at 60m below ground surface, the surface settlements during docking operation was mostly envisaged due to ground water drawdown. As a sensitivity study, ground settlement due to different levels

Table 1. Sensitivity study on ground water drawdown.

Ground water draw down level [m]	Maximum ground settlement [m]
10	23
15	24
20	26
25	27

of simulated ground water table drawdown were studied. Results of the sensitivity study are tabulated in Table 1.

In addition, to assess the level of allowable water ingress during TBM docking operation, a 3D flow model is considered to predict the impact due to different water ingress levels in the docking location. Based on this assessment, a maximum allowable water ingress of 80 litres per minute for a maximum duration of 85 days was set during the design stage. In case of such water ingress, in order to maintain the ground water table, water recharge wells were proposed to be installed with a minimum pumping capacity of 10 m^3 per day within the zone of influence.

All the buildings inside the zone of influence were installed with building settlement markers and continuously monitored during the docking operation. In addition, two piezometers and five ground settlement markers were installed above the docking location to be monitored during the temporary support installation.

3.7 *Face logging*

Hard Rock Slurry TBM (TBM1) reached the docking location first. Articulation jack were retracted to minimum for the cutter head intervention. Upon face inspection, full face moderately to slightly weathered, dark competent sand stone and siltstone (SIII) was observed at the face. There was no over break observed at this stage. Opening around the shield was irregular due to intersection of joints. As can be seen from the face inspection log Figure 8, water ingress was at 7 O'clock position (at the siltstone and sandstone joint) around 2 litre per minute under free air intervention. Overall RQD was estimated to be 50% which translates to an excavation stand-up time of 2 to 3 months (Hoek 2000).

All the roller cutters in the slurry TBM (TBM1) were removed and 8 nos of PVC pipes with pressure gauge to sense oncoming EPB TBM (TBM2) were installed before backfilling the excavation chamber and bulkhead with grouting material. The oncoming TBM2 was mined till the pressure could be sensed from PVC sensors placed from TBM1.

After EPB TBM arrived at the docking location, face inspection from the EPB TBM confirmed slightly to moderately weathered and fractured reddish siltstone and grey sandstone (SII/SIII). Because of intersection of bedding and major joints, there was on over

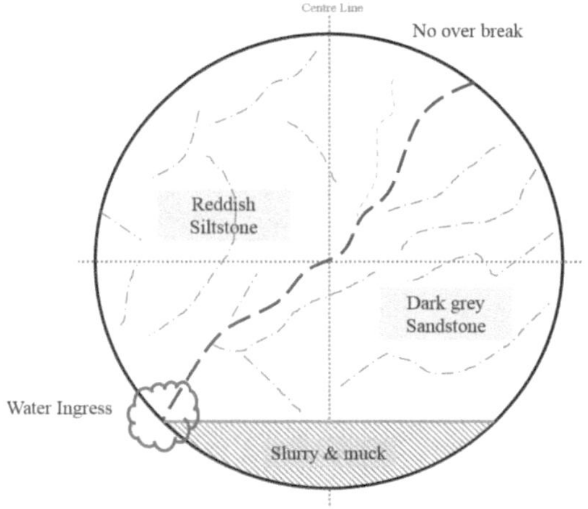

Figure 8. TBM face inspection log.

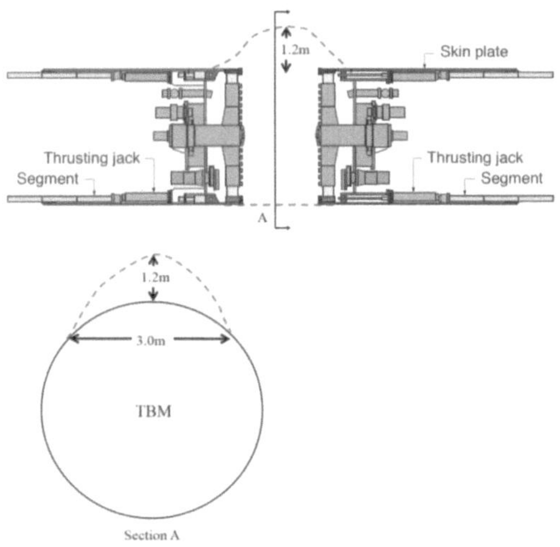

Figure 9. Over break at TBM docking location.

break in the crown area (10 'clock to 2 'clock) as indicated in Figure 9 and 10. The wedge that came off from the crown was about 1.2 m depth and there was no major sign of instability. Water ingress at the time of TBM2 inspection was less than 5 litre per min. Most of which was the water ingress from the shield instead of from the exposed rock face. The exposed rock sides were immediately scaled and a layer of sealing concrete was applied.

3.8 *Grouting*

Polyurethane (PU) resin grout was considered for grouting around the docking region. A two component instant set PU elastomer with low viscosity was used. 42 mm diameter holes were drilled in the pattern as shown in Figure 11 and were grouted using single packer system from both the TBMs to achieve an overlapping water cutoff around the docking area.

Figure 10. Site inspection after shotcrete layer.

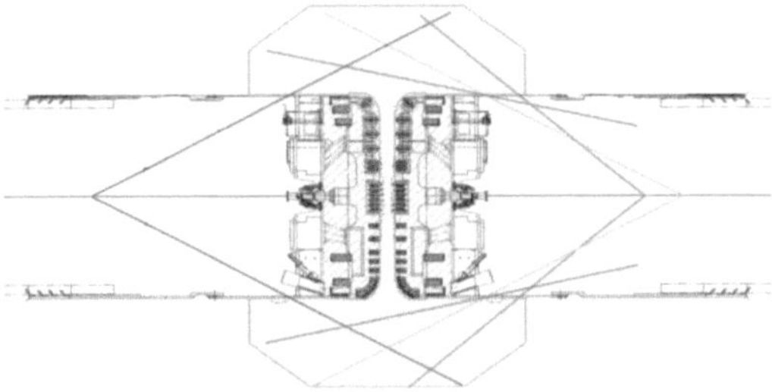

Figure 11. Grouting scheme from TBM.

Chemical used for the purpose of backfilling the excavation chamber was a mix of sodium silicate and sulphuric acid. Sodium silicate grout has been proven to be effective application for water shutoff and plugging (Bauer et al. 2005). Other types of grouts get diluted quickly by surrounding water which results in separation of grout components and loss of all self-supporting characteristics of the mix. On the other hand, sodium silicate grout forms a calcium silicate barrier on the grout surface that limits the dilution of the mix with water and has given good water shutoff for the docking location.

4 CONCLUSIONS

Experience gained during this docking operation indicated that with proper design considerations, continuous design support during execution and effective construction management, mid-tunnel docking can be achieved safely without any effect to surface structures even in a well-developed urban area. Selecting a proper location for TBM docking based on TBM face logging is very crucial in avoiding surface settlements. Grouting from the TBM skin shield and tail skin grouting played a pivotal role in controlling the water ingress level to less than 10 litre per min during docking operation. In terms of environmental impact, safety, disturbance to public, short-term construction period and interference to other underground services, this docking was really stealth and made way for successful completion of tunneling operation.

REFERENCES

Bauer, S. et al., 2005. High-temperature plug formation with silicates. *SPE international Symposium on Oilfield chemistry*, pp. 1–6.

Boresi, A.P., Schmidt, R.J. & Sidebottom, O.M., 1993. *Advanced mechanics of materials*, Wiley New York.

Hoek, E., 2000. Practical rock engineering.

Kavvadas, M., 2005. Numerical Analysis in the Design of Urban Tunnels. (June), pp. 19–24.

Nakamura, H., 1992. Development of shield machine for Mechanical Shield Docking Method. In V.-U. Montanez-Cartaxo, ed. *Towards new worlds in tunnelling*. Balkerma, pp. 461–475.

PMO Singapore, 2013. *A sustainable population for a dynamic Singapore – Population White Paper*

Woon, M.L.K. & Yingxin, Z., 2010. *Geology of Singapore*

Challenges and Innovations in Geotechnics – Zhussupbekov (Ed.)
© 2016 Taylor & Francis Group, London, ISBN 978-1-138-03007-7

Reliability analysis of pile driving formula for concrete driven piles in stiff soil layer: Case study in Chiang Mai province

S. Thongmunee & A. Thongchai
Chiang Mai University, Chiang Mai, Thailand

ABSTRACT: Pile driving formulas, which have been proposed by many researchers, are widely used to predict and control the load carrying capacity of the driven piles. However, efficiency of each formula in predicting the capacity mainly depends on subsoil properties, pile installation data and adopted empirical parameters. In other words, when construction area is different, accuracy of each formula will be changed. The aim of this study is to investigate the appropriate pile driving formula for the area of Chiang Mai province, the largest city in northern Thailand, where the number of construction projects has been increasing rapidly for the last 10 years. In this study, the authors have collected the construction information on concrete driven piles including soil boring log, pile information, installation data and its Dynamic Load Test (DLT). The information was obtained from 7 construction projects around Chiang Mai downtown. Typical soil horizon in all construction sites was found to be alternating layer of stiff silty clay (CL) interbedded with clayey or silty sand (SM or SC). More than 7000 driven piles were installed, and dynamic load tests were performed on 69 selected driven piles in all project. All of the driven pile is square solid concrete pile. Cross section of the driven pile varies from 0.26 × 0.26 m to 0.45 × 0.45 m. Length of the pile varies from 9 to 28 m. Seven pile driving formulas were chosen for predicting the load carrying capacity of the selected DLT piles. The correlation between the predicted capacities and its DLT results were then tested to conduct reliability analysis of the pile driving formulas. The result showed that Gate formula is most reliable formula, based on the higher of R^2 value compared with others for Chiang Mai subsoil.

Keywords: Pile driving formula, Driven pile, Deep foundation, Dynamic pile load test, Stiff soil

1 INTRODUCTION

Driven piles are still widely used as deep foundation in Chiang Mai province, the largest city in northern part of Thailand, due to the advantage of construction cost, construction time, quality of pile and equipment comparing with other foundations such as bored pile or press-in pile.

During installation of the driven pile, pile driving formulas are usually employed as a control in predicting the load carrying capacity by using last 10 blows. The number of pile driving formulas used in Thailand is more than 10 formulas. Validity of each pile driving formula depends on subsoil condition and engineering experience in the local area. So, authors intend to determine the appropriate pile driving formula for our area by comparison between predicted results with pile load test results.

Dynamic pile load test (DLT) is a frequently used method to estimate the load carrying capacity of piles after installation, because its high reliability and efficiency in cost and time. Moreover, the standard testing of Dynamic pile load test has been developed for more reliability by the Department of Public Works and Town & Country Planning (DPT). So in this study, the results obtained from DLT will be used as most reliable value for comparison.

Reliability of pile driving formulas for concrete piles in Chiang Mai area has been studied by many researches (Jitsa-ngiam 2001, Anantasech 2006, Tocharoenying et al 2012, Thongmunee et al 2013 and Tocharoenying 2014). Those authors recommended different appropriate pile driving formula, based on their limited information of construction projects. However, number of construction projects in Chiang Mai province has increased enormously for last 10 years, and more information of the construction projects has been collected.

So, this study focuses on the reliability of pile driving formulas for concrete piles in Chiang Mai by using more information the construction projects. Firstly, the geotechnical properties of Chiang Mai subsoil will be reviewed. The information of 7 construction projects in Chiang Mai, including geotechnical properties, driven pile installation data and its DLT capacity of selected piles, will be collected and summarized. Then, prediction of load carrying capacities (P_u) of the selected piles using 7 popular pile driving formulas will be

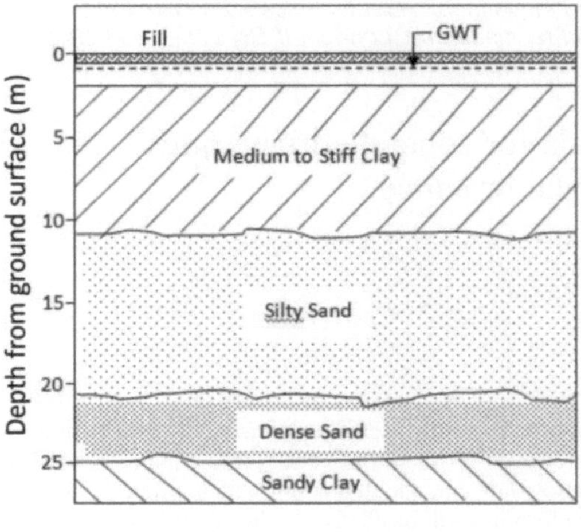

Figure 1. Typical characteristics of Chiang Mai subsoil. (Anantasech 1991).

carried out, and plotted together with its DLT capacities (Q_u). The reliability of each pile driving formula will be discussed based on the comparison.

1.1 *Properties of Chiang Mai subsoil*

Chiang Mai province is the largest city in northern part of Thailand. Geomorphology of northern part of Thailand, including Chiang Mai province, have been studied by many researchers (Anantasech 1991, Anantasech et al 1992, Thanadpipat 1982).

Northern part ofThailand is mountainous area. Chiang Mai city is located in the plain valley between mountains called "Chiang Mai basin". According to the geomorphology of Chiang Mai city, especially downtown area, sedimentation can be classified into 3 types including high terrace sediment, low terrace sediment and flood plain sediment.

High terrace sediment generally consists of gravel and sand with little clay content. Low terrace sediment consists of sand and clay with some gravel. Flood plain area consists of silty sand alternated with silty clay or clayey sand. Depth and thickness of each soil layer are varied depending on location. Also, the percent of sand and clay varies with wide range.

Figure 1 shows typical characteristic of Chiang Mai subsoil (Anantasech 1991). It can be roughly summarized that the 1st layer is usually fill soil with poor geotechnical properties. The 2nd layer is mainly medium to stiff clay interbedded with silty or clayey sand, having 4–20 of SPT N-value. The 3rd layer is mainly silty sand with wide range of fine content, having 30–50 of SPT N-value. The 4th layer is dense sand with few of fine content, having SPT N-value more than 50. And, the 5th layer is sandy clay with 40–50 of SPT N-value.

1.2 *Pile driving formulas used in the study*

In this study, 7 pile driving formulas shown in Table 1 were selected to predict the load carrying capacity of selected driven concrete piles. All the chosen formulas are frequently used by Thai engineers. Two of these formulas were developed by the Engineering Institute of Thailand under His Majesty the King's Patronage (EIT), with some recommended parameters.

Table 1. Pile driving formulas used in the study.

Formula	Equation
ENR.	$P_u = \dfrac{W_r h}{s+c}$
ENR-EIT	$P_u = \dfrac{W_r h}{s+0.9c}$
HILEY	$P_u = \dfrac{e_h W_r h}{s+0.5(k_1+k_2+k_3)} \cdot \dfrac{W_r+n^2W_p}{W_r+W_p}$
HILEY-EIT	$P_u = \dfrac{e_h W_r h}{s+0.5(c_1+c_2+c_3)} \cdot (\dfrac{W_r+n^2W_p}{W_r+W_p})$ $c_1 = \dfrac{1.8P_u L_2}{A}, c_2 = \dfrac{0.72P_u L}{A},$ $c_3 = \dfrac{3.6P_u}{A}$
JANBU's	$P_u = \dfrac{e_h E_h}{k_u s}$ $k_u = C_d\left(1+\sqrt{1+\dfrac{\lambda}{C_d}}\right)$ $\lambda = \dfrac{e_h E_h L}{AEs^2}, C_d = 0.75+0.15\dfrac{W_p}{W_r}$
DANISH	$P_u = \dfrac{e_h E_h}{s+c_1}$, where $c_1 = \sqrt{\dfrac{e_h E_h L}{2AE}}$
GATE	$P_u = 104.5\sqrt{e_h E_h}(2.4-\log s)$

ENR and HILEY with latter of EIT are the pile driving formulas, which some parameters have been recommended by the Engineering Institute ofThailand under His Majesty the King's Patronage (EIT).

In Table 1, P_u is predicted ultimate load carrying capacity obtained from pile driving formulas, W_r is weight of hammer, h is drop height of hammer, s is settlement per blow, c is elastic compression of pile components, e_h is efficiency factor, W_p is weight of pile, n is coefficient of restitution, L is length of pile, A is cross section area of pile, E_h is driving energy produced by hammer ($E_h = W_r h$), E = elastic modulus of pile.

For prediction of driven pile in this study, some parameters will be used from the pile installation data in field. And some parameters will be assumed based on the typical values used in Chiang Mai area.

2 INFORMATION OF CONSTRUCTION PROJECTS

The authors have collected the geotechnical information of 7 construction projects, including subsoil properties and pile driving installation data. All projects are within Chiang Mai downtown area.

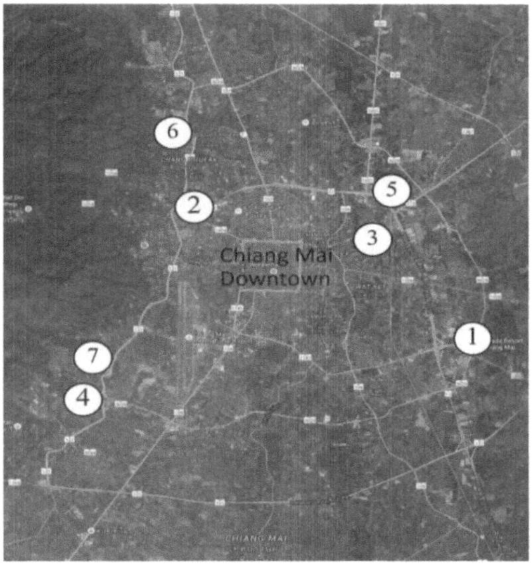

Figure 2. Location of the construction projects compared to Chiang Mai downtown.

The collected pile driving installation data includes weight of hammer, drop high, pile length, pile size, last 10 blow settlements. Detailed information is shown in the following.

2.1 *Locations of the construction project and subsoil properties*

Geotechnical information of 7 construction projects has been collected. Location of each project is shown in Fig. 2. All projects are located within Chiang Mai downtown area.

Project 1 is a large shopping mall, sitting on flood plain area. Square concrete piles with 0.40 × 0.40 m cross section were used in this project.

Project 2 is shopping mall with 2 underground floors, sitting on low terrace area. Square concrete piles with 0.45 × 0.45 m cross section were used.

Project 3 is a small shopping mall, sitting on flood plain area. Square concrete piles with 0.26, 0.30 and 0.35 cross section were used.

Project 4 is a condominium, sitting on high terrace area. Square concrete piles with 0.40 × 0.40 m cross section were used.

Project 5 is a large shopping mall, sitting on flood plain area. Square concrete piles with 0.35 and 0.40 cross section were used.

Project 6 is a convention and exhibition center, sitting on high terrace area. Square concrete piles with 0.40 × 0.40 m cross section were used.

Project 7 is an office building of 16000 m^2 of usable area, sitting on high terrace area. Square concrete piles with 0.40 × 0.40 of cross section were used.

The authors have also collected subsoil properties of 4 construction projects, including project no. 1, 4, 6 and 7. The soil properties in each project were summarized in Table 2.

Table 2. Subsoil properties in some project.

Project No.	Summary of subsoil properties
1	1st soil layer at 0–10 m of depth is silty clay with 20–45 of SPT N-value. 2nd soil layer is silty sand with 40–70 of SPT N-value is at the depth of 10–15 m. from ground surface.
4	1st soil layer is silty sand with 5–12 of SPT N-value at 0–3 m. 2nd soil layer is silty clay with 9–12 of SPT N-value at 3–6 m. 3rd soil layer is silty sand with 12–30 of SPT N-value at 6–8 m. 4th layer is silty clay with 8–10 of SPT N-value at 8–10 m. 5th layer is silty sand with 10–25 of SPT N-value at 10–11 m. 6th layer is silty clay with 20–30 of SPT N-value at 11–15 m. 7th layer is silty sand with 30 of SPT N-value at 15–17.5 m.
6	1st soil layer is silty sand with 5–10 of SPT N-value at 0–3 m. 2nd soil layer is silty clay with 10–20 of SPT N-value at 3–6 m. 3rd soil layer is silty sand with 10–20 of SPT N-value at 6–10 m. 4th layer is silty clay with 10–20 of SPT N-value at 10–13 m. 5th layer is silty sand with 10–30 of SPT N-value at 13–15 m.
7	1st layer is fill soil with variant properties 2nd layer is silty clay with 5–20 of SPT N-value from 0.5 to 9 m. 3rd layer is silty clayey sand with 23–50 of SPT N-value from 9.0 to 15.5 m.

It can be summarized that the SPT N-value increases with increase of depth in all project, and increases significantly at the depth of 10–13 m. Subsoil is mostly silty clay (CL) interbedding with silty sand (SM) at varied depth.

The subsoil properties of these construction projects seem to conform with the typical characteristics of Chiang Mai subsoil (Anantasech 1991).

2.2 *Information of tested piles and its DLT results*

For all projects, more than 7000 of concrete driven piles were installed by drop hammer method (popular method in Chiang Mai area). 69 driven piles were chosen for dynamic pile load tests.

Information of tested piles was summarized in Table 3. All of tested piles is square solid piles with concrete, f'c = 350 ksc. Cross section of piles varies from 0.26–0.45 m., and the most frequency used cross section is 0.40 m. The embedded length vary from between 9 to 28 m. Settlement per blow of the tested piles vary between 1 to 12 mm. The soils below the pile tips are mostly silty sand.

Table 3. Summarized information of tested piles.

Topic	Summary
Number of tested pile	69
Pile type	Square solid pile
Pile material	Concrete (f'c = 350 ksc)
Pile size	0.26, 0.30, 0.35, 0.40* and 0.45 m.
Pile length	9 to 28 m. (10 to 14 m.)*
Weight of hammer	5 to 8 ton (8 ton)*
Drop height	0.4 to 1.6 m. (0.8 m.)*
Settlement per blow	1 to 12 mm.
Soil beneath tip	Sand for 61 pile Stiff clay for 8 pile

*Most frequency used

It should be noted that the information of tested piles in Table 3 will be used for the pile driving formulas, in order to obtain the predicted load carrying capacity (P_u).

The DLT results (Q_u) of the tested pile will be shown together with the predicted capacities obtained from the pile driving formulas (P_u) in next topic.

3 PREDICTION OF LOAD CARRYING CAPACITY OF DRIVEN PILES

Load carrying capacities of the DLT tested piles were predicted by using the pile driving formulas as shown in Table 1. Then, the correlation between the predicted capacity (P_u) by each pile driving formula and the DLT capacity (Q_u) will be investigated to determine the appropriate pile driving formula, based on the statistical parameters.

3.1 *Parameters for pile driving formula*

Parameters used in each pile driving formula were obtained from pile installation data and typical parameters used locally.

The parameters obtained from the installation data were weight of hammer, drop height, pile dimension and settlement per blow.

The typical parameters used in this study were shown in Table 4. The locally used parameters were chosen based on typical values that frequently used in Chiang Mai area.

For ENR and ENR-EIT formulas, empirical constant, c, used was set at 2.54 cm.

Efficiency factor, e_h, of drop hammer installation method was set at 75%. Coefficient of restitution, n, used for Hiley formulas was set at 0.25. Elastic modulus of concrete pile was calculated by using $E = 15100(f'c)^{0.5}$. Strength of concrete at 28 days was set at 350 ksc.

For Hiley-EIT formula, thickness of pile head cushion was recommended by EIT as 0.10 m.

When considering the typical parameters in detail, it was found that only one parameter was used for ENR, ENR-EIT and Gate formulas. These formulas do not take into account the elastic properties of the driven pile.

Table 4. Assumed parameters used in each formula.

Formula	Assumed parameters
ENR.	$c = 2.54$ cm.
ENR-EIT.	$c = 2.54$ cm.
HILEY	$e_h = 0.75$ $n = 0.25$ $E = 15100(f'c)^{0.5}$ k_1 = recommended by Bowles, 1988 $k_2 = P_u L/AE$ $k_3 = 4.50$ mm.(for sand)
HILEY-EIT	$L_2 = 0.10$ m. $n = 0.25$ $e_h = 0.75$
JANBU's	$e_h = 0.75$ $E = 15100(f'c)^{0.5}$
DANISH	$e_h = 0.75$ $E = 15100(f'c)^{0.5}$
GATE	$e_h = 0.75$

Table 5. Summary of Q_u vs P_u in each formula.

Eq. No.	Formula	m	R^2
1	ENR.	0.941	0.584
2	ENR-EIT.	0.899	0.589
3	HILEY	1.002	0.334
4	HILEY-EIT	0.850	−0.219
5	JANBU's	0.592	−0.221
6	DANISH	0.492	0.007
7	GATE	2.131	0.628

Other formulas took into account the elastic properties of driven pile for predicting the load carrying capacity. Moreover, elastic compressions of pile head cushion and soil beneath pile tip were taken into account for Hiley and Hiley-EIT formulas.

3.2 *Prediction of load carrying capacity of driven piles and its comparison with DLT results*

In order to check the accuracy of the prediction of load carrying capacity of driven pile, the predicted capacity (P_u) by each formula were correlated by plotting against its DLT capacities (Q_u).

For each correlation between P_u and Q_u, two simple statistical measuring parameters, including m and R-square (R^2), will be shown together to determine the best fitted formula. m is the slope of fitted regression line. The regression line will be set to intercept the origin (0, 0). R-square is a statistical value for indicating how the P_u vs Q_u correlation closes to the fitted line. So, author will mainly use R-square to determine the best fitted formula.

Table 5 shows the summary of m and R^2 of each formula. It was found that Hiley-EIT and Janbu formulas have negative the R^2 value, indicating bad fitting of regression line. So, the correlation of those formulas will not be shown.

Figure 3 shows the correlation of Q_u and P_u for each pile driving formula. The dash line is the fitted regression line with slope of m. The solid line is the regression line with $m = 1$.

The m value indicates the accuracy of prediction. m is equal 1 indicates the best accuracy of prediction. m larger than 1 indicates that the predicted capacity is less than the DLT capacity. Vice versa, m less than 1 indicates that the predicted capacity is larger than the DLT capacity. So, the m value should slightly greater than 1 to incorporate the safety.

Based on the m value, it is clearly shown that only Hiley and Gate formulas give the m value greater than 1. m value of Gate formula is also larger than 1 and much larger than that of Hiley formula, but is

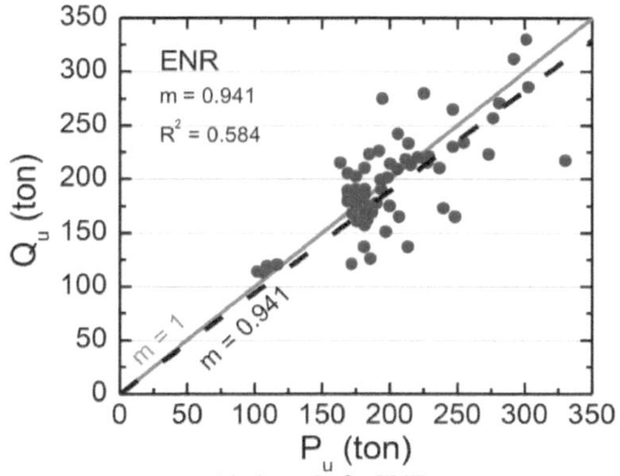

(a) Q_u vs P_u for ENT

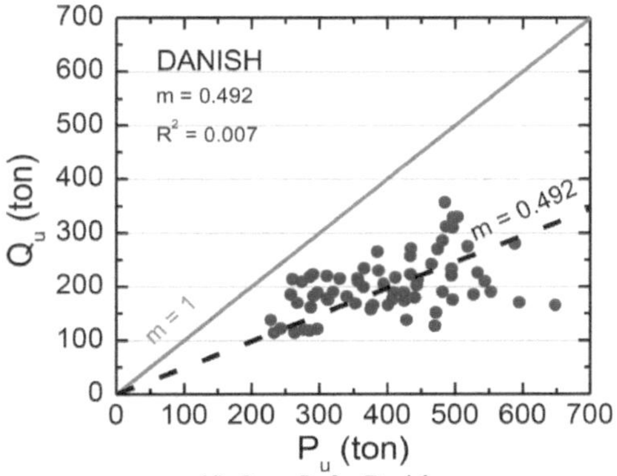

(d) Q_u vs P_u for Danish

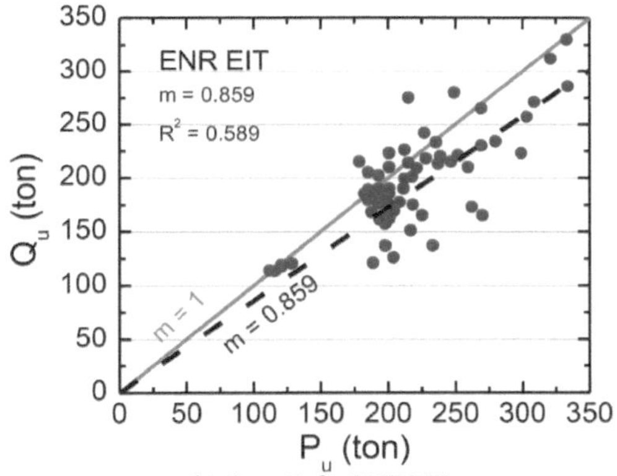

(b) Q_u vs P_u for ENT-EIT

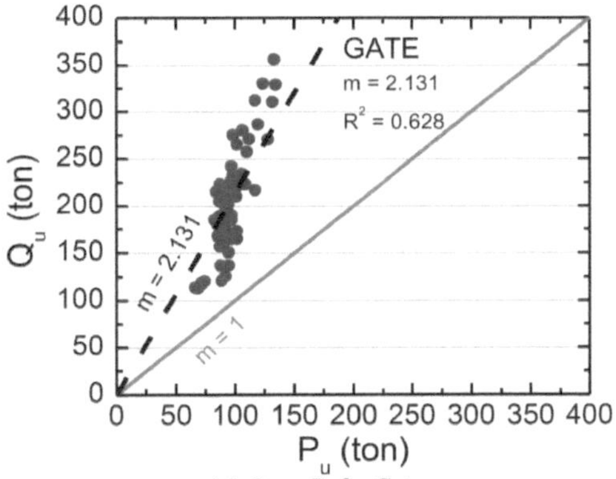

(e) Q_u vs P_u for Gate

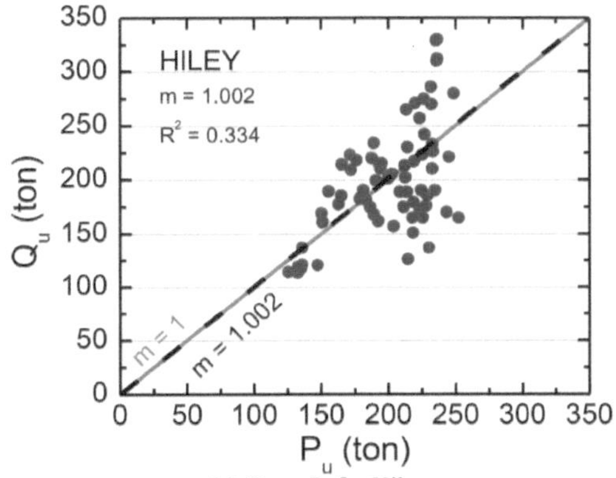

(c) Q_u vs P_u for Hiley

Figure 3. Relationship between dynamic load test capacities (Q_u) against predicted load carrying capacities (P_u) for each pile driving formula.

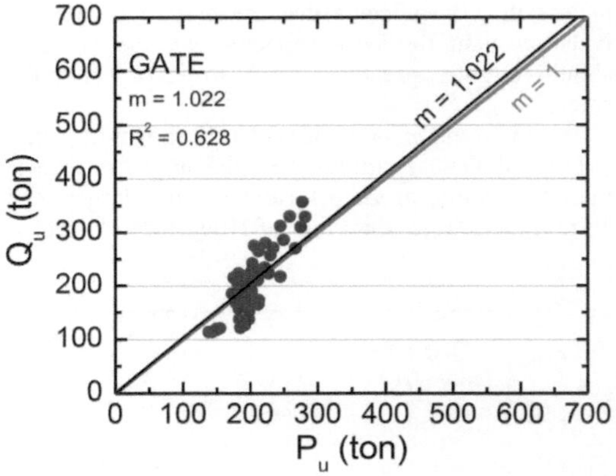

Figure 4. Relationship between dynamic load test capacities (Q_u) against predicted load carrying capacities (P_u) after modification of Gate formula.

has highest R-square value indicating best correlation compared to others.

So, it can be concluded that Gate formula is reliable formula for concrete driven piles in Chiang Mai area, due to the higher of R^2 value compared with other formulas. This conclusion on reliable formula for driven pile in Chiang Mai is in line with the study by Anantasech (2006) recommending the Gate formula.

It can be noted that the adopted elastic compression parameters, such as pile, cushion and soil beneath pile tip, may have influence on the scattering of the data more than efficiency factor. The formulas using fewer parameters will give less scattering.

In order to use the Gate formula for predicting load carrying capacity for Chiang Mai area, it may be to revise an empirical constant of Gate formula by back-calculation, new constant value of 21.79 should be used in place of 104.5. The Gate formula with new constant is shown in Figure 4.

$$P_u = 21.79\sqrt{e_h E_h}(2.4 - \log s)$$

Figure 4 shows the relationship between Q_u and P_u obtained from the Gate formula with revised empirical constant. It was shown that the *m* value was improved to 1.022 and R-square is still the same as in the previous formula.

4 CONCLUDING REMARKS

In this study, the information of 7 construction projects in Chiang Mai, including geotechnical properties, driven pile installation data and its DLT capacity of selected piles, was collected. The number of selected piles is 69. All of the driven pile is square solid pile. Cross section of the driven pile varies from 0.26 × 0.26 m to 0.45 × 0.45 m. Length of the pile varies from 9 to 28 m.

Then, the reliability analysis of pile driving formulas for concrete driven piles in Chiang Mai area has been carried out by making comparison between DLT pile capacities and its predicted capacities.

For Chiang Mai subsoil generally consisting of silty clay (CL) interbedded with silty or clayey sand (SM or SC) at varied depth, it clearly presents that Gate formula is the most reliable formula for concrete driven piles, based on the highest of R^2 value compared with others.

Moreover, it is remarkable that the formula using fewer typical assumed parameters shows less scattering of the predicted capacity.

In order to use the Gate formula for Chiang Mai, an empirical constant of the formula has been revised for good efficiency. However, more information of driven piles and its DLT test results should be collected for increasing the reliability of an empirical constant as well as the pile driving formula.

ACKNOWLEDGEMENT

Authors would like to thank Panupon lekprasert, our master student in department of civil engineering, for his assistance in data collection.

REFERENCES

Anantasech, C. (1991). Geotechnical properties and foundation in Chiang Mai subsoil. *Seminar on foundation and underground construction*, the Engineering Institute of Thailand under His Majesty the King's Patronage (EIT), Bangkok, Thailand, 8.1–8.15. (in Thai)

Anantasech, C., Suttipunta, T. & Boonrawd, A. (1992). Geotechnical properties of stiff Chiang Mai clay. *GEOTROPIKA'92*, Johor Bharu, Malaysia.

Anantasech, C. (2006). Prediction on bearing capacity of driven pile using pile driving formulas for stiff clay layer. *Seminar on technology and innovation for sustainable development*, Faculty of engineering, Khonkaen University, Khonkaen, Thailand, 35–38. (in Thai)

Bowles, J. E. (1988). Foundation analysis and design, *4th New York McGrew-Hill inc.*

Jitsa-ngiam, P. (2001). Prediction on bearing capacity of driven pile in Chiang Mai subsoil. *Master thesis*, Faculty of engineering, Chiang Mai University, Chiang Mai, Thailand. (in Thai)

Thanadpipat, C. Ramingwong, T. and Lerdthersee, S. (1982). Environmental Geology of Chiang Mai. *proc. LANDPLAN 1*, Bangkok, Thailand.

Thongmunee, S., Tocharoenying, T. & Thongchai, A. (2013). Soil parameters in wave equation for estimating loading capacity of driven pile in Chiang Mai stiff clay layer, *Proceeding of 18th National Convention on Civil Engineering*, Chiang Mai, Thailand, pp. GTE 175–181. (in Thai)

Tocharoenying, T. & Thongchai, A. (2012). Reliability of pile driving formula for driven pile in Chiang Mai stiff clay, *Proceedings of the 17th National Convention on Civil Engineering (NCCE17) – Civil Engineering for Integrated Developments*, Udon Thani, Thailand. (in Thai)

Tocharoenying, T. (2014). Loading capacity evaluation of driven piles in Chiang Mai subsoil by wave equation. *Master thesis*, Faculty of engineering, Chiang Mai University, Chiang Mai, Thailand. (in Thai)

Challenges and Innovations in Geotechnics – Zhussupbekov (Ed.)
 ISBN 978-1-138-03007-7

Two-lane vehicle underpass using pipe roof method at Sentosa Gateway Tunnel

S.C. Teo, T.G. Ng, I. Yogarajah, K.X. Woon & H.P. Dang
Golder Associates (Singapore) Pte Ltd

ABSTRACT: A 1.3 km long vehicular tunnel, Sentosa Gateway Tunnel, was proposed to connect outbound traffic from Sentosa Island directly to major roads to cope with growing traffic in the area. Due to the presence of existing buildings in close proximity and carriageways that could not be diverted, a trenchless construction method known as pipe roof method is adopted for the construction of parts of the Sentosa Gateway Tunnel instead of the conventional cut and cover method. This paper presents the design considerations made and the challenges encountered during the construction of the two-lane piped box tunnel. Finite element analysis is conducted prior to the construction of the vehicular tunnel. The settlement of the pipe roof measured during the mined excavation is discussed and compared with the predicted settlements obtained from finite element analysis.

1 INTRODUCTION

As part of the measures to improve the traffic congestion faced by road users, various solutions are being proposed and implemented by the Government of Singapore. Other than providing an efficient rail public transportation system, new roads are often planned to lessen the load of existing roads and to diverge the high traffic volume. As Singapore is a highly-developed and densely-populated country, many of the lands are already occupied by existing structures, which may be in the way the proposed new roads. Often, the roads will have to be aligned as flyovers over existing ones, or sometimes, as underground vehicular tunnels.

The most typical construction method for underground tunnels beneath existing roads involves a combination of staged traffic diversions and cut and cover construction method. Nevertheless, such method is not applicable at all times, mostly due to site limitations such as the presence of existing roads or structures that cannot be diverted or relocated. In such cases, trenchless construction methods such as pipe roof method may be considered.

The pipe roof method has been used for several projects in Singapore. Yogarajah et al (1994), Ng et al (2011) and Ng et al (2016) had discussed the application of this method in several projects carried out in Singapore.

This paper will discuss the construction of a two-lane vehicular underpass using pipe roof method at the Sentosa Gateway Tunnel project in Singapore.

2 SENTOSA GATEWAY TUNNEL

2.1 *Project description*

Sentosa is an island located to the South of Singapore. It has been developed as an island resort and welcomes some twenty million people per year. The gateway to Sentosa is located at an area with very high traffic usage, as a result of several nearby vicinities such as Brani Terminal and Keppel Distripark, a large cargo distribution complex, both under Port of Singapore Authority, Vivocity shopping complex and several high-rise residential developments.

In order to address this congestion issue and to cater to the projected increase in traffic demand, a new underground vehicular tunnel, known as the Sentosa Gateway Tunnel, has been proposed to bridge the outgoing traffic from Sentosa Island directly to major roads which will lead motorists to different parts of Singapore. It will bypass the busy traffic junctions at Sentosa Gateway, Telok Blangah Road and Kampong Bahru Road. The location and layout plan of Sentosa Gateway Tunnel are presented in Figure 1 and Figure 2 respectively.

2.2 *Pipe roof method*

A multitude of construction proposals were studied and reviewed to produce a feasible and efficient excavation design while maintaining the robustness of the system and keeping the resulting movements to a minimum. Diaphragm wall supported by multiple layers

Figure 1. Location of Sentosa Gateway Tunnel.

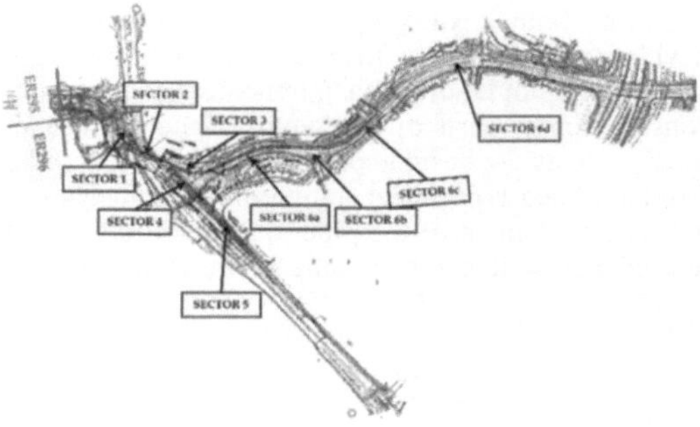

Figure 2. Layout plan of Sentosa Gateway Tunnel.

of steel strutting or concrete slabs was proposed for a great part of the tunnel construction, adopting a conventional cut-and-cover method.

Diaphragm wall could not be installed for two stretches of the tunnel, due to the presence of several existing structures, namely the Mass Rapid Transit (MRT) tunnels underneath, 10 m high natural slope to the West, buildings to the East and a network of roads and flyovers with high traffic volume that inhibits the possibility of traffic diversions. These two stretches, one a single-lane underpass and the other, a two-lane underpass, were proposed to be constructed using trenchless construction approach instead. The pipe roof method was selected as the most suitable trenchless construction approach due to the relatively short length required for tunneling and the shallow overburden pressure. The lengths of these tunnels are approximately 95 m, the longest to be constructed in Singapore, to date.

The topic of discussion for this paper will be limited to wider of the two piped box tunnel, the two-lane piped box tunnel.

3 DESIGN AND CONSTRUCTION CONSIDERATIONS

3.1 *Two-lane piped box tunnel*

The proposed two-lane piped box tunnel (PBT) is 14.5 m wide, 9.1 m high, 95 m long and located approximately 6 m beneath existing ground level. It is situated beneath Telok Blangah Road and West Coast Highway Viaduct, amidst a number of services. Critical structures nearest to the tunnel are highlighted in Figure 3.

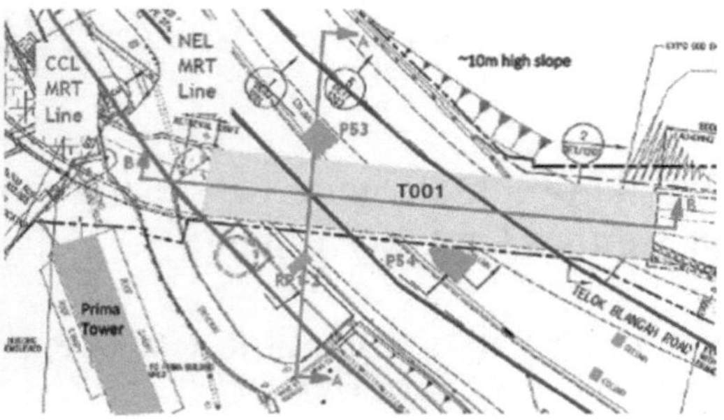

Figure 3. Two-lane piped box tunnel and surroundings.

North-East Line (NEL) tunnels and Circle Line (CCL) overrun tunnel of the existing MRT, are located about 2 m underneath the center of the PBT while existing viaduct columns are present about 10 m away, with the distance of the nearest pile to the edge of the PBT being 2.3 m. Existing natural slopes higher than 10 m are located approximately 10 m away to the West of the tunnel location.

3.2 *General geology and subsurface condition*

In general, the site consists of ~3 m layer of sandy silt fill followed by residual soils with increasing SPT N value from N = 12 to N > 100. Then Jurong Formation sandstone and siltstone of varying weathering grade are encountered below the residual soil from 10 m to 20 m below ground level. Traces of estuarine clay (E), fluvial clay (F2) and fluvial sand (F1) were recorded at locations South of the tunnel, near the Retrieval Shaft. The total thickness of the Kallang Formation ranges from 3 m to 5 m. The soil condition can be broadly categorized into stiff soil at the launching shaft end and relatively weaker soil at the retrieval shaft end.

3.3 *Construction sequence*

Pipe roofing requires launching and receiving shafts for the operation of the micro tunnel boring machine (TBM) used to install the interlocking steel pipes forming the rectangular tunnel box and as the access for the subsequent excavation and casting processes. The launching shaft (LS) is proposed at the northern end of the tunnel and pipe installation will commence from this shaft. The retrieval shaft (RS) for the completed pipes is located at the southern end of the tunnel.

Once the rectangular box is formed, stepped-excavation is then carried out from both shafts. A 3(V):1(H) slope is formed for each excavation step, immediately protected by 100 mm thick of shotcrete. Steel frame set is then installed to support the piped box. Typically, excavation is carried out from a single direction. In order to reduce the construction duration,

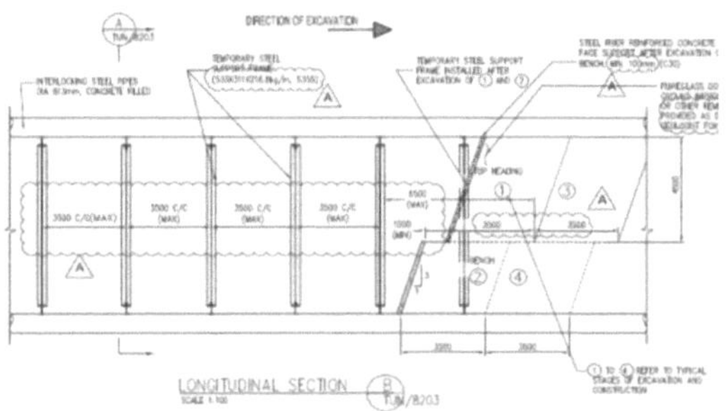

Figure 4. Typical construction sequence.

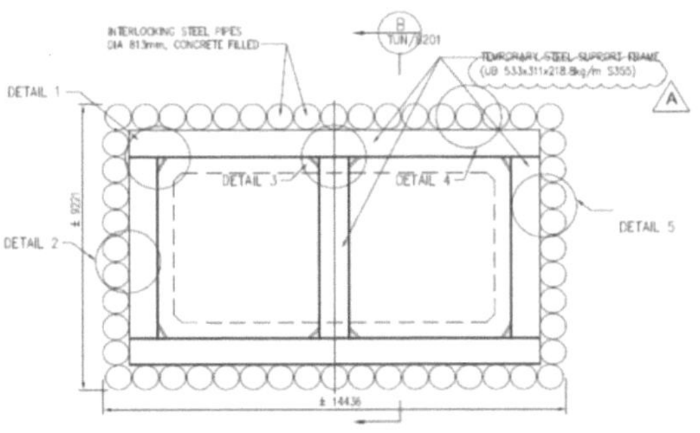

Figure 5. Typical cross section.

a two-directional excavation (excavation advancing from both the shafts) was adopted.

It was observed from the excavation carried out earlier for the one-lane PBT that, apart from time-consuming, the removal of the steel frames also caused additional movement to the stabilized system. Therefore, the steel frames are improvised to be embedded in the concrete lining of the permanent structure. Steel beam depth is reduced from the originally-proposed 900 mm series to 600 mm series for encasement inside the permanent structure. The spacing of the steel frame is reduced correspondingly from 5 m apart to 3.5 m apart.

Figures 4 and 5 shows the typical sequence and section of the PBT excavation.

The whole operation ideally minimizes the impact to surrounding structures and the existing condition at ground level. Each excavation cycle takes about six days.

3.4 *Instrumentation monitoring*

Real-time inclinometer was proposed to be installed in the roof pipe at the center of the tunnel to monitor and assess the settlement that will occur during the construction works, for both the PBTs.

During the one-lane PBT excavation, it was observed that, as the horizontal inclinometer exhibits a cumulative settlement based on one fixed datum point, it does not represent the actual behavior of a two-way excavation. Therefore, in the latter excavation of the two-lane PBT, a series of prisms spaced at 10 m apart

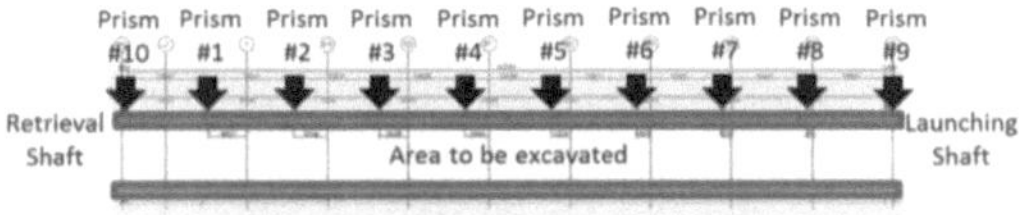

Figure 6. Prisms installed within the pipe to monitor excavation movement.

Table 1. Soil design parameters.

Soil type	Unit Weight kN/m^3	Strength and Modulus Parameters c_u kN/m^2	c' kN/m^2	φ' deg	E' MPa
Fill	20	30	0	30	8.7
E	15	12	0	16	2.6
F1	20	–	0	30	8.7
F2	18	20	0	22	5.2
S:VI	20	–	3	38	70
S:V	22	–	3	32	175
S: IV	22	–	25	30	175
S:III	25	–	50	32	340

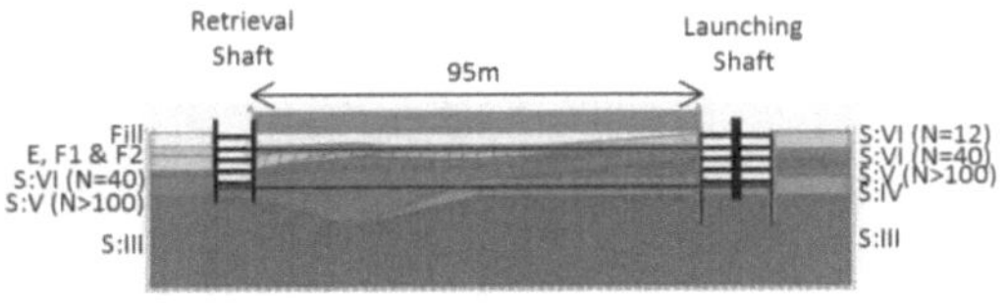

Figure 7. Finite element mesh.

was installed in an adjacent pipe to monitor the excavation movement. Figure 6 shows the layout of the prisms along the pipe.

3.5 *Finite element modelling*

Two-dimensional plane strain analyses were carried out using finite element (FE) method software PLAXIS 2D to model the excavation of the PBT and to predict the ensuing settlement.

Elastic perfectly plastic soil material conforming to the Mohr Coulomb failure criteria was applied for all soil types. The typical soil parameters adopted in the analyses are shown in Table 1.

Groundwater table was modeled at existing ground level. The steel pipe piles were modeled as impermeable plate elements while the steel frame sets are modeled using node-to-node anchors akin to a horizontal shaft excavation. The PBT is assumed to be impermeable, simulating a dry excavation. The actual 2-step excavation of sloped top heading and bench is combined into a single step of sloped excavation, which resulted in an additional excavation up to 0.5 m in accordance with BS8002. The soil layers and FE mesh is presented in Figure 7.

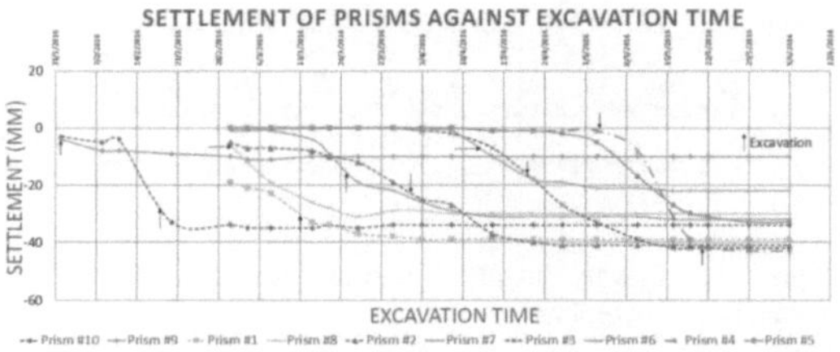

Figure 8. Settlement trends of prisms monitoring the pipe roof.

4 OBSERVATIONS FROM CONSTRUCTION PHASE

4.1 *Pipe roof settlement trends*

The installation of the pipes forming the tunnel box was carried out from October 2014 to December 2015. Fifty-two numbers of pipes were installed. Excavation commenced in late January 2016 from both shafts and was completed in May 2016, i.e. about 4 months.

Figure 8 shows the settlement trends of the pipe roof measured by the prism points. Trends shown in dotted lines are from Prisms No. 10, 1, 2 and 3, used for monitoring of the excavation advancing from the RS. Prisms No. 5 to 9, whose trends are shown in solid lines, are monitoring the relatively-faster excavation advancing from the LS. Prism No. 4 is located approximately at the mid-point of the PBT, where both excavations met.

From Figure 8, it can be surmised that the settlement trends recorded by all ten prism points are similar. All points recorded additional settlement as excavation reached the prisms' location. The settlements measured for the LS side prisms range from 10 mm to 33 mm while those of the RS side recorded larger settlements, ranging from 34 mm to 43 mm. The magnitude is largely dependent on the soil condition at the excavation face, and this observation coincides with the soil information available. The settlement stabilized with the installation of steel frames.

4.2 *Pipe roof settlement profiles*

Figure 9 shows the pipe roof settlement profiles for different stages of construction. The pipe roof settlement increased as excavation advanced from both ends, with larger settlements recorded for the RS end. The settlement profile exhibits an accumulative behavior. It is also observed that preceding excavation steps brought about some advance settlement.

At the RS end, Prism No. 2 breached the Alert Level set at 40 mm after excavation reached 30 m. Despite so, the settlement stabilized and did not increase further with excavation, which is possibly due to the removal of the soft clay layers to an extent of 25 m from the RS, shown by boreholes sunk. At the LS end, the settlement accumulated up to 33 mm recorded by Prism No. 7 and stabilized beyond that. During the last 15 m of excavation, a huge increase in settlement is observed at the mid-point of excavation. This is likely due to the large unsupported pipe span resulting from the removal of the remaining interfacing soil between the two excavation ends.

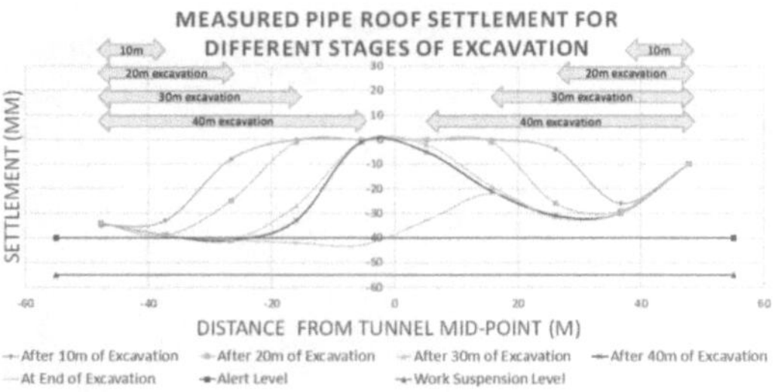

Figure 9. Settlement trends of prisms monitoring the pipe roof.

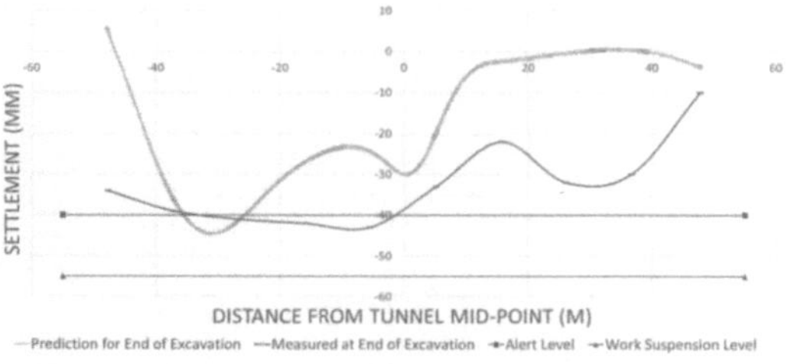

Figure 10. Measured against predicted pipe roof settlement.

4.3 *Comparison of measured and predicted pipe roof settlement*

The comparison of predicted against measured settlement at the completion of excavation stage is presented in Figure 10.

The larger settlement predicted for the RS end was reflected by site measurements. The large increase in settlement at mid-point of the tunnel predicted in the last excavation stage is also observed to be consistent with site measurement.

In terms of magnitude, the maximum measured settlement of the pipe roof matches the predicted settlement, which are 43 mm and 45 mm respectively. However, the location of the maximum settlement did not agree. This is postulated to be due to the presence of a previously undetected abandoned sewer line, which was not considered in modeling.

A divergence in the profile trend is also noted. The settlement predicted in the FE model exhibit minimal rate of change once supports in the form of steel frame sets are provided. This behavior is not demonstrated in the physical construction stage and measured data, as the actual settlement accumulated as excavation progressed.

5 CONCLUSION

Pipe roof method was used to construct two stretches of Sentosa Gateway Tunnel's vehicular underpass, where conventional cut and cover method is infeasible due to the presence of existing structures. Considerable advancements to the typical pipe roof method were carried out, including integration of the temporary support frames into the permanent tunnel linings, a two-directional excavation and a customized monitoring system to monitor such excavation. The excavation of the two-lane piped box tunnel demonstrates settlement behaviors similar to those predicted using two-dimensional finite element software. Nevertheless, several disparities are also observed, and this will hopefully, be addressed by the advancement in geotechnical engineering modeling in the near future.

REFERENCES

Yogarajah, I., Chua, T.S. & Ganeshan, V. 1994. Microtunnelling for Underpass Construction in Built-Up Areas. *2nd National Trenchless Technology Conference.* Sydney, Australia.

Ng, D.C.C., Prasad, P., Tew, B.W., Neo, C.W., Pong, K.F. & Supargo, R. 2011. Geotechnical & Engineering Challenges for Downtown Line Stage 2 C917 & C918 Projects. *Proceedings for Underground Singapore 2011*: 196–207.

Ng, T.G., Yogarajah, I., Woon, K.X., Teo, S.C. & Adalarasu, J. 2016. Trenchless Construction of Underground Infrastructures in Singapore. *Proceedings of the 19th Southeast Asian Geotechnical Conference & 2nd AGSSEA Conference Malaysia 2016*: 149–155.

Challenges and Innovations in Geotechnics – Zhussupbekov (Ed.)
© 2016 Taylor & Francis Group, London, ISBN 978-1-138-03007-7

Analysis of pile foundation behavior on modern and ancient clay bases

A.B. Ponomaryov & E.N. Sychkina
Perm National Research Polytechnic University, Perm, Russia

ABSTRACT: Large areas in many countries are characterized by the spread of ancient deposits. The purpose of this paper is comparison analysis of the characteristics of pile foundations behavior on modern and ancient clays. The following tasks were solved during the study: 1) The analysis of existing works on settlement and bearing capacity of piles on the modern and ancient clay are summarized. 2) Characteristics of experimental sites and methods of field research are described. 3) The experimental data are compared with the results of analytical methods calculations. 4) Based on these results recommendations for forecast piles work on ancient clay in comparison with modern clays and cone penetration test data interpretation in different ages clays are presented.

Keywords: Pile, Foundation, Clay, Settlement, Bearing capacity.

1 INTRODUCTION

1.1 *Review of existing researches*

Numerous studies have shown that the soils of different ages interact with foundations of buildings and constructions in different ways. It is caused by the features the microstructure of the soil, formed as a result of long interaction of various processes: sedimentation, compression and decompression, dehydration, weathering. Large areas in many countries are characterized by the spread of ancient deposits. For example, Vendian clay, Permian and Sochi Formation claystones of European part of Russia (Ponomaryov & Sychkina 2015), the overconsolidation clay and sandstone of Europe (Cooke et al. 1989, Bond & Jardine 1991), North America and Canada (De Ruiter & Beringen 1979, Lehane & Jardine 1994), Japan (Matsumoto et al. 1995). These ancient deposits often lie under the modern sand-clay deposits. However, because of the low values of mechanical properties of modern sand-clay deposits, ancient soils are increasingly being used as the basis for the pile and the pile-plate foundation of heavily loaded buildings and constructions. Existing researches of settlement and bearing capacity corrected the information about the pile behavior on modern sand-clay deposits and increased the accuracy of foundation calculations. However, for the ancient soil a number of questions were not resolved.

Most methods of calculating pile bearing capacity were developed in the 80s and 90s of the last century. The main difference between European methods of calculation and Russian methods is determining coefficients from the cone to the pile. It may be noted among the most common European methods of pile bearing capacity calculation for cone penetration tests in stiff clay:

1. Schmertmann and Nottingham method (Schmertmann 1978, Fellenius 2014);
2. de Ruiter and Beringen method, "Dutch method" or "European method" (Fellenius 2014);
3. Bustamante and Gianeeselli method, "LCPC" method or the "French method" (Bustamante & Gianeeselli 1982);
4. Prince and Wardle method (Price & Wardle 1982).

The main difference between methods is the determining soil resistance under the pile tip and on the lateral surface of the pile.

1.2 *The research aim, object and tasks*

The purpose of this paper is comparison analysis of the characteristics of pile foundations behavior on modern and ancient clays. Object of the research are Early Permian age (P_1) claystones and Quaternary age (Q_{IV}) lean clays.

The following tasks were solved during the study: 1) the analysis of existing works on settlement precipitation and bearing capacity of piles on the modern and ancient clay are summarized. 2) characteristics of experimental sites and methods of field research are described. 3) experimental data are compared with the results of analytical methods calculations. 4) based on these results recommendations for forecast piles work on ancient clay in comparison with modern clays and cone penetration test data interpretation in different ages clays are presented.

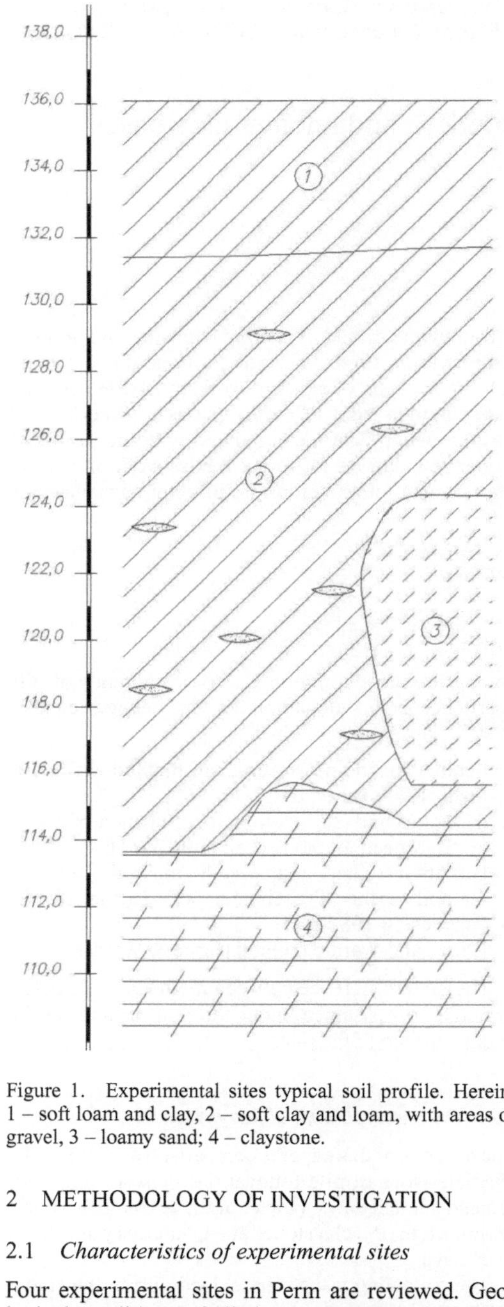

Figure 1. Experimental sites typical soil profile. Herein: 1 – soft loam and clay, 2 – soft clay and loam, with areas of gravel, 3 – loamy sand; 4 – claystone.

2 METHODOLOGY OF INVESTIGATION

2.1 *Characteristics of experimental sites*

Four experimental sites in Perm are reviewed. Geological conditions of all sites are composed of Early Permian age claystones (P_1), which are covered by the modern sand and clay deposits (Q_{IV}). The typical soil profile one of the experimental sites is presented in Figure 1.

The detailed stratigraphy of the soil layers on Figure 1 are given below:

Layer 1: soft, clay and loam clay (Q_{IV}).
Layer 2: p soft clay and loam clay, with areas of gravel up to 25–35% (Q_{IV}).

Table 1. The average values of physical properties of claystones (P_1) and clays (Q_{IV}).

Soil type	density g/cm^3	moisture %	consistency index %
Claystone (P_1)	2.01	20	8
Stiff clay (Q_{IV})	1.98	19	5
Soft clay (Q_{IV})	1.91	27	34

Layer 3: sandy and gravel, from soft to stiff loamy sand (Q_{IV}).
Layer 4: weathered claystone (P_1).

The Table 1 shows some values of the physical and mechanical characteristics of soft, stiff clays and claystone.

The comparison of settlement and bearing capacity of piles on claystones and lean clays with low plasticity on piles was presented in this work.

Much attention is given to the static load tests of driven piles. Pile size was 0.3 m × 0.3 m. Settlement of pile was determined by the results of full-scale static load pile test according to Russian standard documents. Settlement of piles during unloading period was measured for evaluating elastic and inelastic strains.

In the article eleven piles on a different clay deposits are discussed. Including:

- on modern soft clay and loam (Q_{IV}) 4 piles are tested (Pile Nos 1, 2, 52, 69).
- on modern stiff clay and loam (Q_{IV}) 3 piles are tested (Pile Nos 53, 243, 361);
- on Permian claystones (P_1) 4 piles are tested (Pile Nos 403, 407, 587, 592).

Tests for soft clays and loams were performed at 90 days (Pile 1), 96 days (Pile 2), 22 days (Pile 52) and 43 days (Pile 69) after pile driving.

Tests for stiff clays and loams were performed at 19 days (Pile 52), 17 days (Pile 243) and 17 days (Pile 361) after pile driving.

2.2 *Description of methodology*

The comparison of settlement and bearing capacity of piles on claystones and lean clays with low plasticity on piles was presented in this work.

Much attention is given to the static load tests of piles. Pile size was 0.3 m × 0.3 m. Settlement of pile was determined by the results of full-scale static load pile test. Settlement of piles during unloading period was measured for evaluating elastic and inelastic strains. In the paper eleven piles on a different clay deposits are discussed. Including:

- on modern soft clay and loam (Q_{IV}) 4 piles are tested (Pile Nos 1, 2, 52, 69).
- on modern stiff clay and loam (Q_{IV}) 3 piles are tested (Pile Nos 53, 243, 361);
- on Permian claystones (P_1) 4 piles are tested (Pile Nos 403, 407, 587, 592).

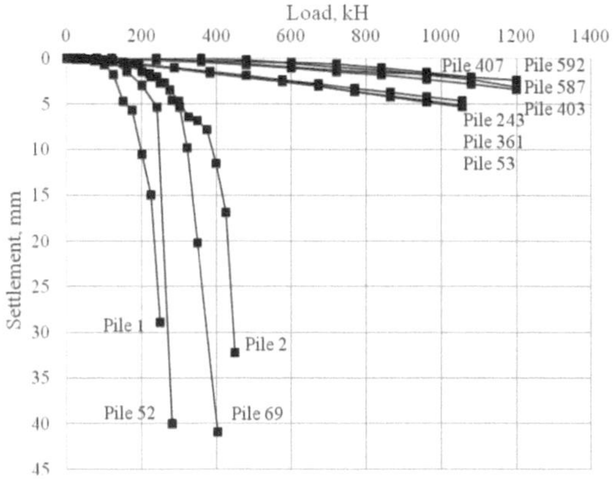

Figure 2. The pile settlement on soft clays, stiff clays and claystones.

Tests for soft clays and loams were performed at 90 days (Pile 1), 96 days (Pile 2), 22 days (Pile 52) and 43 days (Pile 69) after pile driving. Tests for stiff clays and loams were performed at 19 days (Pile 52), 17 days (Pile 243) and 17 days (Pile 361) after pile driving. Pile tests on claystones were performed at 32 days (Pile 403), 20 days (Pile 407), 14 days (Pile 587) and 9 days (Pile 592) after pile driving.

The bearing capacity of pile on claystones was determined by the results of the full-scale static load pile tests and cone penetration tests. The cone penetration tests were carried out in eight points on mentioned experimental sites in Perm. The type of cone – II.

The cone penetration tests results in soft modern clays on this experimental site were taken according to Ponomaryov et al. 2015. The authors (Ponomaryov et al. 2015) analyzed the results of the calculation of the bearing capacity of single pile on soft clays according to the methodology: Schmertmann and Nottingham; de Ruiter and Beringen ("European" or "Dutch" method); Bustamante and Gianeeselli (LCPC or "French method"); Tumay and Fakhro; Aoki and De Alencar and method Eslami and Fellenius.

The results of calculation bearing capacity of pile on claystone according to the calculation methodology in Russian standard document SP 24.13330.2011 (p. 7.2.); and with cone penetration test data according Russian standard document SP 24.13330.2011; Schmertmann and Nottingham method; de Ruiter and Beringen method; Bustamante and Gianeeselli method; Prince and Wardle method are analyzed.

3 RESEARCH RESULTS

3.1 *Pile settlement on different age clay bases*

Figure 2 shows the character of single pile settlement on three varieties of clays: soft clay and loam (Pile Nos 1, 2, 52, 69), stiff clay (Pile Nos 53, 243, 361) and claystone (Pile Nos 403, 407, 587, 592).

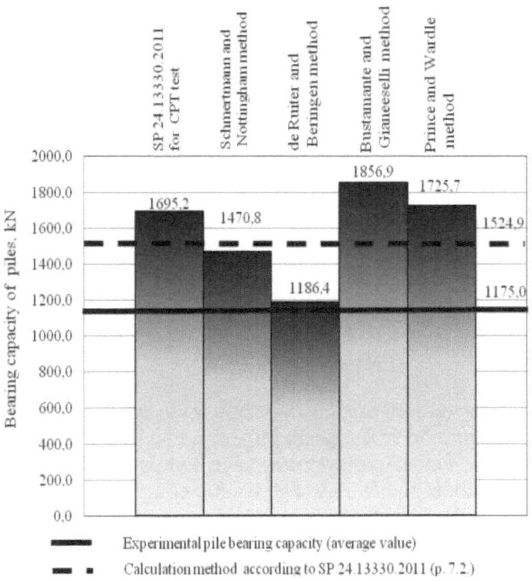

Figure 3. The average values of the calculated and experimental pile bearing capacity on claystones.

Settlement of piles on soft clays significantly increased even for loads of 250–450 kN, the maximum value of piles settlement was 26.10–38.31 mm. Maximum pile settlement on stiff clays was 4.58–5.28 mm for load of 1056 kN. The maximum pile settlements on claystones were still less than for stiff modern clay and amounted 2.17–3.37 mm for load on the pile of 1100–1200 kN. The elastic component of pile settlement on soft clay and loam was 5–12% of the total pile settlement, on stiff clay and loam this value was in the range 97–99%. For claystones elastic deformations were in the range 46–69% of the total pile settlement.

According to authors, it is caused by the cementation bonds in claystones. These bonds are not able to recover after the failure from pile load. Inelastic strain for claystones higher than the inelastic strain for stiff clays without cementation bonds at 30–50%.

3.2 *Pile bearing capacity on different age clay bases*

As noted by Ponomaryov et al. 2015, the most reliable among the bearing capacity calculation methods based on the results of the cone penetration tests in weak clay is the calculation method SP 24.13330.2011 with cone penetration test data from cone type II, method Schmertmann and Nottingham, de Ruiter and Beringen, method Eslami and Fellenius (accuracy within 25% with static testing piles).

The Figure 3 shows the bearing capacity of piles on claystones determined by different methods.

According to Figure 3, the closest results to experimental pile bearing capacity on claystone the calculation by de Ruiter and Beringen method (difference was less than 1%), by Schmertmann and Nottingham method (difference was 25.2%), and calculation method SP 24.13330.2011 p.7.2 (difference was about 30%) showed. Calculation by the method Bustamante and Gianselli, method Prince and Wardle showed overestimated value of pile bearing capacity to 44.3–58% as compared with pile tests on claystones.

The most reliable among the bearing capacity calculation methods is based on the results of the cone penetration tests in claystones de Ruiter and Beringen method, method Schmertmann and Nottingham.

4 RECOMMENDATION AND CONCLUSIONS

The behavior of pile on claystone is close to the behavior of pile on stiff clay and loam. It can be concluded, that the Permian claystones have compressibility less than modern stiff clay. The elastic component of pile settlement on soft clay and loam was 5–12% of the total pile settlement, on stiff clay and loam this value was in the range 97–99%. For claystones elastic deformations were in the range 46–69% of the total pile settlement. It is caused by the cementation bonds in claystones, which are not restored after loading. Character of the interaction between claystone and pile shows that pile under load act as end bearing pile. In summarizing Permian claystone can be a reliable low-compressibility base and is able to take a significant load from the pile foundation.

The most reliable among the bearing capacity calculation methods based on the results of the cone penetration tests in soft clay is the calculation method SP 24.13330.2011 for cone type II, method Schmertmann and Nottingham, de Ruiter and Beringen.

The most reliable among the bearing capacity calculation methods based on the results of the cone penetration tests in claystones is de Ruiter and Beringen method, method Schmertmann and Nottingham. The calculation method SP 24.13330.2011 p.7.2 is also a reliable method.

In summarizing method Schmertmann and Nottingham, method de Ruiter and Beringen for determination bearing capacity shows good agreement with static load pile tests on different age clay bases. These methods can be applied for soft, stiff clays and loams, claystones.

The obtained results can be used to forecast bearing capacity and settlement of piles on clay bases with appropriate justification. According to the authors, this line of research is promising for further study because of the widespread of non-quarter age bases and using them as bases for foundations of buildings and constructions in many countries.

REFERENCES

Bond, A.J, Jardine, R.J. 1991. Effects of installing displacement piles in high OCR clay. *Geotechnique* 41: 341–363.

Bustamante, M., Gianeeselli, L. 1982. Pile bearing capacity predictions by means of static penetrometer CPT. *Proceedings of the 2nd European Symposium on Penetration Testing, ESOPT-II.* Amsterdam 2: 493–500.

Cooke, R.W., Price, G., Tarr, K. 1979. Jacked piles in London Clay: a study of load transfer and settlement under working conditions. *Geotechnique* 29: 113–147.

De Ruiter, J., Beringen, F.L. 1979. Pile foundations for large north sea structures. *Marine Geotechnology* 3: 267–314.

Fellenius, H. 2014. *Basics of Foundation Design.* Revised Electronic Edition. Sidney, Canada.

Lehane, B.M., Jardine, R.J. 1994. Displacement pile behaviour in glacial clay. *Canadian Geotechnial Journal* 31: 79–90.

Matsumoto, T., Michi, Y., Hirano, T. 1995. Performance of axially loaded steel pipe piles driven in soft rock. *Journal of geotechnical and geoenvironmental engineering* 121 (4): 305–315.

Ponomaryov, A.B., Bezgodov, M.A., Bezgodov, P.A. 2015. Comparison of methods for determination of the bearing capacity of driven piles using the cone penetration test in soft clay. *Herald of Perm National Research Polytechnic University, Construction and architecture* 2: 24–39.

Ponomaryov, A., Sychkina, E. 2015. Analysis of strain anisotropy and hydroscopic property of clay and claystone. *Applied Clay Science* 114: 61–169.

Price, G., Wardle, I.F. 1982. A comparison between conep test results and the performance of small diameter instrumented piles in stiff clay. *Proceedings, the 2nd European Symposium on Penetration Testing.* Amsterdam 2: 775–780.

Schmertmann, J.H. 1978. *Guidelines for cone penetration test, performance and design.* U.S. Department of Transportation, Report No. FHWA-TS-78-209, Washington, D.C.

Challenges and Innovations in Geotechnics – Zhussupbekov (Ed.)
© 2016 Taylor & Francis Group, London, ISBN 978-1-138-03007-7

Stress strain behaviour of urban road with underground communication structure

K. Tileu & B. Teltayev
Kazakhstan Scientific Research Highway Institute, Almaty, Kazakhstan

ABSTRACT: The article shows algorithm for determination of displacements, strains and stresses in urban pavement with underground heat pipeline by plane elasticity problem with the use of finite element method. The results for solving of a certain problem, by implementing MATLAB software, are shown in graphical form. The analysis is given for the results obtained. It is found that availability of underground collector in subgrade has a considerable impact on stresses and strains in pavement.

Keywords: pavement, underground reinforced concrete collector, plane problem, finite element method, displacements, strains, stresses.

1 INTRODUCTION

Nowadays the technical level of existing network of urban roads does not always comply with the requirements of safety and comfort for traffic. Early damage of the roads is caused by variation of operational conditions of the roads, and particularly, by the increase of weight carrying capacity for vehicles and their density.

Strains, occurred in pavement, are caused not only by mobile load impact, but also by the impact of natural climatic factors, and particularly, by the impact of temperature on constructive elements.

Strains and stresses in urban pavements, occurred under impact of environment temperature, increase under impact of operation of the heat network, located under carriageway of urban streets and roads. In addition to temperature impact, underground heat networks also cause the impact of physical nature, as the stiffness of reinforced collectors, where steel pipes are mounted for supply of hot and cold water, exceeds essentially the stiffness of subgrade.

The purpose of this work is the investigation of strain stress behaviour of urban pavement with underground reinforced concrete collector under impact of vehicle load by the developed mathematical model.

2 PROBLEM DESCRIPTION

We shall consider cross section of urban road to construct mathematical model for urban pavement with underground reinforced concrete system.

Road structure consists of three-layer asphalt concrete pavement and base course from graded crushed stone, resting on the layer from sand gravel mix, where reinforced concrete system is constructed.

Reinforced concrete system, which is used here, is a serial product and has the following geometrical dimension: total width of double collector $l_k = 2.80$ m and height $h_k = 1.40$ m. Figure 1 shows the schematic form of pavement.

Table after the Figure shows geometrical dimension for constructive layers and thermophysical properties of the materials for constructive elements.

The vertical forces, determined by load value from vehicle wheels and density equal to $q = 0.6$ MPa, are accepted as specified nodal forces for the unsubstantial plane elasticity theory regarding stress and strain behaviour of pavement and subgrade.

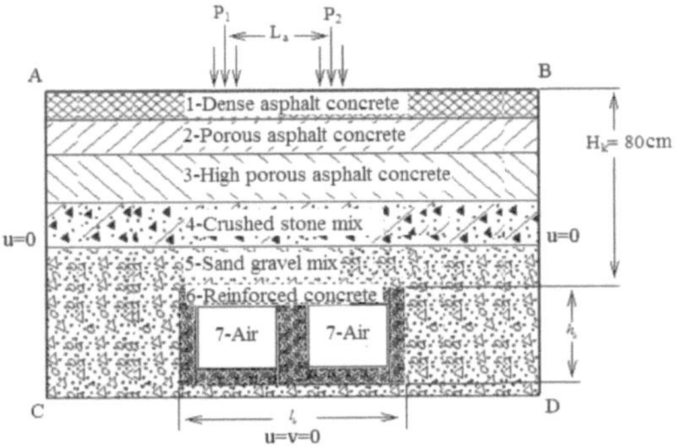

Figure 1. Schematic form of pavement: 1 – dense asphalt concrete; 2 – porous asphalt concrete; 3 – high porous asphalt concrete; 4 – graded crushed stone mix; 5 – sand gravel mix; 6 – reinforced concrete walls of underground collectors; 7 – air inside collectors; L_a – wheel base of vehicle; H_k – depth of laying for reinforced concrete collectors; P_1, P_2 – loads from vehicle wheels; u, v – components for displacement of points at the edges towards coordinate x and coordinate y respectively.

Table 1. Characteristics of constructive elements.

No	Material of constructive element	Elasticity modulus E, MPa	Poisson's coefficient, ν	Density, kg/m^3	Height of layer, m
1	Dense asphalt concrete	3200	0.25	2400	0.05
2	Porous asphalt concrete	2000	0.25	2300	0.10
3	High porous asphalt concrete	1400	0.25	2100	0.12
4	Crushed stone mix	250	0.35	1900	0.33
5	Sand gravel mix	180	0.35	1900	–
6	Reinforced concrete of collector	3000	0.20	2750	1.4
7	Air inside collectors	–	–	–	–

Horizontal displacements ($u = 0$) are not available for profile edges of investigated area by symmetry (Figure 1, lines AC and BD), and all displacements ($u = v = 0$) are forbidden for lower edge of investigated area (Figure 1, line CD).

3 THEORETICAL BASE FOR THE CALCULATION METHOD

Method of minimization for complete potential energy system with effect of stresses and external applied load, which is given by Segerlind (1979) & Zolotaryov (1973), is selected for solving of such problems, when displacements and the values are shown in the edges of the investigated area

Complete potential energy system in the form of matrix will be in the following form:

$$\Pi=\sum_{e=1}^{E}\int_{V^{(e)}}\frac{1}{2}\{U\}^T\left[B^{(e)}\right]^T\left[D^{(e)}\right]\left[B^{(e)}\right]\{U\}dV-$$
$$-\int_{V^{(e)}}\{U\}^T\left[B^{(e)}\right]^T\left[D^{(e)}\right]\left\{\varepsilon_0^{(e)}\right\}dV-$$
$$-\int_{V^{(e)}}\{U\}^T\left[N^{(e)}\right]^T\begin{Bmatrix}X^{(e)}\\Y^{(e)}\\Z^{(e)}\end{Bmatrix}dV- \qquad (1)$$
$$-\int_{S^{(e)}}\{U\}^T\left[N^{(e)}\right]^T\begin{Bmatrix}p_x^{(e)}\\p_y^{(e)}\\p_z^{(e)}\end{Bmatrix}dS-\{U\}^T\{P\}.$$

Formula (1) should be differentiated under $\{U\}$, so that to minimize complete potential energy, and the result should be equal to zero:

$$\frac{\partial\Pi}{\partial\{U\}}=\sum_{e=1}^{E}\left[\int_{V^{(e)}}\left[B^{(e)}\right]^T\left[D^{(e)}\right]\left[B^{(e)}\right]dV\{U\}-\right.$$
$$-\int_{V^{(e)}}\left[B^{(e)}\right]^T\left[D^{(e)}\right]\left\{\varepsilon_0^{(e)}\right\}dV-$$
$$-\int_{V^{(e)}}\left[N^{(e)}\right]^T\begin{Bmatrix}X^{(e)}\\Y^{(e)}\\Z^{(e)}\end{Bmatrix}dV- \qquad (2)$$
$$\left.-\int_{S^{(e)}}\left[N^{(e)}\right]^T\begin{Bmatrix}p_x^{(e)}\\p_y^{(e)}\\p_z^{(e)}\end{Bmatrix}dS\right]-\{P\}=0.$$

Integrals of the formula (2) determine vector of nodal load $\{f^{(e)}\}$ and element stiffness matrix $[k^{(e)}]$. Then the condition for minimization of complete potential energy is recorded in the following form:

$$\frac{\partial\Pi}{\partial\{U\}}=\sum_{e=1}^{E}\left[\left[k^{(e)}\right]\{U\}+\left\{f^{(e)}\right\}\right]=0. \qquad (3)$$

Here matrix $[k^{(e)}]$ is obtained through volume integral as

$$\left[k^{(e)}\right]=\int_{V^{(e)}}\left[B^{(e)}\right]^T\left[D^{(e)}\right]\left[B^{(e)}\right]dV. \qquad (4)$$

and vector of nodal load $\{f^{(e)}\}$ is obtained under formula

$$\left\{f^{(e)}\right\}=-\int_{V^{(e)}}\left[B^{(e)}\right]^T\left[D^{(e)}\right]\left\{\varepsilon_0^{(e)}\right\}dV-$$
$$-\int_{V^{(e)}}\left[N^{(e)}\right]^T\begin{Bmatrix}X^{(e)}\\Y^{(e)}\\Z^{(e)}\end{Bmatrix}dV- \qquad (5)$$
$$\left.-\int_{S^{(e)}}\left[N^{(e)}\right]^T\begin{Bmatrix}p_x^{(e)}\\p_y^{(e)}\\p_z^{(e)}\end{Bmatrix}dS\right]-\{P\}.$$

Elasticity matrix $[D^{(e)}]$ is obtained under formula, which is given by Segerlind (1979), as the strain of the considered mathematical model for pavement occurs in the form of plane strain:

$$[D]=\frac{E(1-\nu)}{(1+\nu)(1-2\nu)}\begin{bmatrix}1 & \frac{\nu}{1-\nu} & 0\\ \frac{\nu}{1-\nu} & 1 & 0\\ 0 & 0 & \frac{1-2\nu}{2(1-\nu)}\end{bmatrix}. \qquad (6)$$

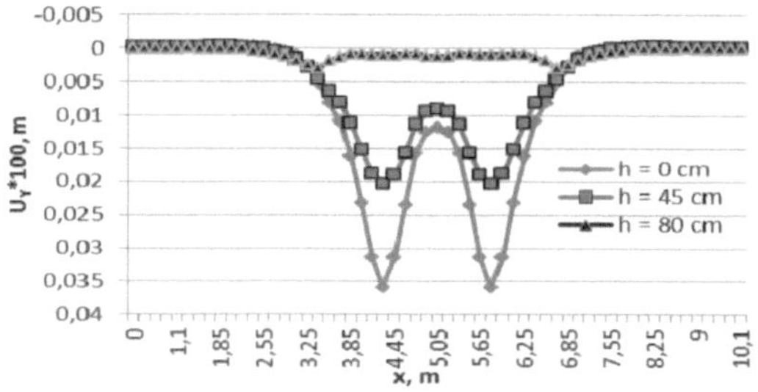

Figure 2. Graphs for variation of vertical displacements u_y in the depths of h = 0 cm, h = 45 cm and h = 80 cm.

Stiffness modulus E and Poisson's coefficient ν are accepted for isotropic materials in the formula (6).

MATLAB code is developed for numerical implementation of mathematical model, by using information taken from Potyomkin (1999), Mathews & Fink (2001).

Figure 2 shows graphs of vertical displacements u_y for points of horizontal sections, located in the depths of h = 0 cm, h = 45 cm and h = 80 cm.

The graphs show centuplicated values for displacements of points.

4 STRESS STRAIN BEHAVIOUR OF PAVEMENT STRUCTURE

Vector of nodal displacements, obtained as the result of solving of the equation (3), has the following form

$$\{U\} = \{u_1\ u_2\ \dots u_{NP}\ u_{1+NP}\ u_{2+NP}\ \dots u_{2*NP}\}, \tag{7}$$

where NP IS A common number of nodes in calculated scheme. Here horizontal displacements of nodes are specified through $u_i, i = \overline{1, NP}$, and vertical displacements of nodes are specified through $u_{i+NP} = v_i,\ i = \overline{1, NP}$.

Equations for relation between strains of finite element points and displacements for nodal points in the form of matrix are given by Segerlind (1979) and shown as follows

$$\begin{Bmatrix} \varepsilon_x \\ \varepsilon_y \\ \gamma_{xy} \end{Bmatrix} = \frac{1}{2A} \begin{bmatrix} b_i & b_j & b_k & 0 & 0 & 0 \\ 0 & 0 & 0 & c_i & c_j & c_k \\ c_i & c_j & c_k & b_i & b_j & b_k \end{bmatrix} \begin{Bmatrix} u_i \\ u_j \\ u_k \\ v_i \\ v_j \\ v_k \end{Bmatrix}, \tag{8}$$

where b_α, $c_\alpha (\alpha = i, j, k)$ are obtained as the result of differentiating of the functions $N_\alpha (\alpha = i, j, k)$ according to coordinates. Here A is the area of element, i, j and k are global numbers of nodes for triangle element.

As the displacements of nodes for each element u_i, u_j, u_k, v_i, v_j and v_k are constant, the strains in all points of finite element, obtained under formula (8), are also constant within each finite element.

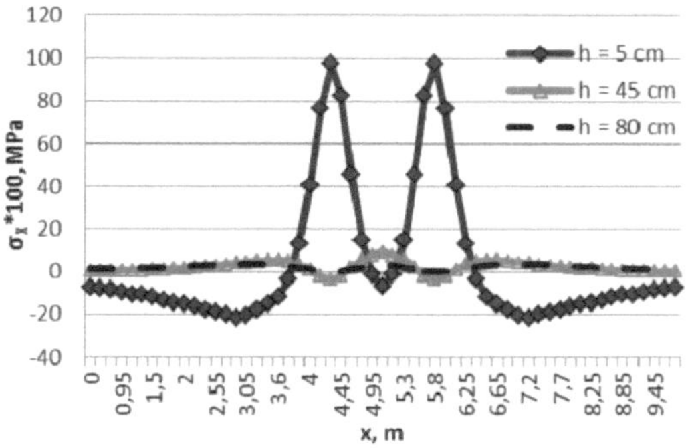

Figure 3. Graphs for variation of normal stresses σ_x in the depths of h = 5 cm, h = 45 cm and h = 80 cm.

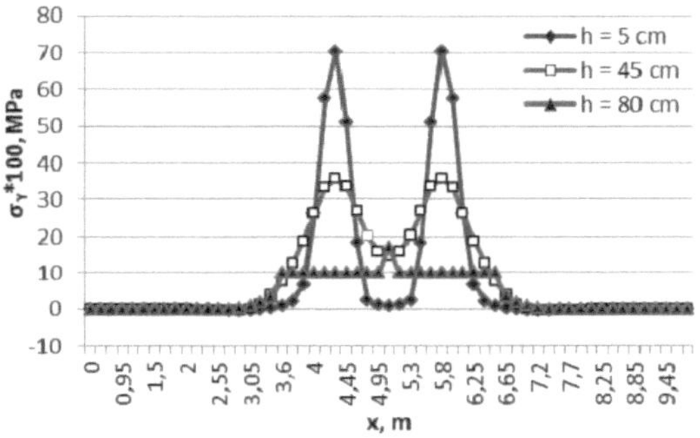

Figure 4. Graphs for variation of normal stresses $\sigma_y$in the depths of h = 5 cm, h = 45 cm and h = 80 cm.

We use Hook's law, as offered by Timoshenko & Goodier (1975), to determine components of stresses σ_x, σ_y and τ_{xy}

$$\sigma_x = E\varepsilon_x, \quad \sigma_y = E\varepsilon_y, \quad \tau_{xy} = G\gamma_{xy}, \tag{9}$$

where $G = \frac{E}{2(1+\nu)}$ is shearing modulus. Here components of strains ε_{xx}, ε_{yy} and γ_{xy} are obtained under equation (8).

The calculation results for components of stress tensor, performed under formula (9) and reduced to the nodes by arithmetic averaging, are shown in the Figures 3–5.

Figures show that external vertical edges for the calculated area are determined accurately, as the gradients for all components of stresses tend to zero near vertical edges of the investigated area, i.e. they reach their natural level.

5 CONCLUSIONS

Analysis of the results obtained allows making the following conclusions:

- impact of collector on stresses and strains increases from top to bottom, i.e. approaching to it;

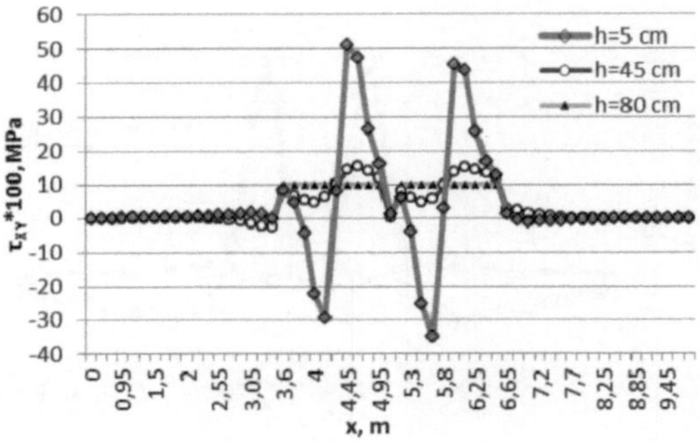

Figure 5. Graphs for variation of shearing stresses τ_{xy} in the depth of h = 5 cm, h = 45 cm and h = 80 cm.

- all components of stresses and also vertical displacements remain constant in the depth of h = 80 cm (surface of underground collector);
- the increase of tensile normal stresses σ_x in the area between the wheels can be explained by interference of impacts from two neighboring wheels. Usually compression stresses σ_x under impact of one wheel of the vehicle, straight under the wheel, are replaced by tensile stresses outside it, which can be observed in the Figure 3;
- the depth of horizontal section essentially influences on the variation value for all components of stresses.

REFERENCES

Mathews D.G., Fink K.D. Numerical computing. The use of MATLAB. 3rd edition: Translation from Emglish – Moscow: Publishing House "Williams", 2001. P. 720.

Potyomkin V.G. System of engineering and scientific computations MATLAB 5.X. In 2 volumes. – Moscow: DIALOG – MIFI, 1999. P. 670.

Segerlind L. 1979. The use of finite element method. Moscow: Mir. P. 392.

Timoshenko S.P. & Goodier J. 1975. Elasticity theory (translation from English), Moscow: Chief editorial board for physics and mathematics literature of Publishing House "Science", P. 576.

Zolotaryov G.S. Test evaluation of slope stability of complex geology by finite element method and model experiment. Moscow: MGU, 1973. P. 227.

Challenges and Innovations in Geotechnics – Zhussupbekov (Ed.)

Development of a reinforced concrete foundation system for renewable energy storage using compressed air technology

Ainur Sabirova
Kazakhstan Geotechnical Society, Kazakhstan

Dichuan Zhang, Jong Kim, Minh Nguyen & Chang-Seon Shon
Nazarbayev University, Kazakhstan

ABSTRACT: The objective of this project is to develop a reliable, safe and economic renewable energy storage media utilizing one of building structural components: a reinforced concrete (RC) foundation system. The proposed RC foundation system, in addition used for transferring structure loads into soils or rocks underground, will be configured to store renewable energy generated from solar panels or wind mills attached to building structures through compressed air energy storage (CAES) technology. The RC foundation system developed from this project can be incorporated as an important component into next generation CO_2 free building structures to improve sustainability of building structures and minimize negative environmental impacts.

Keywords: Reinforced Concrete Foundation System, Compressed Air Technology, Renewable Energy, Sustainable Construction, Green Buildings, Smart Buildings.

1 INTRODUCTION

Most countries in the world including Kazakhstan currently rely heavily on traditional energy such as coal, oil and natural gas energy originated from fossil fuels. However, fossil fuels are non-renewable and have high negative environment impacts. Because of these issues of traditional energy, the use of renewable energy such as solar and wind energy has been highly promoted world widely for residential and commercial purposes, especially in Kazakhstan where our government has identified it as one of key country developments in the coming decades and one of major topics for 2017 EXPO at Astana. The wide utilization of renewable energy leads to challenges of efficient energy generation, reliable energy storage and economic energy transportation. One of popular ways of utilizing the renewable energy is to supply electrical power required for a building directly from energy generated by solar panels or wind mills attached to the building itself (Hayter and Kandt 2011). This method is efficient for the use of the renewable energy in buildings and avoids energy loss from long-distance power transmissions. However, solar and wind energy have an intermittent nature that their availability depends on climate and day-night diurnal cycles (Rugolo and Aziz 2012). This intermittent nature requires efficient methods for energy storage matching the generation of renewable energy with customer demands. For example, solar energy can be generated during day time when the

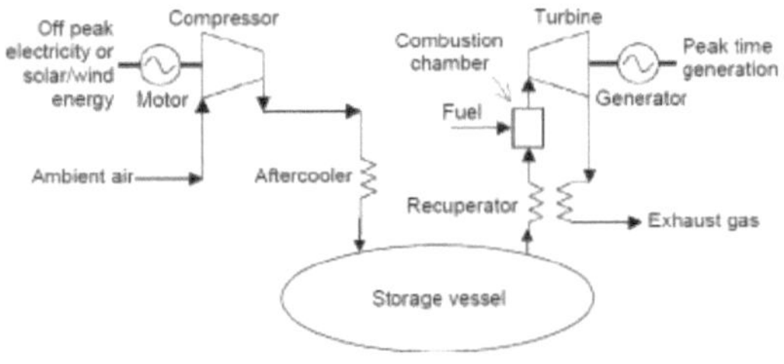

Figure 1. Typical CAES system (after Zhang et al. 2012).

needs of electrical power are not substantial for a residential building while it will not be available during night time when the power demand increases. Several technologies have been developed for energy storage including CAES (See Fig. 1), battery energy storage, electrochemical flow cell system, flywheel storage and hydrogen based energy storage system (DTI 2004). Among them, CAES is one of promising methods with its high reliability, economic feasibility and low environmental impact (Cavallo 2009) (Lund and Salgi 2009). However, current applications of CAES are mainly limited to energy storage for power plants 2 using large scale underground caverns (e.g. McIntosh plant in the USA and Huntorf plant in Germany) due to the challenges for identifying reliable, safe and cost-effective storage media. Although research works have been conducted for the thermal dynamic efficiency

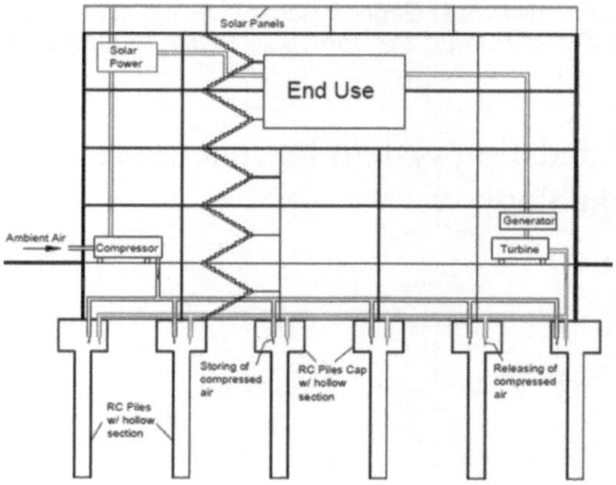

Figure 2. Renewable energy storage using RC foundation system via CAES.

of energy conversions in CAES (Kim et al. 2012), little research has been done for the reliability and safety of the storage media (Zhang and Ahmari 2008). Therefore research works are needed to address these challenges for utilizing the CAES to store renewable solar or wind energy in building structures.

2 THE PROJECT CONCEPT

The concept of proposed research is to store extra renewable energy generated from buildings using the RC foundation system via CAES. This concept is illustrated in Fig. 2 using solar power as an example. The solar power generated from solar panels installed at the building roof will be divided into two parts: one will be directly converted to electric power for immediate end use and the other part will be converted into compressed air with high pressure through a motorized compressor and stored into the foundation system. When the solar power is not available, the compress air stored in the foundation system will be released and converted into electric power by a turbine and generator for end use. To fulfill this concept, it is important to maintain the reliability and safety of the RC foundation system. Therefore this project will mainly focus on developing the RC foundation system which is able to reliably store the compressed air with high pressure as well as safely transfer structural loads. The proposed foundation system has to be configured to contain hollow sections, e.g. waffle matt slab for shallow foundation or thin-wall piles/pile caps (See Fig. 2) for providing energy storage space. However, the use of hollow section will reduce effective area and available strength of the foundation system for transferring loads from the above structure. To maintain the loading transferring capacity, recent developed high strength concrete will be used in the proposed foundation system.

3 SCIENTIFIC NOVELTY AND SIGNIFICANCE

The novelty of this research project is to combine two functions (structural loads transferring and renewable energy storing) into one structural component: the RC foundation system. Rather than building additional storage vessels, this project provides an alternative and cost-effective energy storage media for CAES technology by utilizing the building structure foundation system. In order to meet both energy storage and structural load transferring requirements, innovative high strength and durability material will be investigated and applied to the foundation system proposed in this project. The foundation system will be studies and developed in terms of the structural and geotechnical safety, stability and durability through state-of-art analytical (nonlinear finite element analysis) and experimental (large scale structural and geotechnical tests) integrated research methods.

The proposed research is classified as one of research priority areas and specialized fields set by the Supreme Science and Technology Commission in Kazakhstan: 1. *Energy*: *Renewable energy sources*. The development of proposed foundation system using new high strength and durability materials is related to another research priority area in Kazakhstan: 2. *Deep processing of raw materials and products*: *Technologies of new materials*.

4 UTILIZATION OF FOUNDATION PILES FOR SUSTAINABLE ENERGY

Substantial research works have been carried out for utilizing of foundation piles as ground hear exchangers (so called "energy pile") to transfer and obtain geothermal energy for heating or cooling buildings (Morino and Oka 1994) (Vuataz et al. 2003) (Laloui et al. 2006) (Hamada et al. 2007). The energy pile using the similar dual-function concept proposed in this project has been successfully commercialized in several projects in the UK, such as Neo Bank side, London: One New Change, and Knightsbrigde Palace Hotel. Although the extra loading for the energy pile is originated from thermal-mechanical loads (Amatya et al. 2012) rather than compressed air pressure, the success of the energy pile proves the feasibility of extending functionality of the structural foundation system to more than structural load bearing.

5 NEW MATERIALS FOR RC STRUCTURAL MEMBERS

The normal used concrete typically has a compressive strength around 30 MPa. However recent developed high strength concrete (HSC) can increase the concrete compressive strength up to 120 MPa (Rashid and Manur 2009). This four times increase in the compressive strength allows a 1/4 of reduction of section area of the proposed RC foundations without increase the overall size of the foundation. The high strength concrete has been used for high-rise buildings and large span bridges with detail design guidelines (ACI 363 2010) (FHA Report 2013). HSC can be used to compensate the available area reduction from using hollow sections in the proposed foundation system. However, concrete has generally relative lower tensile strength than its compressive strength regardless of types of concrete such as normal or high strength concrete. Meanwhile the proposed foundation system will incur significant lateral tensile stress from the high air pressure. Such tensile stress can be resisted by the reinforcement placed in the concrete. However it might make the concrete crack and result in leakage of the air pressure. Therefore using HSC will not be sufficient for the proposed foundation system in terms of crack prevention.

In this project, reactive powder concrete (RPC) will be used in the proposed RC foundation to provide sufficient compressive and tensile strength for bearing, crack prevention and air tightness. The RPC originally developed by a French company is characterized by high strength, very low porosity, and high ductility (Richard and Cheyrezy 1994). The properties of RPC are obtained by (i) suppression of weak interfacial transition zone normally developed around the aggregate through improved particle packing, (ii) refinement of the hydrated paste microstructure by use of silica fume and elevated temperature curing, (iii) use of super plasticizer to decrease the water to cementations material ratio, and addition of steel and synthetic micro-fibers. Although RPC is more expensive than regular concrete, its isotropic nature and high ductility make it competitive with steel, consequently resulting in a significant cost advantage for many structural applications. For example, Dowd and Dauriac (1996) reported that RPC beams can be designed with an equal moment capacity to steel beams at comparable mass and cross-sectional dimensions. Richard (1994) and Lee and Chisholm (2006) showed that compressive strength of 200 to 800 MPa, moduli of elasticity of 50 to 60 GPa, and tensile strength of 6 to 50 MPa have been achieved with RPC. Moreover, Song and Wei (2006), Shi and Shi (2000), and Wang et al. (2013) reported that RPC has good durability including anti-chloride and anti-carbonate ion penetration resistance, corrosion resistance, and sulfate-damage resistance due to low permeability and air tightness. Thus, the RPC with high compressive and tensile strength and durability can be effectively used for the proposed RC foundation system.

In general, materials used in reactive powder concrete (RPC) consist of cement, supplementary cementations materials (SCMs), quartz powder, standard sand, and (steel) fiber. The key features of RPC mix design include low water to cementations material ratio and replacement of conventional aggregate with fine quartz sand with particle size of between 150 and 425 micrometer. It may also include different curing temperatures: ambient curing temperature under ambient condition as per conventional site practice and high temperature curing as per pre-cast plant curing practice. Based on several previous researches (Richard and Cheyrezy 1994; Cheyrezy and Richard 1995; Bonneau et al. 1997) and Badillo transform function, the combination of materials for mixture design will be optimized and developed.

6 ANTICIPATED ECONOMIC AND SOCIAL EFFECTS

The proposed RC foundation system, as the major research product, can be incorporated as an important component into next generation CO2 free building structures to increase the effectiveness of the use of sustainable energy. This foundation system provides a cost-effective energy storage media and has potential to improve sustainability of building structures and reduce negative environmental impacts. The proposed RC foundation system can be widely used for many types of structures such as residential, office and industry buildings etc. Moreover the proposed research is expected to generate the following results: (1) new knowledge on behaviors of structural foundations subjected to varying internal pressure; (2) an alternative storage media for CAES; (3) new modeling techniques for the investigation of RC structural foundation systems; (4) significant experimental database for the performance of the RC foundation systems; and (5) new applications of high strength materials in structural engineering.

7 CONCLUSION

The final product of the proposed research is a reliable, safe and cost-effective RC foundation system with dual functionalities: structural load bearing and renewable energy storage. This system can be incorporated as an important component into next generation CO2 free building structures to improve sustainability of building structures and reduce negative environmental impacts. In addition it is expected to: (1) expand the applications of CAES technology and provide alternatives for CAES storage media; (2) advance the knowledge on the behaviors of RC foundation systems under varying air pressure; and (3) promote the use of high strength and durability materials in RC foundations.

REFERENCES

[1] Rashid, M. A., & Mansur, M. A. (2009). Considerations in producing high strength concrete. Journal of Civil Engineering (IEB), 37(1), 53–63.

[2] ACI Committee 363. (2010). Report on high-strength concrete. American Concrete Institute, Technical report: ACI 363R-10.

[3] FHA R&D Committee. (2013). Ultra-high performance concrete: a state-of-the-art report for

the bridge community. the US Federal Highway Administration, Technical report: FHWA-HRT-13-060.
[4] Richard, P and Cheyrezy, M.H. (1994). Reactive powder concrete with high ductility and 200-800 MPa compressive strength, ACI SP 144, 507–518.
[5] Dowd, W.M. and Dauriac, C.E. (1996). Reactive powder concrete. The Construction Specifier 47–52.
[6] Song, S.-M. and Wei, C.-X. (2006). Durability of the Reactive Powder Concrete (RPC). Concrete, (2): 72–73.
[7] SHI, Huisheng and SHI Tao (2006). Research of Chloride Ion Diffusivity in Reactive Powder Concrete with Blast-Furnace Slag, Journal of Tongji University (Natural Science), (1): 93–96.
[8] Wang, Y., An, M.Z., and Han, S. (2013) Durability of reactive powder concrete under the action of sulfate dry-wet cycles, Advances in Civil Engineering and Building Materials, Taylor & Francis Group, London, 125–129.
[9] Bonneau, O., Lachemi, M., Dallaire, E., Dugat, J., and Aitcin, P.C. (1997). Mechanical properties and durability of two industrial reactive powder concretes, ACI Materials Journal, 94(4), 286–290.
[10] Cheyrezy, M.H. and Richard, P. (1995). Composition of reactive powder concrete, Cement and Concrete Research, 25(7), 1501–1511.
[11] Lee, N. P., & Chisholm, D. H. (2006). *Reactive powder concrete*. Branz.
[12] Amatya, B. L., Soga, K., Bourne-Webb, P. J., Amis, T., & Laloui, L. (2012). Thermo-mechanical behaviour of energy piles. Géotechnique, 62(6), 503–519.
[13] Cavallo, A. (2007). Controllable and affordable utility-scale electricity from intermittent wind resources and compressed air energy storage (CAES). Energy, 32(2), 120–127.
[14] Hayter, S. J., & Kandt, A. (2011). Renewable Energy Applications for Existing Buildings. Technical report: NREL/CP-7A40–52172.
[15] Hamada, Y., Saitoh, H., Nakamura, M., Kubota, H., & Ochifuji, K. (2007). Field performance of an energy pile system for space heating. Energy and Buildings, 39(5), 517–524.
[16] Laloui, L., Nuth, M., & Vulliet, L. A. U. R. E. N. T. (2006). Experimental and numerical investigations of the behaviour of a heat exchanger pile. International Journal for Numerical and Analytical Methods in Geomechanics, 30(8), 763–781.
[17] Lund, H., & Salgi, G. (2009). The role of compressed air energy storage (CAES) in future sustainable energy systems. Energy Conversion and Management, 50(5), 1172–1179.
[18] Morino, K., & Oka, T. (1994). Study on heat exchanged in soil by circulating water in a steel pile. Energy and Buildings, 21(1), 65–78.
[19] Rugolo, J., & Aziz, M. J. (2012). Electricity storage for intermittent renewable sources. Energy & Environmental Science, 5(5), 7151–7160.
[20] Vuataz, F. D., Gorhan, H. L., & Geissmann, M. (2003). Promotion of geothermal energy in Switzerland: a recent programme for a long-term task. Geothermics, 32(4), 789–797.
[21] Zhang, L., and Ahmari, S. (2008). Solar energy storage using compressed air in geologic repositories. Progress Report, September- October 2008, Department of Civil Engineering and Engineering Mechanics, University of Arizona, Tucson, AZ.
[22] Kim, Y. M., Lee, J. H., Kim, S. J., & Favrat, D. (2012). Potential and evolution of compressed air energy storage: energy and exergy analyses. Entropy, 14(8), 1501–1521.
[23] Department of Trade and Industry. (2004). Review of electrical energy storage technologies and systems and of their potential for the UK. Technical report: DG/DTI/00055.

Challenges and Innovations in Geotechnics – Zhussupbekov (Ed.)
 ISBN 978-1-138-03007-7

Prospects of using an innovative crushed stone (lightweight, solid, heat-insulating) and an aggregate for concrete structures as well as in the building process of the underground part

Ainur Montayeva
George Washington University, Washington, NW, USA

ABSTRACT: Technical progress in construction is inseparably linked with the development and using of advanced materials. In this direction, particularly relevant inventing new construction materials, more stronger, lighter, cheaper and resistant to aggressive environments of construction materials. One way to increase the resistance of concrete to chloride and sulphate environment is to use the new non-traditional fillers and additives that are capable to reduce the permeability of concrete and the rate of corrosion processes in aggressive environments. Creating a primary protection at the design stage of concrete allows increasing the durability of structures in the building process of underground part. Such measures eliminate the need for secondary protection: corrosion resistance coatings in aggressive environment. Designing a sulfate and chloride resistant concrete are important issues in building foundations on corrosive soil and groundwater, bridges, marine structures, structures of various industrial buildings and structures. The main purpose of this project is: to develop a new type of gravel technology that can be characterized by its thermal insulation properties, lightness, strength, and resistance to aggressive environment; to design a concrete by using innovative gravel and filler that are developed by us. In order to achieve this goal we have developed a new technology of producing artificial gravel. The filler of that new artificial gravel is based on the conversion of a flask that is from the Taskala field of the West Kazakhstan region. Up to 75–80% of chemical and mineralogical composition of this innovative gravel and filler is mainly made up of tiny particles of silica opal-cristobalite. This crystal structure is able to suppress the corrosion process and provide the sulphate resistance to a concrete. The following advantageous characteristics of concrete and concrete are found: due to light weight of innovative gravel, it is easy and safe to mix the concrete mix – without any lumps; provides high speed concrete work. Concrete mixture evenly distributed over the entire surface and fills voids; due to its light weight, the innovative filler and gravel is easily transported through concrete pumps.

Keywords: innovative crushed stone, construction, concrete, siliceous rock – flask, sulphate resistant, foundations.

1 INTRODUCTION

Technical progress in construction is inseparably linked with the development and using of advanced materials. In this direction, particularly relevant inventing new construction materials, more stronger, lighter, cheaper and resistant to aggressive environments of construction materials. The task is achieved due to realization of the following defining factors:

- Involvement of nonconventional raw material resources with development of technological solutions of their deep processing;
- Development new compositions of the raw systems and materials providing their multifunctionality, durability and energy efficiency of technology;
- Modification of the existing raw systems, materials and design taking into consideration providing demanded physical, mechanical, deformation properties and chemical resistance;
- carrying out complex scientific and experimental and test works on using of new composite materials instead of the traditional used materials and in the construction of buildings and facilities;
- creation of the international consortium, the uniting scientists and experts in the field of construction (geotechnicians, architects, specialists in building constructions, experts of construction materials science) for the purpose of examination of competitiveness, an assessment of novelty and technical and economic justification developed innovations;
- carrying out joint scientific and experimental works of scientists and experts in the field of construction directed, on increase of seismic stability, energy efficiency, decrease in a material capacity, and also development complex technological solutions of building of the construction objects conforming to requirements of "green construction".

One of the way to increase the resistance of concrete to chloride and sulphate environment is to use the new non-traditional fillers and additives that are capable to reduce the permeability of concrete and the rate of corrosion processes in aggressive environments.

Creating a primary protection at the design stage of concrete allows to increase the durability of reinforced concrete structures. Such measures in many cases eliminate the need for secondary protection: corrosion resistance coatings in aggressive environment.

Designing a sulfate and chloride resistant concrete are important issues in building foundations on corrosive soil and groundwater, bridges, marine structures, structures of various industrial buildings and structures.

Now there is a wide range of scientific developments and the offered chemical and mineral additives for increase of corrosion resistance of concrete [1–6]. However, the proposed additives currently used in preparation of concrete in Kazakhstan are imported from the countries of the near and far abroad and differ by high cost. Therefore is the task of development of the production technology of domestic additives and materials for improvement of quality of concrete of different function becomes actually.

The main purpose of this project is:

- To develop a new type of gravel technology that can be characterized by its thermal insulation properties, lightness, strength and resistance to aggressive environment.
- Development of a new type technology of filler, giving an astringent system sulphate resistance
- To design a concrete by using innovative gravel and filler that are developed by us
- Design of the modified composition of concrete, with the using of innovative crushed stone and a filler developed by us;
- Research of their physical and mechanical properties;
- Detection of primary characteristics of concrete mix and concrete for the purpose of their recommendation for construction of underground parts of the building and construction.

2 METHODS AND EXPERIMENTAL PROCEDURES

In order to achieve this goal we have developed a new technology of producing artificial gravel and the filler of that new artificial gravel is based on the conversion of a flask-breed that is from the Taskala field of the West Kazakhstan region.

Flasks of Taskala—field – is light microporous rocks, mineral composition, which is mainly composed of very small (less than 0.005 mm) particle opal-cristobalite silica (75–85%).

For receiving crushed stone the molding flasks at first is exposed to crushing by means of gear crushers and to a dry screening by means of a roar. The received fractions (5–10, 10–20, 20–40) is exposed to high-speed roasting on specially developed mode (the innovative patent No. 30783). The small fractions which have remained after a dry screening are exposed to a grinding in a mill to a specific surface of 3000 g/cm^{2}which are used as a filler as a part of the knitting system of concrete mixes.

Distinctive features of the developed technology are the fact that in one technological production line of crushed stone will reach release of three categories of crushed stone concerning their average density, durability and coefficient of heat conductivity. At the same time at the request of the customer it is possible to make crushed stone of average density from 650 to 1000 kg/m^3, durability on compression from 50 MPa to 120 MPa and heat conductivity coefficient from 0,1–0,12 W/мхK.

For design of composition of concrete, for the purpose of their assessment suitability for underground part the building and a construction used crushed stone, possessing the following properties:

- average density – 950–1000 kg/m^3;
- durability on compression of 100–120 MPas;
- coefficient of heat conductivity of 0,1–0,12 W/мхK.

The filler was applied as an additive in cement part of concrete.

Preparations of concrete mix and receiving concrete cubes (10 kh10 kh10 sm) and their tests are made by a standard technique.

To assess the suitability of the concrete blocks on the basis of innovation gravel and filler for use in the construction of the underground part of buildings and structures separately produced concrete tests to sulphate resistance.

For determination of sulfate resistance of concrete used the accelerated method of determination of corrosion resistance of concrete in the environments characterized by sulphatic aggression with concentration of aggressive ions no more than 2000 mg/l. The method is based on comparison of speed of absorption of aggressive ions of SO_4^2 by the examinee the concrete and especially dense concrete of the increased sulfate resistance prepared from a Portland cement.

For determination of sulfate resistance of concrete on a usual Portland cement were made samples of cubes with sizes of 5 × 5 × 5 cm. The hardening of cubes was carried out by the regime, appropriate heat treatment regime.

The composition of concrete was selected so that to provide brand on water tightness of not less W8. On each test is made on 6 samples of one batch.Samples have placed in an exsiceator and have filled in with the sodium sulfate solution amount of 5 l prepared at the rate of 2,8 g of waterless sulfate of sodium ((Na_2 SO_4) in 1 l of the distilled water.

After installation of samples in aggressive solution, in terms, to the corresponding 1, 3, 6, 9 and 12 weeks, from each exsiceator selected a pipette of test of aggressive solution for definition of ions of SO_4^{2-} in amount of 100 ml.

Ions content of SO_4^{2-} was defined in each test and in the initial aggressive solution prepared for test of cubes.

For determination of content in solution of sulphatic ions in amount of 100 ml have been added 1 ml of the

Table 1. Results of physical and mechanical properties pilot scale batches of artificial crushed stone.

No	Indicator name	ND on a test method	Norm indicator on ND	Actually received results
1	Grain composition: the full remains on sets, % on weight d,	ST RK 1213–2003	90–100	97,96
	0,5 d (d + D),		30–60	48,98
	D		до 10	0
	1,25 D		до 0,5	0
2	Content of grains of a lamellar (flatness) and needle shaped form, %	ST RK 1213-2003	35	4,60
3	Break ability, not below	ST RK 1213-2003	800	M1200 (5,26%)
4	Rub-off, not less	ST RK 1213–2003	И-3	И–1 (21,29%)
5	Content of soft rocks grains, %, no more	ST RK 1213–2003	5	2,5
6	Frost resistance, not below	ST RK 1213–2003	F25	F15 (2,9%) F25 (0.6)
7	Maintenance of dust-like and clay particles, %, no more	ST RK 1213–2003	1,0	0,2
8	Content of clay in lumps	ST RK 1213–2003	0,25	Not found
9	Bulk density, kg/m^3	ST RK 1213–2003	Not normalized	0,75

Table 2. Physical and mechanical properties of concrete products on the basis of artificial crushed stone.

Name products	Durability at compression, MPa	Density, kg/m^3 W /мК	Heat conductivity,	Water absorption, %
Base blocks	43,2	1100–1200	0,75–0,8	16–17
Easy heat-insulating constructional blocks	10–12	800–950	0,75–0,8	23–25
Borders	17–17,5	1100–1200	0,75–0,8	10–15
Paving slabs	15–16	1100–1200	0,75–0,8	10–15

concentrated HCl on a tile almost to boiling. In 25 ml of 2,5% – BaCl solution heated to boiling have added in stirring the studied solution and left on a warm tile on 3 h, having covered a glass with the hour glass investigated by solution. After cooling of solution to room temperature, it was filtered through the dense filter. The sediment was quantitatively transferred to the filter and washed out warm water before disappearance in washing waters of reaction to a chlorine ion (on nitrate silver). The filter with a deposit was transferred to a crucible, dried up in a drying cabinet and dry out at a temperature of 800–900°C up to the constant weight.

According to this method the absorption curve by results of pilot studies lies below a reference curve, concrete is considered sulfate-resistant and it can be used for production of construction designs working in sulphatic environments with concentration ions of SO_4^{2-} no more than 2000 mg/l.

3 RESULTS AND DISCUSSION

Chemical and mineralogical the composition of innovative crushed stone and a filler to 75–80% is presented generally from the smallest particles of disgraces-crystobalite of silicon dioxide. Such crystal structure is capable to suppress corrosion process, which explain sulfate resistance of the offered concrete. Below we present physical and mechanical properties of innovative crushed stone and concrete on the basis of innovative crushed stone and filler below.

The received results of experimental studies are presented in table 1 and 2.

4 CONCLUSION

1. Developed the technology of a new type of crushed stone, differing in lightweight, durability, resistance to hostile environment and possessing heat-insulating properties. The technological decision is confirmed with the innovative patent;
2. Established sulfate resistance of innovative crushed stone and filler as a part of concrete, through of chemical and mineralogical composition of innovative crushed stone and a filler of the silicon dioxide presented generally from the smallest particles of disgraces-cristobalite silica (till 75–80%);
3. Designed the modified optimum composition of concrete, with using of the innovative crushed stone

and filler are developed by us and their physical and mechanical properties are investigated;

4. The analysis of the received physical and mechanical properties of concrete has shown that the weight of designs is lost almost twice in comparison with usual concrete and keeps at the same time high heat-insulating properties;
5. The following primary characteristics of concrete mix and concrete are revealed:

   - due to ease of innovative crushed stone concrete mix on their basis, significantly easier and more reliably mixes up – without formation of lumps;
   - High speed of concrete works is provided. Concrete mix is evenly distributed on all surfaces and fills emptiness;
   - Due to application of a natural fine filler water tightness increases;
   - The offered innovative crushed stone and filler not lead to decrease in properties of other applied additives for concrete or properties of mix;
   - The offered innovative crushed stone and filler not make an adverse effect on properties of concrete during solidification;
   - The need for concrete breakers available. Only at increased requirements to a consistence to use the means increasing fluidity;
   - due to the filling easy and viscosity, it is ideally suited for floor designs and creation of other plain surfaces;
   - due to the filling easy innovative crushed stone and filler is easily transported by concrete pumps.

REFERENCES

1. Batrakov, V.G. (1987). Increase of durability of reinforced concrete additives – modifiers. Concrete and reinforced concrete. 7.
2. Takhirov, M. K. (1998). Concrete with additives the acetone-formaldehyde of pitches. M.
3. Shevyakov, V.P. (1984). Design of protection of building constructions of the chemical companies against corrosion. M.
4. Solomatov, V. I. & Silyaev V.P. (1987). Chemical resistance of composite construction materials. M.
5. Rybyev, I.A. (2002). Construction materials- science. M. The higher school.
6. Velichko, I.G. (2014). Structure and main properties of construction materials. M.

Challenges and Innovations in Geotechnics – Zhussupbekov (Ed.)
© 2016 Taylor & Francis Group, London, ISBN 978-1-138-03007-7

The analysis of the piling tests on construction site "The future of the free country"

A.R. Omarov, A.Zh. Zhussupbekov, A.D. Tulegulov, G.A. Zhukenova & G.K. Tanyrbergenova
Department of Civil Engineering, Eurasian National University, Astana, Kazakhstan

ABSTRACT: In this article used loading tests for large diameter and deep drilling piles in Astana, Kazakhstan. Diameter bored piles which have diameter more than 820 mm and 13.5 m depth have been executed at the construction site of New Monument in Astana. Also are considered some of the requirements of the designer of piled foundations and to what extent the accessible testing methods comply with these requirements. Finally, are introduced some recommendations are made for testing methods suitable for problematical ground conditions of Kazakhstan. Traditionally, pile load tests in Kazakhstan are carried out using static and dynamic load test methods. Static pile load test is the most reliable method to obtain the load-settlement relation of piles. Most of the static pile load tests are performed using reaction systems. Furthermore, cost and time for the static pile load test are relatively high compared to the dynamic pile load testing. Therefore, the number of pile load tests in construction site is limited to 2 or 5 piles in usual of constructed piles in Kazakhstan.

Keywords: Pile, test, load, bearing capacity, settlement

1 INTRODUCTION

The monument is called "The Future of the Free Country" (see Figure 1). The project was performed in a simple, concise and dynamic format. The total height of the monument is 76 m, symbolizing 25 years of independence of the Republic of Kazakhstan, is located on the square by the cardinal points. At the top, there are gently curving shapes that are turning into an octahedron with its silhouette that is a symbol of the kazakh slogan: "Segiz Kyrly Bir Syrly".

Figure 1. "The Future of the Free Country" monument in Astana.

In choosing a stylistic solution of the object, universal and international form of strict geometric vertical has been used. A light silver color of the proposed titanium material corresponds to futuristic horizons and is interpreted as the further movement of the country along the path of creation. The monument, facing the sky, 25 facets is 25 years of independence, important events in the history of modern Kazakhstan.

The target of this tests was obtaining of bearing capacity of piles on problematical soils ground of construction.

2 CONSTRUCTION SITE AND GEOTECHNICAL CONDITIONS

To construct of Megaprojects safety and qualitatively we use deep piling foundations with large diameter. Therefore piling foundations must by to design correctly.

Typical geotechnical conditions of the construction site of "The Future of the Free Country" monument which mentioned before are represented by the following soils:

Soil 1 – loam with detritus is covered by top soil and fill-up soil in the depth of 0.20–0.40 m. the thickness of layer varies from 3.6 to 8.1 m. On the field description loam is brown, carbonated, and with middle coarse sand band which is thickness equal to 5 cm and with 5–10 cm thickness loamy sand layer.

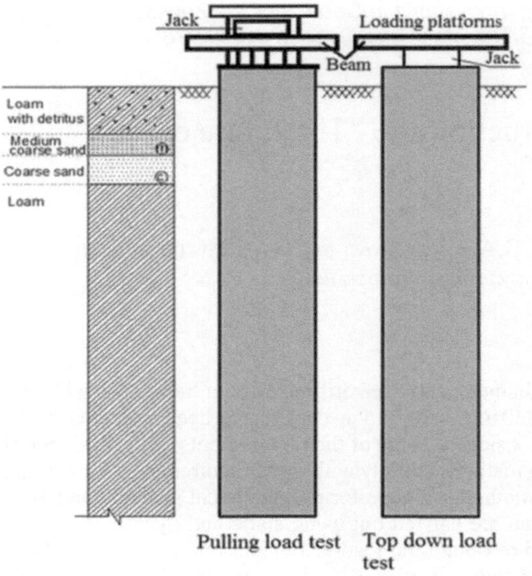

Figure 2. Geological conditions of construction site and details of testing piles.

Soil 2 – middle coarse sand is brown, with loamy soil which is thickness equal to 5–10 cm and water-saturated. The middle coarse sand thickness is 0.60–4.50 m.

Soil 3 – Coarse sand thickness is 1.0–3.9 m. Coarse sand is described by semi-gravel, color is brown, and with middle coarse sand band which thickness equal to 5–10 cm, water-saturated and include 10 percent gravel fragment.

Soil 4 – loam is deep-brown and mahogany color find include debris and detritus up to 10 percent.

Geotechnical condition of construction site and details of piles are shown in Figure 2 and Table 1.

A borehole was drilled at the location of the test pile in the case of the site monument is called "The Future of the Free Country".

Figure 3 shows the results of the odometer tests, axial stress σ_a versus axial strain ε_a. The one-dimensional deformation modulus, $E_c = \Delta\sigma_a/\Delta\varepsilon_a$, was estimated from the stress range from 0.1 to 0.3 MPa.

3 THE PROCEDURE OF PILE TESTSING

SLT one of the more reliable field tests in analyzing pile bearing capacity. SLT should be carried out for driving piles after the "rest" and for bored piles after achievements of the concrete strength, by more than 80%.

According to requirements of Kazakhstan Standard – SNiP RK 5.01-03-2002 and SNIP RK 5.01-01-2002, (2002) – ultimate value of settlement of the tested pile is determined as and depending on category of construction is equal to 16 or 24 mm. The last argument shows conditional character of SLT method.

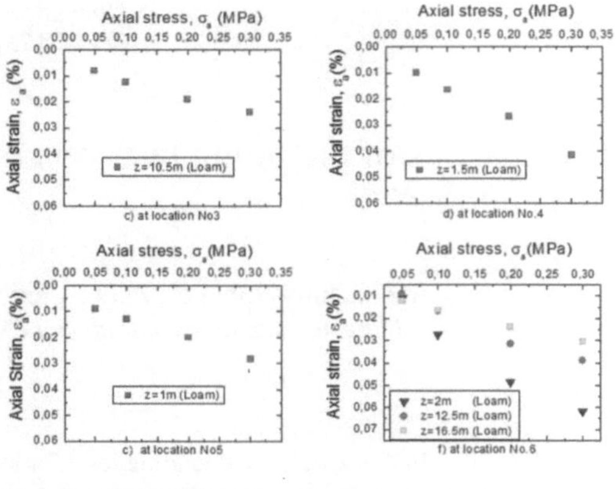

Figure 3. Results of odometer tests.

Table 1. Physical – mechanical properties of soil ground.

Soil #	Soil name	Characteristic of ground						
		ρ, g/sm³	C, kPa	φ	E, MPa	e	I_L max	S_r
Soil 1	Loam aQ II-IV;	1,97	21	31	13	0,60	0,71	0.60
Soil 2	Medium coarse sand aQII-IV;	1,62	2	35	17,0	-	-	-
Soil 3	Coars sand aQII-IV;	2,00	-	-	23,0	-	-	-
Soil 4	Loam e(C1);	2,00	35	33	16,0	0,66	0,04	0.83

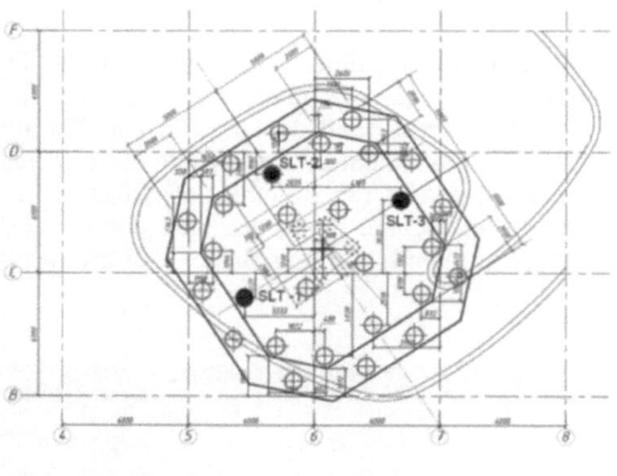

Figure 4. Testing location on the construction site of "The Future of the Free Country" monument.

According to Kazakhstan Standard 1% of constructed piles on construction site must be tested by SLT, but at least 2 Vertical Static Load Tests and in a site must be done.

Field test was performed in 21 days after bored pile had been prepared. Static Loading Test was carried out in accordance with requirements of GOST 5686-94 "Soils. Field test methods by piles" and GOST 20276-99, (1999).

3.1 *Vertical top down load test*

Experienced bored piles with a length of 13.5 m, diameter 820 mm. Static load test carried out in accordance with GOST 5686-94, (1994). Load test

Figure 5. Top down static load tests on the site of "The Future of the Free Country" monument.

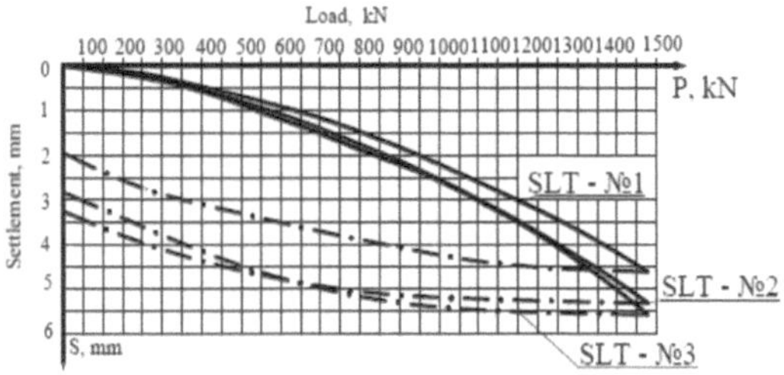

Figure 6. Results of vertical top down load test from SLT-1 to SLT-3.

on pile amounted to 1477 kN. Tests carried out after reaching the strength of the concrete piles more than 80%.

Layout and design considerations of the bored piles are presented in Figure 5. The load test set up and the platform pile layout are shown in Figures 5.

Graph of relationship between piling settlement S from load P and changing of pile displacement S on time t was constructed on result of static load test. The results are presented in Figure 8. Bearing capacity F_d on result of field static load test is determined by formulation:

$$F_d = \gamma_c \frac{F_{u.n}}{\gamma_g}, \tag{1}$$

where, γ_c – coefficient of working condition, which received during the load activity equal $\gamma_c = 1$;

$F_{u.n}$ – guideline value of pile limited resistance;

γ_g – safety factor on ground, which equal 1.2;

As a result of static load test, bearing capacity of bored pile is 1477 kN corresponding to SLT-1 2.5 mm settlement, SLT-2 and SLT-3 (See Figure 6)

Allowable bearing capacity of the piles with an allowance for safety factor ($FS = 1.2$) equal to 4667 kN.

3.2 *Vertical pulling load test*

One pile number No 3* tested static, stepwise increasing pulling loads, in steps of 52.75 ÷ 105.5 kN

Figure 7. Vertical pulling load test on the construction site of "The Future of the Free Country" monument.

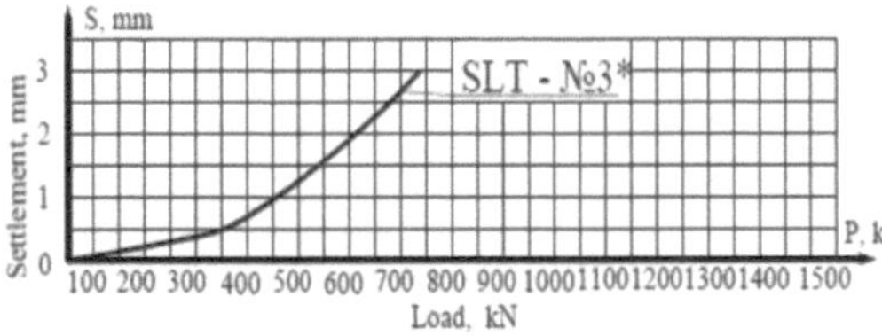

Figure 8. Results of vertical pulling load test SLT-3.

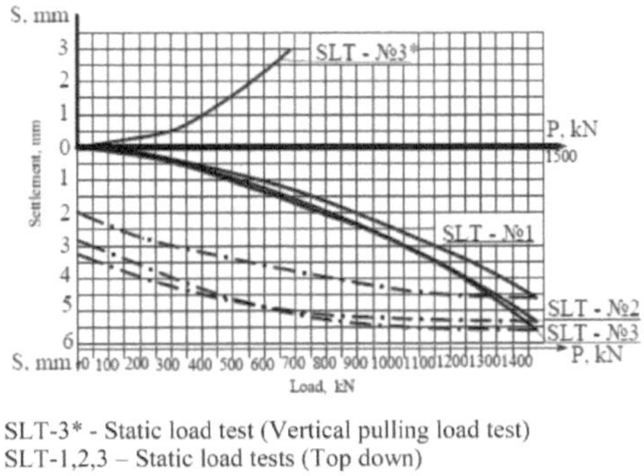

Figure 9. Results of static load testing.

(the first three stages on 105.5 kN and further by 52.75 kN). Maximum load was increased to 686 kN, and the piles of settlement soil was 2.98 mm. The load test set up and the platform pile layout are shown in Figures 7. Result of vertical pulling load test shown in Figure 8.

3.3 *Results of the pile tests*

The maximum load applied on the pile was equal to 1477 kN and the corresponding settlement of the pile was 4.62, 5.53 and 5.59 mm (see Figure 9).

CONCLUSIONS

Existing pile foundation Standards practiced in Kazakhstan are out-of-date and are in urgent need for modernization. This paper presented very short descriptions of coming changes to the concept of Kazakhstan pile foundation design.

In construction site in site monument "The Future of the Free Country" Station the static loading tests of piles No. 1, 2 and 3.

Test bored pile diameter 820 mm was constructed up to 348,56 ÷ 348,68 m level, with a depth of 13.5 m.

The vertical static loading (top down) test was successfully employed to proof load on the pile up to 1477 kN, with a maximum displacement of 4.62, 5.53 and 5.59 mm and vertical static loading (vertical pulling load).

Allowable bearing capacity of the piles with an allowance for safety factor ($FS = 1.2$) equal to 1253 kN and vertical pulling load test 572 kN.

These investigations are important for understanding of soil-structure interaction especially of behavior of boring piles with 820 mm diameter on soil ground conditions of "The Future of the Free Country" in Astana.

REFERENCES

GOST 5686-94, (1994), Methods for field testing by piles. Gersevanov Research Institute of Bases and Underground Structures (NIIOSP), Moscow.

GOST 20276-99, (1999), Field methods for determining the strength and strain characteristics. Gersevanov Research Institute of Bases and Underground Structures (NIIOSP), Moscow.

SNIP RK 5.01-03-2002, (2002), Pile foundations. KAZGOR Design Academy, Almaty.

SNIP RK 5.01-01-2002, (2002), Soil grounds of the buildings. KAZGOR Design Academy, Almaty.

Challenges and Innovations in Geotechnics – Zhussupbekov (Ed.)
© 2016 Taylor & Francis Group, London, ISBN 978-1-138-03007-7

The pile anchor base with use of the extending mixes

G.G. Kamshibaev, L.V. Gorshkova & K.T. Sakanov
S. Toraighyrov Pavlodar State University, Kazakhstan

ABSTRACT: The short review of the existing methods of application of the anchor bases is provided in article. The way of the device of anchor piles which represent the hollow metal pipes having vertical longitudinal sections at the basis is offered, the cavity of a tubular pile is filled with the extending mix, during reaction of hydration there is an increase in volume of mix, the widened basis of a pile and the condensed soil around a pile is formed. The definition formula increase in diameter of hollow part of a tubular pile as a result of her expansion is removed.

The most widespread devices of the anchor bases are of great importance in construction is stretched – intense designs in which most often the pulling-out efforts perceived by anchor devices work in the place of an label-stop.

For the first time the problem of anchoring has arisen in the middle of the last century during construction of big suspension bridges. The big stretching efforts in the bearing elements of these bridges were transferred to soil by means of anchor devices of big weight, also at construction of dams the method of fixing of the bearing cables by their anchoring in rocky massifs was used. Further the problem has gained development due to the need of development of anchor devices for retaining walls and embankments, with the device of the bases for masts of power lines and other high-rise constructions. Anchor piles are widely applied to strengthening as tape and column bases.

In the field of the anchor bases till today there are no comprehensive researches. Due to the increasing distribution it is stretched intense designs of research in this area have acquired considerable relevance.

Successful permission of the tasks connected with the device of the anchor bases will allow to increase profitability of new constructive decisions.

Anchors are subdivided into two categories depending on final appointment:

- the temporary anchors holding any construction (temporary or constant) certain time and becoming unnecessary after carrying out the corresponding stage of construction works;
- the constant anchors which are a component of designs and providing stability of a capital construction during the long time provided by the project.

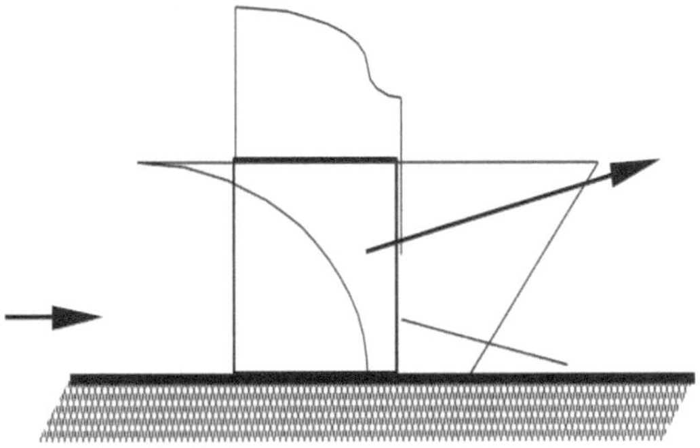

Figure 1. Equally effective wind loading.

The anchor devices perceiving the pulling-out efforts are necessary in all cases when the body weight of a construction is less than a possible wind suction.

The weight of designs of modern buildings usually is less than this size even if this is not about such extremely easy designs as it is stretched intense. In figures 1 and 6 are shown equally effective wind load of constructions of various form. In more flat constructions (fig. 2) equally effective is directed more abruptly up, in particular in case of dome-shaped buildings (fig. 3).

High buildings test the overturning efforts (fig. 4) so to the base from a windward side the maintaining efforts are used, and from down wind side the bases work for compression.

In the flat building equally effective a suction it is directed abruptly up in this connection the base of a construction on all area comes off the basis. The wind suction in the building opened from all directions to a wind stream (fig. 6) is especially big.

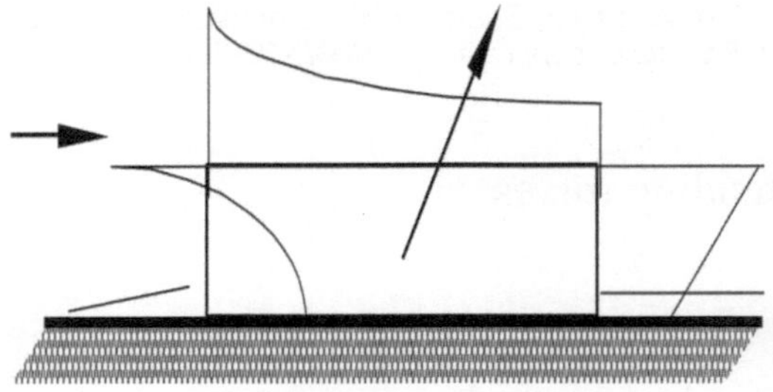

Figure 2. Equally effective wind loading in flat constructions.

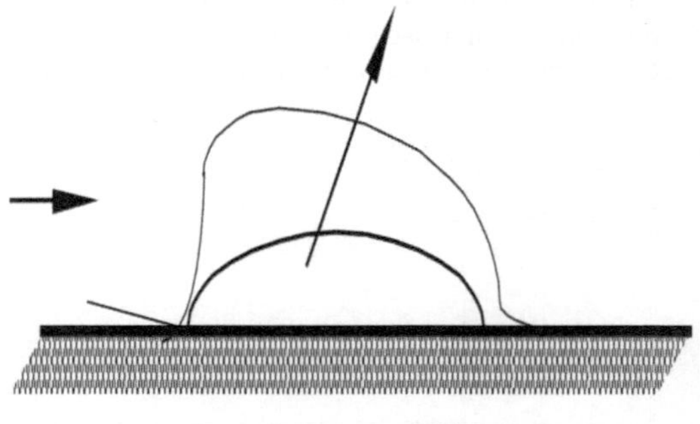

Figure 3. Equally effective wind loading in dome-shaped buildings.

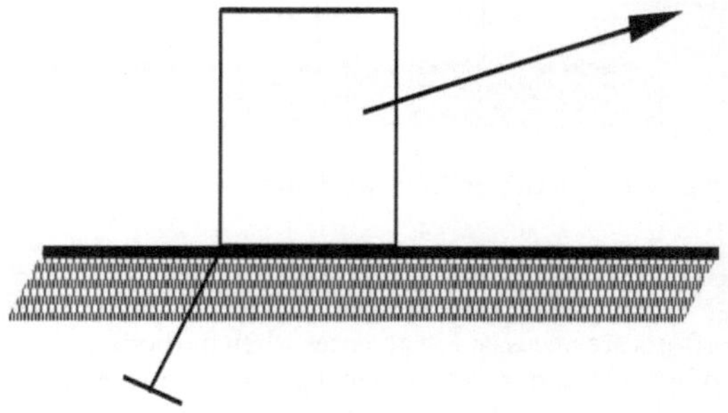

Figure 4. Equally effective wind loading in high buildings.

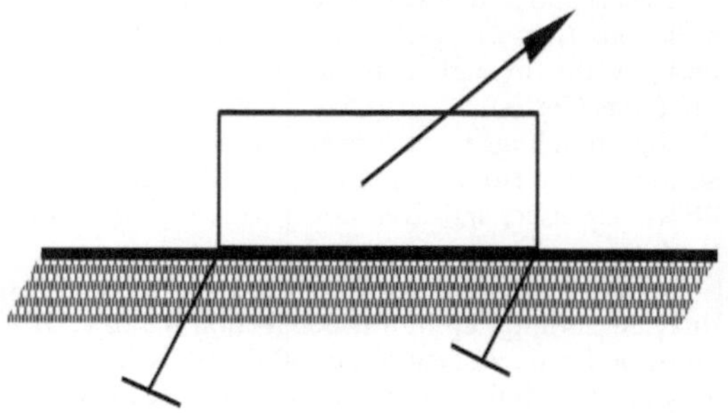

Figure 5. Equally effective wind loading in buildings of average number of storeys.

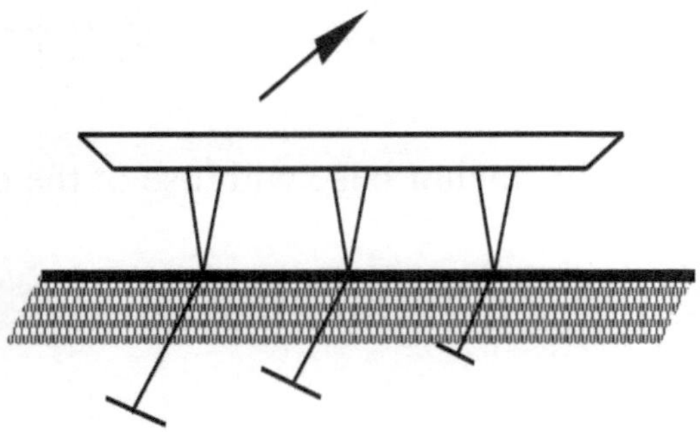

Figure 6. Equally effective wind loading in open constructions.

In such buildings it is necessary to count each separate support on wrest.

Wind loading isn't constant neither in size, nor in the direction. She causes dynamic loading of the bases. As settlement the maximum wind loads get out to a construction.

The internal structure of dynamic wind loadings is still a little studied.

The speed maximum often follows directly a minimum which value makes only 30% of a maximum. At the same time the period of change of loading makes from 3 to 20 sec. Also isn't constant so danger of resonant rocking, generally, is small if only this is not about some special cases of aerodynamic instability. Strong fluctuations of speed of wind define dynamic character of the wind loading arising and disappearing during 1–2 sec.

There is not clear in what degree these dynamic blows are absorbed by the building at the expense of his weight and internal elasticity and what part of loading is transferred to the bases.

Fluctuations of loading on which the anchor bases have to pay off in general are much higher, than in case of the "usual" bases working for compression. Considerable size also constant maintaining efforts, for example have efforts of a preliminary tension or effort from snow loading.

In studying by authors of a question of application of anchor piles and the carried-out patent search it has been established that in practice of foreign and domestic construction the following views of the anchor bases are applied:

- freely lying gravitational anchors the perceiving stretching efforts. The weight which is freely lying on the earth perceives the effort directed up until his size doesn't surpass the weight of the anchoring massif (the ballast anchor devices having now a wide circulation);
- Screw anchors which are screwed in in soil like a drill (are widely applied in the USA) the bearing ability of such anchors with a diameter from 10 to 35 cm, are buried at 2 m and more, fluctuates ot10 to 80 kN;
- expansion anchors are designed in the form of harpoons.

During clogging of an expansion anchor in soil it develops so that has the minimum resistance. At wrest the anchor automatically reveals and his resistance to wrest sharply increases.

The revealing anchors, represent a core from a pipe with the operty plate fixed on it sharnirno which can turn and be established in soil across the direction of the pulling-out force.

The needle anchors which are applied when the considerable stretching efforts from a trailing covering have to be transferred directly to the basis in many points. The anchor needle is carried out from a steel pipe with perforation on the lower end. After deepening of a needle through openings in soil cement mortar or other solutions curing and strengthening soil is squeezed out.

Membrane anchors in the form of the stretched flexible panels brought to soil; first, to provide reliable fixing of dams, secondly, to arrange a waterproof veil on the way of the filtered water.

The gravitational anchors perceiving the shifting efforts. At a tilt angle of the stretched cable of smaller 30 degrees application of earlier considered types of anchors is inexpedient. For perception of inclined efforts gravitational anchors, i.e. the anchors working due to the friction caused by a body weight are used.

Similar functions carry out anchor piles and plates. Piles are long since applied to perception of almost horizontal stretching efforts used under corners of 0–10 degrees, piles can be hammered vertically or obliquely.

Advantage of piles is that they are easily entered into soil. The pile hammered into undisturbed soil has considerably big resistance, than the same dug pile.

The anchor plate usually settles down normally to the direction of the stretched cable, in soil anchor plates are dug in.

Authors offer a new way of the device of anchor piles which represent the hollow metal pipes having vertical longitudinal sections of a certain length (lp) at the basis. Place the extending material which represents powdery mix, usual gray color, nonflammable, inexplosive, the bulk volume density of 12,5 – 13,5 kN/m^3, water requirement of 30% in the basis of a hollow pile. The volume mass of solution of working mix – 21,7 kN/m^3. The developed pressure after a zatvoreniye water to 30 MPas. The main component of powder a calcium oxide, her contents reaches 70–90% of powder lump.

Mechanical tension in working mix results from increase in its volume due to reaction of hydration in the constrained conditions.

Increase in tension in working mix happens about two months, and, for the first days the working tension of mix reaches about 70% of maximum. Other 80% of hobby of tension occurs gradually within two months.

Thanks to pro-saws there is an expansion of the lower end of a pile, as a result receive a pile with the widened lower end, besides around disclosure there is an additional consolidation of soil, pile resistance in

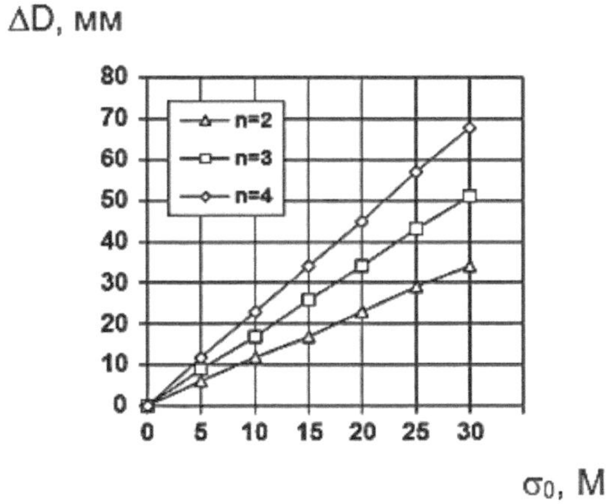

Figure 7. Change of diameter of a tubular pile depending on tension in working mix.

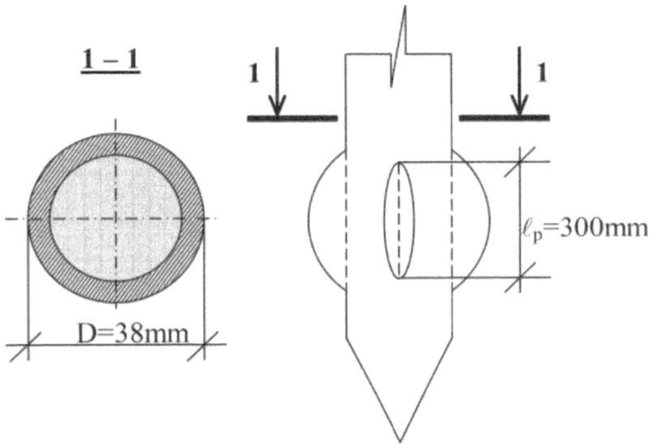

Figure 8. Scheme of change of diameter of a hollow pile.

turn increases by the pulling-out loadings. If necessary it is possible to fill an internal cavity of a pile with concrete mix.

Increase in diameter of hollow part of a tubular pile of ΔD, in the location of longitudinal sections as have shown experiments, it is expedient to define from expression:

$$\Delta D = \frac{\sigma_0 \cdot \pi \cdot r \cdot l_p^4 \cdot n}{96 \cdot E \cdot I_{\text{тр}}},$$

where: σ_0– tension caused by expansion of working mix, kN/m^2;

r – internal radius of hollow part of a tubular pile, m;

l_p– length of a longitudinal section of hollow part, tubular pile, m;

E – module of elasticity of material of hollow part of a tubular pile, kN /m^2;

I_{Tp} – minimum moment of inertia of a tubular pile, m^4;

n – quantity of longitudinal sections (not less than 2) hollow piles, piece.

Change of diameter of hollow part of a tubular pile from tension caused by expansion of working mix at $l_p = 300$ mm and various quantity of propil (n) is given in fig. 7.

REFERENCES

CR and R RK 5.01-03-2002 Pile bases. / Committee on affairs of construction of the Ministry of the industry and trade of the Republic of Kazakhstan – Astana, 2002. – 82 p.

Sorochan E.A. Bases of industrial buildings. – Moscow: Stroyizdat, 1986. – 303 p.

Ganichev I.A. Device of the artificial bases and bases.– Moscow: Stroyizdat, 1981. – 542 p.

Ground improvement (TS3)

Challenges and Innovations in Geotechnics – Zhussupbekov (Ed.)

Role of considering non-constant soil thermal parameters in the simulation of geothermal heat storage systems in the vadose zone

T. Başer
Department of Structural Engineering, University of California, San Diego, La Jolla, CA, USA

Y. Dong
Colorado School of Mines, Golden, CO, USA

N. Lu
Department of Civil and Environmental Engineering, Colorado School of Mines, Golden, CO, USA

J.S. McCartney
Department of Structural Engineering, University of California, San Diego, La Jolla, CA, USA

ABSTRACT: This study focuses on understanding the role of thermal properties of soils (thermal conductivity and specific heat capacity) in the performance of geothermal heat exchanger arrays used to store heat in the vadose zone. Although the impacts of degree of saturation and temperature on the apparent thermal conductivity of soils has been widely studied, the same is not true for the volumetric heat capacity of soils. To investigate the role of the heat capacity, a three-dimensional (3D), transient finite element model was built in COMSOL to consider the representative field conditions as well as coupled heat transfer and water flow processes in the unsaturated soil within a soil-borehole thermal energy storage (SBTES) system. The numerical analyses were performed considering two cases: using constant thermal properties and using thermal properties that vary with changing degree of saturation. Two models were considered to predict volumetric heat capacity of different soils with changing degree of saturation and were compared with laboratory measurements. Results indicate that using constant thermal properties leads a difference in temperature distribution in the heat exchanger arrays. The proposed prediction equation in this study for volumetric heat capacity was more successful than the common prediction equation.

Keywords: SBTES systems, Thermal properties, Numerical analysis

1 INTRODUCTION

Soil-Borehole Thermal Energy Storage (SBTES) systems are used to store heat collected from renewable sources so that it can be used later for heating of buildings (Sibbitt et al. 2012; Zhang et al. 2012, Başer & McCartney 2015). They function in a similar way to conventional geothermal heat exchange (GHE) systems, where heat is transferred from a source to a sink via circulation of fluid through a series of closed-loop heat exchangers. However, they differ from GHE systems in that the heat is injected or extracted continuously over the course of a season into the borehole heat exchanger array. Further, the borehole array in a SBTES system is overlain by a hydraulic barrier to retain pore water within the subsurface and a thermal insulation layer to minimize heat losses to the atmosphere (Başer et al. 2015, 2016).

The more the SBTES systems become popular, the greater effort has been given to determine the thermal properties of soils to model the heat storage and transfer through soils. Because the heat transfer and storage in a soil is governed by its thermal characteristics, thermal properties are necessary for modeling the heat transfer in soils (Abu-Hamdeh 2003).

The main goal of this paper is to understand the impact of the specific heat capacity on spatial temperature distributions in heat exchanger arrays installed in different types of soils in unsaturated conditions, considering coupled heat flow and thermally induced water flow. Thus numerical analysis were performed on unsaturated soils to present the importance of the thermal properties in heat transfer modeling in heat exchanger arrays. The second phase of study focused on the prediction of volumetric heat capacity of different soils with volumetric water content. Some prediction equations were used and the results were compared with the laboratory measurements.

2 BACKGROUND

Several field and numerical studies have established that SBTES are proven to be efficient at storing heat in the subsurface (Sibbitt et al. 2012, Zhang et al. 2012, Başer et al. 2015). However, a better understanding of the heat transfer processes in these systems is required as the heat transfer and the heat storage in the SBTES arrays are governed by the thermal properties; thermal conductivity and the specific heat capacity. It has been proven by various researchers that thermal properties of soils change with degree of saturation (Duarte et al. 2006, Smits et al 2012).

A recent study by Lu and Dong (2015) evaluated a new equation based on the soil-water retention curve (SWRC) that can consider the impact of changes in degree of saturation on the thermal conductivity. Although other thermal conductivity models have been developed in the past, this new equation employs the parameters from the widely-used van Genuchten (1980) SWRC model. Although they focused on the thermal conductivity, they also found that the specific heat capacity varied with degree of saturation. They evaluated several soils and found a good match between their predictive model and the measured thermal conductivity values as a function of the degree of saturation.

The volumetric heat capacity (C or VHC) describes the amount of heat (thermal energy) required to increase the temperature of a certain volume of material by 1°C, but without undergoing a phase transition. In this sense, it can be used to reflect the heat storage in a volume of material at a certain temperature. It is different from the specific heat capacity (C_p) in that the VHC is a 'per unit volume' measure of the relationship between thermal energy and temperature of a material, while the specific heat is a 'per unit mass; measure. VHC can converted to specific heat capacity by multiplying it by the total density of the soil (ρ). The total heat (Q) content of a system possessing the volumetric heat capacity, and being at an absolute temperature (T), is expressed by as follows:

$$Q = \int C dT \tag{1}$$

Once the heat is transferred into the ground, the ability of the soil to retain heat under the typical subsurface boundary conditions is dependent on both its heat capacity and thermal conductivity. A conceptual evaluation of heat storage in soils is given in Figure 1. From point *a* to *b* is the heating period corresponding to heat energy being absorbed in the soil at the point being monitored. The time for the heating period is denoted ΔT1. The time period ΔT2 corresponds to the period where external heat is no longer being put in the ground but energy is still being received from

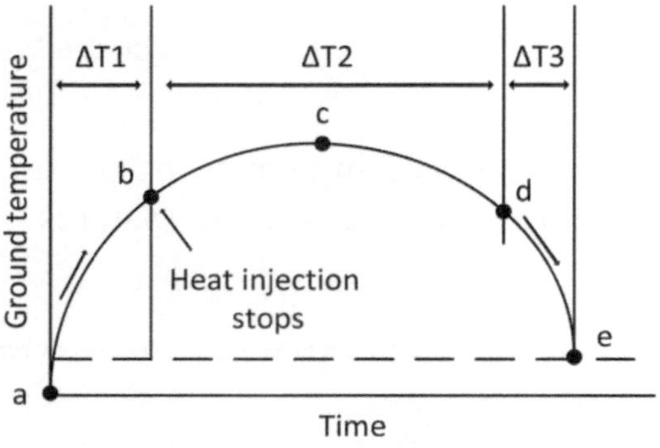

Figure 1. Conceptual heat storage capacity of soils (Adapted from Fang and Chaney 1983).

point *d* to *e* at the same depth. The thermal storage capacity of the soil is then can be defined by the area under the curve. During ΔT3 the dissipation stage of the heat happening. From a practical point of view, the ideal soil for use in thermal storage applications will have a small ΔT1 and a large ΔT2.

3 NUMERICAL MODEL

3.1 *Model formulation*

A transient three-dimensional finite element model was built in COMSOL to predict temperature distributions inside and outside of the thermal heat exchanger arrays. The model was developed considering both heat transfer and water flow since when the unsaturated soils is heated water flows due to the decrease in density. The other driving mechanism for water flow is the alteration of surface tension with temperature. Thermally-induced vapor flow is not considered in this study.

Water flow in unsaturated soils, assuming that air pressure in the pores equal to atmospheric pressure, is governed by Richards' equation. Using this equation, mass balance in unsaturated soils can be expressed as follows (Bear 1972):

$$n\frac{dS_w}{dP_c}\frac{\partial \rho_w P_c}{\partial t} + \nabla \cdot \left(\frac{-\rho_w k_{rw} k_{\text{int}}}{\mu_w} (\nabla P_w + \rho_w g) \right) = -Q_m \tag{2}$$

where n = porosity of the soil; S_w = wetting (water) degree of saturation (dim.); ρ_w = density of water (kg/m^3); μ_w = dynamic viscosity of water (Pa·s); P_c = capillary pressure ($P_c = P_{nw} - P_w$) (kPa); t = time (s); k_{int} = intrinsic permeability of soil (m^2); k_{rw} = relative permeability of water (dim.); g = gravitational acceleration (m^2/s); and Q_m = mass storage (kg/m^3) which is assumed to be zero in this study.

Heat transfer in unsaturated soils is governed by the combination of Fourier's law for heat conduction and Newton's law for heat convection. For these two

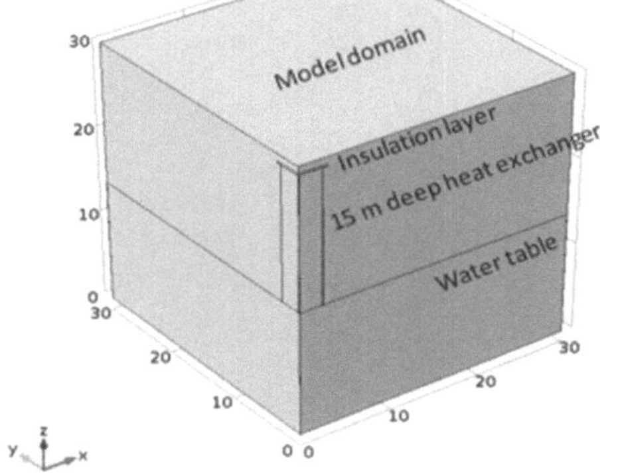

Figure 2. Model geometry.

modes of heat transfer, the governing equation for heat transfer is porous media can be expressed as follows:

$$(\rho C_p)\frac{\partial T}{\partial t} + \nabla \cdot (\rho C_p u_w T) = \nabla \cdot (\lambda \nabla T) + Q \quad (3)$$

where ρ = total density of the soil (kg/m^3); C_p = specific heat capacity of the soil at constant pressure (J/(kgK)); u_w = Darcy velocity; T = absolute temperature (K); λ = apparent thermal conductivity of the soil (W/(mK)); and Q = heat source.

3.2 *Model geometry and boundary conditions*

The model geometry consists of an SBTES array of 15 m-deep vertical borehole geothermal heat exchangers installed in a deep homogeneous soil layer in an array having a width of 30 m and a depth of 30 m. A 0.1 m-thick insulation layer was placed on top of the heat exchangers, which was covered by a layer of site soil. The details of the model geometry is given in Figure 2. Symmetry was used in configuring the model geometry, with one of the geothermal heat exchangers at the corner of the array, and the other two are spaced at a distance of 2.5 m from the center in orthogonal directions.

3.3 *Material properties*

The entire domain is assumed to be a uniform and isotropic soil layer. The properties of Hopi silt were considered for the soil layer. To perform coupled heat transfer and water flow analysis the thermal and hydraulic properties are needed. These properties differ for the different type of soils as they are dependent on the grain size, pore size distribution, and degree of saturation of the soil. Dong et al. (2014) used a transient water release and imbibition method (TRIM) which was modified to include measurement of the thermal conductivity function (TCF) and volumetric heat capacity function (VHCF) in conjunction with concurrent measurement of the soil water retention curve (SWRC) and hydraulic conductivity function (HCF). The modified TRIM test works in a way that two pairs of dielectric and thermal needle sensors embedded in the soil specimen monitor continuously spatial and temporal variation of degree of saturation, thermal conductivity, thermal diffusivity, and volumetric heat capacity during drying and wetting processes. For SWRC van Genuchten (1980) fitting model was used. The soil hydraulic and thermal properties are given in Figures (3a) to 3(b). The saturated hydraulic conductivity of Hopi silt is 5.3 × 10^{-7} (m/s).

Lu and Dong (2015) proposed a closed form equation for the thermal conductivity function (TCF) based on soil water retention regimes to predict the thermal conductivity. This equation allows prediction of thermal conductivity as a function of degree of saturation incorporating van Genuchten (1980) parameters. TCF is expressed as follows:

$$\frac{\lambda - \lambda_{dry}}{\lambda_{sat} - \lambda_{dry}} = 1 - \left[1 + \left(\frac{S}{S_f}\right)^m\right]^{1/m-1} \quad (4)$$

where λ_{dry} and λ_{sat} are the dry and saturated thermal conductivities, respectively, S is the actual degree of saturation, S_f is the degree of saturation at which the funicular regime is onset, m = defined as the pore fluid network connectivity parameter for thermal conductivity that also could be related to the pore-size parameter n in the van Genuchten (1980) SWRC model. Volumetric heat capacity can be assumed to varying a similar way with the degree of saturation as the thermal conductivity. In this case, the saturated and dry thermal conductivities can be replaced with the volumetric heat capacities for soil in saturated and dry conditions. Accordingly, the Lu and Dong (2015) TCF can be converted to a VHCF:

$$\frac{C_p - C_{p\,dry}}{C_{p\,sat} - C_{p\,dry}} = 1 - \left[1 + \left(\frac{S}{S_f}\right)^m\right]^{1/m-1} \quad (5)$$

There are another prediction methods exist in the literature. One of the most common prediction method for volumetric heat capacity is expressed by the following equation (Abu-Hamdeh 2003):

$$C_v = \rho_d \left(c_s + Snc_w\right) \quad (6)$$

where ρ_d = the dry density of the soil, n = porosity, and c_s and c_w are the specific heats in J/kg/^{o}C of dry soil particles and soil water. Usually, the contribution of air can be neglected because of negligible mass of gaseous phase. Volumetric heat capacity of Hopi silt was predicted both VCHF and Abu-Hamdeh (2003) and the results were compared with the laboratory measurements in Figure 4.

Although the fit is not perfect, the newly proposed VHCF provides a better prediction of the VHC better than Abu-Hamdeh (2003) equation. Overall Abu-Hamdeh (2003) equation gives a relationship between

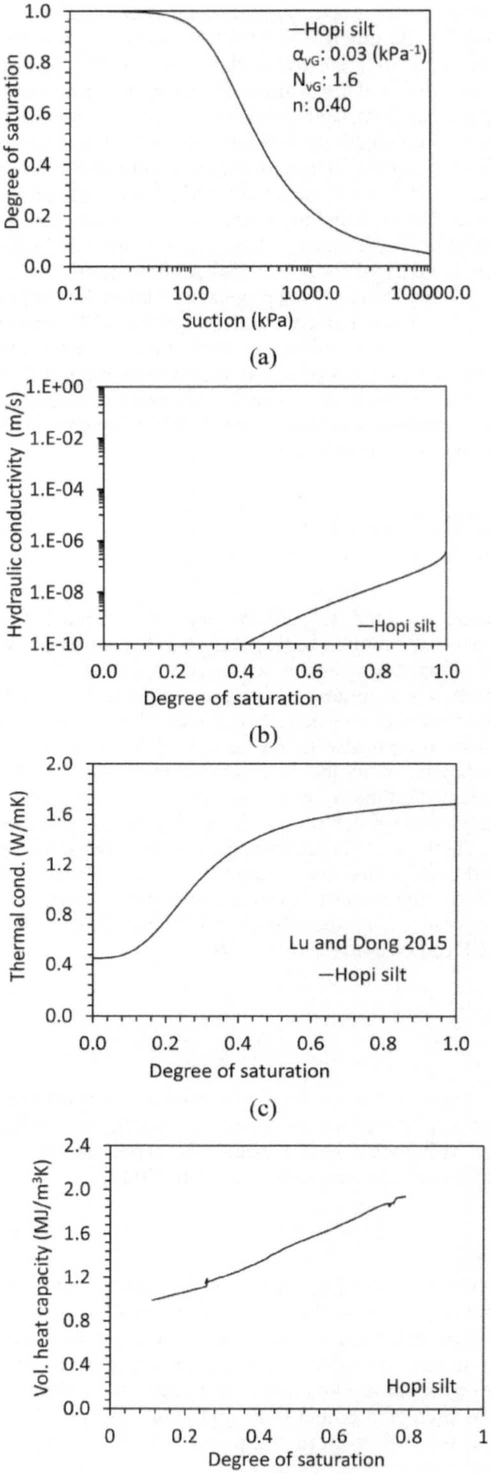

Figure 3. Hydraulic and thermal properties of Hopi silt used in the analyses (a) SWRC; (b) HCF; (c) TCF; (d) C_v vs degree of saturation.

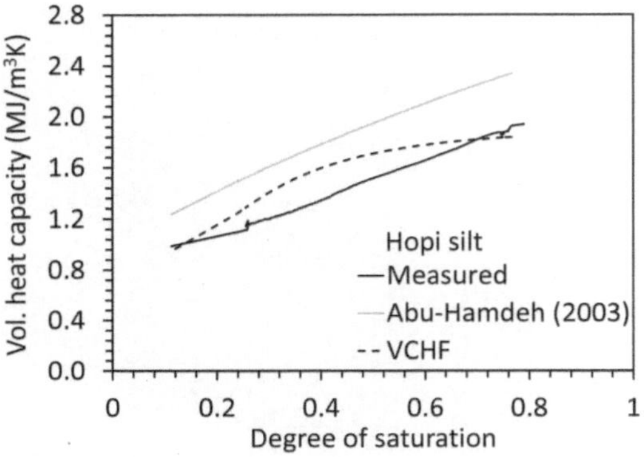

Figure 4. Comparison of predicted and measured heat capacities.

degree of saturation and VHC. It is not always the case, in fact, thermal properties are proven to follow the trend of the water retention regimes (Lu and Dong 2015).

3.4 *Initial and boundary conditions*

For heat transfer, a constant heat flux of 30 W/m was applied at the borehole boundaries for a period of 90 days. A sinusoidal temperature function was applied at the top assuming that maximum and minimum daily air temperatures are 25°C and 10°C. The bottom temperature was fixed to 12°C because of the groundwater. For water flow, zero flux was assumed for all boundaries except the bottom boundary, where a constant total head of 14 m was applied.

The initial temperature of the domain was assumed to be uniform and equal to 12°C. This is equal to the mean annual air temperature in San Diego, CA, and represents the transition profile between the hot and cold seasons of the year. The water table was assumed to be at a depth of 16 m, coinciding with the bottom of the heat exchangers. The initial conditions for the profiles of degree of saturation and suction with depth correspond to hydrostatic conditions.

After the implementation of the initial and boundary conditions, the system of partial differential equations (1) and (2) in three-dimensional domain was simultaneously solved using the COMSOL Multiphysics software to understand the effect of non-constant thermal properties on the heat transfer in unsaturated soils. The simulated domain has a volume about 10 times that of the heat exchanger domain in order to minimize boundary effects. Also initial simulations verified that there was no boundary effects.

4 ANALYSIS

In heat transfer modeling efforts specific capacity is one of the governing parameters. In most of the heat transfer modeling studies it is assumed to be a constant parameter. This assumption is valid when the soil is either nearly dry or saturated. On the other hand when the soil in the heat exchanger array is unsaturated thermal properties change with depth due to the degree of saturation. To understand the effect of specific heat capacity spatial and temporal distributions of the temperatures inside (x = y = 1.25 m) and outside (x = y = 3.75 m) of the heat exchanger array were determined considering three cases. First simulation was run considering the actual laboratory measurements while the second case considered the Abu-Hamdeh (2003) prediction values. Lastly, in the third case analysis was performed assuming that soil is having a uniform degree of saturation of 0.6 along the heat exchangers and the associated thermal properties with this particular degree of saturation were used in the model. The temperature distributions were plotted in Figures 5(a) to 5(c).

It was observed that although not significant, there are differences in the temperature distributions in the soil after 90 days of heating period when the actual values and the predictions from the model of Abu-Hamdeh (2003) were compared. The maximum temperature in the array was observed to be 44°C when the actual thermal properties were used while this value was 40°C in the case of the model of Abu-Hamdeh (2003). This is because the model of Abu-Hamdeh (2003) over-predicting the VHC led to a decrease in the temperature inside and outside of the heat exchanger array. On the other hand in the case of constant thermal properties there was only 1°C difference between the model of Abu-Hamdeh (2003) and the constant parameters in the maximum temperatures of inside and outside of the array. The reason might be the selecting the constant thermal parameters associated with the degree of saturation of 0.6. Although not given, because the newly proposed VHCF predicted the measured volumetric heat capacity better than the Abu-Hamdeh (2003) model, the temperature distribution in the array was quite similar. These results concluded that the usage of representative thermal properties will result in proper heat transfer modeling.

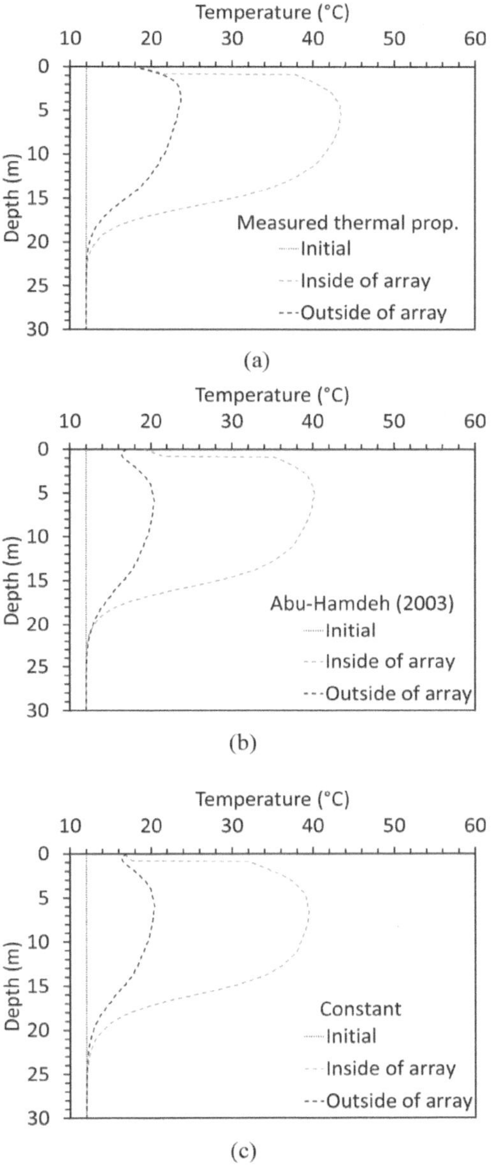

Figure 5. Temperature profiles inside (x, y = 1.25 m) and outside (x, y = 3.75 m) of the array (a) using the newly-proposed VHCF; (b) with Abu-Hamdeh (2003); (c) Constant thermal properties.

5 CONCLUSIONS

The effect of non-constant thermal properties on volumetric heat capacity in heat transfer modeling and the prediction approaches of volumetric heat capacity were investigated in this study. A 3D transient finite element model was built to understand the effect of specific heat capacity in unsaturated soil profiles. A particular soil properties those of Hopi silt was used in the coupled heat transfer and water transfer numerical modeling. Two prediction equations for volumetric heat capacity were investigated and compared with the actual laboratory measurements. These predictions then were employed in the numerical model. Initial numerical results concluded that using constant thermal properties in heat transfer modeling in the vertical heat exchanger arrays may result in different temperature distribution than the actual when the SBTES systems are designed. The temperature distributions in the SBTES array using the newly proposed VHCF were in good agreement than the results obtained using the common prediction equation.

ACKNOWLEDGEMENTS

Funding from National Science Foundation (NSF 1230237) is much appreciated. The opinions are those of the authors alone and do not reflect those of the sponsor.

REFERENCES

Abu-Hamdeh, N.H., 2003. Thermal properties of soils as affected by density and water content. Biosystems Engineering, 86(1). pp. 97–102.

Başer, T. & McCartney, J.S. 2015. Development of a full-scale soil-borehole thermal energy storage system. Proc. Int. Foundations Conference and Equipment Exposition (IFCEE 2015). ASCE. pp. 1608–1617.

Başer, T., Lu, N., & McCartney, J.S. 2015. Operational Response of a Soil-Borehole Thermal Energy Storage System. J. Geotech. Geoenviron. Eng., 10.1061/(ASCE) GT.1943-5606.0001432, 04015097.

Başer, T., McCartney, J.S. & Lu, N. 2016. Impact of a thermo-hydraulic insulation layer on the long-term response of soil-borehole thermal energy storage systems. GeoChicago 2016. ASCE. 1–10.

Bear, J. 1972. Dynamics of Fluids in Porous Media. Dover, Mineola, N. Y., 764 p.

Dong, Y., Lu, N., Wayllace, A., & Smits, K., 2014. Measurement of thermal conductivity 4 function of unsaturated soil by using a transient water release and imbibition method. Geotechnical Testing Journal, Vol. 37, No. 6, 2014, pp. 1–11.

Duarte, A.P.L., de Campos, T.M.P., Araruna Jr., J.T., Rocha & Filho, P., 2006. Thermal properties of soils. Unsaturated Soils 2006, pp. 1707–1718.

Fang H.Y. & Chaney R.C., 1983. Mechanics of soil-heat interaction and its thermal storage capacity, Recent Advances in Engineering Mechanics and Their Impact on Civil Engineering Practice, ASCE, edited by W.F. Chen and A.D.W. Lewis, ASCE, Vol. 2, 1983, p. 786–789.

Lu, N. & Dong, Y. 2015. A closed form equation for thermal conductivity of unsaturated soils at room temperature. Journal of Geotechnical and Geoenvironmental Eng. 141(6), 04015016.

Sibbitt, B., McClenahan, D., Djebbara, R., Thornton, J., Wong, B., Carriere, J., & Kokko, J. 2012. The performance of a high solar fraction seasonal storage district heating system – Five years of operation. Energy Procedia, 30: 856–865.

Smits, K.M., Sakaki, S.T., Howington, S.E., Peters, J.F., & Illangasekare, T.H. 2013. Temperature dependence of thermal properties of sands across a wide range of temperatures (30–70°C).VadoseZone Journal, doi:10.2136/vzj2012.0033.

van Genuchten, M.T. (1980). "A closed-form equation for predicting the hydraulic conductivity of unsaturated soils." Soil Sci. Soc. Am. J., 44(5), 892–898.

Zhang, R., Lu, N. & Wu, Y. 2012. Efficiency of a community-scale borehole thermal energy storage technique for solar thermal energy. Proc. GeoCongress 2012. ASCE. 4386–4395.

Challenges and Innovations in Geotechnics – Zhussupbekov (Ed.)
 ISBN 978-1-138-03007-7

Strengthening of pile foundation under dynamic loads by high-pressure injection

M.L. Nuzhdin
SPEC Company "O&F", Novosibirsk, Russia

L.V. Nuzhdin
Novosibirsk State University of Architecture and Civil Engineering (Sibstrin), Novosibirsk, Russia

ABSTRACT: Here is described the method for strengthening (and construction) of the pile foundation for the machines under dynamic loads or vibration-sensitive equipment, allowing the reduction of their vibration level. It involves injecting of cement-sand mixture into the ground under the foundation frame foot for hardening and involving additional volume of soil mass for collaboration with the foundation. This leads to reducing the performance problem of the pile foundation vibration under dynamic loading and in case of base excitation. The injection is performed at several points situated between piles or around the perimeter of the foundation frame. The penetration of injectors is performed through the specially provided holes, i.e. injecting conductors. The parameters of injecting works (the number of injection points and their placing according to the plan, high-altitude position of injection horizons, required volume of injected mortar, the sequence of injection, etc.) are set depending on the pile foundation construction, geotechnical conditions of the site, dynamic equipment operating mode and other factors.

1 INTRODUCTION

It is known that the parameters of the foundation vibration produced by dynamic loads or base excitation directly depend on the foundation rigidity and damping parameters as well as on the weight of the vibrating system composed of the foundation, the machine or constructions installed on the foundation and the "adjoined soil body" involved in the vibration. From technical perspective in general for reducing vibrations it is efficient to increase the foundation rigidity and the vibrating system weight. The use of pile foundation suffices for this purpose.

However, it should be noted that permanent loads are relatively low for pile foundations under machines with dynamic loads, engineering units, vibration-sensitive equipment and special constructions; so the number of piles even on very soft soil is set structurally based on the size of the foundation frame appointed in accordance with the dimensions of the placed equipment. In the process of construction of pile foundations under machines with dynamic loads and in the course of their operation, all permanent loads are transmitted to piles, as a result of which the soil under the foundation frame foot remains unconsolidated. In the process of vibration, coupling of the foundation frame foot with soil can be violated, and a microgap can be formed between them. As a result of this situation stiffness and damping parameters of the foundation are determined only by the relevant characteristics of piles in position. There is no resistance of the soil along the foundation frame foot when it is vibrating, the soil mass of the inter-pile space almost does not collaborate with the foundation.

There are a great number of scientific papers and engineering solutions dedicated to the problem of including of the soil in inter-pile space into dynamic loads and increasing the weight of the vibrating system. However, most of the proposed methods with certain advantages have some obvious drawbacks. First of all, it is the complexity of the proposed designs and labor intensity of their implementation at relatively low efficiency.

2 CONSTRUCTION OF PILE FOUNDATION WITH SUBSOIL CONSOLIDATION BY HIGH-PRESSURE INJECTION OF CEMENT-SAND MIXTURE

The construction method of pile foundations used under dynamic loading with the consolidation of the soil in the inter-pile space by high-pressure injection of cement-sand mixtures has been developed and introduced into construction practice in the Scientific Research Laboratory (SRL) of Dynamics of Bases and Foundations of the Novosibirsk State University of Architecture and Civil Engineering (Sibstrin).

The method involves injecting cement-sand mixture under the foundation frame for the purpose of

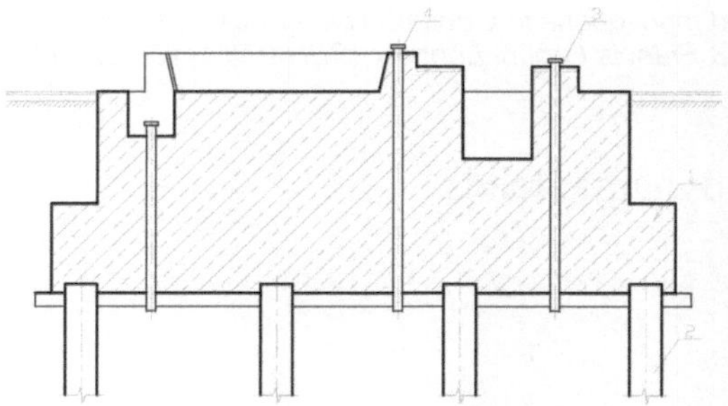

Figure 1. Layout of injection conductors in the foundation frame body: 1 – foundation frame body; 2 – pile; 3 – injection conductor; 4 – removable cover.

soil compaction, elimination of possible gaps between the foundation frame foot and the soil, and the inclusion of soil foundation into work under dynamic loads. Performing the complex of injection works provides the increase of overall stiffness of the pile foundation base and allows increasing the weight of the vibrating system, which reduces the foundation vibration level.

Also, if necessary, the solidifying mortar can be directly injected into inter-pile space, resulting in forming of solid injectable bodies interconnecting piles and, possibly, the foundation frame. This contributes to a further increase in the weight of pile foundation and stiffness of pile footing.

The mixture is injected in several injection points that are located between the piles and/or around the perimeter of the foundation frame (Fig. 1). The number of points and their location depend on the dimensions of the foundation frame, presence of technological space, etc. In this case, the distance between adjacent points in the average shall not exceed 2.5–3.0 m. The mixture is injected under the pressure, exceeding the structural strength of the soil. This leads to discontinuity of soil mass with the formation of tension cracks that are filled with the cement and sand mixture, followed by its hardening and formation of solid injection bodies. At each point the injection is carried out on one or several injection horizons; the number and high-altitude position of which depend on the site geotechnical conditions, the type of piles, the pile foundation structure and dynamic equipment operating mode. The injection process is controlled by changing the value of injection pressure and volume of the injected mixture. Injection work are usually stopped when the supply pressure is increased by 40–60% compared to the steady-state operating pressure, which indicates that the gaps have been eliminated – filled with cement and sand mortar and soil consolidation under the foundation frame foot. Furthermore, the works are completed when the volume of the injected mortar reaches 2.0–2.5 m^3 in one injection point on the same horizon.

Injection of mortar into inter-pile space under the foundation frame is carried through injectors, submerged in the pre-made holes in the body of foundation frame – injecting conductors.

Figure 2. Installation of the injecting conductor in the body of the pile foundation frame before concreting.

Conductors are industrially manufactured metallic or plastic pipes with the inner diameter sufficient to immerse injectors with minimum clearance. Injecting conductors are installed in the body of the foundation frame in the process of its concreting – they are fastened by binding wire or welding (for metal pipes) to reinforcement bars or formwork (Fig. 2). Conductors are normally arranged vertically with a slight elevation of one end over the top edge of the foundation frame and embedment of the other end into the subsoil under its foot sole or base concrete (Fig. 3).

The proposed method can also be used for strengthening of previously constructed pile foundations. In this case, even if the size of the foundation frame in plain view is relatively small, it is advisable to inject mixtures through injectors, submerged into the body of the foundation frame. This is due to inhomogeneous condition of the soil under the foundation frame after pile installation, the possibility of uncontrolled distribution of mortar in the soil beyond the foundation and as a consequence, the lack of the required soil consolidation in the inter-pile space under the foundation frame foot after injection of the designed mixture volume.

If it is impossible to make holes in the foundation frame body for injectors immersion, it is required to develop a special, more complex technology of injection works performance, which includes several stages of injection or execution of preliminary activities for

Figure 3. Foundation frame with conductors after concreting of its bottom part and formwork removal.

making subsoil screens, preventing the uncontrolled distribution of the injected mixture.

The proposed method has been used for reinforcement of several full-scale pile foundations, operated under dynamic loading. In particular, the subsoil of pile foundation of VPL-200-400 was reinforced to reduce the high level of vibrations. The foundation frame with the plain view dimensions of 1.4 × 2.45 m and 0.6 m in height was based on 4 of precast concrete piles S6-30 of 6.0 m. The foundation soils are represented by water-saturated sand clay of the consistency from sloppy to high-plastic. Weight of the installation is 45 kN, the power is 144 kW, and the operating frequency is 2980 rpm. The vibration parameters were registered before and after subsoil reinforcement. The measured amplitude of the VPL-200-400 foundation vibrations exceeded the limit values upon a number of indications, including sanitary and hygienic norms. About 1.0 m^3 of plasticized cement-sand mixture was injected into the upper zone of the inter-pile space for foundation reinforcement. As a result of the works performed, the vertical component of the vibration amplitudes decreased 1.2–1.6 times depending on the frequency range, while the horizontal vertical component decreased 1.5 ... 3.0 times and more (Fig. 4). After the reinforcement performed, the foundation vibration parameters became less than limit values.

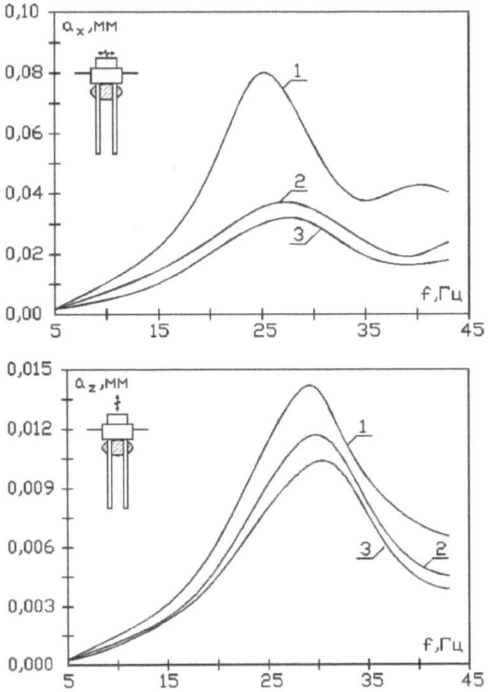

Figure 4. Reducing the pile foundation vibrations after the subsoil reinforcement by high-pressure injecting of the cement-sand mixture; horizontal (left) and vertical (right) components of vibrations: 1 – before reinforcement; 2 – immediately after the mixture injection; 3 – after its solidification.

3 CONCLUSION

Strengthening of pile foundation under dynamic loads by high-pressure injection of cement-sand mixtures through conductors for subsoil consolidation significantly (up to 1.5–2.0 times and more) reduce the level of vibrations from the dynamic loading, including base excitation. This is achieved by relatively low cost and technological simplicity of works.

REFERENCES

Nuzhdin, M.L. 2003. Prevention of uneven settlement of foundation plate using high-pressure injections. Experience of construction and reconstruction of buildings and structures on weak soils. *Materials of International Scientific and Technical Conference Arkhangelsk*, 119–122.

Nuzhdin, M.L. 2004. Application of injection conductors to prevent occurrence of excessive tilts of foundation plates using a method of group high-pressure injection. *Geotechnical problems of construction of large-scale and unique objects. Works of International Geotechnical conference. Almaty*, 351–354.

Nuzhdin L.V. and Nuzhdin M.L. 2009. *Patent for invention No. 2352723 Russian Federation*, Method to adjust vertical position of buildings and structures on foundation slab. Bulletin No. 11.

Nuzhdin, M. L. and Nuzhdin, L.V. 2010. Experience of application high-pressure injection method for erection buildings on plate foundation. From Research to Design in European Practice: *Proc. of full papers of the XIV-th Danube-European conf. on Geotechnical Engineering.* Bratislava: SUT, (on CD-ROM insert. – 6p.)

Nuzhdin, M.L. and Nuzhdin, L.V. 2011. Improvement of reliability of high-rise buildings on plate foundation. Proc. of the 14th *Asian Regional Conf. on Soil Mechanics and Geotechnical Engineering. Hong Kong*, (on CD-ROM insert. – 4p.)

Challenges and Innovations in Geotechnics – Zhussupbekov (Ed.)
© 2016 Taylor & Francis Group, London, ISBN 978-1-138-03007-7

New designs of the combined tape bases providing fuller use of the bearing ability of the basis

T.A. Krakhmal'ny, S.I. Evtushenko & M.P. Krakhmal'naya
Platov South-Russia State Polytechnic University (NPI), Novocherkassk, Russia

ABSTRACT: Results of experiments on studying of work of the tape bases with a broken outline of the support plate are given in article. A test objective – development of the new economic tape bases during which work in soil there is "an arch effect". Big efficiency of these designs is caused by increase in a ratio of perimeter of the base to its area (it leads to increase in coefficient of subgrade resistance); change of a contour of a regional zone that leads to emergence in the basis of "arch effect" and to fuller use of the bearing ability of foundation of the bases; besides, it is less zone of plastic deformations of the bases under the bases with a broken outline of support plates, than under the equivalent continuous bases.

1 INTRODUCTION

In recent years in SRSPU (NPI) experiments on studying of work of the tape bases with a broken outline of the support plate are made. A test objective – to develop the new economic tape bases during which work in soil there is "an arch effect". It is reached due to fuller use of the bearing ability of the basis and without decrease in the bearing ability of the base.

2 GETTING STARTED

Tests are carried out by the measuring machine MF-1 developed by professor Murzenko Yu.N., intended for model tests of the bases. She treats the category of vertical test measuring machine with the hydraulic drive. Three independent hydraulic jacks established on a frame over a tray allow to create loadings to 500 kN (i.e. 50 tons) on one jack. The subgrade is modelled by medium-grained dry sand of average density. The sizes of a sandy tray – 3.0 × 3.0 of m. Depth of filling with sand of – 1.8 m.

Loading is transmitted through three metal load traverses and put by steps on 30 kN then is kept by a constant 15 min. for stabilization of deformations in the basis. After each step of loading readings of sensors and measuring devices are taken. Sediment is measured by a progibomer 6-PAO it is also controlled by graphing of dependence of sediment on loading on a diagrammatic device of the MF-1 measuring machine (Murzenko (1965, 1969, 1972)).

The list of the base is controlled by a level. Unicomponent mesdoza measure normal tension in a contact layer of the sandy basis. Linear deformometra of D2 measure relative deformations ε_y, and shift deformometra of SD – γ_{xy}. Poll of sensors is carried out

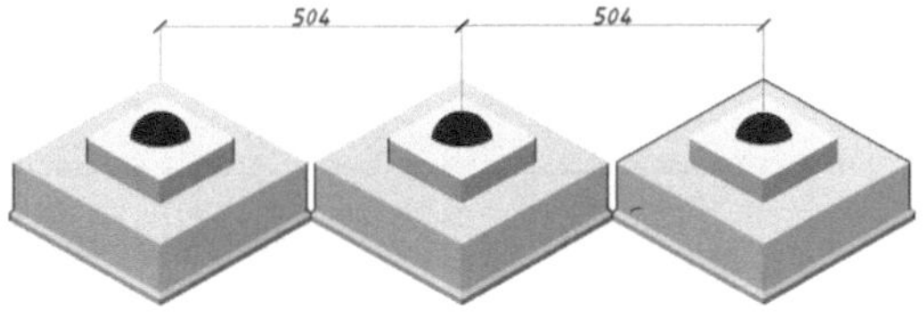

Figure 1. Scheme of placement of stamps with turn on 45°.

by the operator of personal computer of use of a complex of programs (Evtushenko & Krakhmal'ny (2004)) which allows to carry out automatic processing of indications of tensometric devices in real time. Experience comes to an end with either destruction of model, or destruction of the sandy basis, with a sand vapor.

3 RESEARCH INFLUENCE ON THE SUPPORTING CAPACITY AND SETTLEMENT OF SANDY FOUNDATION OF TURN OF STAMPS ON THE CORNER 45°

In the first series of experiences influence on the supporting capacity and settlement of the sandy basis of turn of stamps on a corner 45° has been considered.

The scheme of placement of stamps is shown in figure 1. In this series three experiments for obtaining frequency of indications have been made.

In the first experience (experience 1.1) the supporting capacity has made 391 kN (39.1 tons), that corresponds to average footing contact pressure of 1040 kPa. The maximum settlement of an average stamp has made 7.63 mm. 1.2 supporting capacity of the sandy basis are equal in the second experience to 394.0 kN (39.4 tons). Average footing contact

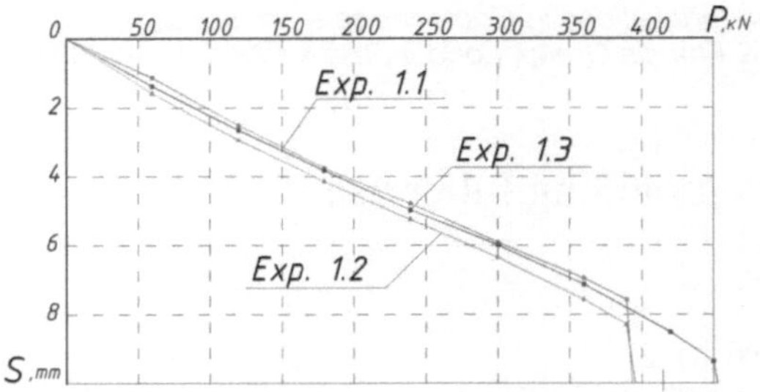

Figure 2. Schedules of dependence of settlement of an average stamp on loading in the second series of experiments.

Figure 3. The condensed core of the sandy basis (experience 3).

pressure of 1042 kPa. Settlement is equal to 8.18 mm. In experience 1.3 supporting capacity of the basis have appeared more, than in previous and has made 457.6 kN (45.7 tons) that corresponds to average footing contact pressure of 1217 kPa. In this experience resultant settlement of sandy foundation has also appeared more previous and has made 9.1 mm. Schedules of dependence of settlement of the central stamp on loading for the second series of experiments are given in figure 2.

When dismantling the basis of the third experience photofixing of the condensed core arising under the bases has been executed. The condensed core of the sandy basis, had a difficult outline in space. Appearance of a core is given in figure 3.

4 RESEARCH OF THE SUPPORTING CAPACITY OF THE BASIS AT THE BROKEN OUTLINE OF THE SUPPORT PLATE WITH THE INTERNAL ANGLE OF CUTS 60°

In the following experiences the supporting capacity of the basis at a broken outline of the support plate of the base with the internal angle of cuts was investigated 60°. The scheme of a base plate is submitted in figure 4. The total area of a sole of the base is equal to 0.492 m^2.

The model of the strip foundation was established on the center of a tray on the sandy basis.

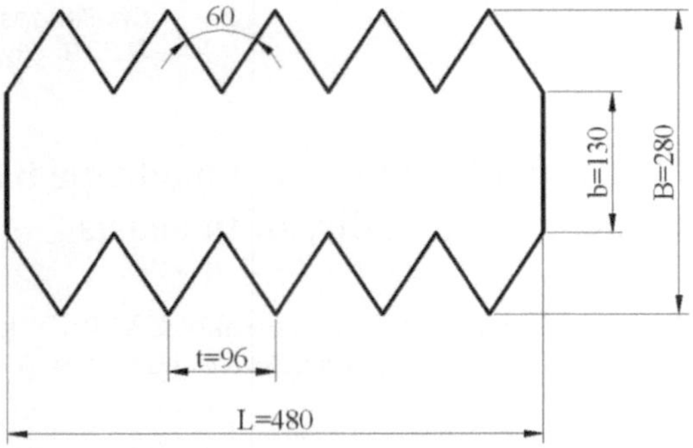

Figure 4. The sizes of a base plate with the internal angle of cuts 60°.

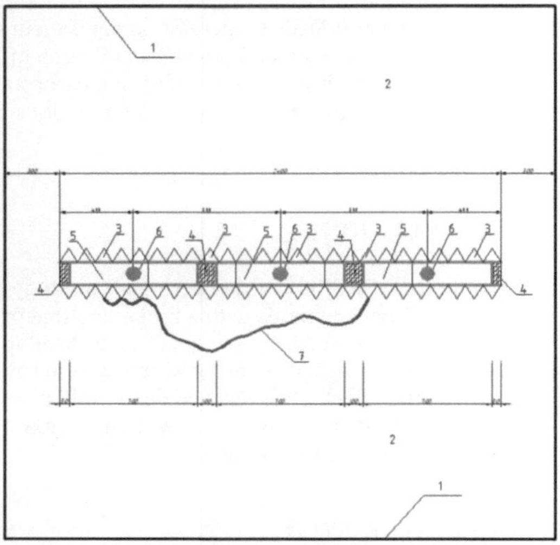

Figure 5. A look and the sizes of model in a tray 1. MF-1 measuring machine tray. 2. Sandy basis. 3. A plate with a broken outline of a regional zone. 4. The board distributing loading. 5. Load traverse. 6. Spherical support of a jack. 7. Basis vapor zone.

The general view of model in a tray of the measuring machine MF-1 is presented in figure 5.

In these researches four experiments have been made, first of which (experience 2.1) I was unsuccessful and at calculation of the average supporting capacity wasn't considered.

In the second experience (experience 2.2) the supporting capacity of the sandy basis has made 541.2 kN (54.12 tons) that corresponds to average footing contact pressure of 1100 kPa. The maximum settlement was measured on the center of model by means of a progibomer type 6-PAO and has made 9.10 mm.

In the following experience (experience 2.3) the supporting capacity was lower and has made 534.0 kN (53.4 tons) that corresponds to average footing contact pressure of 1085 kPa, and the maximum settlement has appeared more, than in the previous experience, and has made 9.62 mm.

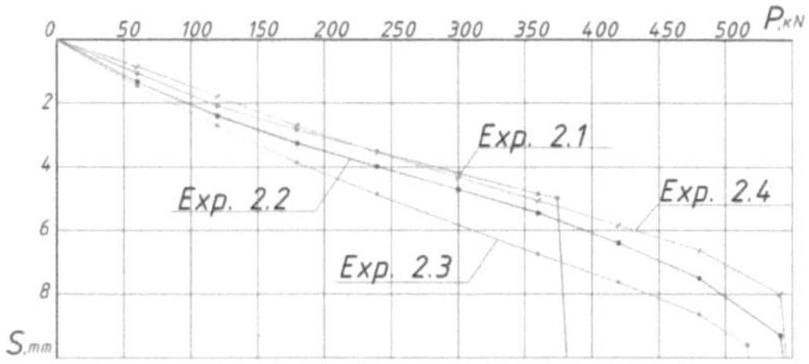

Figure 6. Schedules of dependence of settlement on loading.

In the last experience (experience 2.4) the greatest supporting capacity of 558.4 kN (55.84 tons) at the smallest settlement of 8.4 mm has been received. Average footing contact pressure has made 1135 kPa.

All made experiments came to an end with destruction of the basis that was followed by a sand vapor from under model and falling of loading on a power measuring instrument. The dispersion of the received values makes no more than 2.5% and is explained not by uniformity of consolidation of the sandy basis.

Schedules of dependence of settlement of model of the base on loading have been constructed according to indications of a progibomer 6-PAO and given in figure 6.

Average value of the supporting capacity of the basis by results of three experiences makes 545 kN (54.5 tons), and average footing contact pressure – 1107 kPa.

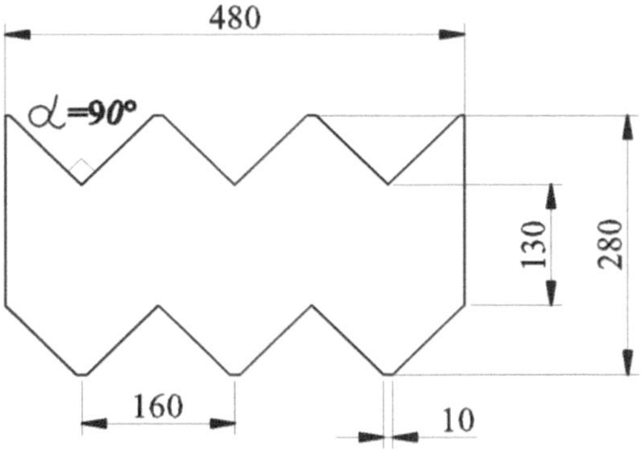

Figure 7. The scheme of a base plate with an internal corner 90°.

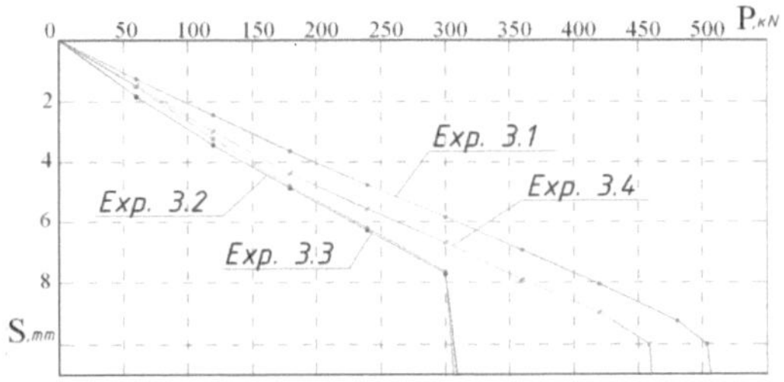

Figure 8. Schedules of dependence of settlement on loading.

5 RESEARCH OF THE SUPPORTING CAPACITY OF THE BASIS AT THE BROKEN OUTLINE OF THE SUPPORT PLATE WITH THE INTERNAL ANGLE OF CUTS 90°

Further the internal angle of triangular cuts – 90° has been changed. Five plates (figure 7) model the strip foundation of long 12 m. The area of all model remained invariable - 0.492 m^2.

In the first experience (experience 3.1) the supporting capacity of the basis was 510 kN (51.0 tons) that corresponded to footing contact pressure of the basis of 1036.5 kPa. Extreme settlement has made 9.96 mm. In this experience the greatest supporting capacity of model has been received.

In the following experience (experience 3.2) the supporting capacity of 384 kN (38.4 tons) has been received that corresponded to footing contact pressure of the basis of 780 kPa. The maximum settlemen of model has made 7.60 mm. Capsizing of model – at increase in uneven settlemen became result of experience, there was an eccentricity on extreme load a traverse, as has led to her falling.

In the third experience (experience 3.3) there was a similar situation with falling traverses. However, the maximum load has been recorded at the size of 387 kN (38.7 tons) that corresponds to footing contact pressure of the basis of 787 kPa. The largest settlemen of model of the base has made 7.64 mm.

For control the fourth experiment (experience 3.4) which has shown results similar with the first has been made. The extreme supporting capacity of the basis has made 471 kN, with a pressure to footing contact pressure of the basis of of 957 kPa. The maximum settlemen of the base, is comparable to the settlemen of the first experience and has made 9.98 mm. Schedules of dependence of settlemen on loading in all four experiences are given in figure 8.

Average value of the supporting capacity of model of the strip Foundation with the internal angle of cuts 90° by results of two experiences brought to exhaustion of the supporting capacity of the basis is equal to 490 kN that corresponds to footing contact pressure of the basis of 996 kPa.

6 ANALYSIS AND GENERALIZATION OF RESULTS OF EXPERIMENTAL STUDIES

Experimentally, in experiences with rigid square stamps, an established fact of increase in the supporting capacity of the basis at increase in perimeter of model of the strip Foundation. The supporting capacity of foundation of the strip Foundation at turn of stamps on a corner 45° on average increases by 18%, in comparison with linear installation of elements.

In experiences the difficult spatial form of the condensed kernel in the basis of a rigid stamp is recorded. With growth of depth, a zone of consolidation of the basis of three stamps are crossed, and the greater volume of the sandy basis gets into gear.

Researches on definition of an optimum internal angle of cuts are conducted. The strip Foundation with the internal angle of cuts in 60°, in comparison with the strip Foundation with the internal angle of cuts 90°, at the identical area of both models has supporting capacity the reasons 12.0% more.

The proposed constructive solutions of the extended and strip Foundation are fixed by patents for useful models (Murzenko et al. 2003, 2005a, b, c).

REFERENCES

Evtushenko, S. I. & Krakhmal'ny, T.A. 2004. Model of the tape base with a broken outline of the support plate.

Murzenko, Yu.N. 1965. The test measuring machine MF-1 for pilot studies of the bases and the bases.

Murzenko, Yu.N. 1969. Metodik of pilot studies of collaboration of the bases and the squeezed basis at static loading.

Murzenko, Yu.N. 1972. Experimental and theoretical researches of power influence of the bases and sandy basis.

Murzenko, Yu.N. & Evtushenko, S.I. & Anishchenko, E.Yu. & Arkhipov, D.N. & Krakhmal'ny, T.A. 2005a. Tape base: Patent for the useful U1 50552 RU 7 E 02 D 27/01 model. – No 2005119951/22; patent application 6/27/2005; publication 1/20/2006, Bulletin No 02.

Murzenko, Yu.N. & Evtushenko, S.I. & Murzenko, A.Yu. & Krakhmal'ny, T.A. 2005b. Retaining wall: Patent for the useful U1 49543 RU 7 E 02 D 29/02, E 02 B 3/06 model. – No 2005121969/22; patent application 7/11/2005; publication 11/27/2005, Bulletin No. 33.

Murzenko, Yu.N. & Evtushenko, S.I. & Murzenko, A.Yu. & Krakhmal'ny, T.A. 2005c. Tape base: Patent for useful model U1 55386 RU 7 E 02 D 27/01 model. – No 2005138664/22; patent application 12/12/2005; publication 8/10/2006, Bulletin No 22.

Murzenko, Yu.N. & Evtushenko, S.I. & Skibin, G.M. & Anishchenko, E.Yu. & Arkhipov, D.N. 2003. Tape base: Patent for the useful U1 32139 RU 7 E 02 D 27/01 model. – No 2003108928/20; patent application 4/3/2003; publication 9/10/2003, Bulletin No 25.

Challenges and Innovations in Geotechnics – Zhussupbekov (Ed.)
© 2016 Taylor & Francis Group, London, ISBN 978-1-138-03007-7

Properties of collapsible soil—a case study

Aisulu Zhakulina
Department of Civil Engineering, Karaganda State Technical University, Karaganda, Kazakhstan

Yerken Akhmetov
ACCL, Abu Dhabi Plaza Project, Astana, Kazakhstan

Ayan Akhmetov
Faculty of Physics and Engineering, National Research Tomsk State University, Tomsk, Russia

Altay Zhakulin
School of Trade and Economics, Guangdong University of Foreign Studies, Guangzhou, China

ABSTRACT: According to engineering-geological and hydrogeological zoning, territory of Kazakhstan can be divided into following parts: Folded areas of Southern Kazakhstan (Tang-Shan; Zaisan; Altai; intermountain, piedmont lowlands); Turan plateau (Piedmont valleys of Chu depression and plains of Shymkent and Karatau), where predominantly collapsible soils form a sufficient stratum. Because of complex regional engineering-geological prospecting and following generalization, research came up with the following; mainly collapsible soils represent lower quaternary wind deposits, loess loams. The analysis reveals that majority of structures in developed areas are built on quaternary soils. Quaternary soils are notable for the variability of physical properties, complexity, and diversity of mechanical characters. Therefore, collapsible soils were chosen as a target of the following research. Mountainous area around Almaty city taken as an experimental zone. The underground water of 40 meters deep is untapped in this region. However, laboratory tests reveal characteristics of loam as following: density – 15–16 kN/m^3; porosity – e > 0.9 and minor mechanical properties. The value of the total settlement is 8.8–73.51 cm. According to the complex seismic zoning map, Almaty city and neighbour areas are adjacent to seismic district "III-B-2". Section of given area represent in massive highly porous loess loam rested on gravel. According to seismic properties, following soil has the third category, because of high porosity coefficient (e > 0.9). Preliminary retaining wall design reveals that all northern slope prone to landslide. According to physical properties in full saturation stage, the most unfavourable against slope stability is collapsible loam. Retaining wall stability analysis carried out in PLAXIS 2D reveal that slope in a saturated state prone to landslide. For retaining wall stability, physical and mechanical properties in saturated state influence greatly.

1 INTRODUCTION

Surface level fluctuates within +910.000 to +985.835 meters above sea level; therefore, the difference in elevation is 69.835 meter. A geologic-lithological formation mainly include lower quaternary aeolian deposits. It is loess collapsible loams (Q_I) and upper quaternary alluvial-proalluvial deposits (arQ_{III}), which subsequently represented by top soil (Q_{IV}), loam soil and cobble soil. Brownish loam soil (Q_I) 21 meter deep is collapsible below that is not. Loam soil penetrated down to 40 meters (Zhakulin, 2015). Cobble soil with sand inclusions observed on North side and have following structure: pebbles 10–15%, cobble 50–55%, gravel 10–15%, and crushed stone 15–20%. Loess collapsible soils are prone to cause a slope slide under partial or full soil saturation due to sufficient design parameters degradation (cohesion, angle of internal friction and modulus of deformation). It is necessary to analyse collapsible soil slope's stress-strain state, and estimate stability and bearing strength of retaining structure for motorway security.

2 DEFECTS ANALYSIS OF RETAINING WALL

In the aftermath of various tests – slope, reinforced concrete retaining wall and storm water drainage system reveal numerous defects, predominantly, during the design stage. Following inaccuracies lead to the thread of motorway security breach. Monolithic reinforced concrete retaining walls with natural stone cladding has a height of 4.0–20.0 m, width of 0.6 m and bearing on plate 1.2 × 2.4 m and rigidly fixed to pile foundation. Slope subjected to atmospheric and storm water precipitations during a year. Discovered defects point to unacceptable cracks and inclination of retaining wall along the road (Figure 1).

Figure 1. Damaged part of retaining wall.

Table 1. Physical mechanical properties.

Designation	Units	Initial	Design
I_p	–	0.080	–
W	–	0.170	–
ρ	g/cm^3	1.50–1.60	–
ρ_s	g/cm^3	2.70	–
ρ_d	g/cm^3	1.47	–
e	–	0.84–1.1	–
S_r	–	0.55–0.61	1.0
I_L	–	–0.25	–
c	KPa	35	21
φ	degree	22	16
E	MPa	15	4.9

3 RESULTS OF COLLAPSIBLE SOIL INVESTIGATION

According to the investigation, following physical-geological phenomena observed on south regions of Kazakhstan: rainwash, gullying, settlement, seismic activity. Slope formed from loam soils with following properties: brownish colour, from hard to low plasticity, carbonized, up to a depth of 15.5–21.0 m collapsible. Physical-mechanical properties given in Table 1. Underground water is untapped at depth of 44.0 m. However, laboratory tests state following characteristics: density – 15–16 kN/m^3; porosity – e > 0.9 with low mechanical properties. According to compression tests, loam discovered at a depth of 15.5–21.0 m during the saturation lose its strength characteristics up to 50% (Figure 2). Initial settlement pressure deviates from 0.028 until 0.361 MPa (0.112). Coefficient of relative settlement under unit pressure of 0.05 MPa fluctuates within 0.001 to 0.056 MPa (0.014); under unit pressure of 0.1 MPa range is 0.001–0.064 MPa (0.023); under unit pressure of 0.2 MPa range is 0.001–0.105 MPa (0.406); under unit pressure of 0.3 MPa range is 0.019–0.113 MPa (0.059) (Figure 3). Calculations represent a value of total settlement is 8.8–73.51 cm. It is the Second type of soil against the collapsibility.

According to a complex seismic map of Almaty piedmont region and adjacent ones, investigation site

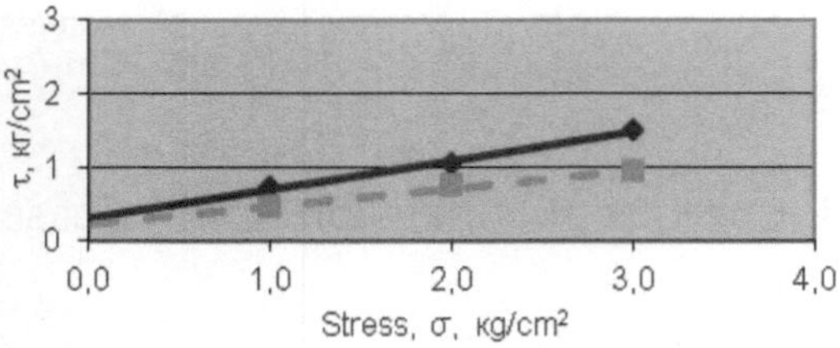

Figure 2. Graphics of strength parameters identification before and after water saturation.

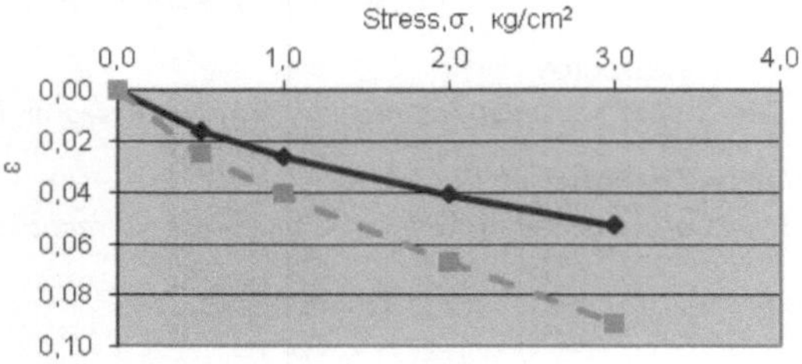

Figure 3. Graphics of deformational soil properties before and after water saturation.

located in seismic region "III-B-2" (upper piedmont step). Elevation of this step represented by thick mass (40 to 80 m) of highly porous loess loams rested on cobble. Subsequent to the analysis of seismic properties of soil they belong to the third category, because of high porosity (e > 0.9) coefficients.

4 RETAINING WALL DESIGN ANALYSIS

During stress-strain analysis of slope, stability interacting with retaining wall collapsible soil behaviour investigated. Herewith, slope's collapsible soils are characterised by physical-mechanical properties variability (porosity reduction, deformation modulus, cohesion, the internal angle of friction); stress condition change, due to relocation after saturation and strain tensor change, due to elastoplastic and viscoplastic deformations propagation (Maslov N.N. 1982, Malyshev M.V. 1994). Target stress state occurs under saturation process in collapsible soil, which differs from initial one. The problem was set in the elastoplastic environment using the Coulomb-Prandtl model. It has an elastic behaviour of the environment at a stress below the yield point, and equivoluminar plastic flow at stress at the yield point. Coulomb formula describes the stress at the yield point (Morgenstern, N.R. & Price, V.E. 1967, Terzaghi, K. and Peck R. B. 1967):

$$\sigma_{max} = S + \lambda\sigma_{min}, \quad (1)$$

where $\lambda = \cot^2(\pi/4 - \phi/2)$ – passive pressure coefficient; $S = 2C\cot(\pi/4 - \phi/2)$ – uniaxial compressive strength; σ_{max}, σ_{min} – maximum and minimum primary stresses.

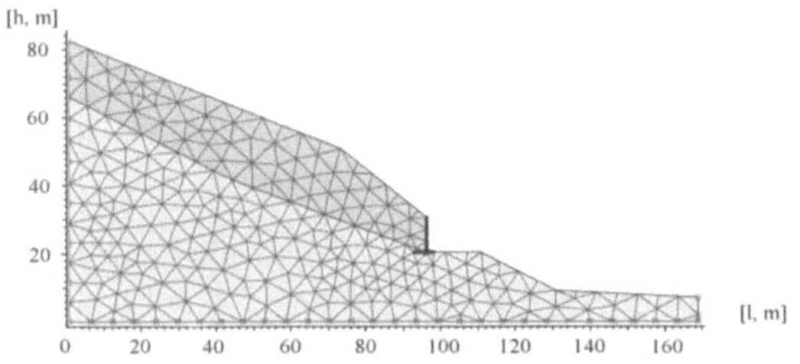

Figure 4. Calculation scheme.

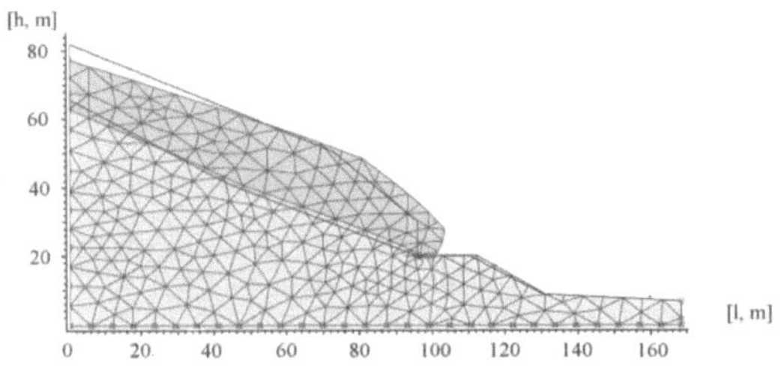

Figure 5. Deformed mesh (extreme total displacement $162.42^* 10^{-3}$ m).

At the region of yield criterion:

$$\sigma_{min} = -T, \tag{2}$$

where T – tensile strength is taken as $C/5$.

After the rupture propagation under stress $\sigma = -C/5$ further tensile strength analysis taken as zero ($T = 0$). The elasto-plastic solution obtained by the finite element analysis and initial stress method using Newton-Raphson iteration approach with constant stiffness matrix. However, with alternating load vector replenished during iteration process by "initial stresses" in plastic elements. Tolerance in FEM consists of discretization error, conditional object substitution with an infinite number of degrees of freedom of the model with a finite number of degrees of freedom, and rounding errors in performing computations in software (Brinkgreve R.B.J. et al. 1997, Fadeyev A.B. 1987, Paramonov V.N. 2012). Because of numerical analysis for the designated problem, we obtained: (Figure 3. Calculation scheme) distortion of finite element mesh (Figure 4), horizontal and general deformation of the slope under design characteristics change of collapsible soils (Figure 5 and 6), and the isolines of maximum shear stress (Figure 7) and the trajectory of collapsible soil movement (Figure 8) under stress-strain state modification of the slope on the retaining wall. Result analysis of numerical solution for modelling the interaction of slope and retaining wall shows that the most dangerous development of maximum horizontal deformation at the wall is due to shear deformations. The range of shear deformations covers large areas in slope. Collapsible soils settle under its own weight during water saturation, which leads to an initial value of cohesion, the angle of internal friction and soil modulus degradation, due to stress-strain states change. The path of collapsible

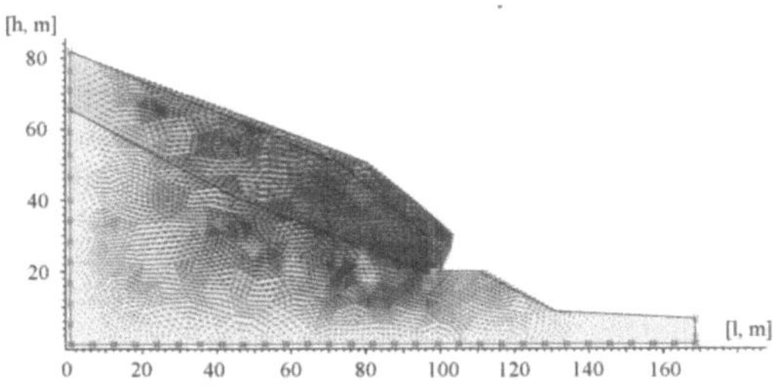

Figure 6. Horizontal displacements (U_x) (extreme – 0.162 m).

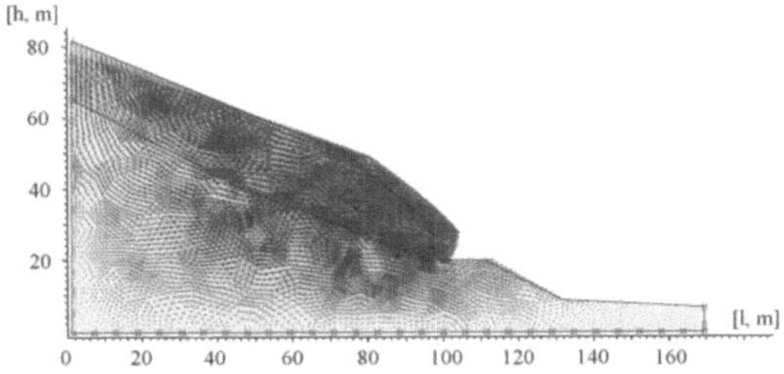

Figure 7. Total displacements (U_{tot}) (Extreme $U_{tot} = 0.165$ m).

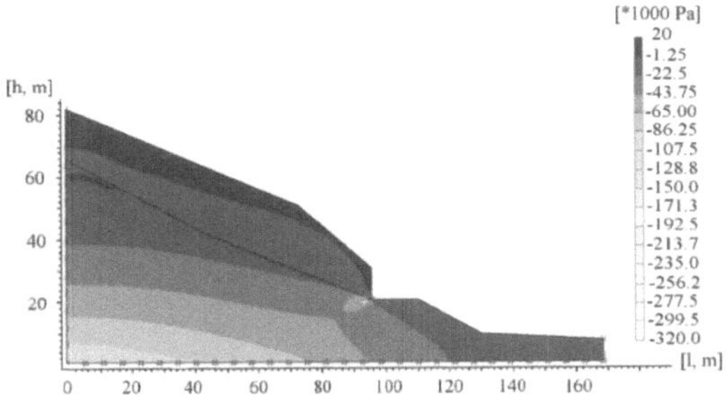

Figure 8. Horizontal total stresses.

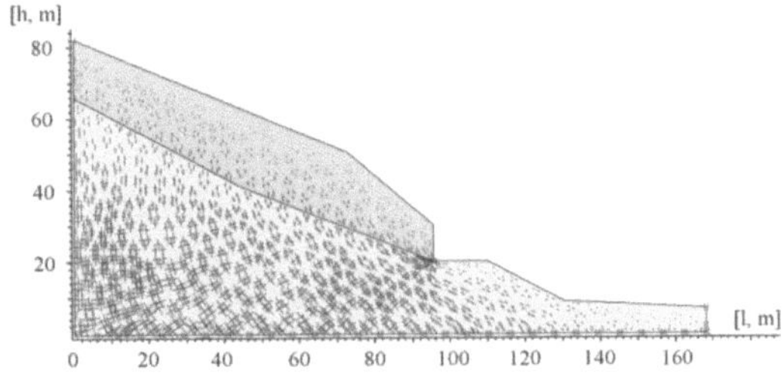

Figure 9. Total stresses (Extreme total principal stress -483.25^*10^3 Pa).

soil particles under the stress-strain state modification perform that movement of the soil particles takes place on the border in between the collapsible loam and non-collapsible loam. Calculations show that the retaining wall at the slope is not stable and the soil sliding occurs in the range of collapsible soil presence.

5 CONCLUSIONS

According to engineering-geological and hydrogeological zoning, territory of Kazakhstan can be divided into following parts: Folded areas of Southern Kazakhstan (Tang-Shan; Zaisan; Altai; intermountain, piedmont lowlands); Turan plateau (Piedmont valleys of Chu depression and plains of Shymkent and Karatau), where predominantly collapsible soils form a sufficient stratum. Because of complex regional engineering-geological prospecting and following generalization, research came up with the following; mainly collapsible soils represent lower quaternary wind deposits, loess loams.

Lower quaternary aeolian deposits, represented by loess-like loam (Q_I) is a part of the geological and lithological structures.

As the result of experimental tests, reinforced concrete retaining wall and drainage system survey, numerous defects that were committed in the process of design and construction revealed. In return, this damage involves a threat to the safe operation of the road and facilities.

According to the results of compression tests, loam that lies 15.5–21.0 m deep under saturation exhibits collapsible soil properties. The initial subsidence pressure varies from 0.028 to 0.361 MPa (0.112). Calculations show that the total value of the settlement is 8.8–73.51 cm.

Retaining wall stability design carried out by numerical approach reveal that slope under saturation prone to landslide. Results of numerical analysis during the interaction of slope and retaining wall modelling show that the most dangerous development of critical horizontal deformation at the face of retaining wall due to shear deformations at the slope.

REFERENCES

Brinkgreve R.B.J et al. 1997. PLAXIS, *Version 8.* Balkema.

Das, M. Braja. 1993. "Principles Geotechnical Engineering." *Third Edition. PWS Publishing Company*, Boston, p. 672.

Fadeyev A. B. 1987. Finite element analysis in geomechanics. "*Heдрà*". Moscow. p. 224.

Maslov N.N. 1982. Basics of engineering geology and soil mechanics. Moscow.

Malyshev M.V. 1994. Strength and stability of foundations. Moscow.

Morgenstern, N.R. & Price, V.E. 1967. A numerical method for solving the equations of stability of general slip surfaces. *Computer Journal,* 9, p. 388–393.

Paramonov V.N. 2012. FEM in nonlinear geomechanical problems. "Geostructure". *Saint Petersburg.*

Tarzaghi, K. and Peck R. B. 1967. Soil Mechanics in Engineering Practice. *Wiley*, New York.

Ukhov S.B. et al. 2002. Soil mechanics, basements and foundations. Moscow.

Zhakulin A.S. 2015. Foundation soil deformation under water saturation. *LAP Lambert Academic Publishing.* Saarbrucken.

Challenges and Innovations in Geotechnics – Zhussupbekov (Ed.)
© 2016 Taylor & Francis Group, London, ISBN 978-1-138-03007-7

Improvement of soft soil characteristics through application of dry concrete mix pillars

Tatiana Tronda
Belarusian National Technical University, Minsk, Belarus

ABSTRACT: The territory of Belarus is characterized by nonhomogeneous ground conditions. In some cases a soft soil layers can reach tens of meters in depth. Application of piles and other foundations is rather difficult and expensive in such complicated conditions. Alternatively dry concrete mix pillars can be applied for improvement of upper soft water-saturated soils. Application and field investigation results are described in the paper. Improvement of physical and mechanical characteristics was achieved for all kinds of soil. According to comparative calculations the cost of foundations on improved base can be 32% lower than the cost of conventional foundations.

Keywords: soft soils, improvement, characteristics, dry concrete mix pillars

1 INTRODUCTION

One of the effective improving methods for soft water-saturated soils is reinforcing upper layers by dry concrete mix pillars (Figure 1). It allows to drain and to compact natural soils at the same time. As a result strength and deformation characteristics of natural soils improve (Tronda 2013, Тронда 2014).

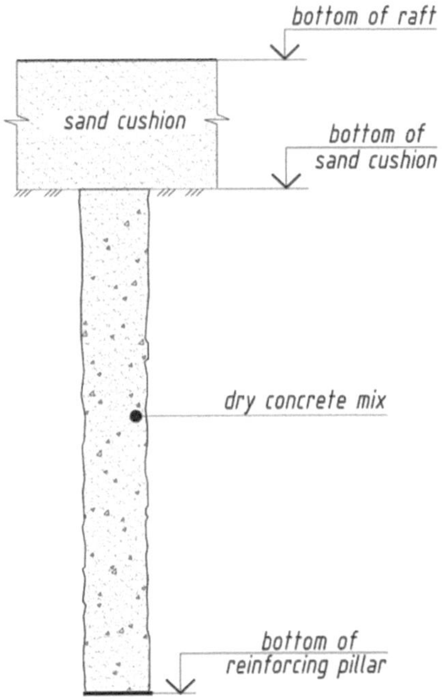

Figure 1. Draft of dry concrete mix pillar.

2 TECHNOLOGY

Reinforcing pillars are fulfilled in pressed wells (Figures 2, 3). Wells are made by well-casing with closed end. Well-casing is put into by vibrating hammer to design level. Then well-casing is filled with dry concrete mix and pulled out by vibrating hammer compacting concrete mix. Well pressing provides compacting of surrounding soils and dry concrete mix drains surrounding water-saturated soils improving its properties and speeding up the consolidation process.

Mass arrangement of dry concrete mix pillars is produced in two steps. In the first step each second pillar

Figure 2. Top of dry concrete mix pillar after filling the well.

Figure 3. Application of dry concrete mix pillars at the site.

is produced in staggered-order. Second step pillars are produced in three days after the first step. Then over reinforcing pillars sand cushion is applied to distribute loads from upper structures or building.

3 APPLICATION

Dry concrete mix pillars were applied at the site in Kolodischi. Since such kind of constructions has never been used in Belarusian practice and no regulatory documents have any reference of them, its availability was checked by series of experiments such as static penetration and static load tests.

Table 1. Soils and their characteristics at the site.

Soil	γ_{II}, kN/m^3	c_{II}, kPa	φ_{II}, °	E, MPa
1. Peat soil	12.9	–	–	–
2. Soft sandy clay	19.9	15	15	4.3
2a. Medium-strength sandy clay	20.2	22	18	11
3. Strength sandy clay	20.4	32	23	35
4. Medium-strength clay sand	21.0	12	19	8
5. Strength clay sand	21.2	18	28	37
6. Medium-strength medium sand	–	1	35	25
7. Medium-strength sandy clay	20.5	20	17	10
8. Sandy clay with organic inclusions	19.1	24	18	14
9. Medium-strength medium sand	–	1	35	27
10. Peat soil	14.0	31	7.5	9
11. Peat	11.3	13	5.7	15
12. Medium-strength medium sand	–	1	35	23
13. Medium-strength clay sand	21.2	15	24	16
14. Medium-strength sandy clay	21.0	25	20	17
15. Medium-strength fine sand	–	2	31	19

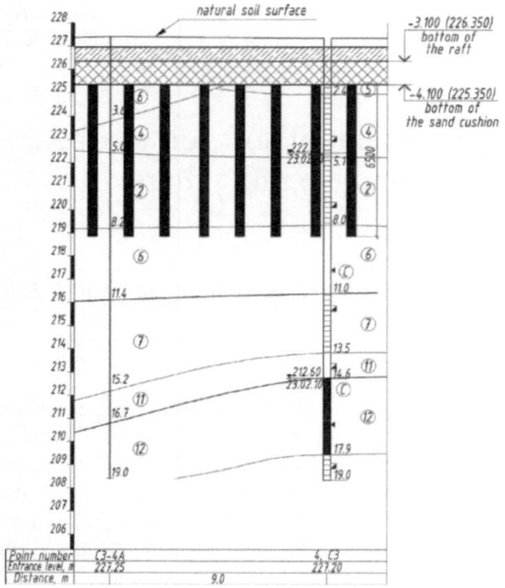

Figure 4. Fragment of engineering-geological section.

Ground conditions at the site are complicated (Figure 4, Table 1).

There is a layer of soft sandy clay under the base of designed foundations. The layer reaches a depth of 3.0 . . . 8.5 m. A layer of medium-strength sand, clay soils with organic inclusions and peat lie below. Then medium-strength sand is located at a depth of 12.5 . . . 16.7 m.

Application of sectional driven pile foundations was proposed as the first variant due to the big thickness of soft soils. Piles would be united by a raft under this project. The length of piles would be from 25 to 33 m. As an alternative variant the cast-in-place reinforced raft over sand cushion and reinforcing pillars were proposed. Cost-performance comparison of variants showed that the cost of raft with improved base is 32% lower than the cost of sectional driven pile foundations.

Reinforcing pillars were made according to the technology of shaking pressed piles with 426 mm diameter, 1.5 m step and 4 . . . 7 m length. The arranging of 1.0 m thick sand cushion was fulfilled layer by layer by shaking dynamic method.

4 TEST RESULTS

4.1 *Static penetration tests*

For the quality control of soil compaction between pillars the static penetration was carried out before and after pillar arrangement.

One of static penetration points is shown at Figure 5. Zones of influence under pillars noticed on the graphs come to 3 . . . 7 diameters of pillar.

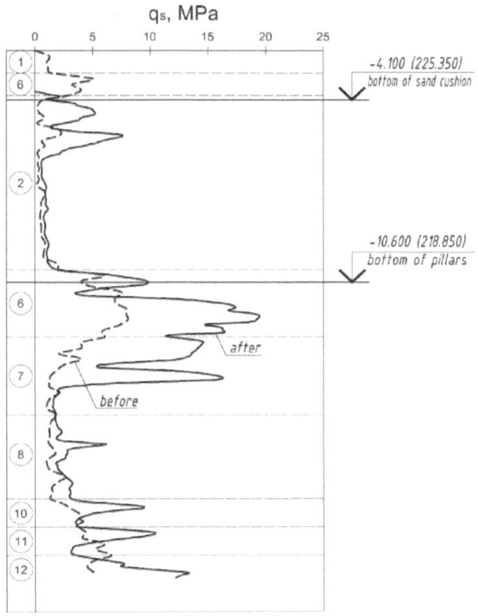

Figure 5. Toe resistance q_s, MPa before and after.

Table 2. Change of toe resistance q_s, MPa and Young modulus E, MPa for each layer at the site.

	q_s, MPa		E, MPa	
Soil	before	after	before	after
Within the length of pillars				
2	0.8	6.9	4.8	42.2
2a	3.3	5.5	20.2	33.5
4	1.5	6.5	8.6	37.3
6	4.3	7.4	19.7	33.8
Below the pillars				
6	6.1	12.0	28.1	55.1
7	2.0	4.3	12.4	26.0
8	1.9	3.4	10.9	19.2
10	3.6	4.8	9.3	12.2
11	4.6	4.9	14.4	15.5
12	4.6	7.0	21.0	32.3

Toe resistance q_s, MPa of soil within the length of pillars increased in 2.6 times, from 2.4 MPa to 5.7 MPa. Toe resistance of soil within the length of well increased in 2.1 times, from 3.2 MPa to 6.7 MPa.

Changes of toe resistance and Young modulus for each layer at the site are shown at Table 2.

Toe resistance and Young modulus calculated according to results of static penetration improved for all kinds of soil and increased in 1.1...9.8 times depending on the kind of soil and its location. The increase of toe resistance and Young modulus averaged 4.3 times within the length of pillars and 1.6 times below the pillars.

For soft soils within the length of pillars physical and mechanical characteristics were received (Tables 1, 4).

Table 3. Change of moisture content w, %, void ratio e and saturation S_r.

	w, %		e		S_r	
Soil	before	after	before	after	before	after
2	21.5	18.3	0.64	0.56	0.92	0.86
4	15.5	14.6	0.49	0.47	0.9	0.81

Table 4. Change of friction angle φ, °, cohesion c, kPa and Young modulus E, MPa.

	φ, °		c, kPa		E, MPa	
Soil	before	after	before	after	before	after
2	15	21	15	31	4.3	42.2
4	19	26	12	17	8.0	37.3

Physical and mechanical characteristics improved for soft sandy clay and medium-strength clay sand within the length of pillars. For sandy clay moisture content was reduced by 3.2%, void ratio in 1.14 times, saturation by 0.06, and friction angle was increased by 6°, cohesion in 2.1 times, Young modulus in 9.8 times. For clay sand moisture content was reduced by 0.9%, void ratio in 1.04 times, saturation by 0.09, and friction angle was increased by 7°, cohesion in 1.4 times, Young modulus in 4.7 times.

4.2 *Static load tests*

Static load tests were carried out after arrangement of pillars and compaction of sand cushion in order to determine Young modulus. Reinforced soils with pillars were tested under the slab 4.0 m^2. The results by three points showed increase of Young modulus on average up to 40 MPa.

5 CONCLUSIONS

Application of dry concrete mix pillars is effective and allows reaching significant soft soil improvement. Improvement of physical and mechanical characteristics was achieved for different kinds of soil.

Results of static penetration showed the decrease of moisture content by 0.9...3.2%, void ratio in 1.04...1.14 times, saturation by 0.06...0.09, and the increase of friction angle by 6...7°, cohesion in 1.4...2.1 times, toe resistance and Young modulus in 1.1...9.8 times depending on kind of soil and its location.

According to static load tests Young modulus reached 40 MPa.

The cost of foundation with improved base by dry concrete mix pillars was 32% lower than the cost of usual pile foundations.

REFERENCES

Tronda, T. 2013. Actual experience with the geomassif in complicated soil conditions. *Advances in soil mechanics and Geotechnical engineering; Proc. 5th iYGEC 2013, Vol. 2, Paris, 31 August – 1 September 2013*. Amsterdam: IOS Press.

Тронда, Т.В. 2014. Упрочнение слабых грунтов элементами из сухой бетонной смеси. *Современные геотехнологии в строительстве и их научно-техническое сопровонсдение; сб. трудов менсдунар. науч.-тех. конф., часть I, Санкм-Петердург, 5–7 февралR 2014 г.*: 221–224. Санкм-Пемердург: СПбГ АСу.

Challenges and Innovations in Geotechnics – Zhussupbekov (Ed.)
© 2016 Taylor & Francis Group, London, ISBN 978-1-138-03007-7

Laboratory modeling of soil dam reinforced by geosynthetic material

R.E. Lukpanov
Department of Civil Engineering, Eurasian National University, Astana, Kazakhstan

ABSTRACT: The paper presents the results of the laboratory modeling of soil embankment which had been subjected to the static loading, as well as horizontal and vertical strains. The research had made by the necessity of making engineering solutions for the reconstruction of the existing dam, located on the territory of a metallurgical plant in the Karaganda region, Kazakhstan. Model tests conducted on a scale of 1:40. In the scopes of research program, two models of soil dam were tested: without reinforcing elements and with reinforcement elements presented by geosynthetic (geogrid). The test results are presented in graphical and tabular form. The qualitative and quantitative analysis of the results is presented in this paper.

Keywords: soil dam, geosynthetic, equivalent material

1 INTRODUCTION

The object of the research is a dam of the thermal power plant slurry tank staying on the huge metallurgical plant of Karaganda region, Kazakhstan. Nowadays slurry tanks are out of functionality, nevertheless soil structures are required proper control and service as the objects of special dangerous.

Periodic accidents often lead to serious consequences – an abnormality on the power plant, environmental pollution, flooding large areas, etc [1].

Since 1957, there have been many repair work to strengthen the slope of the dam. The main decision was refilling of the dam slope broken structure. This decision did not give proper result, periodic restoration measures of the dam lead to economic losses to the thermal power plant. There is a question arose about the dam reconstruction with the use of technologies to maximize serviceability of the dam. Figure 1 shows the collapse of the slope of the dam recorded during survey in 2006.

During geological investigation and monitoring of the dam it was revealed: ways of a possible filtration, hydrogeological structure of foundation and dam soil, tendency of soil movements, mechanical and physical properties of soil.

The general reason of the soil dam collapse is a suffusion effect, particularly reasoned by the groundwater and meltwater migration, removal of the foot's soil, and appearance of the tensile or shear stresses exceeding the limit of tensile or shear strength of the soil.

Previous solutions of the foundational soil reconstruction are not effective as long as the risqué of dam soil movement after reconstruction will still exist.

Figure 1. Land sliding of the dam (recorded in June 2006).

This project includes additional preventive measures for the stability of dam with using reinforced elements as geogrid. The initial condition for the slope stability analysis is more potential direction of soil movement (Figure 2).

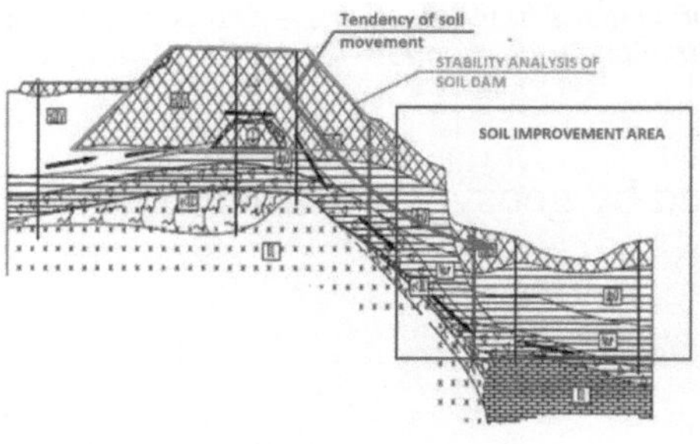

Figure 2. General concept of dam stability.

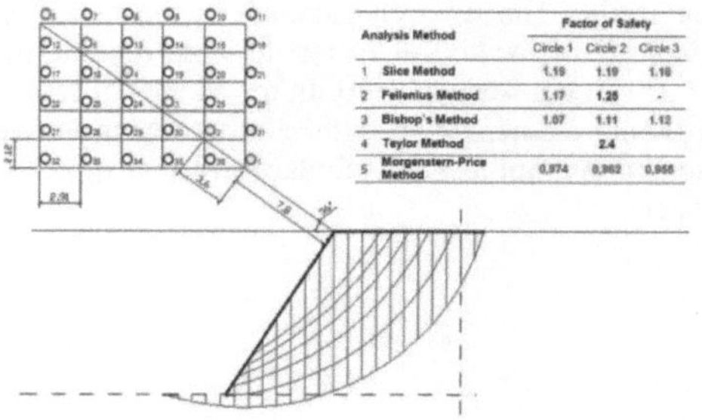

Figure 3. Slope stability analysis results.

Forecast of slope failure is very important part of design and so preventing possible failure, collapse or crumble of slope, seems to be very important part of stability analysis concern to the critical slip circle definition. Forecast of slope failure had been made by different methods and results of critical slip circle analysis shown in Figure 3. The factor of safety is less than required by standard Fs = 1.2, that is means that the dam stays in unstable condition [2].

2 LABORATORY MODELLING

The model materials are chosen by Newton's law of the dynamic similarity with taking into account simultaneous action of gravity and internal stresses. To definition of required parameters of equivalent materials the law of dynamic similarity is transforming to the following form:

$$N_m = \frac{i}{J} \cdot \frac{\gamma_m}{\gamma_n} \cdot N_n \qquad [1]$$

Where γm, γn – unit weight of model and natural soil, kN; i, J – linear dimension of model and natural dam, m; Nm, Nn – value corresponding to different characteristics of the model and nature.

For the dam model materials (equivalent material) takes a mixture composed of 97% of fine quartz sand and 3% of spindle oil by weight. Spindle oil allows modeling cohesive soil. Before the test all the walls of stand had been oiled for prevention friction effect with soil.

Table 1. Characteristics of natural and model dam.

Soil and model material	γ, (kN/M^3)	C, (kPa)	φ(^{deg})	E, (MPa)	ν
Natural soil (Loam)	19,0	38	38	27	0,35
Equivalent material	17,7	0,90	21	0,26	0,25

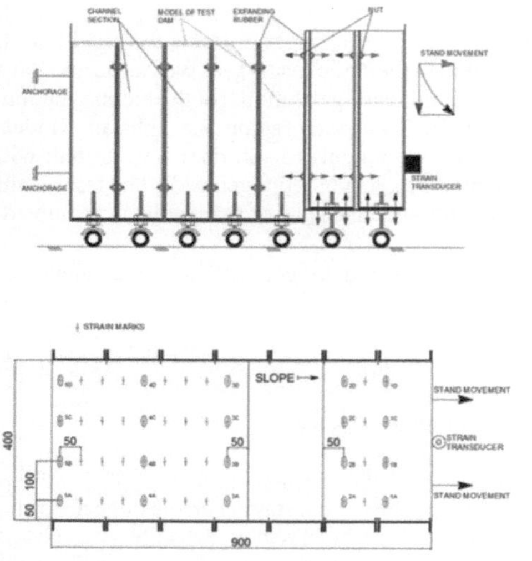

Figure 3. Model stand.

The properties of natural soil dam and equivalent material of model are presented in Table 1.

The axial strength of one rod of natural geogrid is 51 kN/m, then axial strength of 1 m reinforcement is follow: 51·5 = 255 kN/m (where 5 is a number of rods per 1 m of natural dam). Final parameters of the model geogrid are follow: diameter of the rod is 0.8 mm, cross section is 0,524 · 10-6 m, and axial strength of equivalent 1/40 m of model is EA = 100,48 kN/m (where 1/40 is a scale of natural to model dam [3]).

The scheme of testing stand is presented in Figure 4. The stand preseted by the group of steel elements, and the joints of this elements presented by hinges. The foundation soil and filling soil of dam had been made by consequent interlaced soil compaction. The sickness of layer up to 5 cm, the density of each layer had been controlled (13 g/cm3). Reinforced dam include 3 layers of reinforced net made by polyethylene with a holes 1x1 cm. The dimensions of the model dam are follows: height is 430 mm that is correspond to the natural dam height of 17.2 m and angle of slope is 45 degree. After preparation of model dam strain transducer had been fixed on the front of stand to control stand movement. Then strain marks had been installed on the top of dam and on the foundational soil with a view of strain monitoring (Figure 3).

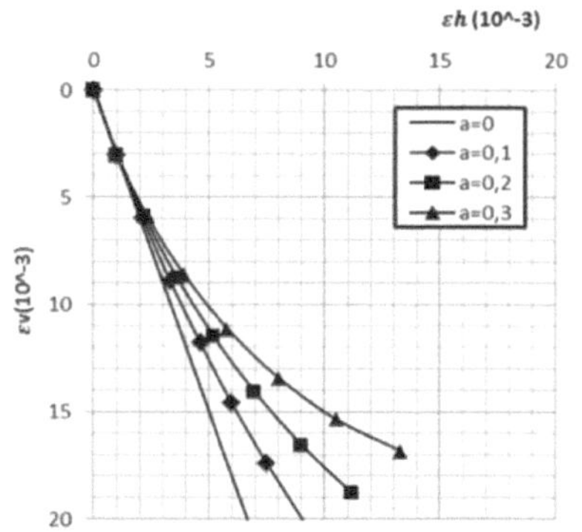

$$\varepsilon_h^n = \varepsilon_h^1(1 + a(n-1)) + \varepsilon_h^{n-1}$$

$$\varepsilon_v^n = \varepsilon_v^{n-1} + \sqrt{(\Delta\varepsilon_h^{n-1})^2 + (\Delta\varepsilon_v^{n-1})^2 - (\Delta\varepsilon_h^n)^2}$$

ε_h^n - horizontal strain of n step;
ε_h^1 - vertical strain of the 1 step;
ε_v^n - vertical strain of n step;
$\Delta\varepsilon_h^n$ - increment of horizontal strain of n step;
$\Delta\varepsilon_v^n$ - increment of vertical strain of n step;
a – intensity factor of increment horizontal strain to the vertical (for a = 0, the ratio of the horizontal to the vertical movement is a constant)
n – number of step.

Figure 4. Definition of horizontal and vertical increments.

Table 2. Increments of horizontal and vertical strain.

Step	Horiz. Strain ε_h^i, %	Vert. Strain ε_v^i, %	Total displ. l_{total}^i, mm
1	$1 \cdot 10^{-3}$	$3{,}0 \cdot 10^{-3}$	12
2	$2{,}3 \cdot 10^{-3}$	$5{,}9 \cdot 10^{-3}$	25
3	$3{,}9 \cdot 10^{-3}$	$8{,}6 \cdot 10^{-3}$	37
4	$5{,}8 \cdot 10^{-3}$	$11{,}1 \cdot 10^{-3}$	49
5	$8 \cdot 10^{-3}$	$13{,}4 \cdot 10^{-3}$	62
6	$10{,}5 \cdot 10^{-3}$	$15{,}3 \cdot 10^{-3}$	74
7	$13{,}3 \cdot 10^{-3}$	$16{,}8 \cdot 10^{-3}$	86

Totally two series of tests had been made: for reinforced and unreinforced dam.

During the washing out the soil on the dam base the voids are appearing at the foot, hence the trajectory of the initial displacement of the dam soil more vertical (by the principle of the path of least resistance) Namely, in the beginning destruction of existing strength has a place and vertical movement exceed horizontal, then destructed soil creeping character, where prevail horizontal movement. The first predetermined step or displacements (both vertical and horizontal) characterized by the tendency of the resultant displacement (total displacement) to vertical (instantaneous collapse of dams due to of destruction of structural strength), then the ratio of vertical to horizontal displacement changes.

Thus, trajectory of predetermined displacements should be finding from the conditions of prevailing horizontal movement at each subsequent stage, wherein total displacement remains constant for each steps. To meet the above conditions it was suggested simple equations for increments of horizontal and vertical movement, the essence of which is described in Figure 4. The increments of horizontal and vertical strain by step are presented in Table 2.

Figure 5. Experimental stand of dam modeling.

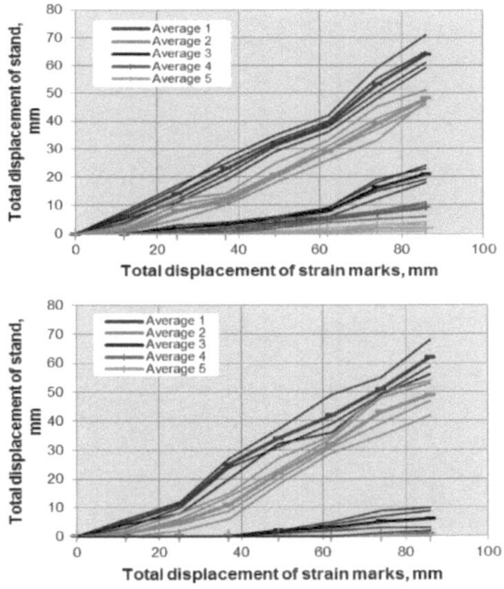

Figure 7. Results of the test.

The experimental stand of reinforced and unreinforced dam model is presented in Figure 5.

The results of strain marks displacement presented in Figure 6. Proection of each marks displacement (and average from 1 to 5, see Figure 3) to horison presented in Figure 6A for unreinforced and 6B for reinforced dam. Comparison of reinforced and unreinforced dam displacement presented in Figure 7.

From the diagrams is follow that reinforced elements essentially resist to horizontal movement of soil dam remaining dam is stable condition.

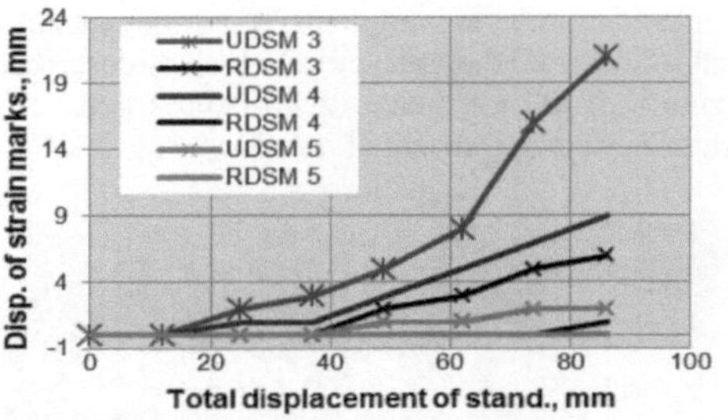

UDSM – unreinforced dam strain marks,
RDSM – reinforced dam strain mark

Figure 8. Results of the test.

CONCLUSION

Two series of test for reinforced and unreinforced soil dam performed. Model dam had subjected to the potential movement of foundation soil which described by gradual increase of lateral strain. Maximum value of soil dam displacement occurred on the crest 21 mm (strain mark 3) for unreinforced and 6 mm for reinforced model. Strain analysis of each strain marks do not shown critical points of strain distribution, the dam is slowly moved toward the stand movements, no any crumble effects was observed. In any case this model test shown retaining effect of reinforced elements, which should to be recommended for the following strengthening of existing soil dam.

REFERENCES

SN 522–79 Instruction for design and construction hydrogtechnical constructions on reclaimed territories.

SP 21.13330.2012 Buildings and constructions on reclaimed and sediment soil.

Tanaka T., Zhussupbekov A., Aldungarova A. and Lukpanov R.. Model test on the stability of the dam model with horizontal and radial deformations of the subgrade. 6th International geotechnical symposium on Disaster Mitigation in Special Geoenvironmental Conditions. January 2015, Indian Institute of Technology, Madras, Chennai, India, p. 375–378.

Challenges and Innovations in Geotechnics – Zhussupbekov (Ed.)
© 2016 Taylor & Francis Group, London, ISBN 978-1-138-03007-7

Estimation of liquefaction potential: A case study in San Francisco, California

A. Mussa, E. Satekenova, D. Zhang, S. Thiyyakkandi & J.R. Kim
Nazarbayev University, Astana, Kazakhstan

ABSTRACT: One of the main issues with the infrastructure construction in seismic regions is liquefaction. This paper reports on a case study of liquefaction evaluation of a proposed building site in San Francisco, California. The risk of liquefaction was assessed considering the following site specific factors: ground water table, liquefaction hazard maps, soil type, soil density, and peak ground acceleration. The analysis revealed that the selected region has low risk of liquefaction. However, it is suggested that the uncertainty associated with the estimation of corrected SPT N value (e.g., spatial variability, other human errors, etc.) should be accounted by incorporating an additional site-specific reduction factor in the estimation of Cyclic Resistance Ratio (CRR) of the soil.

1 INTRODUCTION

Liquefaction is a phenomenon of losing the strength of soil mass during an earthquake, which is a serious concern of infrastructure development in seismic prone areas. It typically occurs in saturated granular soil condition; where the pore water pressure increases significantly during earthquake, which causes the soil particles to separate from each other (i.e., effective stress becomes zero) leading to the failure of soil mass. Therefore, site-specific liquefaction susceptibility analysis should be performed for each construction site in the seismic region before the design phase of a project. This paper presents the preliminary liquefaction potential assessment for a proposed 12 stored building in San Francisco, California, which is historically a highly seismic zone. Standard Penetration Test (SPT) based approach is considered in the present study. The influence of spatial variability of SPT blow counts (N) on the liquefaction assessment is also investigated.

2 SITE DESCRIPTION

The proposed construction site is located in San Francisco Bay Area. The soil exploration information from an adjacent site reported by ENGEO Inc. was used for the preliminary liquefaction assessment. Simplified soil profile and relevant properties used for the present analysis are given in Table 1.

As shown from Table 1, soil condition was characterized as predominantly clay and silt with a thin layer of sand at approximately 11.5 m depth.

Table 1. Soil profile (ENGEO, 2012).

WT	Depth (m)	Soil Type	Unit weight (kN/m^3)	Cu (kN/m^2)
	2.5	Clay	19.8	129
∇	2.3	Clay	16.8	31
	3	Clay	20.6	122
	3	Clay	17.6	64
	0.6	Sand	20.8	–
	15	Clay	19.8	104
	22	Silt	20	–

3 LIQUEFACTION ASSESSMENT PROCEDURE

There are a number of methods for assessing liquefaction potential in granular soils and most of them rely on empirical correlations and graphs (Youd et al., 2001; Seed et al., 2003; Idriss and Boulanger, 2008). The factor of safety (FS) against liquefaction can be estimated as the ratio of cyclic stress ratio (CSR) to cyclic resistance ratio (CRR), where CSR is a function of earthquake loading (peak ground acceleration) and CRR depends on the soil properties (soil strength). The most commonly used SPT based liquefaction approaches for granular soils are Youd et al. (2001) method, Seed et al. (2003) method and Idriss and Boulanger (2008) method.

In case of fine grained soil, Wang (1979) reported that any clayey soil containing 15–20% particles (by weight) smaller than 0.005 mm and with water content to liquid limit ratio (w_c/LL) > 0.9 is liquefiable. Later, Seed and Idriss (1982) added one more criterion that liquid limit (LL) should be less than 35. These criteria are commonly known as "Chinese criteria". In 2006,

Bray and Sancio showed that "Chinese criteria" is not conservative and stated that the fine grained soil can liquefy if plasticity index (PI) < 12 and w_c/LL > 0.85. In addition, clayey silts and silty clays of moderate plasticity ($12 < PI < 18$) and w_c/LL>0.8 can also liquefy under cyclic loading. However, soils with PI > 0.8 are generally not susceptible to liquefaction.

Recently, California Department of Transportation (Caltrans) has proposed a procedure for liquefaction assessment based on past studies discussed above (Caltrans Geotechnical Manual, 2014). There are 6 main steps through which the estimation of preliminary screening of soil liquefaction is performed.

a. Ground Water Table
b. Liquefaction Hazard Maps
c. Age of Soil
d. Soil Type
e. Soil Density
f. Peak Ground Acceleration

Following is the procedure for liquefaction analysis as per Caltrans Geotechnical Manual (2014):

Step 1: Determine the ground water evaluation. The site is potentially liquefiable if ground water table (GWT) is within top 15.2 m (50 ft.).
Step 2: Identify soil type: sand or fine grained soil.
Step 3: Apply correction for SPT blow counts using equations suggested by Youd et al. (2001).

$$(N_1)_{60} = N_m C_N C_E C_B C_R C_S \tag{1}$$

where,

N_m is measured SPT blows.

C_N is depth correction factor: $C_N = \left(\frac{P_a}{\sigma'_{vo}}\right)^{0.5}$, where $P_a = 1$ atm and σ'_{vo} is effective overburden pressure.

C_E is hammer energy correction factor:

$C_E = ER\%/60\%$, where ER is rod energy ratio (for auto trip hammer: 80–140%).

C_B is borehole diameter correction factor (present case, C_B is 1).

C_R is rod length correction factor:

$C_R = 0.05L + 0.6$ for $4m < L \leq 6m$;

$C_R = -0.0004L^2 + 0.017L + 0.83$ for $6m < L < 20$ m;

$C_R = 1.0$ for $L \geq 20m$ and $L \leq 4m$, where L is depth.

C_S is correction factor for samplers with or without liner (present case, C_S is 1).

Step 4: Estimate Cyclic Stress Ratio (CSR) using the following equation (Youd et al., 2001):

$$CSR = 0.65 a_{max} (\sigma_o/\sigma_o') r_d \tag{2}$$

where, σ_o, σ'_o = total and effective vertical overburden stresses, a_{max} = peak horizontal acceleration, r_d = stress reduction coefficient:

$r_d = 1 - 0.00765z$, where, z is depth in meters.

Step 5: Apply correction to $N1_{60}$ (step 4) for fine content using the following equation (Youd et al., 2001):

$$(N1)_{60cs} = \alpha + \beta (N1)_{60} \tag{3}$$

where, $(N1)_{60cs}$ = the blow count corrected with fine content, and α, β = coefficients that are function of fines content. For example, %Fine larger than 35%, $\alpha = 5$ and $\beta = 1.2$ (Xiao, 2015).

Step 6: Determine Cyclic Resistance Ratio (CRR) with the equation given below (Youd et al., 2001).

$$CRR_{7.5} = \frac{1}{34-(N1)_{60cs}} + \frac{(N1)_{60cs}}{135} + \frac{50}{[10((N1)_{60cs})+45]^2} - \frac{1}{200} \tag{4}$$

Step 7: Calculate Magnitude Scaling Factor (MSF) (Youd et al., 2001).

$$MSF = 10^{2.24}/M_w^{2.56} \tag{5}$$

where, M_w is usually 7.6.

Step 8: Finally, calculate the factor of safety (FS) against liquefaction.

$$FS = \left(\frac{CRR_{7.5}}{CSR}\right) MSF \tag{6}$$

If the Factor of Safety is less than 1, it can be considered that the soil is susceptible to liquefaction affect.

4 PRESENT LIQUEFACTION ANALYSIS

The Ground Water Table in the proposed site is 3.26 m below the ground surface. According to the first step in the Caltrans procedure, the site may be susceptible to liquefaction. Therefore, further analysis (steps 2 through 7) for liquefaction is required.

In case of clay and silt layers, PI is greater than 18 and therefore, these layers can be considered non-liquefiable as per Bray and Sancio (2006). Accordingly, the remaining steps were only performed for sand layer.

In order to estimate the cyclic stress ratio (CSR), peak horizontal ground acceleration (a_{max}) is to be determined first. *The proposed site* is categorized as class D, *and* peak ground acceleration (PGA_M) *was estimated* using Equation 1.

$$PGA_M = F_{PGA} PGA \tag{7}$$

where, PGA_M = peak ground acceleration adjusted for site class effects, PGA= peak ground acceleration, and F_{PGA} = site coefficient that can be obtained from ASCE 7–10. For the location of proposed building, PGA is 0.6g, and F_{PGA} is 1 (ASCE 7–10). Consequently, PGA_M is 0.6.

Maximum spectral acceleration for the proposed site was calculated using "Equivalent-linear Earthquake site Response Analyses" (EERA) excel add in. For the analysis, the parameters such as time step, desired maximum acceleration, maximum frequency cut-off, and number of point for FFT should be entered for a particular earthquake. The present study considered Loma Prieta earthquake occurred in Northern

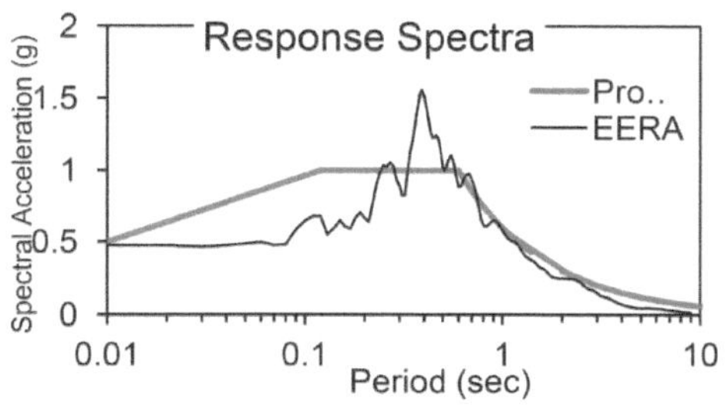

Figure 1. Response spectra.

California with magnitude 6.9 Richter scale in 1989. The required soil parameters included soil type, layer thickness, unit weight, shear wave velocity, and location of water table. Ratio of critical damping was assumed as 5%. Using these information response spectra was obtained from EERA software, which was then scaled in order to fit design response spectra of the proposed site. Figure 1 shows a comparison of design response spectra with that obtained using EERA. Note that the design values for ground conditions of the proposed site were based on ASCE 7-10. As can be seem from Figure 1, the maximum spectral acceleration is equal to 1.56 g.

Corrected SPT N value, $(N_1)_{60}$ was estimated for sand layer using Equation 1 (Youd et al., 2001) and found to be 35. However, as per Caltrans Geotechnical Manual (2014), granular soils with $(N_1)_{60} > 30$ and %fines ≥ 5 are not liquefiable. Therefore, in the present case, sand layer is also not liquefiable.

As can be seen from Equations 1 and 3, corrected SPT N value, $(N_1)_{60CS}$, depends on several correction factors and fine content in addition to the spatial variability of measured N value at a specific site. This uncertainty in the estimation of actual $(N_1)_{60CS}$ can significantly influence the cyclic resistance ratio (CRR) estimation and hence the factor of safety against liquefaction. Therefore, a probabilistic approach considering $(N_1)_{60CS}$ as a random variable may be used for site-specific liquefaction assessment. Alternatively, an additional reduction factor (ϕ) may be applied to the estimated CRR value to incorporate the uncertainty of associated with the estimation of $(N_1)_{60CS}$.

5 CONCLUSION

This paper presents a case study of the liquefaction evaluation of a proposed 12 storied building site in San Francisco, California. Liquefaction assessment procedure recommended by the Caltans Geotechnical Manual (2014) was used. As per the analysis, the site is not susceptible to liquefaction. However, considering the uncertainty involved in the estimation of corrected SPT N value (e.g., spatial variability, errors in the estimation of fine content, various correction factors, etc.), it is suggested to apply an additional site-specific reduction factor to the estimated CRR value.

ACKNOWLEDGMENTS

This paper is a part of the undergraduate capstone project by Adilzhan Mussa, Elnara Satekenova, Dinmukhambet Raiymbekov, Umut Bektimirova, and Meruyert Zhunussova in the Department of Civil Engineering at Nazarbayev University. Dr. Dichuan Zhang and Prof. Kim are the supervisor and co-supervisor of this project. The authors thank Dr. Sudheesh Thiyya Kkandi for his valuable comments on this project.

REFERENCES

American Society of Civil Engineers (ASCE). (2010). Minimum Design Loads for Buildings and Other Structures. Reston: ASCE.

ENGEO. (2012). Geotechnical Exploration. [Online]. Available from http://www.ci.milpitas.ca.gov/_pdfs/geotechnicalreport.pdf. [Accessed 25th March 2016].

Caltrans Geotechnical Manual. (2014). Liquefaction Evaluation. Available from: http://www.dot.ca.gov/hq/esc/geotech/geo_manual/page/Liquefaction_Evaluation_Dec2014.pdf. [Accessed: 29th October 2015].

Earthquake and environment (TS4)

Challenges and Innovations in Geotechnics – Zhussupbekov (Ed.)
© 2016 Taylor & Francis Group, London, ISBN 978-1-138-03007-7

Seismic bearing capacity of footings in c-ϕ soil by using a non-associated flow rule

D. Chakraborty
Indian Institute of Technology (IIT), Kharagpur, India

ABSTRACT: Limited investigations have been reported in the literature for finding out the seismic bearing capacity of foundations. Also, in order to find out the seismic bearing capacity of foundations, mostly the soil mass has been assumed to follow an associated flow. In general, frictional soils do not follow the associated flow rule strictly. In this paper a plane strain lower bound limit analysis technique in combination with the finite element and nonlinear programming is used to investigate the effect of considering a non-associated flow rule on the seismic bearing capacity of a strip footing. The bearing capacity factors (N_c, N_q and N_γ) with changes in soil friction angle (ϕ) and seismic acceleration coefficient (k_h) for different values of dilative coefficient (η) are computed for rough strip footings. The analysis clearly reveals that, the bearing capacity factors under the seismic loading decrease drastically with a decrease in the magnitudes of η.

Keywords: Bearing capacity, Seismic coefficient, Failure, Limit analysis, Non-associated flow rule

1 INTRODUCTION

With the consideration of an associated flow rule few researchers have obtained the seismic bearing capacity of shallow foundation (Sarma & Iossifellis, 1990; Budhu & Al-Karni, 1993; Richards et al., 1993; Kumar & Rao, 2003; Choudhury & Subba Rao, 2006; Kumar & Chakraborty, 2013). With the consideration of a non-associated flow rule in the energy dissipation method the solutions have been provided by Yang et al. (2007) for seismic bearing capacity of foundation on slope. However, there is no research work available in literature which provides the lower bound solutions on the seismic bearing capacity of foundation with the consideration of a non-associated flow rule. It should be mentioned here that the lower bound theorem of the limit analysis is based on the consideration that the material follows an associated flow rule. The dilation angle considered in case of an associated flow rule is usually greater than that experienced by frictional soils (Drescher & Detournay, 1993). The purpose of this paper is to find out the influence of dilation angle (ψ) on the lower bound solutions for the seismic bearing capacity of a strip footing in a general cohesive-frictional (c-ϕ) soil mass. The results are presented in terms of design charts with (i) the dilative coefficient $\eta = \psi/\phi = 0.0$, 0.5 and 1.0, (ii) soil friction angle (ϕ) between 10° and 4°, and (ii) seismic acceleration coefficient (k_h) between 0. and 0.4.

2 PROBLEM DEFINITION

It is required to find out the influence of dilation angle (ψ) on the bearing capacity factors of a rigid strip footing of width B in a c-ϕ soil under the pseudostatic seismic loading by incorporating a non-associated flow rule in the lower bound finite element limit analysis

3 ANALYSIS

The plane strain lower bound limit analysis formulation, in combination with finite element (as per Sloan, 1988) and nonlinear programming, is used in this paper. It should be mentioned here that Sloan (1988) utilized the linear programming technique. Each node is associated with three basic unknown stress variables (σ_x, σ_y, τ_{xy}). The magnitude of the collapse load is maximized subject to a number of (i) the equality constraints developing from the satisfaction of (a) the equilibrium conditions, (b) the stress boundary conditions, and (c) the continuity of the normal and shear stresses along the stress discontinuities; and (ii) the inequality constraints developed from the nonlinear Mohr-Coulomb yield criterion at all the nodes. The nonlinear optimization method, as presented by Krabbenhoft & Damkilde (2003), is adopted in this paper.

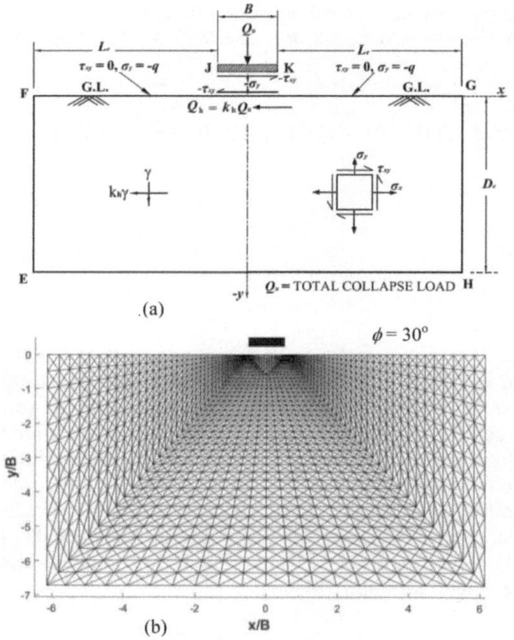

Figure 1. (a) Selected domain for analysis and the corresponding boundary conditions; and (b) finite element mesh utilized in the present study for $\phi = 30°$.

3.1 *Incorporation of non-associated flow rule*

In case of a frictional material if ψ is not equal to ϕ, as per the theory of plasticity it can be stated that the material follows a non-associated flow rule. This ψ value may vary between 0 and ϕ. Similar to the work of Chakraborty (2016), for non-associative materials following the Mohr-Coulomb yield criterion, the effect of ψ is taken into consideration with the help of following expressions (Drescher & Detournay, 1993):

$$\tan\phi^* = \tan\phi \frac{\cos\phi\cos\psi}{1-\sin\phi\sin\psi} \quad (1a)$$

$$c^* = c\frac{\cos\phi\cos\psi}{1-\sin\phi\sin\psi} \quad (1b)$$

where ϕ^* and c^* are the modified friction angle and cohesion.

3.2 *Mesh details*

The chosen domain and the related stress boundary conditions are indicated in Figure 1a. The D_e and L_e of the domain are selected in such way that the values of the failure load remain more or less constant even if a bigger size of the domain is utilized. Three-noded triangular elements are used to discretize the entire domain. Typical finite element mesh in case of $\phi = 30°$ used for the present study is presented in Figure 1b.

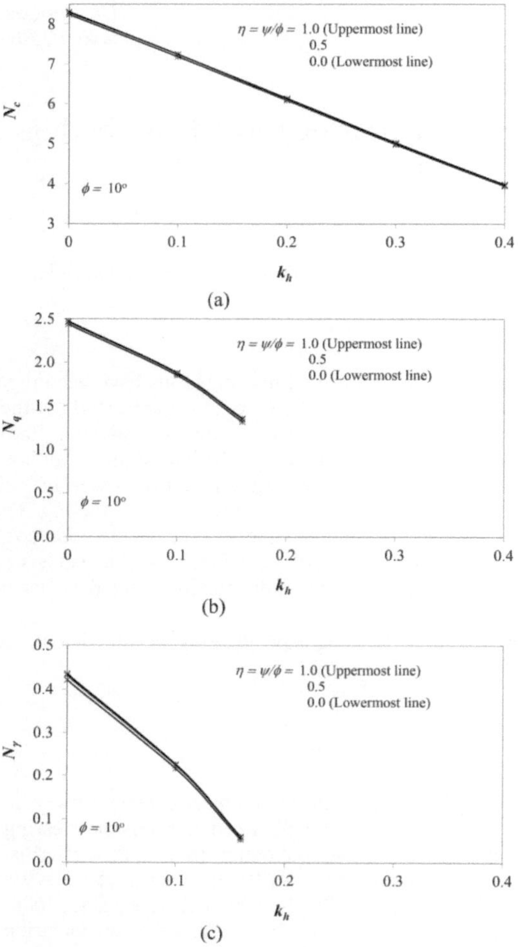

Figure 2. For $\phi = 10°$ with the changes in η value the variation of (a) N_c with k_h; (b) N_q with k_h; and (c) N_γ with k_h.

It should be mentioned here that for greater ϕ value a bigger domain is used.

Note that for incorporating the effect of pseudo-static horizontal seismic forces, the formulation of Kumar & Chakraborty (2013) is adopted in this paper.

3.3 *Final form of the optimization procedure*

After satisfying the stress boundary conditions, the corresponding global matrices and vectors are founded. The nonlinear programming problem can then be presented as:

Maximize the objective function: $\boldsymbol{u}^T\boldsymbol{\sigma}$ (2a)
Subjected to

(i) linear equality constraints: $\boldsymbol{M\sigma} = \boldsymbol{d}$ (2b)

(ii) nonlinear inequality constraints: $\boldsymbol{f}(\boldsymbol{\sigma}) \leq \boldsymbol{0}$ (2c)

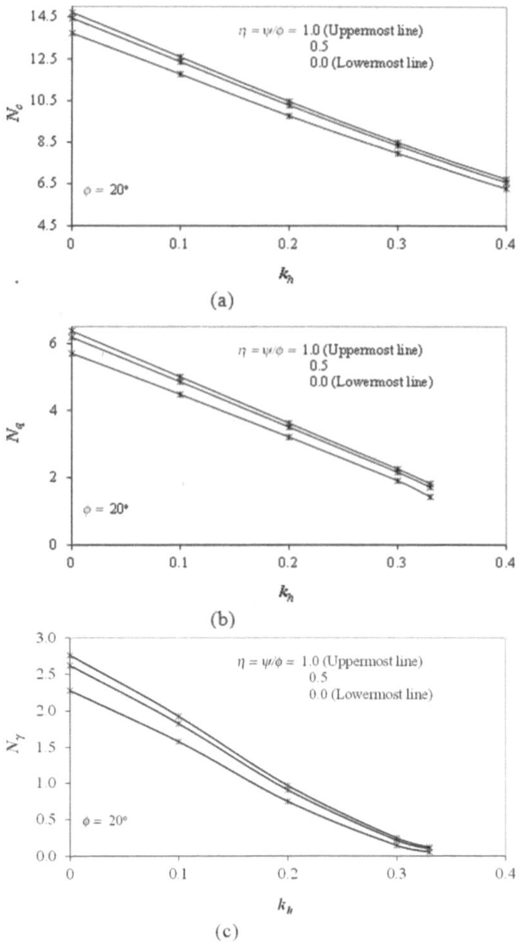

Figure 3. For $\phi = 20°$ with the changes in η value the variation of (a) N_c with k_h; (b) N_q with k_h; and (c) N_γ with k_h.

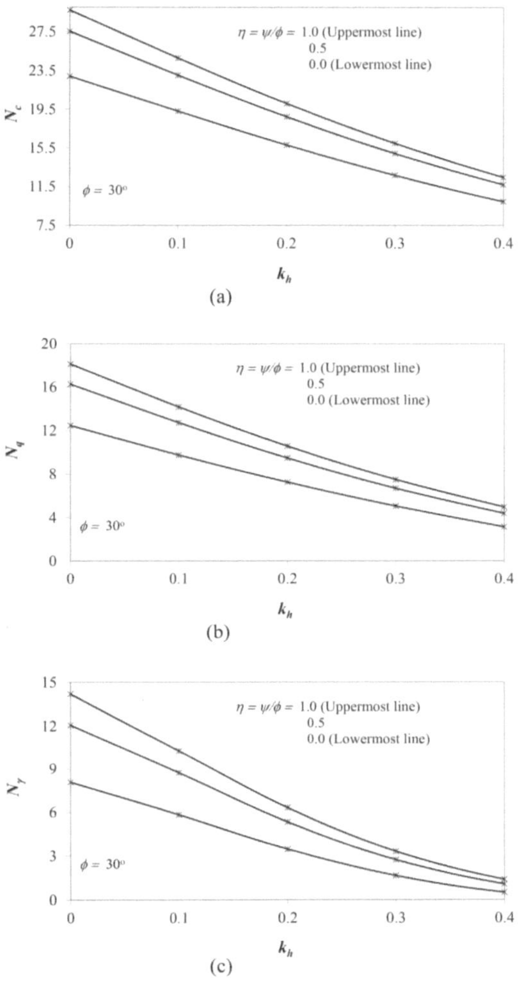

Figure 4. For $\phi = 30°$ with the changes in η value the variation of (a) N_c with k_h; (b) N_q with k_h; and (c) N_γ with k_h.

Here $\boldsymbol{\sigma}$ indicates a vector comprising of all the nodal stresses. For complete understanding of the nonlinear optimization methodology one can refer Chakraborty & Kumar (2013). For carrying out the present research study, a computer code is developed in *MATLAB*.

4 BEARING CAPACITY FACTORS

After finding out the value of ultimate collapse load (Q_u), the magnitude of N_c, N_q and N_γ on account of the components of cohesion of soil (c), surcharge pressure (q), and unit weight of soil (γ), respectively, are derived by using the following expression

$$\frac{Q_u}{B} = cN_c + qN_q + 0.5B\gamma N_\gamma \tag{3}$$

5 RESULTS AND COMPARISONS

Computational results are derived for different values of (i) k_h ranging from 0.0 to 0.4, (ii) ϕ ranging from 10° to 40°, and (iii) η ranging from 0.0 to 1.0. The results are presented here.

5.1 *The variation of the bearing capacity factors with k_h for the changes in the dilative coefficient*

The variations of N_c, N_q and N_γ with various magnitudes of k_h with the changes in η for $\phi = 10°$, 20°, 30° and 40° are provided in Figures 2, 3, 4 and 5, respectively. The analysis clearly reveals that for a given ϕ, the factors N_c, N_q and N_γ decrease with a decrease in the magnitudes of η. As expected, the effect of η is found to be increasing significantly with an increase in ϕ value. The results also clearly indicate that for a

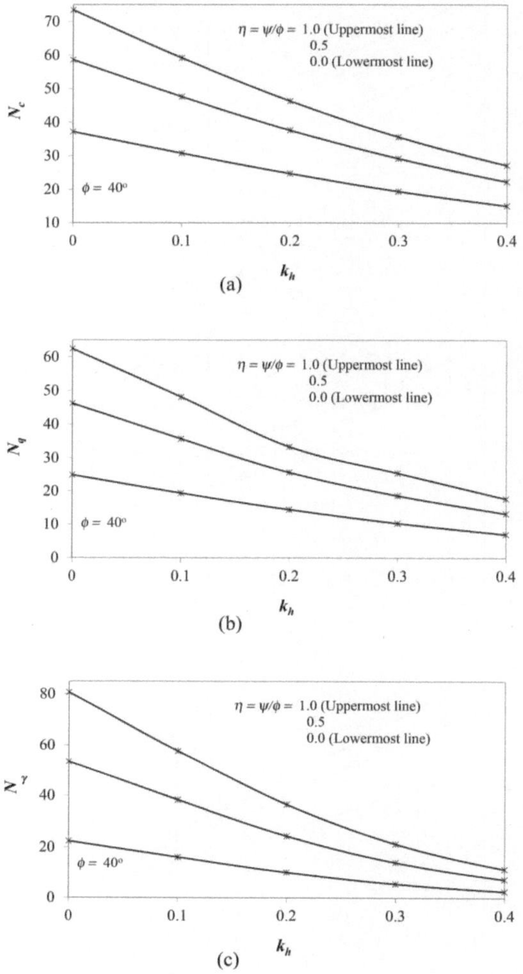

Figure 5. For $\phi = 40°$ with the changes in η value the variation of (a) N_c with k_h; (b) N_q with k_h; and (c) N_γ with k_h.

particular magnitude of η the factors N_c, N_q and N_γ decrease considerably with an increase in the value of k_h. The rate of decrease in the magnitude of the factors N_c, N_q and N_γ is found to be maximum for $\eta = 0.0$ and minimum for $\eta = 1.0$. The maximum effect of η can be observed in case of N_γ. For $\phi = 40°$ and $\eta = 1.0$, with an increase in k_h value from 0.0 to 0.4 the magnitude of N_γ decrease by 85.8%. Whereas, for $\phi = 40°$ and $\eta = 0.0$, with an increase in k_h value from 0.0 to 0.4 the magnitude of N_γ decrease by 88.6%. It clearly shows the significance of incorporating a non-associated flow rule in the analysis.

5.2 *Comparison*

Under static condition (i.e., $k_h = 0.0$) for $\eta = 1.0$ (i.e., with the consideration of an associated flow rule), the presently obtained values of N_c, N_q and N_γ are compared with the solutions of (i) Chakraborty (2016) based on the lower bound limit analysis with finite element and nonlinear programming, and (ii) Yin *et al.* (2001) based on FLAC (Fast Lagrangian Analysis of Continua) analysis. The comparisons are provided in Figures 6a, b, c. Note that the present solutions compare quite well with the existing solutions. The magnitudes of N_c, N_q and N_γ obtained by Yin *et al.* (2001) become little higher than that of the present solutions. As expected the present solutions compare very nicely with the solutions of Chakraborty (2016).

Similarly, in case of $k_h = 0.0$ with $\eta = 0.0$, the values of N_c, N_q and N_γ are compared with the solutions of (i) Chakraborty (2016), and (ii) Yin *et al.* (2001). The comparisons are provided in Figures 6d, e, f. The comparison indicate that the magnitudes of N_c, N_q and N_γ obtained by Yin *et al.* (2001) become little higher than that of the present solutions. As expected the presently obtained solutions are found to be comparing very nicely with the solutions of Chakraborty (2016).

It should be mentioned here that based on the energy dissipation method Yang et al. (2007) provided the solutions for slope angle varying between 5° and 25°. As a result of that it is not possible to compare the present lower bound solutions for horizontal ground with the solutions of Yang et al. (2007).

6 FAILURE PATTERNS

The state of stress at a point with respect to the yield is presented as a ratio a/d; where

$$a = (\sigma_x - \sigma_y)^2 + (2\tau_{xy})^2 \tag{4a}$$

$$d = \{2c\cos\phi - (\sigma_x + \sigma_y)\sin\phi\}^2 \tag{4b}$$

Here $a/d = 1$ denotes that the point is in a state of shear failure, and $a/d < 1$ denotes that the point is in a non-plastic state. For $\phi = 30°$, at the time of finding out N_γ, the failure patterns associated with $\eta = 0.0$ and $\eta = 1.0$ for $k_h = 0.0$, $k_h = 0.1$, and $k_h = 0.2$ are provided in Figure 7. It can be noted that the failure patterns become gradually more and more asymmetric with an increase in the k_h value. Also, with a decrease in η value from 1.0 to 0.0, there is a decrease in the extent of the plastic zone can be noted.

7 CONCLUSIONS

The influence of dilation angle on the seismic bearing capacity of a strip footing in a general cohesive-frictional soil mass by incorporating a non-associated flow rule in the lower bound limit analysis is studied in this paper. The results clearly indicate that for a given ϕ, the bearing capacity factors decrease with a decrease in the magnitudes of η. In case of greater ϕ values, this rate of decrease becomes more noticeable.

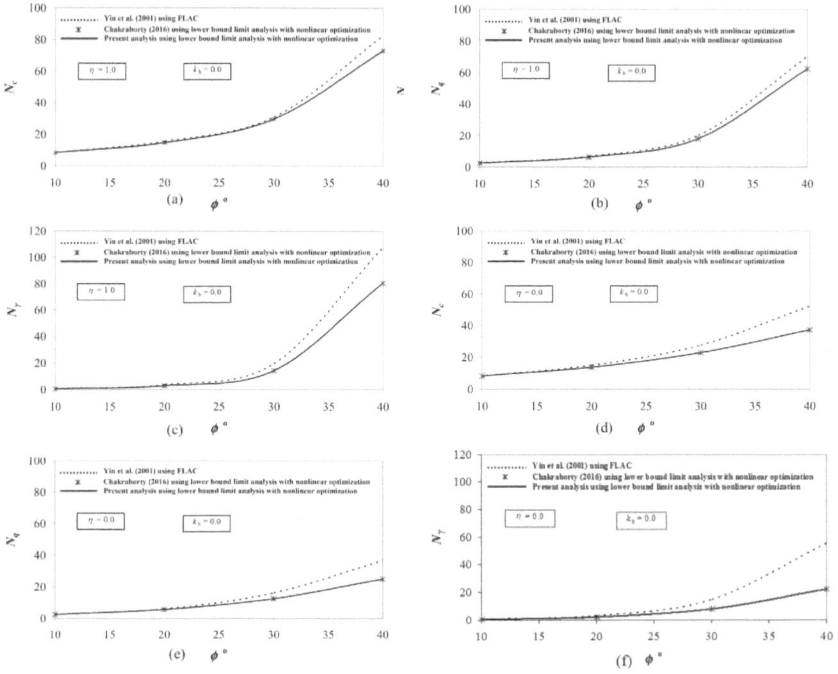

Figure 6. A comparison of the present results with those available in the literature for $k_h = 0$ with (a) $\eta = 1.0$ in case of N_c; (b) $\eta = 1.0$ in case of N_q; (c) $\eta = 1.0$ in case of N_γ; (d) $\eta = 0.0$ in case of N_c; (e) $\eta = 0.0$ in case of N_q; and (f) $\eta = 0.0$ in case of N_γ.

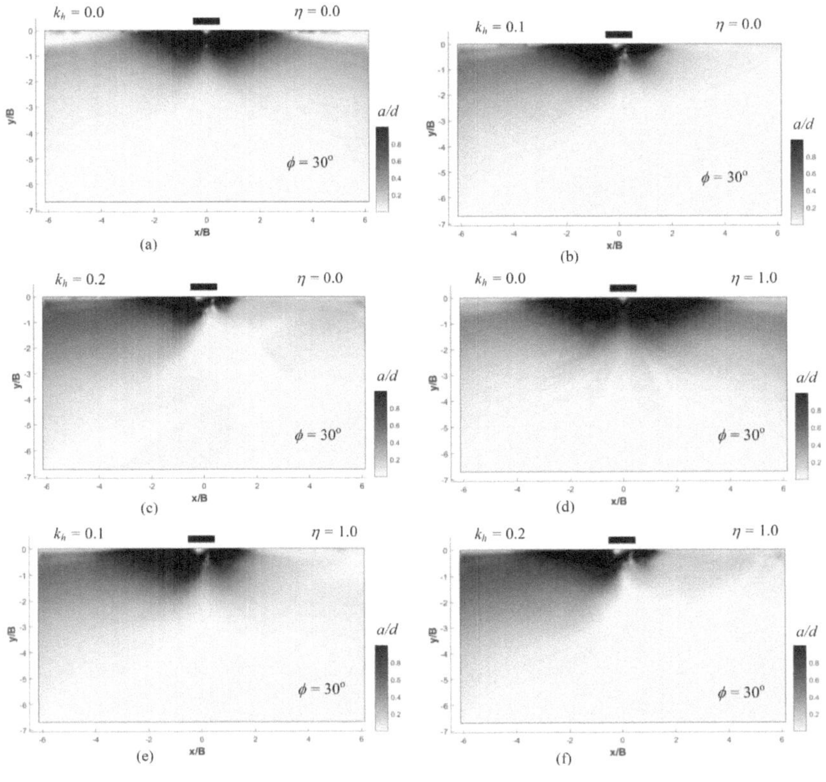

Figure 7. Failure pattern for $\phi = 30°$ with (a) $\eta = 0.0$, $k_h = 0.0$; (b) $\eta = 0.0$, $k_h = 0.1$; (c) $\eta = 0.0$, $k_h = 0.2$; (d) $\eta = 1.0$, $k_h = 0.0$; (e) $\eta = 1.0$, $k_h = 0.1$; and (f) $\eta = 1.0$, $k_h = 0.2$ obtained after computing N_γ for rough footing base.

The results also clearly indicate that for a particular magnitude of η the factors N_c, N_q and N_γ decrease considerably with an increase in the value of k_h. The rate of decrease in the magnitude of the factors N_c, N_q and N_γ is found to be maximum for $\eta = 0.0$ and minimum for $\eta = 1.0$. The maximum effect of η can be observed in case of N_γ. It is anticipated that the design charts presented in this paper will be quite useful for the practicing engineers.

REFERENCES

Budhu, M. & Al-Karni, A. 1993. Seismic bearing capacity of soils. *Géotechnique* 43(1): 181–187.

Chakraborty, D. 2016. Bearing capacity of strip footings by incorporating a non-associated flow rule in lower bound limit analysis. *International Journal of Geotechnical Engineering, Taylor and Francis* 1-5: DOI: 10.1080/19386362.2016.1142272.

Chakraborty, D. & Kumar, J. 2013. Bearing capacity of foundations on slopes. *Geomechanics and Geoengineering: An International Journal* 8(4): 274–285.

Choudhury, D. & Subba Rao, K.S. 2006. Seismic bearing capacity of shallow strip footings embedded in slope. *International Journal of Geomechanics, ASCE* 6(3): 176–184.

Drescher, A. & Detournay, E. 1993. Limit load in translational failure mechanisms for associative and non-associative materials, *Géotechnique* 43(3): 443–456.

Krabbenhoft, K. & Damkilde, L. 2003. A general non-linear optimization algorithm for lower bound limit analysis. *International Journal for Numerical Methods in Engineering* 56(2): 165–184.

Kumar, J. & Chakraborty, D. 2013. Seismic bearing capacity of foundations on cohesionless slopes. *Journal of Geotechnical and Geoenvironmental Engineering, ASCE*, 139(11): 1986–1993.

Kumar, J. & Rao, V.B.K.M. 2003. Seismic bearing capacity of foundations on slopes. *Géotechnique* 53(3): 347–361.

Richards, R., Elms, D.G. & Budhu, M. 1993. Seismic bearing capacity and settlement of foundations. *Journal of Geotechnical Engineering Division, ASCE* 119(4): 662–674.

Sarma, S.K. & Iossifelis, I.S. 1990. Seismic bearing capacity factors of shallow strip footings. *Géotechnique* 40(2): 265–273.

Sloan, S.W. 1988. Lower bound limit analysis using finite-elements and linear programming. *International Journal for Numerical and Analytical Methods in Geomechanics* 12(1): 61–77.

Yang X., Guo N., Zhao L., & Zou J. 2007. Influences of nonassociated flow rules on seismic bearing capacity factors of strip footing on soil slope by energy dissipation method. *Journal of Central South University of Technology* 14(6): 842–847.

Yin, J., Wang, Y. & Selvadurai, A. 2001. Influence of nonassociativity on the bearing capacity of a strip footing, *Journal of Geotechnical and Geoenvironmental Engineering, ASCE* 127(11), 985–989.

Challenges and Innovations in Geotechnics – Zhussupbekov (Ed.)
 ISBN 978-1-138-03007-7

Application of local bentonite as buffer and backfill material in low-level radioactive waste disposal facility

F. Teng
Sinotech Engineering Consultants, Taipei, Taiwan

ABSTRACT: Bentonite has been extensively used as buffer and backfill materials in artificial engineering barriers system in radioactive waste (Radwatse) disposal facility. Due to the low hydraulic conductivity of bentonite, the hazardous radionuclide in the radwastes could be retarded for hundred or thousand years, which ensures radioactivity of radionuclide degrades to a relatively low level. It provides a safe solution for the disposal of poisonous radwastes. Bentonite material used in disposal facilities could be imported or produced locally. This study aims to evaluate the feasibility of adopting bentonite mined in Taiwan. In this work, bentonite of Wyoming imported from the USA and Zhishin clay produced in Taiwan were selected. The buffer and backfill material were made by mixing bentonite with different aggregates. Tested samples was compacted into size of ϕ50 mm × h20 mm under different compaction energies. Soil properties tests, swelling tests, and hydraulic conductivity tests were performed to evaluate the performance of buffer and backfill materials. Based on the test result, the feasibility of using local bentonite as buffer and backfill materials was evaluated in this study. The results showed that the local bentonite was only suitable for backfill materials.

Keywords: Bentonite, Radwaste, Disposal

1 INTRODUCTION

Taiwan has 40 years of experience in the research and peaceful use of nuclear energy. Nuclear power contributed 5144 MW of capacity by 3 active nuclear power plants (NPPs) and 6 reactors, which makes up 18.6% of its electricity generation as of 2014. Radioactive wastes (radwastes) were generated from nuclear power plant (NPP) operations, decommissions of nuclear facilities, as well as medical, agricultural, industrial, and research usage. According to the existing regulations and laws on the categorization of radwaste, with the exception of spent nuclear fuel, all other wastes were categorized as low-level radwaste. Those low-level radwastes were cemented and stored in zinced steel drums within the storage facilities for final disposal. The amounts of low-level radwastes stored in Taiwan are listed in Table 1.

Final disposal strategies of radwaste are long-term isolation and containment of wastes away from biosphere. To achieve the isolation and containment function, a concept of underground disposal facility with multi-barriers is employed for the final disposal of low-level radwaste in Taiwan. The multi-barrier system consists of both engineering and natural barriers including solidified waste, container, buffer, backfill, engineering structure, and host rock, as illustrated in Figure 1. Among the multi-barrier system, the back fill and buffer material with low hydraulic conductivity and high absorbability provide the containment function that retards the transportation of toxic nuclides in low-level radwaste to biosphere.

After strict siting process, two recommended candidate sites for final disposal facility were announced by the Ministry of Economic Affairs. Site characteristics of the 2 recommended candidate sites have been preliminarily studied during site selection process. The host rocks of the 2 sites are argillite and granite, respectively. The buffer and backfill material is basically mixture of bentonite and host rocks.

Bentonite material used in disposal facilities could be imported or produced locally. This study aims to evaluate the feasibility of adopting bentonite mined in Taiwan. Bentonite of Wyoming imported from the USA and Zhishin clay produced in Taiwan were selected as test materials. The buffer and backfill materials were made by mixing bentonite with different aggregates from the 2 recommended candidate sites.

Soil properties tests, swelling tests, and hydraulic conductivity tests were performed to evaluate the performance of buffer and backfill materials. Based on the test result, the feasibility of using local bentonite

Figure 1. Illustration of the multi-barrier system for low-level radwaste disposal.

Table 1. Current amount of low-level radwaste in Taiwan.

Facility	Amounts (drums)
Chinshan NPP	43,551
Kuosheng NPP	52,974
Maanshan NPP	8569
Lan-Yu interim storage facility	100,277

As of Nov. 2015.

Table 2. Host rock properties for both recommended candidate sites.

Items	Site-D	Site-W
Rock	Argillite	Granite
Unit weight (kN/m^3)	26.39–27.27	25.80–26.78
Water content (qu (MPa)	1.2–47.1	100–250
Young's modulus (GPa)	9.3–62.9	39.8–63.0

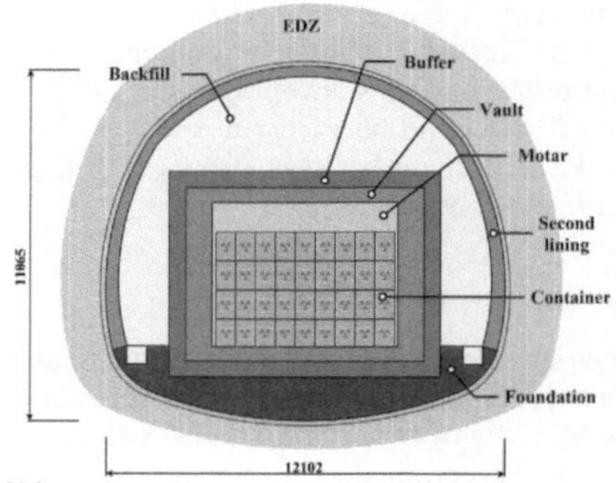

Figure 2. Conceptual design of EBS for Site-D.

as buffer and backfill materials was evaluated in this study.

2 CONCEPTUAL DESIGN OF ENGINEERING BARRIER SYSTEM

Design considerations for the engineering barrier system (EBS) include long-term safety, mechanical stability, and operational safety. For the low-level radwaste contains higher concentrations of nuclides with longer half-lives, low-permeability buffer mater is deployed surrounding the disposal vault to prevent groundwater penetration. Figure 2 and Figure 3 show the conceptual design of EBS for both recommended candidate sites (Taiwan Power Company, 2016). The difference in dimension and geometry between these two designs is mainly due to the property of host rock. Related host rock properties are listed in Table 2.

2.1 *Buffer material*

The buffer would be composed mainly of bentonite which expands upon absorbing water to seal any

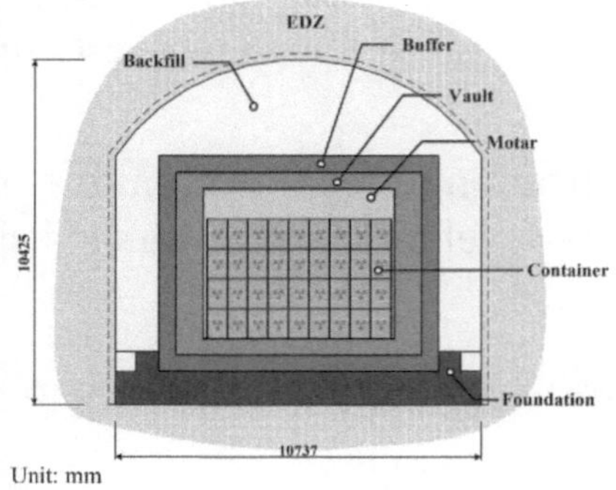

Figure 3. Conceptual design of EBS for Site-W.

water permeable gaps, thus ensuring the safety of the disposal facility. Bentonite also exhibits different properties in saltwater and freshwater environments. The design should consider groundwater properties in the vicinity of the disposal tunnels.

2.2 *Backfill material*

Backfilling would mainly be used to fill up any remaining spaces within the disposal tunnels. A mixture of bentonite and aggregate would be employed for the disposal of the low-level radwaste contains higher concentrations of nuclides with longer half-lives.

3 TEST MATERIAL

The buffer and backfill materials are made of bentonite and aggregates with different ratios. The properties of bentonites are introduced below.

3.1 *Bentonite*

Two bentonites were used as test materials in this study, which are bentonite of Wyoming imported from the USA (MX-80) and Zhishin clay produced in Taiwan (Zh). The properties of MX-80 were extensively studied by many researchers, such as mineral structures (SKB, 2009), long-term behavior (Herbert et al. 2004), bentonite mineralogy (Posiva, 2004) etc. The MX-80 is sodium bentonite which is commonly used for sealing the nuclear repository. Compared to the MX-80, the Zh clay is more like calcium bentonite which was mined in southeastern Taiwan. Performance properties of Zh clay as engineering material in nuclear repository was not clearly investigated. The basic properties for MX-80 and Zh clay tested in this study are listed in Table 3.

3.2 *Buffer and backfill material sample*

Buffer and backfill material samples were made of bentonite and aggregates. Aggregates from the two recommended candidate disposal sites, i.e. granite and argillite, were crushed and mixed with bentonite. The mixture was then compacted under various energies, which resulted in samples with different initial unit

Table 3. Basic properties of MX-80 and Zh claye.

Items	MX-80	Zh clay
Specific weight	2.69	2.66
%: >74 μm	2	28
%: 2–74 μm	33	46
%: <2 μm	65	26
Water content %	14.1	7.3
PL	52	17
LL	430	67
PL	378	50
Activity	5.37	1.78

Table 4. Samples notation for different bentonite and aggregate.

		Bentonite type		
		Zh clay(Z)	MX-80(B)	Mixed clay(M)
Aggregate	Taitung argillite (T)	TZ	TB	TM
	Kimen granite (K)	KZ	KB	KM

weight. The sample dimension is ϕ50 mm × h20 mm for the tests.

4 TEST PLAN

The test plan aims to investigate the performance of local bentonite as buffer and backfill material in radwaste disposal facilities. Samples made of different bentonite and aggregates (Table 4), different ratio of bentonite, and different compaction energies (Table 5) were used in one-dimensional swelling test (ASTM D4546) and hydraulic conductivity test (ASTM D5856). The sample notation, for instance 50TZ5M, denotes that the 50% Taitung argillite mixed with 50% Zh clay was compacted under 5 MPa stress. And the Mixed clay (M) represents the bentonite used in producing the sample is 50% of Zh clay and 50% of MX-80.

5 TEST RESULT

5.1 *Swelling behavior*

The buffer and backfill materials are used to seal possible voids in the disposal facility. The swelling property of the buffer and backfill material under different configuration is a key parameter for the future design. Results on different bentonite are show in Figures 3. The higher bentonite/aggregate ratio yields in stronger swelling potential. The local bentonite, Zh clay, has less swelling potential compared to the MX-80. Swelling strain for Zh clay samples are less than 50%, nevertheless the MX-80 has 100% to 500% swelling strains depends on the bentonite ratio. For sample with the same bentonite ratio, different aggregate, i.e. argillite and granite, has minor effect on the swelling behavior for buffer and backfill material.

Table 5. Samples notation for different ratio of bentonite and compaction energy.

Compaction Stress	Bentonite ratio 50%	75%	100%
100 kPa	50XX100K	75XX100K	100XX100K
500 kPa	50XX500K	75XX500K	100XX500K
1 MPa	50XX1M	75XX1M	100XX1M
5 MPa	50XX5M	75XX5M	100XX5M
10 MPa	50XX10M	75XX10M	100XX10M

Note 1. XX is the notation for different bentonite and aggregate, e.g. TB or KZ etc.
Note 2. Notations for 10% and 30% are omitted in the table due to the limitation of table width.

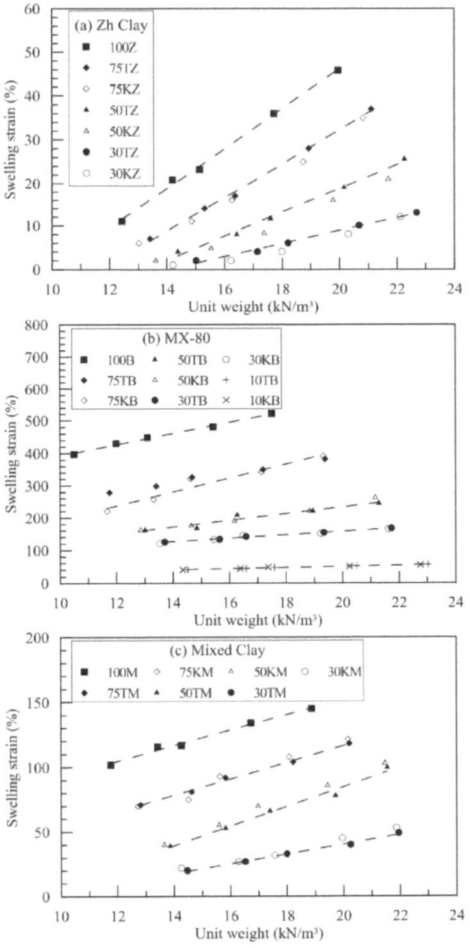

Figure 3. Swelling strain versus unit weight (a) Zh clay, (b) MX-80, (c) Mixed clay.

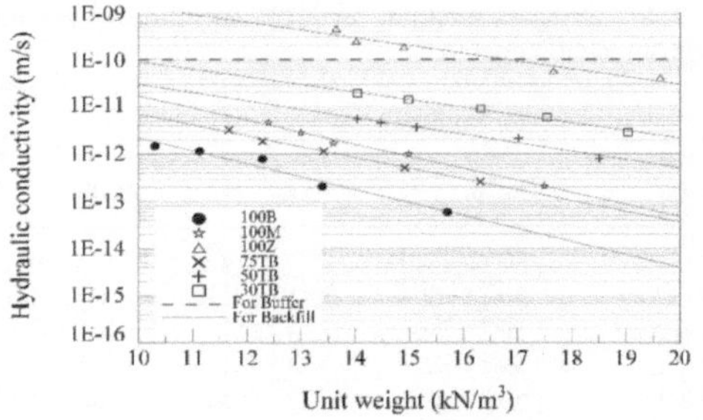

Figure 4. Hydraulic conductivity versus unit weight.

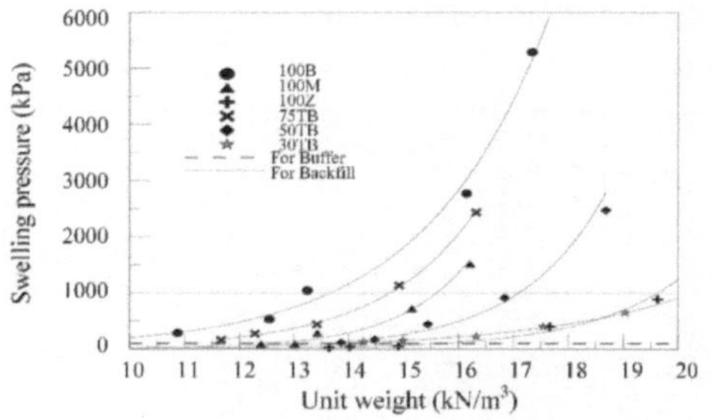

Figure 5. Swelling pressure requirement for buffer and backfill material.

5.2 *Hydraulic conductivity*

The buffer and backfill material are used to retard the transportation of toxic nuclides with water flow during the disposal. The hydraulic conductivity of the buffer and backfill material under different configuration is also a key parameter for the future design. Results on different bentonite are show in Figure 4. The local bentonite, Zh clay, has higher hydraulic conductivity compared to the MX-80.

6 DISCUSSION

The hydraulic conductivity and swelling pressure are indices used in design of buffer and backfill material. Requirements of hydraulic conductivity for buffer and backfill material are 1E-12 and 1E-12 m/s, respectively. For Zh clay sample shown in Figure 4, only pure clay sample can satisfy the hydraulic conductivity requirement of backfill material. The MX-80 has better performance on retarding transportation of nuclides than Zh clay and mixed clay. Samples with only 30% of MX-80 can satisfy the hydraulic conductivity requirement of backfill material. And for hydraulic conductivity requirement of buffer material, at least 75% of MX-80 with proper compaction energy is needed.

Requirements of swelling pressure for buffer and backfill material are 1 MPa and 0.1 MPa, respectively. Figure 5 shows the swelling pressure measured on selected bentonite samples. According to the results, design of buffer and backfill material, such as bentonite type, bentonite ratio, compaction energy, could be suggested.

7 CONCLUSION

Bentonites are commonly used in radwaste disposal facilities as buffer and backfill materials. The function of bentonites are retarding the transportation velocity of toxic nuclides from disposal tunnels to biosphere. Thus, the swelling property and hydraulic conductivity of bentonites are the key design parameters. Due to the consideration of construction cost, local bentonite are preferred in producing the buffer and backfill materials. This study evaluated the feasibility of applying local bentonite, i.e. Zh clay, in buffer and backfill material in a disposal facility. The performance of Zh clay is compared to MX-80. Soil properties tests, swelling tests, and hydraulic conductivity tests were performed to evaluate the performance of buffer and backfill materials. The hydraulic conductivity test results showed that the local bentonite was only suitable for backfill materials. Samples with 100% Zh clay can yield a hydraulic conductivity of 1E-10, which satisfied the requirements of hydraulic conductivity for backfill material. The requirements of hydraulic conductivity for buffer material can only be satisfied with MX-80 and mixed clay (50% Zh and 50% MX-80).

ACKNOWLEDGEMENT

Technical supports from Taiwan Construction Research Institute (TCRI) and advices from Mr. C.-L. Yuan at Sinotech Engineering Consultants (SEC) are gratefully acknowledged.

REFERENCES

ASTM, D4546.
ASTM, D5856
H.-J Herbert, J Kasbohm, H.C Mooga, K.-H Henning, 2004, Long-term behaviour of the Wyoming bentonite MX-80 in high saline solutions, *Applied Clay Science*, Volume 26, Issues 1–4, August 2004, Pages 275–291.
Posiva, 2004, bentonite mineralogy, working report
SKB, 2009, Structure and forces in bentonite MX-80, *Technical report, TR-09-06*
Taiwan Power Company, 2016, Evaluation of Technical Feasibility for Low Level Radioactive Waste Disposal, *LLWD1-SC-2016-01-V01*

Challenges and Innovations in Geotechnics – Zhussupbekov (Ed.)
© 2016 Taylor & Francis Group, London, ISBN 978-1-138-03007-7

Geotechnical approaches on the seismic risk assessment for architectural heritages at historical sites using field and centrifuge tests

Heon-Joon Park & Dong-Soo Kim
Department of Civil and Environmental Engineering, KAIST, Daejeon, Korea

ABSTRACT: This paper describes a research methodology for the seismic risk assessment of architectural heritages at historical sites from a geotechnical engineering viewpoint. Field seismic tests, site-specific ground response analysis, and geotechnical centrifuge model tests with an earthquake simulator were used to determine the geotechnical characteristics of Cheomseongdae, which is a representative of Korean architectural heritage. This artifact is known to be the oldest astronomical observatory in Asia. According to historical records, the Gyeongju area where Cheomseongdae is located has suffered from numerous medium-scale earthquakes. Based on site investigation and ground response analysis of scenario earthquakes, the ground around Cheomseongdae is highly susceptible to site amplification. In addition, Cheomseongdae is a masonry structure that is apparently vulnerable to horizontal dynamic loads such as earthquakes. However, features such as the filler of the lower half, irregularly shaped inner stones that can induce high frictional resistance, eight long horizontal tie stones inside the artifact, and a grid of interlocking headstones are recognized to increase the structure's resistance to horizontal dynamic loads despite its appearance. Through field and centrifuge tests, a geotechnical approach to the seismic risk assessment of architectural heritages at historic sites is established. The test results verify our ancestors' "seismic design" technology.

1 INTRODUCTION

Cultural heritages reflect their temporal backgrounds and history in their entirety. Thus, we have a responsibility to preserve them safely and transmit them to posterity. In particular, architectural heritages are exposed to natural and manmade disasters, so the need for systematic safety management and research to prevent disasters and preserve structures has arisen. Because all architectural heritages are found on the ground, a geotechnical engineering approach was developed in this study for the seismic risk assessment of architectural heritages in historical areas. Cheomseongdae, which is one of the most famous and representative architectural treasures, was considered as a case study. Field site characterization, including boring, downhole seismic tests, and surface wave tests, was performed near the Cheomseongdae ground. Based on the results of the site investigation and site-specific ground response analyses, the surface motions during scenario earthquakes were obtained. The geotechnical centrifuge tests were performed on 1/15-scale Cheomseongdae models at 15 g, and the surface motions were used as input earthquake motions.

2 CHEOMSEONGDAE

Cheomseongdae is a star-gazing platform that is known to have been constructed in the 7th century AD and the oldest astronomical observatory in Asia. Despite various natural and manmade disasters such as wars that occurred over the past 1300 years, its form has been maintained without much damage up to the present. The materials and original form at the time of the construction have been preserved, and the structure has a beautiful appearance characterized by a harmonious mingling of curves, straight lines, circles, and squares.

Cheomseongdae is a stone masonry structure that was built from nearly four hundred stones and is apparently vulnerable to horizontal dynamic loads such as earthquakes (Figure 1). The main body is formed from 373 stone members in 27 layers and piled up in a bottle-shaped smooth curve on a square stereobate made of processed granite. Two layers of interlocking headstones are placed on top of the main body. The entire structure is 9.17 m in height (NRICH 2009).

There is a small entrance or window at the southern part from the 13th–15th layers. From the 1st–12th layers of the main body (lower half of structure, lower part of entrance or window), the inside is filled with large and small pieces of rubble and soil. From the 13th layer upward, the inside is empty.

The interior is composed of roughly touching stones and irregularly shaped stones, unlike the smoothly faced exterior. This interior shape increases the contact area and roughness between stone materials and increases the frictional resistance inside Cheomseongdae.

Figure 1. The typical photos of exterior and interior Cheomseongdae: (a) filler beneath the 13th layer; (b) inner irregular shaped stones; (c) long shaped horizontal tie stones of 19, 20, 25 and 26th layer; (d) grid of interlocking stones and half-lap joint.

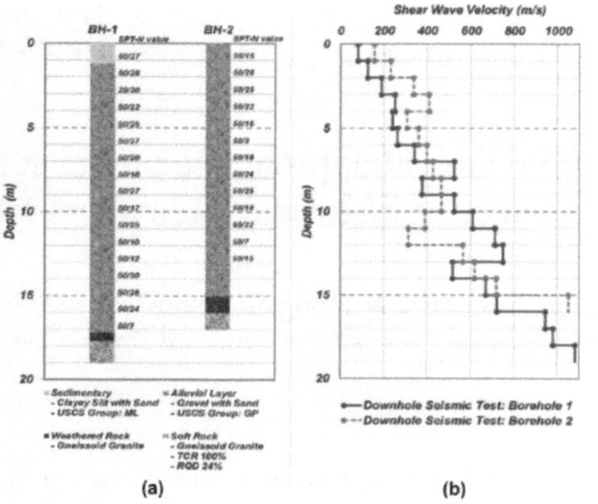

Figure 2. Site characterization for cheomseongdae ground: (a) boring result; (b) downhole seismic test result.

The eight stones of the 19th, 20th, 25th, and 26th layers and the headstones of the 28th and 29th layers are long horizontal tie stones. The 19th layer has two long horizontal tie stones in the shape of rectangles that penetrate the cylindrical plane in the east and west directions. The 20th layer has two long horizontal tie stones inside the cylindrical plane in the north and south directions. The 21st–24th layers have stone materials formed by the same method as that used for the 12th–18th layers. The 25th layer again has two long stones that penetrate the cylindrical plane in the north and south directions, and the 26th layer has another two long stones resting in the east and west directions (perpendicular to the 25th layer) on the 25th layer.

A flagstone rests on the 27th layer, which is lopsided to the east. The top part of the 28th and 29th layers is a grid of interlocking stones in a square shape. The 28th layer has two long horizontal tie stones laid in the east and west directions, and another two long horizontal tie stones are inserted in the north and south directions through a partial cutting of the former long stones. The 29th layer has two long horizontal tie stones laid in the north and south directions, and on top of these stones another two long stones are connected in the east and west directions by a half-lap joint. This grid of interlocking headstones increases the resistance to horizontal dynamic loads such as earthquakes.

According to historical records, the Gyeongju area, where Cheomseongdae is located, suffered from numerous medium-scale earthquakes after the construction of Cheomseongdae. These historical records note damage from earthquakes, such as "The ground of Gyeongju shook and residences collapsed and one hundred people died" in 779 AD and "After a major earthquake around Gyeongju, a stone pagoda at Bulguksa Temple was on the verge of collapse" in 1036 AD. It is possible but not probable that Cheomseongdae underwent significant repairs during its life, even without earthquakes. Nevertheless, it has survived over 1300 years and still preserves its shape today despite the numerous medium-scale earthquakes in these seismic records.

In order to understand this curiosity, a site investigation and ground response analysis of the ground of Cheomseongdae were performed to evaluate the site-specific earthquake ground motion, and centrifuge model tests with an earthquake simulator were conducted to evaluate the seismic response of Cheomseongdae. The centrifuge tests were performed on two models: an exact duplicate of the original Cheomseongdae and one without the long parts. The results of the centrifuge tests on the two Cheomseongdae models were used to evaluate and verify our ancestors' seismic design technology (Park & Kim 2014).

3 FIELD SITE INVESTIGATION AND SITE-SPECIFIC GROUND RESPONSE ANALYSIS

In the previous study, field investigations were performed on architectural heritage sites in the Gyeongju area (Park et al. 2012). In the present study, the results from two borehole and downhole seismic tests were used for the surrounding ground of Cheomseogdae. Boreholes were drilled down to the bedrock depth at two spots approximately 20 m to the north and south of Cheomseongdae, and a standard penetration test and downhole seismic test were performed to identify the subsoil profile and stiffness distribution.

According to the borehole test results, a 15–17 m thick alluvial layer is present in the ground surrounding Cheomseongdae consisting mostly of gravel with sand, as shown in Figure 2(a). The bedrock is at a depth of approximately 17 m, and boring was terminated when soft granite rock was reached.

At the two boreholes, downhole seismic tests were conducted to derive the shear wave velocity profile, as shown in Figure 2(b). The downhole method measures the time for body waves to travel from the source on the surface to receivers at different depths in a single borehole. The results confirmed that sandy gravel was present at a depth of 1–16 m. In addition, downhole seismic tests were used to obtain the stiffness distribution of the ground with the depth (Kim et al. 2002, Park et al. 2013).

Considering the importance of the ripple effect on the collapse of the architectural heritage structure in this study, input rock outcrop accelerations between 0.044 and 0.329 g were considered. In the Korean seismic design guide for new buildings, the design rock outcrop accelerations at Gyeongju for inducing collapse in the event of an earthquake are 0.220 g for a return period of 2400 years and 0.154 g for a return period of 1000 years.

Based on the site investigation results for the ground at Cheomseongdae, site-specific ground response analyses for scenario earthquakes were performed by using a one-dimensional equivalent linear scheme of the SHAKE program (Schnabel et al. 1972). The equivalent linear analysis provided relatively reliable results when the strain level caused by an earthquake was less than 1%, which is a typical range in moderate seismic zones such as Gyeongju. One artificial synthetic earthquake, three strong motion recordings from strong seismic regions (i.e., El Centro, Hachinohe, and Ofunato earthquake motions), and one weak motion recording were used as input ground motions. In order to reflect the characteristics of regional seismicity, an earthquake acceleration waveform recorded in Gyeongju was also applied to derive site response data.

The results of the one-dimensional ground response analysis found that the ground around Cheomseongdae showed an amplification of 1.70–2.16 times at peak ground acceleration (PGA), and the natural period of the soil layers was 0.21–0.23 s. The results of the site investigation and ground response analysis illustrated that the ground near Cheomseongdae is highly susceptible to site amplification nearby and resonance with the Cheomseongdae structure itself. The ground response analysis results were also applied to derive incident design seismic waves that can occur on the ground surface of Cheomseongdae and were used as input motions for dynamic centrifuge tests.

4 CENTRIFUGE MODEL TEST

A centrifuge test that simulates the gravity field of the N g state for a 1/N scale model is a good alternative to numerical analysis and the 1 g shaking table test. The stress conditions of a scaled model in a centrifuge are the same as those in a full-scale prototype. This equivalent stress condition makes it possible to simulate real seismic behavior by using a reduced-scale model. In the previous study (Park & Kim 2013), centrifuge tests with the "modelling of models" procedure were used to check the general scaling factor for the dynamic behavior of stone block structures. The test confirmed that the seismic behavior of the reduced model can replicate that of the given stone structure in the study. Thus, the dynamic centrifuge test was proposed as a means of assessing the seismic risk of stone pagodas to evaluate their safety against future earthquakes.

This study was conducted at a centrifuge facility of the KAIST Geotechnical Centrifuge Center (Kim et al. 2013a). The centrifuge model test is based on the principle of creating scaled-down models with identical stresses and strains at homologous points in the prototype and model structures and subjecting them to accelerated gravitational fields (Schofield 1980). This centrifuge can also excite seismic events by spinning at a desired centrifugal acceleration. An electrohydraulic servo-type earthquake simulator was used for the centrifugal test in this research. This earthquake simulator provides a shaking acceleration of 40 g with no payload and 20 g with a 700 kg payload, which is equivalent to a PGA of 0.5 g in the prototype when excited at a centrifugal acceleration of 40 g (Kim et al. 2013b).

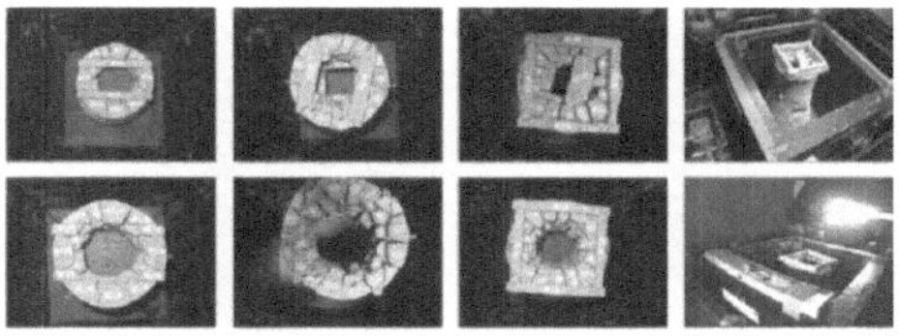

Figure 3. The difference between the Cheomseongdae Model 1 (upper) and Model 2 (lower): (a) 19th layer; (b) 25th layer; (c) 28th layer, lower part of the headstone; (d) Models in the rigid box.

In this test, a centrifugal acceleration of 15 g was induced vertically on a 1/15-scale model of Cheomseongdae. The model size was determined in order to consider the fit in the container of the earthquake simulator in the centrifuge, the anticipated bandwidth of the response frequency, and the feasibility of model building.

By using the scaling factors to fit the model size, time, and frequency conditions to 1/15 of those of the prototype, actual phenomena that could happen at Cheomseongdae in the event of an earthquake can be explained. The appropriate scaling factor for dynamic events was defined, and actual earthquake situations could be simulated in the centrifuge model. The measured frequency, amplitude, and length were converted to the prototype scale. The excitation frequency for the models was 40.0–300 Hz. Based on the 1/15 scaling factor, this is equivalent to a frequency range of 2.67–20.0 Hz for the prototype structure.

In order to reproduce the current appearance of Cheomseongdae, each stone member of the 1/15-scale model was produced by hand. As noted previously, Cheomseongdae is composed of nearly four hundred stones; in particular, the eight long horizontal tie stones of the 19th, 20th, 25th, and 26th layers and headstones of the 28th and 29th layers are longer than the others, so they should help increase the resistance to horizontal dynamic loads. In this study, the dynamic centrifuge tests were executed on two 1/15-scale models: one of the original Cheomseongdae (Model 1) and another where 12 stones from the 19th, 20th, 25th, 26th, and 29th layers, which were considered to provide the most seismic resistance, were excluded (Model 2). Figure 3 shows the differences between the two models and their fabrication process.

Dynamic centrifuge tests were successfully performed on the two Cheomseongdae models in two

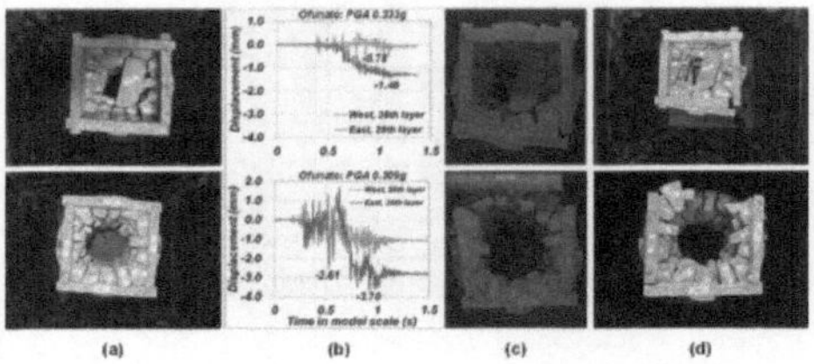

Figure 4. The pictures of top part and displace measurement for Model 1 (upper) and Model 2 (lower): (a) before testing; (b) displacement of 28th layer as the result of image tracking of high speed camera for the return period 1000 years earthquake level; (c) captured images of high speed camera when the upper headstone had fallen out; (d) after testing.

excitation directions. The appearances after the damage from the earthquake were compared. The amplitude of the waveform increased starting from the seismic waves with a low acceleration level. For the earthquake with a return period of 500 years, the displacement of each model was visible with the high-speed camera. For Model 1, the headstones were observed to move when the earthquake level had a return period of 1000 years. For Model 2, the lower part of the headstone and other lower parts fell off at the same excitation level. None of the stone material of Model 1 fell off even with the strong motion of an earthquake with a return period of 2400 years.

Figure 4 compares the top part and measured displacement with a high-speed camera of Models 1 (upper) and 2 (lower). Figure 4(a) shows pictures taken before the tests. After the excitation level of the Ofunato earthquake was increased to a return period of 1000 years (0.333 g PGA), the lateral displacement of the top part was measured by image tracking with the high-speed camera. For Model 1, a maximum displacement of 1.46 mm occurred on the eastern side of the 28th layer with an earthquake duration of 1 s and acceleration of 15 g. At that time, the headstones and top part of the 27th–29th layers were observed to move slightly. For Model 2, a maximum displacement of 3.70 mm took place (0.309 g of PGA), as shown in Figure 4(b), and the northern headstone of the 28th layer fell, as shown in Figure 4(c). If the displacements were scaled up to the actual Cheomseongdae, a 15-s earthquake would cause maximum displacements of 21.9 and 55.5 mm, respectively. Figure 4(d) shows the appearance after the upper headstone fell off. Thus, the long stones and upper headstone of the Cheomseongdae prototype greatly affected the seismic performance of the entire structure.

5 CONCLUSION

A geotechnical approach to the seismic risk assessment of architectural heritages in historic site that uses field and centrifuge tests is proposed, and a case study was performed on Cheomseongdae. Based on close observation of the inner surface of Cheomseongdae, the filler beneath the 13th layer, inner irregular shaped stones, and the positioning of the long horizontal tie stones, there is reasonable confidence that this structure has considerable resistance to horizontal dynamic loads such as earthquakes. After site investigation and site-specific ground response analysis, centrifuge model tests with an earthquake simulator were performed on models of Cheomseongdae. Differences in seismic behavior between the two Cheomseongdae models were observed, and the upper headstone and long stones appear to play an important role in making the structure resistant against seismic motions. In the masonry stone structure, these stones increased the seismic resistance. Through the field and centrifuge tests, a geotechnical approach to seismic risk assessment for architectural heritages at historical sites was organized and the test results verified our ancestors' seismic design technology.

ACKNOWLEDGEMENT

This research was supported by Basic Science Research Program through the National Research Foundation of Korea (NRF) funded by Ministry of Education (grant number 2014R1A6A3A04056405). The authors acknowledge the KREONET service provided by Korea Institute of Science and Technology Information.

REFERENCES

Kim, D.-S., Chung, C.-K., Sun, C.-G. & Bang, E.-S. 2002. Site assessment and evaluation of spatial earthquake ground motion of Kyeongju. *Soil Dynamics and Earthquake Engineering* 22(5): 371–387.

Kim, D.-S., Kim, N.-R., Choo, Y.W. & Cho, G.-C. 2013a. A newly developed state-of-the-art geotechnical centrifuge in Korea. *KSCE Journal of Civil Engineering* 17(1): 77–84.

Kim, D.-S., Lee, S.-H., Choo Y.W. & Perdriat J. 2013b. Self-balanced earthquake simulator on centrifuge and dynamic performance verification. *KSCE Journal of Civil Engineering* 17(4): 651–661.

National Research Institute of Cultural Heritage (NRICH). 2009. *Assessment of damage for Gyeongju Cheomseongdae.* in Korean.

Park, H.-J., Kim, D.-M., Kim, K.-S., Ahn, H.-Y. & Kim, D.-S. 2012. Noninvasive geotechnical site investigation for stability of Cheomseongdae. *Journal of Cultural Heritage* 13(1): 98–102.

Park, H.-J., Kim, D.-S. & Kim, D.-M. 2013. Seismic risk assessment of architectural heritages in Gyeongju considering local site effects. *Natural Hazards and Earth System Sciences* 13(2): 251–262.

Park, H.-J. & Kim, D.-S. 2013. Centrifuge modelling for evaluation of seismic behaviour of stone masonry structure. *Soil Dynamics and Earthquake Engineering* 53: 187–195.

Schnabel, P.B, Lysmer J. & Seed, H.B. 1972. *SHAKE: a computer program for earthquake response analysis of horizontally layered sites*; Report No. EERC72-12. Earthquake Engineering Research Center, University of California, Berkeley.

Schofield, A.N. 1981. Cambridge geotechnical centrifuge operations. *Géotechnique* 30(3): 227–268.

Challenges and Innovations in Geotechnics – Zhussupbekov (Ed.)
 ISBN 978-1-138-03007-7

Liquefaction susceptibility analysis based on SPT and microtremor

N. Shrestha
Nepal Geotechnical Society, Kathmandu, Nepal
Institute of Engineering, Tribhuwan University, Pulchowk, Lalitpur, Nepal

S.K. Jha
Institute of Engineering, Tribhuwan University, Pulchowk, Lalitpur, Nepal

ABSTRACT: This paper outlines the liquefaction susceptibility analysis performed at residential building that suffered a ground failure after 7.8 Mw Gorkha (Nepal) earthquake. Five boreholes were drilled with depth varying from 10 m to 40 m where Standard Penetration Tests (SPT) have been performed. Liquefaction susceptibility in each borehole has been analysed using deterministic approach and probabilistic approach for Peak Ground Acceleration (PGA) of 0.3g and moment magnitude of 7.8 Mw. The results of analysis show that liquefaction susceptibility is comparatively higher at borehole location BH05 at a depth range of 3 m to 4.5 m where silty clay was found with lower plasticity index, higher fine contents and having higher water table. Ground vulnerability index (Kg) values were calculated using Horizonal-to-Vertical fourier spectral ratio (HVSR) technique through microtremor survey to compare the results. The results show that microtremor based liquefaction susceptibility analysis can also be a good alternative indicator to find liquefaction susceptibility.

Keywords: Liquefaction susceptibility, SPT, Deterministic approach, Probabilistic approach, Microtremor, HVSR, Ground vulnerability index

1 INTRODUCTION

The study area is located at latitude 27°36′45.18″N and longitude 85°21′29.60″E in Godawari, Lalitpur which lies in the central region of Nepal. The site lies in the southern peripheral part of Kathmandu Valley which is characterized by lake deposits of Pliocene to Pliestocene in age (Yoshida and Gautam 1998). The Bagmati, Bishnumati and Manahara river drained the Kathmandu Valley with an outlet in the southeast of the valley through the chobhar gorge which flows dissecting the mountain of the lesser Himalaya range in the southeast before it finally joins the Ganges river in India. This phenomenon caused erosion that resulted in formation of plateau where the project site is located.

In Nepal, there have been numerous devastating historical earthquakes recorded since 1255 AD and Kathmandu Valley has been jolted severely. Dynamic loading caused by earthquake abruptly increases pore pressure in soil, decreases effective stresses and considerably enhances liquefaction potentiality. This phenomenon occurs easily in lacustrine soil (UNDP /MOHPP 1994) of Kathmandu Valley. On April 25, 2015, central Nepal was hit by devastating earthquake (Gorkha earthquake) of 7.8 Mw. Central to eastern regions of Nepal including Lalitpur and capital city Kathmandu faced varying levels of damages during this earthquake. More than 40% of house damages in Kathmandu Valley were found to be caused by differential settlement of the land in Kathmandu Valley lacustrine sediments (Dahal 2015). This paper presents a liquefaction susceptibility analysis as a case study for a building located in lacustrine sediments of Kathmandu Valley where gradual crack expansion in ground floor was reported after Gorkha earthquake.

Borehole logging and standard penetration tests performed at 5 borehole locations with varying depth of 10 m to 40 m showed that upper layer of soil is generally composed of silty clay with little fine grained soil whereas mostly black clay is found at depth greater than 10 m which evidently is the deposition of lake of Kathmandu Valley. Liquefaction susceptibility analysis has been done based on SPT by deterministic and probabilistic approach followed by microtremor study carried out for each borehole locations where ground vulnerability index (Kg) values have been determined from HVSR graph.

2 METHODOLOGY

Bray & Sancio (2006) found that loose soils with a $PI < 12$ and moisture content $>85\%$ of the Liquid Limit (LL) are susceptible to liquefaction which correlates with soil sample obtained from 3 to 4.5 m depth at BH05 borehole location with $PI = 7.6$, moisture content of 40.14% and liquid limit of 38.8%. Furthermore, SPT values based analysis have been performed by deterministic approach and probabilistic approach followed by microtremor analysis as discussed below.

2.1 *Deterministic approach*

The factor of safety (FS) against liquefaction in terms of CSR (Cyclic Stress Ratio) and CRR (Cyclic Resistance Ratio) are defined as:

$$FS = \frac{CRR_{Mw=7.5,\sigma_v'=1atm}}{CSR_{Mw=7.5,\sigma_v'=atm}} = \frac{CRR_{Mw,\sigma_v'}}{CSR_{Mw,\sigma_v'}} \tag{1}$$

where $CRR_{Mw=7.5,\sigma'v=1atm}$ is the cyclic resistance ratio (CRR) for earthquakes of magnitude 7.5 and effective overburden pressure of 1 atm (101.3 kPa), is the capacity of soil to resist liquefaction; $CSR_{Mw=7.5,\sigma'v=1atm}$ is the normalized cyclic stress ratio (CSR) for earthquakes of magnitude 7.5 and an effective overburden pressure of 1 atm (101.3 kPa). Cyclic stress ratio is given by (Boulanger & Idriss 2014):

$$CSR_{Mw=7.5,\sigma_v'=1atm} = 0.65\frac{a_{\max}}{g}\frac{\sigma_v}{\sigma_v'}\frac{\gamma_d}{MSF}\frac{1}{K_\sigma} \tag{2}$$

where σ_v is the total vertical stress; σ_v' is the effective vertical stress; a_{max} is the peak horizontal ground surface acceleration; g is the acceleration due to gravity; γ_d is the nonlinear shear stress mass participation factor (or stress reduction factor); MSF is the magnitude scaling factor; K_σ is the correction factor for effective overburden. Liquefaction is predicted to occur if $FS \leq 1$, and no liquefaction is predicted if $FS > 1$. However, still there are many uncertainties that needs to be considered which are discussed in probability approach. Stress reduction factor, magnitude scaling factor, overburden correction factor and K_σ is given by Youd et al. (2001):

$$\gamma_d = \frac{(1-0.4113z^{0.5}+0.04052z+0.001753z^{1.5})}{(1-0.4177z^{0.5}+0.05729z-0.006205z^{1.5}+0.0012z^{2})} \tag{3}$$

where z = depth below ground surface in meters.

Cyclic Resistance Ratio (CRR) can be obtained from the corrected blow count (N1)60 using empirical correlations proposed by Seed et al. (1985):

$$CRR_{7.5} = \frac{1}{34-(N_1)_{60cs}} + \frac{(N_1)_{60cs}}{135} + \frac{50}{[10(N_1)_{60cs}+45]^2} - \frac{1}{200} \tag{4}$$

$(N1)_{60cs}$ is corrected $(N1)_{60}$ to an equivalent clean sand value given by Youd et al. (2001) based on Fine Contents (FC). $(N1)_{60}$ is the SPT blow count normalized to an overburden pressure of approximately 100 kPa and a hammer energy ratio of 60% expressed as (Youd et al. 2001):

$$(N_1)_{60} = NC_NC_EC_BC_RC_S \tag{5}$$

where N = measured standard penetration resistance; C_N = factor to normalize N to a common reference overburden stress (1 atm); C_E = correction for hammer energy ratio; C_B = correction factor for borehole diameter; C_R = correction factor for rod length; C_S = correction for samplers with or without liners. The correction factors given by Youd et al. (2001) are adopted in this paper.

2.2 *Probabilistic approach*

First order second moment (FOSM) method is used to find the probability for liquefaction for the study site in each borehole locations. The mean and coefficient of variation of cyclic stress ratio are given by Jha et al. (2009):

$$\mu_{CSRN} = 0.65\frac{\mu_{a\max}}{g}\frac{\mu_{\sigma v}}{\mu_{\sigma v'}}\frac{\mu_{\gamma d}}{\mu_{MSF}} \tag{6}$$

$$V_{CSRN}^2 = V_{a\max}^2 + V_{\sigma v'}^2 + V_{\sigma v}^2 - 2\rho_{\sigma v\sigma v'}V_{\sigma v'}V_{\sigma v} \tag{7}$$

where μ and V represent the corresponding mean and coefficient of variation (COV: ratio of standard deviation to mean), respectively, and $\rho'_{\sigma v\sigma v}$ represents the correlation coefficient between total and effective stress which is assumed to be negligible in this paper. PGA is assumed to be 0.3 g and mean of MSF is obtained by taking average value of lower and upper bounds of MSF. The COVs of MSF and γ_d are assumed to be 0.05 and 0.1 respectively. Similarly, COVs of other parameters are assumed with reference to Christian and Baecher (2015).

COV (V) for CRR can be calculated as (Jha et al. 2009):

$$V_{CRR} = \frac{\Delta CRR}{2\mu_{CRR}} \tag{8}$$

$\Delta CRR = CRR(\mu(N_1)60 + \sigma(N_1)60) - CRR(\mu(N_1)60 - \sigma(N_1)60)$ The risk for liquefaction in terms of liquefaction probability P_L can be obtained from reliability index by:

$$P_L = 1-\varphi(\beta) \tag{9}$$

where φ is the standard normal cumulative probability. If Z = CRR-CSR, due to inherent uncertainties in estimating CSR and CRR, treating both as variables, hence liquefaction performance function will also be a random variable. The lognormal reliability index, β_{LN}, is defined by Duncan (2000):

$$\beta_{LN} = \frac{\ln\left[\frac{\mu_{FS}}{\sqrt{1+V_{FS}^2}}\right]}{\sqrt{\ln(1+V_{FS}^2)}} \tag{10}$$

where μ_{FS} is the mean value of the factor of safety, V_{FS} is COV (σ_{FS}/μ_{FS}) of the factor of safety.

2.3 *Microtremor analysis*

At each borehole locations, microtremor survey was performed using a portable velocity sensor NewPIC, manufactured by System and Data Research (SDR) Co. Ltd., Japan. Microtremor data were recorded for 300 seconds at a sampling frequency of 100 Hz. Base line correction is done in each velocity-time histories and divided into 15 windows of 2048 samples. Transient signals were minimized from visualization by removing windows of high transient signals to carry out Fast Fourier Transform (FFT) to obtain Fourier spectra. Then, the obtained Fourier spectra were smoothed by the Parzen window of bandwidth 0.5 Hz (Paudyal et al. 2012). The average spectral ratio of the horizontal to vertical component of vibration (H/V) in each window was derived from Equation 11 given by Delgado et al. (2000):

$$H/V = \sqrt{\left(F_{NS}^2 + F_{EW}^2\right)/(2F_{UD}^2)} \tag{11}$$

where, F_{NS}, F_{EW} and F_{UD} are the Fourier amplitude spectra in the north-south (NS), east-west (EW), and vertical (UD) directions, respectively.

H/V spectral ratio (HVSR) for the particular point was then obtained by averaging all those spectral ratios. Nakamura (1997) proposed the vulnerability index "K_g value" as a means to determine the extent of liquefaction derived from strains of ground and structures:

$$K_g = \frac{A_p^2}{F_p} \tag{12}$$

where A_p is peak amplification factor corresponding to predominant frequency, F_p. In this paper, K_g values are determined for each borehole locations to predict the liquefaction potentiality.

3 RESULTS AND DISCUSSIONS

Water table has been found to be at shallower depth of 1.1 m in BH05 borehole location whereas it is deeper in other borehole locations with mostly SPT value greater than 30, characterized as non-liquefiable soil. Therefore BH05 is investigated with special consideration. Considering 7.8 Mw and PGA of 0.3g, factor of safety against liquefaction by deterministic approach has been found to be less than 1 at a depth of 3 m to 4.5 m as shown in Table 1.

The result is further supported by probabilistic approach in which 58% to 81% probability of liquefaction potentiality at the same range of depth has been found in BH05 borehole location as shown in Table 2. Besides BH05, factor of safety for BH04 borehole location at a depth of 15m has been found as 1 with $P_L = 57\%$ where sandy soil with pebbles was logged. $P_L > 15\%$ is considered vulnerable to liquefaction (Jha et al. 2009) Ground vulnerability index (K_g) values have been calculated using Equation 12 (Table 3) with the help of H/V graph obtained from microtremor study as shown in Figure 1.

Ground vulnerability index (K_g) value at BH05 as 12 is reasonably greater than values obtained at other borehole locations which is comparable with the result of SPT based deterministic and probabilistic analysis except at borehole location, BH03.

Water table at BH03 is at 9 m depth with soil of higher plasticity index, lower moisture content

Table 1. Result of liquefaction analysis from deterministic approach for BH05 borehole location.

Depth (m)	γ (kN/m^3)	$(N_1)_{60cs}$	FC (%)	FS
1.5	23.1	28.82	48	1.67
3	23.1	24.72	69	0.99
4.5	23.1	22.11	71	0.79
6	23.1	34.78	65	–
7.5	18	53.75	19	–
9	18	47.31	7	–
10	18	46.54	9	–

Table 2. Result of liquefaction analysis from probabilistic approach for BH05 borehole location.

	Mean (μ)	COV	Depth (m)	β_{LN}	P_L(%)
$(N_1)_{60cs}$		0.108	1.5	1.147	13
MSF	0.891	0.2	3	−0.202	58
K_σ		0.07	4.5	−0.862	81
σ_v/σ'_v		0.05	6		
a_{max}		0.1	7.5		
γ_d		0.2	9		
CSR		0.235	10		

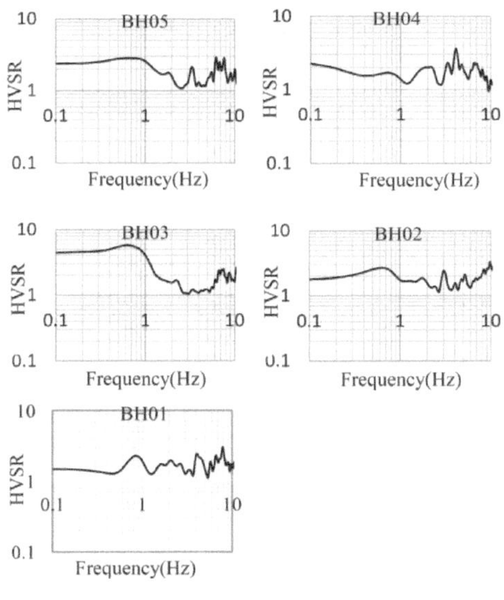

Figure 1. HVSR graph for each borehole location.

Table 3. Ground vulnerability index (K_g) values.

	BH01	BH02	BH03	BH04	BH05
A_p	1.7	2.6	6	1.3	2.9
F_p	0.75	0.65	0.65	0.85	0.7
K_g	3.85	10.40	55.38	1.99	12.01

and cohesive nature. Generally, cohesive soil is not considered vulnerable to liquefaction. Therefore, the result suggests that consideration of water table and existence of liquefiable soil in site should be taken into account while evaluating liquefaction potentiality by ground vulnerability index (K_g) values using microtremor study.

4 CONCLUSIONS

Even though FS $\geq$ 1, probability of liquefaction susceptibility may be higher or lower depending upon uncertainties in SPT performance, procedure and random measurement error that may vary from 15% to 45% as defined by Phoon & Kulhawy (1999).

Especially at top layer of soil, below water table, liquefaction susceptibility analysis based on microtremor survey can also be a good alternative indicator to find liquefaction susceptibility. Considering water table and liquefiability of soil from various parameters (PI, moisture content, LL etc), microtremor based ground vulnerability index (K_g) values are comparable with the result based on SPT for finding liquefaction susceptibility. However, more research needs to be done for better understandings.

ACKNOWLEDGEMENTS

The Author expresses sincere gratitude to Dr. Ranjan Kumar Dahal, Dr. Netra Prakash Bhandary and Dr. Indra Acharya for providing technical support and Nepal Geotechnical Society for the organizational nomination for participating in the 8th Asian Young Geotechnical Engineers Conference.

REFERENCES

Bolton Seed, H., Tokimatsu, K., Harder, L. F., & Chung, R. M. 1985. Influence of SPT procedures in soil liquefaction resistance evaluations. *Journal of Geotechnical Engineering* 111(12): 1425–1445.

Boulanger, R. W., & Idriss, I. M. 2014. CPT and SPT based liquefaction triggering procedures *Report No. UCD/CGM-14/01 Center for Geotechnical Modeling (Davis USA: University of California, Davis)* 134 *Department of Civil and Environmental Engineering Report.*

Bray, J. D., & Sancio, R. B. 2006. Assessment of the liquefaction susceptibility of fine-grained soils. *Journal of geotechnical and geoenvironmental engineering* 132(9): 1165–1177.

Christian, J. T., & Baecher, G. B. 2015. Sources of uncertainty in liquefaction triggering procedures. *Georisk: Assessment and Management of Risk for Engineered Systems and Geohazards*: 1–8.

Dahal, R.K. 2015. Engineering Geological Issues after Gorkha Earthquake 2015 in Nepal-a preliminary understanding. *10th Asian Regional Conference of IAEG.*

Delgado, J., Casado, C. L., Giner, J., Estevez, A., Cuenca, A., & Molina, S. 2000. Microtremors as a geophysical exploration tool: applications and limitations. *Pure and applied geophysics* 157(9): 1445–1462.

Duncan, J. M. 2000. Factors of safety and reliability in geotechnical engineering. *Journal of geotechnical and geoenvironmental engineering* 126(4): 307–316.

Idriss, I. M., & Boulanger, R. W. 2010. SPT-based liquefaction triggering procedures. *UC Davis.*

Jha, S. K., & Suzuki, K. 2009. Reliability analysis of soil liquefaction based on standard penetration test. *Computers and Geotechnics*. 36(4): 589–596.

Nakamura, Y. 1997. Seismic vulnerability indices for ground and structures using microtremor. *World Congress on Railway Research in Florence, Italy.*

Paudyal, Y. R., Bhandary, N. P., & Yatabe, R. 2012. Seismic microzonation of densely populated area of Kathmandu Valley of Nepal using microtremor observations. *Journal of Earthquake Engineering* 16(8): 1208–1229.

Phoon, K. K., & Kulhawy, F. H. 1999. Characterization of geotechnical variability. *Canadian Geotechnical Journal*. 36(4): 612–624.

UNDP/MOHPP 1994. Seismic hazard mapping and risk assessment of Nepal. *United Nations Development Programme and Ministry of Housing and Physical Planning, Government of Nepal.*

Yoshida, M., & Gautam, P. I. T. A. M. B. A. R. 1988. Magnetostratigraphy of Plio-Pleistocene lacustrine deposits in the Kathmandu Valley, central Nepal. *Proceedings of Indian National Science Academy* A 54: 410–417.

Youd, T. L., et al. 2001. Liquefaction resistance of soils: summary report from the 1996 NCEER and 1998 NCEER/NSF workshops on evaluation of liquefaction resistance of soils. *Journal of geotechnical and geoenvironmental engineering* 127(10): 817–833.

Challenges and Innovations in Geotechnics – Zhussupbekov (Ed.)
 ISBN 978-1-138-03007-7

Foundations on artificial bases stabilized by Vertical Armo-Elements (VAE) with a seismic-protective damping layer

Z.A. Khasanov & A.Z. Khasanov
State Architectural and Civil Engineering Institute "M. Ulugbek", Samarkand, Uzbekistan

I.I. Usmankhodjaev
YPUU "Qishloq Qurilish Loyiha", Uzbekistan

ABSTRACT: The paper examines the issues of bases from the compacted soil pads, pile foundations and bases stabilized by vertical reinforcing elements. The results of the modeling of vertical armo-elements on loess subsiding soils and foundation pads are given in the paper. The methods of determining of the basic calculating parameters of bearing armo-element capacity and foundation pad are presented by the authors. The results of practical use of this method of bases and foundation construction in soil conditions of Uzbekistan are also given in the paper.

When design and constructing buildings and structures on loss subsiding soils in seismic conditions, the designers often face the problems of lowering non-uniform sediments, subsidence and increasing the rigidity of the base. The use of the most common method of arrangement of artificial bases in the form of compacted ground pads does not fully meet the requirements of reliability and lead to the rise in the cost of a zero-cycle erection in construction. Besides, the production of earth work is closely connected with seasonality i.e. oversaturation of soil moisture in winter – spring periods or on the contrary, its disadvantage in summer period of the year. In the process of buildings exploitation due to water leakage from communication network foundation pads are saturated by water whech leads to the lowering of its stiffness and increase visco-plastic properties. Experts in geotechnics know what exactly such properties of soils, despite their compactness may reduce their reliability during earth quakes.

The alternative methods of arrangement to prepare artificial bases, particular, the method of the tamped foundation trenches, ground piles, the method of presoaking and etc., in Uzbekistan, due to technical, technological and specific characteristics, these methods practically were not used in construction. Because of the cramped city conditions the method of soil compactness by heavy tampers began to apply rarely.

The alternative variant in urban construction is still the method of construction bases and foundations with the use of pill foundations.

However, as practice shows, for small-story buildings of no more than 5–9 floors the use of pile foundations leads to the rise in the cost of a zero-cycle construction. Even in soil conditions with highly location of groundwater and weak water – saturated soils, instead of driven piles which had been traditionally used before, the method of compacted ground pads construction and continuous or solid slab foundations were gradually begun to use. Moreover, in this case, as the material for ground pads, finely crushed stone or grave preparation are used.

In recent years, alongside with ground pads in construction, the method of a base preparation, using vertical armo-elements (VAE) have been widely used.

Vertical armo-elements from the compacted or strengthened soils are fundamentally different from rigid reinforced concrete piles. In our opinion, there are the following differences between them.

Reinforced concrete piles in conjunction with the raft foundation make up simple foundations under a building or structure. At calculation of pile foundations, practically almost all external loads are distributed to the piles. Ground, -cement – grounded and concrete piles are not a construction and relate to the elements of foundation soils strengthening, which perform the function of its reinforcement. The VAE unlike reinforced concrete piles do not work on the bending loads and allow to redistribute the stresses in the solid mass of grand. The VAE allow to regulate contact pressure and stiffness (deformability of a ground base under foundation). The VAE unlike conventional piles on working conditions in the soil do not fall on stable and hanging piles. This is due to the fact that the VAE are relatively small solid and accordance with it, they are much more deformable than reinforced concrete piles (the strength of the VAE material is much higher than soil strength of reinforced concrete). For this reason, under the influence of external loads in the base, there is more extensive stress dissipation with the depth. At the initial moment of loading, the external load is distributed in the foot of pile grid foundation. At further increasing of loads the VAE themselves are gradually joined in works, through which grounds are included in deeper layers of base. There is also a significant difference between the looking conditions at the contact of the side surface of the VAE and the surrounding soil. For example, for cementsoil piles the distinct side service does not exist and the diameter of the piles is approximately take from the center to conventional borders, equal to the diameter of the screw (a working body) or the zone of water – cement jet influence (injection technology). At the "contact, there is" sticking hardening of surrounding soil to the soil cement pile and therefore interaction terms are changing.

As our researches have shown, when calculating the bearing capacity of soil cement piles on loess soils, it is necessary, to consider not friction but also hanging down of a certain soil volume, located at the same distance from the wall of an armo-element. The area of influence depends on the structural and physical properties of soils and ranges (2–5) D. In the process of subsidence due to overhanging, there is a reduction of the gravitational pressure of soils that prevents subsidence deformation. For this reason at designing bases, strengthened by vertical elements the distance between them should be taken for calculation.

The deformability of bases under foundation besides others depends on the Thickness of a damping layer. Of fundamental importance is the correct selection of the mechanical model of the reinforced soil material. In particular, the strength of both compacted and fixed ground obeys to the Kulon-Mohr theory of strength the VAE unlike cohesive soil and nonecohesive compacted soils characterized by highly structural strength and is characterized by the increased adhesion. The strength limit of these materials is defended similarly to the prism strength of concrete, according to the results of uniaxial compression. The strength of the VAE corresponds to prism strength of material which must be at least 700 kPa. The calculated model of such materials can be adopted as elastic-plastic one.

The longitudinal stiffness both of cement soil and concrete VAE is considered provided when conditions are performed: the ratio of the diameter to its length is D/L ≥ 0.05.

In dependence on volume of loads and the power of the weak subjected to soils reinforcement partial (on the fixed depth) or reinforcement of the entire depth of the spread weak soils.

Reducing the impact of reinforced concrete raft foundations of strength and longitudinal rigidity, the

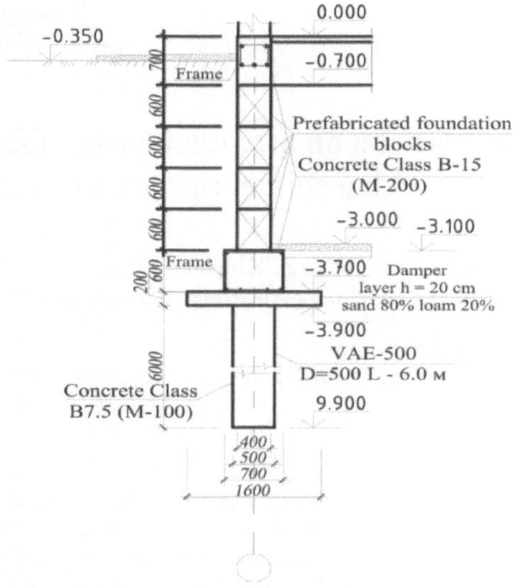

Figure 1. The cross section of the raft foundation and VAE.

VAE are archived by a damping sandy layer. In practice, the thickness of thise layer is taken 15–25 cm (Fig. 1). The width of the damping layer is assumed to be equal to 1,4*b, where "b" Is the width of the pile grid.

In [1.2], the engineering methods of calculation are shown to determine the bearing capacity and deformability of the reinforced VAE. For example for loess soils with physical properties as:

Density $\gamma = (1.60 - 1.8)\,\mathrm{g/cm^3}$; ratio of porosity is e = 0.88 – 0.95; humidity is $W = (17 \div 19)$ %, and the number of plasticity $I_p = (6 \div 8)$ with deformation and strength characteristics of:

$$\frac{E}{E_{sat}} = \frac{(70 \div 100)\text{мПа}}{(40 \div 50)\text{мПа}}; \ \varphi = 22 \div 28^0$$

and $c = (20 \div 30)\,\kappa\Pi a$.

Approximately the bearing capacity of a single concrete VAE with the length of More than 3m can be determined by empirical formula:

$$F_d = \frac{A * Lk}{m} \quad (1)$$

L – respectively the cross area and the length of the VAE. For example, for the same type of loess soils in Uzbekistan, coefficient $K = 0.5$ and $m = 0.003\,\mathrm{m^3/KH}$. The distance between the VAE is determined by calculating the expression:

$$L_{BA\text{Э}} = \frac{F_d - A * P_{sl}}{q - P_{sl} * B} \quad (2)$$

Where P_{sl} – initial subsidence pressure (kPa) or calculated resistance of a weak ground base: q-external distributed load on the pile grid and B-is the width of the pile grid.

The calculation of pile grids is made on the model of Winkler-Pasternak as a beam on an elastic foundation with two parameters C1 and C2, coefficient of sub-grade reaction (elasticity) are determined by the known expression [2], as the average pressure ratio P under the title block to the sagging S.

$$C_1 = \frac{P}{S},\ C_1 = \frac{E}{(1-\mu^2)\omega\sqrt{A}}$$

or

$$C_1 = \frac{\pi}{2}\frac{E}{(1-\mu^2)B\ln(4\alpha)} \tag{3}$$

The VAE in these calculations, both elastic deformable supports and installed along the beam with L_{VAE} step.

The deformability (rigidity) of elastic supports is determined with the help of rigidity coefficient K as the ratio of the force N to the displacement (settlements) S, according to the expression:

$$k = \frac{N}{S} \tag{4}$$

Coefficient of the VAE rigidity is more exactly determined by the results of experiments in natural conditions, by static loads with measurements of the VAE displacement. The deformability (rigidity) of the VAE depends on the diameter, the length of the pile and deformation characteristics of soil. The damping layer from sand is considered as an elastically-plastic body.

Table 1. shows the numerical results of foundation calculation on too considered variants.

In seismic areas, a sandy layer among others acts as a damper. It allows to reduce the impact of shear loads transmitted to the building. Since, a sandy layer of the damper obeys the low of elastic-plasticity under shock loads, it at the moment loses its original strength and transmits into another more stable equilibrium position. In thus case, it is expected to reduce the thickness of the damper, lowering contact stresses under the raft and increasing the loads on the VAE. In essence, besides other functions, mentioned above, the VAE serves the role of a short-term support at impact of dynamic forces, occurring during earthquakes. The construction of the VAE join and a raft are presented in fig.1 with the aim to reduce the coefficient value of friction and permeability, sandy soils of the damping layer are mixed with loam (15–20% on composition).

Table 1.

Name	Foundations on a natural base	Foundations – rafts; bases – the VAE	Coefficient of efficiency
Vertical Displacement Max/Min mm	54/46	21/13	2.5/3.5
Q = (K H)	140	163	0.85
My (K H m)	650	300	2.2

To reduce concentration of contact stress, occurring at the level of the VAE cap, a supporting concrete slab is provided, it is also serves as a leveling layer. The note of the VAE level is controlled by using the level. The difference of notes should note exceed 1–2 cm.

By this technology more than one hundred buildings from 2 to 7 floors buildings have already built. As surveys showed, such buildings, even in the case of accidental soaking of a soil base are exploited without damages. Over the last 5 years of operation, such buildings were also suffered 4–6 magnitudes of earth quakes. As the surveys of buildings after earth quakes showed, damages of buildings, where the bates stabilized with the use of the VAE were applied, they were much lower than in buildings build on compacted ground pads.

Technical – economic comparison of ground pads and traditional strip foundation have shown savings of resources and construction cost of zero cycle erection of a buildings make up to 20–35%.

For comparison, when constructing buildings from 3 to floors, cross-section of rafts does not exceed 60 and 70 cm. The advantage of this method, besides the above-mentioned, is lack of the influence of seasonal prevalence (winter and spring rainy seasons of the year) and water saving, needed to achieve optimal moisture in dry hot climate.

1 CONCLUSION

On the result of simulation of the VAE behave our and as well as on the results of field observations of buildings we can draw the following conclusions:

1. Stabilization of grounds by VAE allows to reduce significantly the emergence of uneven saggings and increase the bearing capacity of a weak base, composed of loess soils.
2. For soil conditions of the 1-st and 2-nd types on subsidency ($H_{sl} < 10 - 15$ m In dependence of operating loads: the length of the VAE can be taken by calculation, but not less than 3–4 m).
3. The approximate calculation of bearing capacity of the concrete VAE with the diameter D and the length L for the described soils can be determined by expression (1)
4. In case of application of cement-soil piles, it is expedient to concrete the upper head to the depth of 1–2 m by lean concrete. Since, it is in thus area, maximum horizontal deformation came into existence.
5. The presented calculation model of the VAE interaction and the surrounding it soil and foundation grills with coefficient of sub-grade reaction allows to make calculation with the use of the known software packages

On the results of the carried out researches, the authors have prepared the text-book on designing bases stabilized by the VAE and advice on application of thus method can be found at 140147 Republic of Uzbekistan, Samarkand, Lolazor 70, Samarkand state Architectural and Civil engineering Institute, LLC "Geofundamentproject" or at the State Comity for Architecture and Construction of the republic of Uzbekistan or e_mail: uzssmge@gmail.com.

REFERENCES

Khasanov A.Z., Khasanov Z.A. Bases and foundations on loess subsiding soils. Tashkent, 2006 IPDT "Uzbekistan".

Khasanov A.Z., Usmonkhodjaev I.I., Khasanov Z.A. Text book on designing converted bases in the form of vertically reinforced soil by bored concrete piles from soil materials for low use buildings on subsiding loess and no subsided clay soils. State Comity for Architecture and Construction of the republic of Uzbekistan. Tashkent 2010. 37 pp.

Designers handbook. Base, foundations and underground structures. M., Stroizdat, 1985-480 pp.

Gorodetsky E.B. and others. Method of finite elements: theory and numerical implementation. Program complex Lira-SAPR.

Software package Plaxis 7.2 Netherland. www.plaxis.nl.

Challenges and Innovations in Geotechnics – Zhussupbekov (Ed.)
© 2016 Taylor & Francis Group, London, ISBN 978-1-138-03007-7

Monitoring and protection of the foundations of buildings of dangerous production plants on seasonally heaving ground

S.A. Kudryavtsev
Far Eastern State Transport University, Khabarovsk, Russia

V.V. Kogay
NIIOSP named after N.M. Gersevanov – Research Centre of Construction JSC, Moscow, Russian Federation

A.A. Moldazhanova
Eurasian National University "L.N.Gumilyov", Astana, Kazakhstan

I.V. Shestakov & A.V. Kazharzkii
Far Eastern State Transport University, Khabarovsk, Russia

ABSTRACT: The need for monitoring the technical condition of buildings and structures of hazardous production facilities requires the creation of a set of measures to ensure to maximize the information to take into account the various factors affecting the safe operation of the facilities.

1 INTRODUCTION

Buildings and facilities located on the territory of hazardous production facilities (GRO) and related also subject to monitoring and inspection of technical condition. But at the moment such events mainly concerns the smoke and ventilating stacks, buildings, pumping stations and substations. To a lesser degree, this concerns a lightweight construction of foundations containers, equipment and heat exchangers. This results in evaluation of the technical condition of the objects only visually accessible components. Meanwhile, during the period of service is located on the foundations of the equipment, as well as technological procedures (filling and emptying) produced during their operation is possible manifestation of the deformation of foundations and bases, associated with a change in the load, which can vary by tens of times. The equipment can be taken out of service for a long period in unfavorable period. In addition to the operation of the physical and mechanical properties of the soil may be subject to change their properties due to penetration into the thickness of the soil storage equipment washing results.

2 METODS OF MONITORING OF BUILDINGS AND STRUCTURES

Since the operation of buildings and structures, as well as their bases and foundations receive damage and are subject to change physical characteristics under the influence of various factors. As measures to control, the state of the main structural elements of buildings and structures arranged for assessing the current technical condition. The main disadvantage in the diagnosis and monitoring of buildings is the lack of technologies for integrated assessment of soil-building system in the three-dimensional dynamic setting and its linking with an integrated security system for monitoring and life support systems of the entire object.

It is worth noting that as a monitoring to determine the soil pellet was used downhole magnetic extensometers, automatic extensometers fixed for the determination of total sediment monitoring system and the differential precipitation (DSM) for the automatic and continuous long-term monitoring of buildings layered sediments. The horizontal displacement of soil are monitored by satellite methods. Expansions of the brand in the form of GPS-receivers are fixed on the ground or on an end surface of the end of the anchor, which lay at a depth of interest to researchers. Alternatively, control the horizontal displacement of soil – tool downhole monitoring with portable inclinometer probes automatic or Incline-metric systems. The water level in wells and pore pressure are monitored piezometers.

However, the main obstacle to the widespread introduction of these systems is their cost. As an alternative measure may be called the use of ultrasound and seismic-acoustic equipment. The use of the above equipment makes great certainty be defined as the technical condition of building structures and mechanical properties of soils.

3 MONITORING USING RESEARCH GEOACOUSTIC

In 2015, for the reconstruction of a catalytic reforming unit of OJSC "Khabarovsk oil refinery" was carried out technical inspection of the foundation state capacity E-2 due to changes in the technological scheme. Changing flowsheet assume periodic filling capacity, as well as operation with incomplete filling, in connection with this load, from the tank to the foundation betray varies from 52 tons to a filled condition of 11 tons – in the absence of liquid. When operating without liquid tank available foundation deformation during freezing thaw from the impact forces of frost heaving. In this regard, it was necessary to perform the assessment of the technical condition of the foundation, evaluate the physical and mechanical properties of the soil, the development of measures to reduce the forces of frost heaving, no meaningful change in the foundation design. Survey subject to the pile foundation grillage with circular cross-section, as well as its base. The foundation is made of reinforced concrete design class B15 strength. In the installed foundation reinforcement cages with a diameter of rods 6 and 8 mm, the reinforcing steel grade A-I and A-II. The foundation raised up to 200 mm in height from the zero mark of the planning figure. This made it possible to determine the lateral dimensions of the grillage, which amounted to and Ø2880 mm.

Because the work was carried out on the existing installation, as well as because of the cramped conditions, there was no possibility of developing pits, evaluation of technical condition of foundations below the earth's surface, as well as the physical and mechanical properties of the soil carried out by seismic-acoustic sounding foundation and seismic sounding soil method refraction. When conducting seismic-acoustic sensing used seismic "seismologist-20" Type of used sensors – geophones GX-20DX SUPER (manufacturer OYO-GEOIMPULSE), vertical and horizontal, with a natural frequency of 10 Hz and an upper limit of bandwidth to 2000 Hz, with improved characteristics especially for the 24-binyh data collection systems. For excitation of the desired signal, used artificial source – a sledge hammer, 5 kg, equipped with piezoelectric transducer. Attempts were made vertically and horizontally protruding parts bases depending on the type of target waves. Seismometers were set based on the dimensions of the surveyed foundations. The nearest to the place of impact geophones placed him at a distance of no more than 0.2 m.

During the work it was determined the strength of the above-ground part of the concrete foundation, which amounted to 34 MPa. Available in working documentation design scheme of the subject regarded

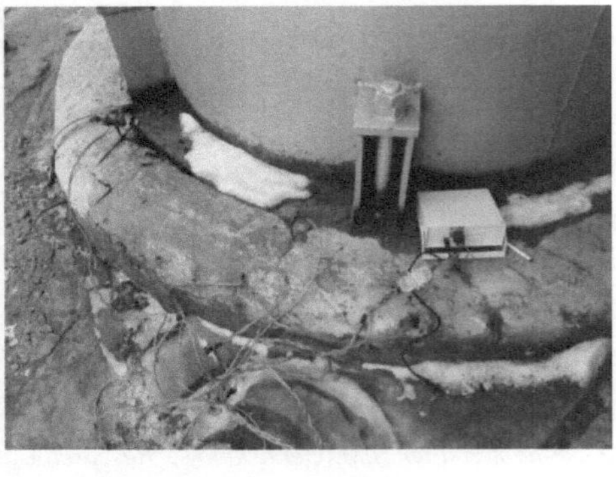

Figure 1. Placement of geophones scheme.

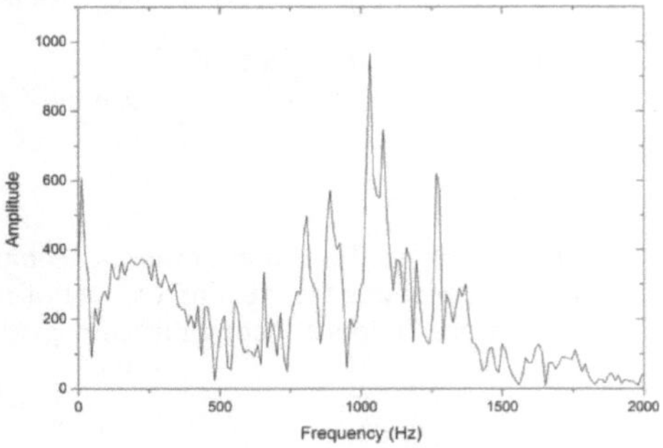

Figure 2. Fourier spectrum of longitudinal vibrations to the foundation.

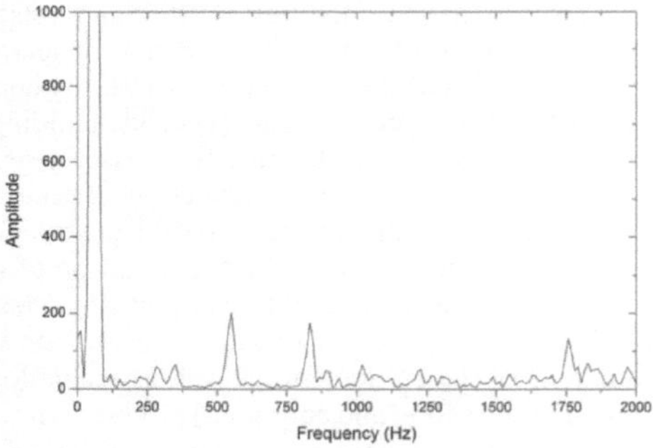

Figure 3. The Fourier spectrum of the transverse oscillation on the foundation.

as the foundation of the first approximation to the actual size, which had to be set in the seismic-acoustic research.

However, due to a long recording length (0.68 milliseconds) has its portion which can be identified with longitudinal and transverse dimensions of the foundation. In the fragment shown in figure 4 with the visible portion of a constant period of 0.002 millisecond that the longitudinal wave velocity Vp = 4000 m/s in reinforced concrete class B12.5 corresponds to the size shown in figure 5 given in the documentation.

Figure 2 and 3 show the Fourier spectra obtained at sounding foundation.

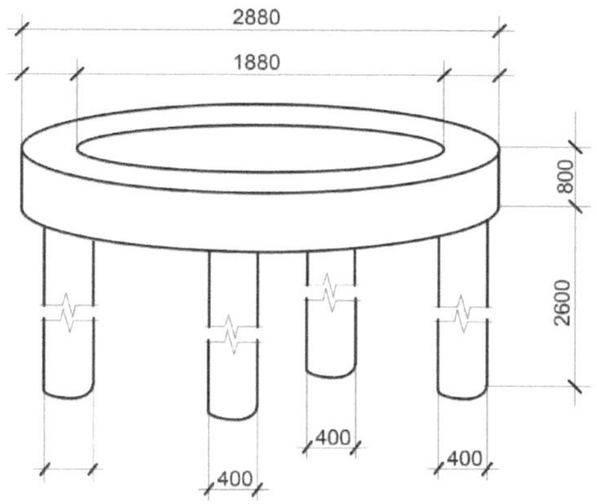

Figure 4. The resulting scheme.

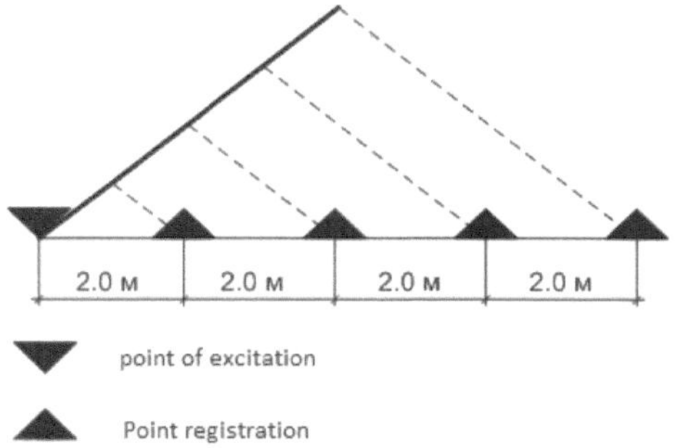

Figure 5. Observing system when developing seismic profiles.

In processing the data took into account the resonance peaks, expressed in the spectra of longitudinal waves of 1033 Hz and the transverse waves – 833 Hz. Method for determining the vertical and transverse dimensions described above. The resulting circuit is shown in figure 4.

Seismic surveys provided for sensing ground stratum, in order to determine the velocity of elastic waves and the associated functional or correlative basic physical and mechanical properties of soils. The work performed used a variant of seismic – sensing on direct waves on profiles, broken within the surveyed facilities.

Performing sensing by observing the system shown in figure 5 conducted through the concrete floor, with an average thickness of 10 cm. For direct access to the ground, in the screed reams holes with a diameter of 1.2 cm, which is inserted in special polypropylene liners, which are installed geophones. The excitation of seismic waves and carried through a hole in the concrete, sledgehammer blows on rebar diameter of 1.0 cm.

This arrangement allowed the measurements virtually eliminate the problem of inversion of seismic waves, and receive their speed related only directly

Table 1. Statement of the measurement of soil characteristics in the sole of the surveyed foundations.

Point number	Density ρ, g/cm^3,	Specific clutch C, kPa	Angle of internal friction φ, degree	total deformation module eE_0, MPa
FD	1,62	10	18	8

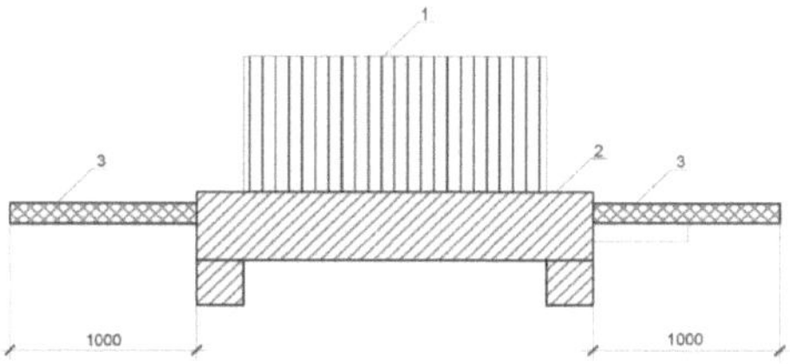

Figure 6. Insulation installing scheme. 1- capacity E-1; 2- the foundation; 3 thermal insulation.

from the soil base (Bendat J., Piersol A. 2000, . Dummer A.E., Tyurin I.M., Varnavskiy V.G. 1991, White J. E. 1986).

When calculating the physical properties of the soil (because of its heterogeneity across the grounds on which the foundations studied), paragraph 4, was used (Kudryavtsev S.A. 2004), where in tables 6 and 8 are all necessary to determine the dependence of the shear characteristics.

At the site was inspected object defined by three points of the subject of the foundation sole, which was required to determine the specific soil adhesion, density, angle of internal friction and modulus (White J. E. 1986). To this were worked seismic profiles, according to the schemes, as shown in figure 5.

At the base of the foundations is loam therefore used to calculate the characteristics of the relevant correlations according to (Kudryavtsev S.A. 2004).

As lying in the bottom of the soil refers to heaving in accordance with GOST 25100-2011, numerical simulation using software FEMmodels Thermoground product was carried out. Numerical modeling was performed to determine the effectiveness of the insulation, as a measure to reduce the impact forces of frost, as interference in a constructive foundation scheme and conduct large-scale excavation at the site is not possible. insulation arrangement is shown in figure 6.

4 ESTIMATED STUDY USING NUMERICAL MODELING

Different types of diagnostics are mostly local in nature, based on the visual strength and the process control with the selection of material samples and studying the constructive execution of the individual components by opening nodes, protective layers, and the development of pits for the study of the actual state of the bases. Making local control, estimating focusing on particulars, you cannot understand the mechanism and the main reason of the object of vulnerability to possible loads. The same local approach exists for the design of buildings, that is, soils and foundations are considered separately from the building itself, there is no single approach to the calculation of the system soil-building as a single system. A large part of software products used in the calculation and numerical simulation of the foundations of buildings and structures allow us to estimate the current state as a result of visual, laboratory and instrumental control, but they do not allow to simulate the dynamic picture, on the basis of which it is possible to make long-term prognosis, and as a result develop most effective recommendations.

Program complex FEM models developed by Geotechnics St. Petersburg, led by Professor V.M.Ulitskogo (Kudryavtsev S.A. 2004, Kudryavtsev S.A., Paramonov V.N., Sakharov I. 2014). An integral part of the program is "Thermoground" to solve the thermal problems of the finite element method. In the creation of the module "Termoground" participated K.G.Shashkin, V.N.Paramonov, I.I.Saharov and S.A. Kudryavtsev (Kudryavtsev S.A. 2004, Kudryavtsev S.A., Paramonov V.N., Sakharov I. 2014). The process of calculating the freezing, frost heaving and thawing solved in two stages. Initially it solved the problem of temperature distribution and humidity with phase transitions "water-ice" in the range of negative temperatures. As a soil model using the model proposed Y.A.Kronikom in which the thermal and thermodynamic characteristics of soil are functions of temperature (Kudryavtsev S.A. 2004, Kudryavtsev S.A., Paramonov V.N., Sakharov I. 2014). Thawing processes are performed according to the method GOST19706-74 and sensitive proposals M.F. Kisileva [5, 6].

The principle of operation of the module "Termoground" based on a set of factors into account: unsteadiness, physical and thermal nonlinearity work soils during the freezing and thawing in the spatial setting. A rigorous solution of these tasks involves consideration of phase transformations of water in the range of negative temperatures, as well as the presence of moisture in the freezing zone, migrating to the freezing front. It is this proximity to moisture during the freezing front of groundwater is a major contributor to the deformation of frost heaving and deformation subsequent thawing (Kudryavtsev S.A., Paramonov V.N., Sakharov I. 2014).

The amount of water in the soil during freezing element in "Termoground" software module is installed when the thermophysical calculations. At the same boundary conditions of heat and ground water the natural moisture and the amount of moisture in the introduced element when migrating determine the provisions of the amount of water in the soil element. If the natural humidity is the same, the individual for moisture in the element determined thermal characteristics of the ground (this determines the temperature distribution) and suction characteristics – different magnitude of migration flows, corresponding to different types of soils.

In three-dimensional space, the temperature distribution of the process in the range of negative temperatures for transient thermal regime is expressed by the equation:

$$\rho_d(C_{th(f)} + L_0 \frac{\partial W_w}{\partial T})\frac{\partial T}{\partial t} = \lambda_{th(f)}(\frac{\partial^2 T}{\partial x^2} + \frac{\partial^2 T}{\partial y^2} + \frac{\partial^2 T}{\partial z^2}) + q_v, \quad (1)$$

where Sth (f) – the specific heat of melting and frozen ground; ρ_d – density of dry soil; T – temperature; t – time; The L0 – specific heat of phase transitions "water – ice" per unit of weight; $\lambda_{th(f)}$ – thermal conductivity or melting of frozen ground; x, y, z – coordinates; qV – power internal heat sources; Ww – unfrozen water content in the soil.

The system of finite-element equations of heat conduction problem can be obtained by minimizing the corresponding functional on the set of functions that satisfy the boundary conditions. From the point of view of the variational solution of 1 with the boundary conditions 1, 2, 3 and 4 kind of equivalent to finding the minimum of functional

$$\chi = \frac{1}{2}\int_V \lambda_{f,th}\left[\left(\frac{\partial T}{\partial x}\right)^2 + \left(\frac{\partial T}{\partial y}\right)^2 + \left(\frac{\partial T}{\partial z}\right)^2 - 2\left(q_v - C_{f,th}\frac{\partial T}{\partial t}\right)T\right]dV + \int_{S_q} q_n T dS + \int_{S_\alpha} \alpha\left(\frac{1}{2}T - T_\infty\right)T dS \quad (2)$$

that leads to a system of ordinary differential equations written in matrix form as follows:

$$\left[C_{f,th}\right]\frac{\partial T}{\partial t} + \left[\lambda_{f,th}\right]T + \{F_k\} = 0 \quad (3)$$

where $[C_{f,th}]$ is the matrix of soil heat capacity in the frozen and thawed sosotoyanii; T – vector of nodal temperatures; t – time; $[\lambda_{f,th}]$ – The thermal conductivity of the soil matrix in the frozen and thawed condition; {Fk} – vector of the right parts of the system resolution.

Unknown function approximated to the temperature T at time t in the elements and in the whole region under consideration shape functions N (x, y, z):

$$T = \sum_{i=1}^{n}\{N(x,y,z)\}_i^T\{T(t)\}_i = [N]\{T\} \quad (4)$$

The matrix elements of the heat capacity of the system has the form

$$\left[C_{f,th}\right] = \sum_{i=1}^{n}\left[C_{f,th}{}^e\right] = \sum_{i=1}^{n}\int_{V^e} C_{f,th}\rho[N]^T[N]dV \quad (5)$$

A matrix element of the system thermal conductivity is of the form:

$$\left[\lambda_{f,th}\right] = \sum_{i=1}^{n}\left[\lambda_{f,th}{}^e\right] = \sum_{i=1}^{n}\int_{V^e}\lambda_{f,th}[B]^T[B]dV + \int_{S_\alpha}\alpha[N]^T[N]dS \quad (6),$$

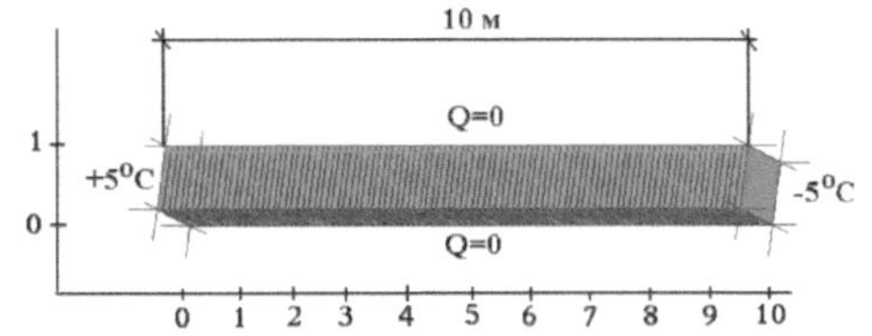

Figure 7. An example of a three-dimensional computational model for the distribution of the heat flow software module "TERMOGROUND".

$[C^e_{f,th}]$ – The heat capacity of the matrix finite element; $[\lambda^e_{f,th}]$ – The thermal conductivity of the matrix of the finite element; [N] – matrix finite element shape functions; [B] – the matrix of derivatives of functions of the form of the final element of the coordinates; $S\alpha$ – the surface area through which heat exchange; α – coefficient of heat transfer surface.

The vectors of the nodal heat gains are as follows:

$$\{F_k\}_i \sum \int_{S_a} N_i (q_v - \alpha \cdot T_a) \partial S \quad (7)$$

The solution of the differential equation 2 can be obtained by a finite-difference scheme. The most simple scheme – the left finite difference:

$$[C]\frac{T_n - T_{n-1}}{\Delta t} + [R]T_n = 0 \quad (8)$$

where Tn – current temperature of a discrete point in time;

Tn-1, – the temperature in the previous discrete time.

Hence we finally arrive at the permissive system of finite-element equations:

$$([C] + \Delta t[R])T_n = [C]T_{n-1} \quad (9)$$

The system of equations (8) is a self-starting, because at the time t0 the temperature field is known and is equal to the specified value.

Unlike previously had theoretical solutions, developed a software module "TERMOGROUND" numerical simulation of the freezing process, frost heaving and thawing adequately describes the processes of heat transfer, moisture, strain and stress of frost heaving and thawing when changing internal and external conditions in the spatial setting. Precision thermal and stress-strain calculations are largely determined by the accuracy defined physical, mechanical, thermal characteristics, the speed of the atmosphere changes in temperature and the level of groundwater. Therefore, their purpose for practical calculations in specific soils it is advisable to carry out according to the engineering and geological surveys using field and laboratory methods. (Kudryavtsev S.A., Paramonov V.N., Sakharov I. 2014).

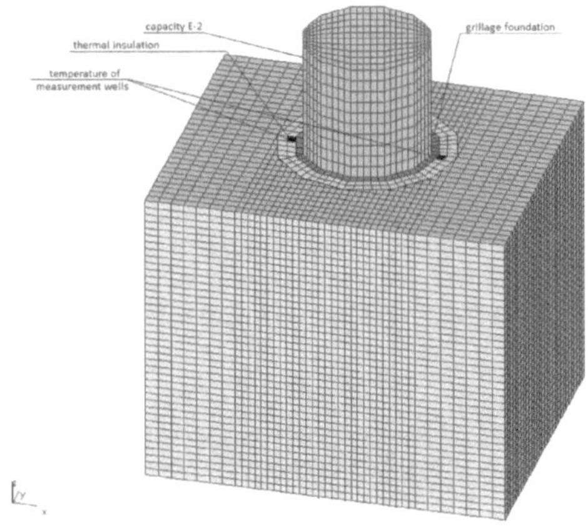

Figure 8. Design scheme thermophysical task.

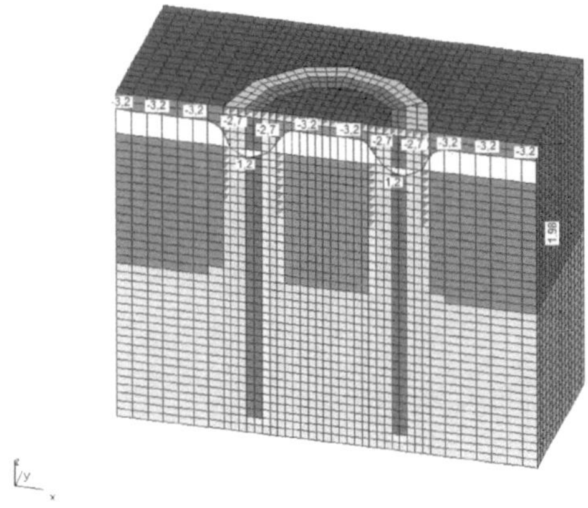

Figure 9. Diagrams of temperature distribution and areas of freezing and thawing the base of pile foundations catalytic reforming capacity in the month of February.

Design scheme to address the thermophysical problem is presented in figure 8.

Originally it was the numerical simulation at ambient temperature characteristic of the month of February for the city of Khabarovsk, as maximizing the depth of soil freezing occurs in February.

The second step was the implementation of a numerical simulation using a heater circuit for the foundation raft foundation, with the placement of insulation in the level of the surface. As polystyrene insulation 300 mm thick. Numerical simulation was performed for February, to allow comparison of temperature and determine the depth of frost penetration. numerical simulation results are shown in figure 9.

Figure 9. Diagrams of temperature distribution and areas of freezing and thawing the base of pile foundations catalytic reforming capacity in the month of February.

As can be seen from the numerical simulation results in the case of discharge or insufficient filling capacity in winter is possible deformation of foundations associated with underutilized foundation. At the same time, the use of insulation allows the contour of raft foundation can reduce the negative impact of the forces of frost heaving due to conservation temperature indicators directly under the foundation, in which the possible onset of frost heaving.

In general it can be noted that for hazardous production facilities is the most appropriate monitoring using non-destructive inspection techniques together with software allows to carry out numerical simulations to assess the current condition and forecasting capabilities.

5 CONCLUSION BASED ON THE RESULTS OF THE WORK PERFORMED

1. Application of seismic sounding methods, with a permanently installed geophones to rapidly detect the appearance of possible defects and damage to the foundation, as well as to evaluate possible changes in the physical and mechanical characteristics of the grounds.
2. Numerical modeling by finite element method of freezing processes, frost heaving and thawing allows prognosis in the long term, and to date, allowing to solve engineering problems related to the reconstruction and repair of buildings and structures during the operation.

REFERENCES

Bendat J., Piersol A. 2000. Analysis and Measurement Procedures, 3rd Edition. 624 p. New York: John Wiley & Sons.

Chmielewska V.K., Lyakhovitskii F.M., Yaschenko Z.G. 1989. Engineering geophysics, Nedra. Moscow.

Gerdalt L., Sheriff R. 1987. Seismic. In 2 that's, that 1, 448 p., translated from English, (eds.)Mir. Moscow.

Dummer A.E., Tyurin I.M., Varnavskiy V.G. 1991. The geological structure and geotechnical conditions of Khabarovsk and its surroundings. Khabarovsk.

Kudryavtsev S.A. 2004. Settlement and theoretical basis of designing and building structures in a freezing heaving ground, ð. 39. St.- Petersburg.

Kudryavtsev S.A., Paramonov V.N., Sakharov I. 2014. The freezing and thawing of soil practical examples and finite-element calculations, 248 p. St. Petersburg.

White J. E. 1986. Excitation and propagation of seismic waves. Translated from English. Goldin S.V.& Pavlova OV. Nedra,216 p. Moscow.

Zalmanzon L.A. 1989. Fourier transform s, Walsh, Haar and their use in the management of communication and other fields. Nauka, 496 p. Moscow.

Challenges and Innovations in Geotechnics – Zhussupbekov (Ed.)
© 2016 Taylor & Francis Group, London, ISBN 978-1-138-03007-7

Hazardous geological processes in the northern Sikhote-Alin

S. Kvashuk, P. Iazvenko & V. Iazvenko
Far Eastern State Transport University, Khabarovsk, Russia

ABSTRACT: Nortern Sikhote-Alin is characterized by its unique geology and consequently difficult engineering-geological conditions. At present, due to the development of the region, its transportation infrastructure in the first place, further study if hazardous exogenous geological processes is very important.

The authors have modernized the method of quantitative evaluation of denudation processes and implemented it on testing sites located in Nortern Sikhote-Alin. Additionally they have studied characteristic gravitation processes of the line and analyze here one of the catastrophic accidents.

1 INTRODUCTION

The territory discussed is located in the Northern part of Sikhote-Alin Range – mountain terrain within the Pacific Belt (Fig. 1).

Tectonic structure of the region is influenced by the Central Sikhote-Alin regional fault and its multiple local offset faults (1).

Block structure of the Amur River Region was created by inherited neotectonic Mesozoic and more aged tectonic dislocations. The kinematics of the north-east strike old-age active faults is most often referred to as of normal and thrust pattern sometimes having strike-slip component commensurable with vertical displacements. This factor is the leading one to stipulate engineering-geological conditions of structures on hard rocks because it rules high rate of fracturing of the rocks. Sikhote-Alin zone is a 7 MSK-64 intensity earthquake prone area. The territory of the study has seasonal sporadic permafrost.

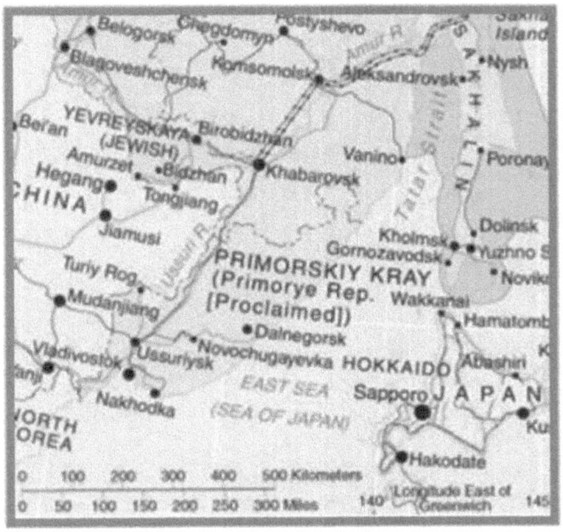

Figure 1. Map of the studied area.

North Sikhote-Alin enters Sovgavan super-basin as well as Primorsky and Sikhote-Alin hydro-geological massifs. Fractured soil, fractured sheet and, more rarely, fractured veiny and pore-soil underwater are connected with the top rocks of the super-basin. In the fractures of the weathering zone of foundation strata waters and the mountain framework waters of dislocations are formed.

The important trait of the territory is its complex hydrographic network which is represented by the river systems of Tumnin and Gur. The character of the flow of these rivers is rapid mountain-like. The contour of the regions' river pattern is very complex which makes construction and operation of engineering facilities cumbersome.

Climate of the territory is severe and typically continental, of monsoon type, forming in the mean latitudes of Asia continent outskirts, in the conditions of close interaction of land and ocean. Specific geomorphology of the region where mountainous landscape is combined with vast isolated planes defines its exclusive contrast of climate. During the winter as cold cyclones move in from central parts of Siberia and Arctic towards the ocean they are being impeded by the mountain ranges located along the coast which makes the temperatures very low throughout the entire territory. The ranges weaken the advection from the sea during summertime and increase the rates of the vertical temperature gradients.

The geological conditions discussed define wide spread of hazardous exogenous geological processes in the region. They are rock falls, block falls, debris falls, frazil phenomena, plane and concentrated wash-out, high rates of water inflow in cuttings, underground structures, etc.

2 FORMULATION OF THE PROBLEM

Railroad sections in the passing part of Sikhote-Alin are characterized by hazardous gravitational phenomena - rock falls, block falls, debris falls.

The territory began to be developed 80 years ago which had stipulated low mechanization, prevalence of manual work and massive use of rock blasting. Dips of the slopes along the line did not correspond modern design and construction codes for a long time. In some sections of the railroad dip angles of slopes exceed 45–60 degrees by the slopes' elevations exceeding 50 meters and rock traps widths up to 1 meter, thus slopes of the cuttings characterized by critical values of main design parameters are not safe for operation of the line.

Wide spread of rock-falls along the line can be explained by the peculiarities of its local sections:

- high rate of difference in elevation of landscape;
- complex tectonics;
- high fracturing of the tectonic origin;
- presence of hydrostatic head from the slope massif.

The main factor that explains the gravitation processes is the massive tectonic fracturing. The Dip of the fractures to the cutting slopes is often unfavorable. Sporadic watering can be noticed in the slopes. There is an additional factor that explains rock-falls: filling of the fractures by sandy or clayey aggregates which in its turn nfluences stability of individual blocks. Rock blasting affected fracture opening during the construction as well.

Rapidly changing activization factors of rock-falls are mostly but often technogenous. They include dynamic forces of heavy trains, catastrophic blasts, accidents, natural disasters, etc.

It should be noted here that the greatest rock masses were observed in the cuttings of mountain pass part of the railroad in 1960s when they reached 800–1200 m^3. Even though these days such rock falls are not observed this is not the evidence of the elimination of the problem. For instance, one boulder 1 m in diameter and weighing 1.5 ton may reach 80 km/hr velocity of its falling out on the track when the width of the trapping bench is insufficient.

The development of rock-falls is influenced by active processes of weathering. In this connection the necessity to monitor slopes of the cuttings and semi-cuttings is up to date, which in its turn defines the necessity to estimate their average rate of degradation – rate of denudation.

3 RATE OF DENUDATION DEFINITION

The authors have developed digital photogrammetry system which allows to estimate the denudation processes using 3D work model of the landscape created automatically with specially worked out software.

The essence of the method is that the model of the land is being built from the shots and outward orientation of the model. Phototriangulation offers to define with individual shots vertical location elevation of base points which are necessary for the creation of 3D digital model of the area.

Digital photogrammetry system was developed by the authors to build 3D digital model of the area which consists of the variety of software products and specially created additional scenarios aimed to simplify the system for photogrammetric reconstruction (2).

The process of the creation of 3D model consists of the two main stages:

1. Field stage.

The main task here is to obtain multiple photographs of the landscape of a single location during different periods of time based on which digital model of the territory is built.

2. Laboratory works consist of two parts:

a) Processing of the obtained photographs and getting the model of the territory is carried out automatically using the developed photogrammetric digital system.
b) Immediate computation of denudation processes velocities.

Here the digital models of different times are being processed by comparison of the areas of design sections. Fixation of changes for a certain period and computation of the denudation rate on the site are performed.

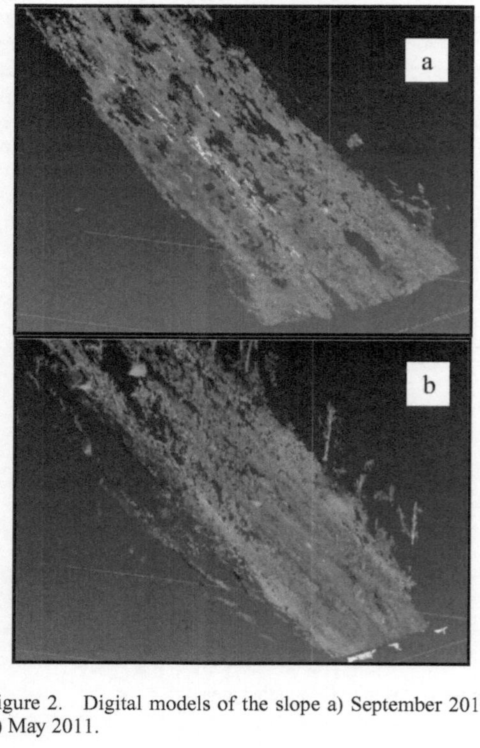

Figure 2. Digital models of the slope a) September 2010, b) May 2011.

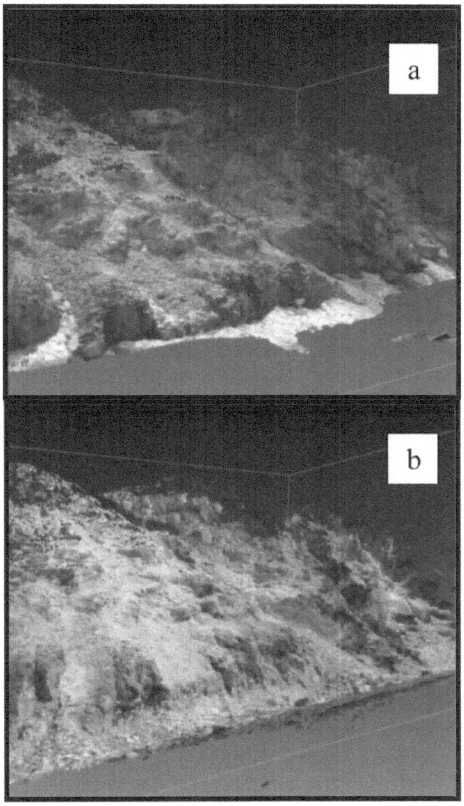

Figure 3. Digital models of the slope a) May 2011, b) June 2012.

Precision of the results obtained is estimated by the comparison of the design model to the main poins of slope's micro-topography measured during tachymetric survey using range finder.

Rate of denudation processes, among other things, was estimated on two sites composed of shalestones (Fig. 2) and andesites (Fig. 3).

3.1 *Rate of denudation estimation in shalestones*

As a result of the processing of the photographs taken during September 2010 and May 2011 field work with the use of the software described above the 2 run-time digital models of the slope were obtained (Fig. 2).

The models were overlapped and divided by cross-sections at 1 m spacing for fixation of the measurements.

The slope is composed of highly fractured shalestones and degrades irregularly. There are zones on the cross-sectional profiles with shifted debris, i.e. 2010 slope contour is above 2011 one and zones with crushed-out rock where 2010 slope line is beneath the 2011 one.

Volume of the crushed-out rock was not taken into account for the final result because it wouldn't provide necessary precision due to some features of the site,

Table 1. Rate of denudation definition.

Type of rock	Volume of material brought down V_2, m^3	Interval between surveys T, mo.	Denudation rate m^3/mo.	Average denudation rate m^3/mo. per 1 m^2 of slope
Andesites	13,9	13	1,07	0,05
Shalestones	288,2	9	32,02	0,70

Figure 4. View of the rock-fall prone site near Gurskaya Station: a) before base excavation, b) after base excavation.

firstly, the slope continues up into natural slope thus some debris might had fallen from it and, secondly, railway maintainers regularly remove rock material from the trapping bench. Conclusion that can be made from this fact is that most reliable denudation rate on the site one can get from the outcrop wall retreat volume computation.

The summary volume of the material brought down within design cross-sections reached 288,2 m^3

3.2 *Rate of denudation estimation in andesites*

On the section composed of andesites the studies were done in 2011 and 2012, field measurements resulted in run-time digital models of the slope (Fig. 3).

The rock-fall prone slope is represented by irregularly fractured kaino-andesites. Practically monolythic or little fractured blocks are found at the base of the slope after its bottom excavation, the top part as opposed is represented by crashed rock. 2011 survey of the site was performed in the middle of May with residual snow at the slope's base as it is seen in the resulting model. 2012 studies were carried out in June which explains more detailed outlines of the slope base. The authors took into account the site's specific characteristics to increase the precision of denudation rate estimation. The rate computation was done for middle part of the slope where changes are most evident (analogues to the preceding site).

The summary volume of the material brought down within design cross-sections reached 13.9 m^3.

Final indices of denudation rate are summed up in Table 1.

The difference between the values of denudation is stipulated by genesis and condition of rocks.

Developed software complex allows to get a runtime digital 3D model of an area with enough high precision allowable for the solution of multiple tasks by geological survey and engineering-geological prospecting. It is possible to monitor different objects demanding constant control with this method.

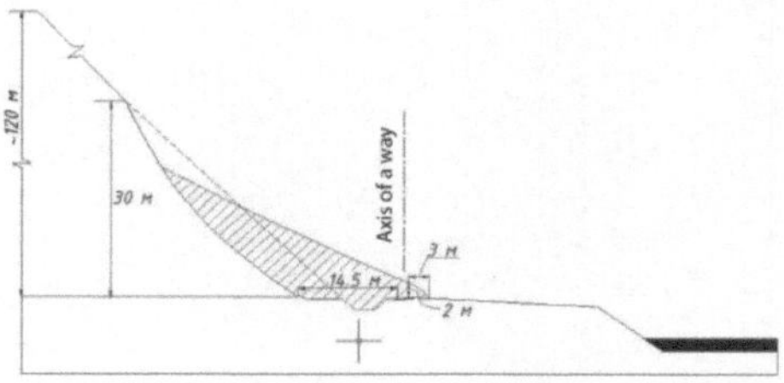

Figure 5. Scheme of the crashing-out near Gurskaya Station.

4 AN EXAMPLE OF HAZARD ROCK-FALL NEAR GURSKAYA STATION

An evident example of hazardous rock-fall occurred near Gurskaya Station. The reason for this was an erroneous design decision – cutting of bottom of the slope to build a trap bench at its base. Fig. 5 shows the site before and after the base cut respectfully.

It is paradoxical that cutting of slope bases is one of the main reasons of rock-falls, i.e. a trigger for gravitational processes' activization.

On March 3 2011 at 16:30 weak rocks crushed out, filled a ditch and spread onto the track with total volume of 5k cubical meters (Fig. 4b and 5).

First momentary stage of the rock-fall lasted for 3 minutes and was video-taped. During the following hours and days individual fragments, boulders and blocks were falling out.

The authors have carried out engineering-geological mapping after this event presented in this paper.

The section is located in the area with difficult engineering-geological conditions which were underestimated during design of trap bench and consequently led to this catastrophe.

Gurskaya station area is 7 MSK-64 intensity earthquake prone area and the interface (along a sub-latitude fault) of two terrains of different ages. Several kilometers to the north from the rok-fall there is Surkumski fault which is an offset of Central Sikhote Alin slip fault (Fig. 6).

The rocks of which this one-sided cutting is composed are shalestones, sandstones and their interbedding, weathered and crushed.

The rocks are significantly fractured, average number of fractures per 1 meter length varies from 10 to 35.

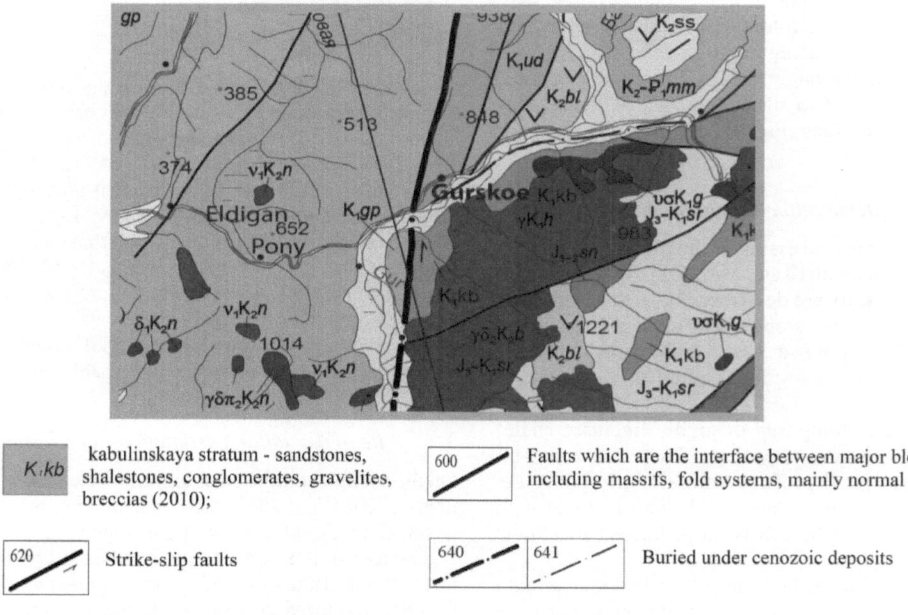

Figure 6. Fragment of the area geological map.

The dip of fractures in relation to the track is unfavorable because vector of the dipping is directed towards the track. The crack opening reaches 1 centimeter. Some cracks are filled with sandy-clayey material which is unfavorable for stability of individual blocks. Depth of the fractures vary from 0.3 to 5 meters. There is evidence of periodic watering on the slopes.

Vegetation is dense mainly represented by birches and larches with developed rooting system which is a crack opening factor.

The angle of the undermined slope plane within the section studied varies from 60 to 80 degrees with its height 20–40 m. The entire height of the slope reaches 120 m.

5 CONCLUSION

The object studied is located in geologically and engineering-geologically unique conditions of North Sikhote-Alin. Complex fault tectonics defining intensive fracturing and block structure of rocks, watering of rock massifs, severe continental climate, which are the factors of intensive physical and chemical weathering of rocks and high rate of denudation, stipulate active development of gravitation processes in the studied region. The developed software complex and methodology of denudation rate estimation are efficient, precise and reliable method to monitor of hazardous process in mountain areas, allowing to design measures for stabilization of hazardous slopes in the right time.

The example of catastrophic rock fall analyzed in the paper showed the high price of engineering mistakes and underestimation of difficult engineering-geological conditions in design of industrial and transportation facilities.

REFERENCES

http://www.roebuckclasses.com/maps/placemap/russia/russianfareast.jpg

Kvashuk S.V., Koltun P.A., Iazvenko V.O. Digital Photogrammetry System in Engineering Prospecting/ S.V. Kvashuk, P.A. Koltun, V.O. Iazvenko //Voronezh: Proceedings of 2nd international scientific-practical conference "Ecological geology: theory, practice, regional problems", 2011 PP. 352–355. (in Russian)

Numerical and analytical modeling (TS5)

Challenges and Innovations in Geotechnics – Zhussupbekov (Ed.)
© 2016 Taylor & Francis Group, London, ISBN 978-1-138-03007-7

Three-dimensional finite element analysis for stabilizing piles used in homogeneous slopes

I-Hsuan Ho
Harold Hamm School of Geology and Geological Engineering, University of North Dakota, Grand Forks, ND, USA

ABSTRACT: This paper presents the analytical methods of slope-stabilizing piles using the three-dimensional (3-D) Finite Element (FE) analysis with the Strength Reduction Method (SRM). This 3-D finite element model is employed to overcome the limitations observed in 2-D FE analysis. The method adopted in this paper can avoid the assumption of soil movement and the pressure distribution along the piles subjected soil movement. The spacing effect of the pile is considered in the 3-D model, the S/D (S: center to center, D: diameter of pile) ratio, equal to 4.0, is found to be equivalent to the single pile stabilization. The middle portion of the slope is identified as the optimal location to place the piles. The proper length of the pile, which can be used to stabilize the slope, is also examined using 3-D FE analyses. It is concluded that L/H greater or equal 0.70 is recommended (L: pile length, H: slope height). The failure mechanisms of the pile-slope system subjected to the pile locations, pile head conditions and pile length are each discussed. The contact pressure, shear force and moment along the piles are presented to illustrate the pile stabilizing mechanism herein.

1 INTRODUCTION

Because they can be easily set up, piles used to stabilize slopes are widely adopted. The advantages of piles used to stabilize slopes have been well documented in the previous literatures. Basically, the stability for the pile-slope system must take into account the slope stability and pile's capacity.

1.1 *Analytical methods*

The analytical methods proposed by other authors are mainly categorized into three types: (1) pressured-based (Ito and Matsui 1975; Hassiotis *et al.* 1997; He *et al.* 2015), (2) displacement-based, (Poulos 1995; Chen and Poulos 1997; Jeong *et al.* 2003; Galli and Prisco 2012) and (3) numerical methods (Goh and Wong 1997; Won *et al.* 2005; Jenck 2009; Kourkkoulis 2012, Ho, 2015). The representative pressured-based method was the theory proposed by Ito and Matsui (1975) where the soil is squeezed between piles in plastic-equilibrium, without considering the soil arching between piles. The assumptions of the method are that the pile is rigid and the length is infinite, and subsequently, it is improper to apply to the finite length and flexible pile. Meanwhile, the soil arching effect is also considered to be one of the important factors that may induce the substantial effect (Kourkoulis 2012). The second method proposes consideration of the lateral movement of the soil, and the piles working as passive piles to stabilize the slopes. The drawback of the displacement method can be concluded as follows: (1) only the single pile was considered, and (2) the distribution of the limiting soil pressure applied on the passive piles is unclear.

The numerical methods such as finite element method and finite different methods became popular due to drastic improvement of computing tools, which are capable of conducting a coupled analysis of the pile-soil interaction. The two-dimensional finite element analysis has been widely employed and applied to geotechnical engineering application in the past several decades. Due to the computational limitations and convenience, several results of the 2-D FE technique have been adopted for solving problems in engineering. However, the 2-D results were found to be more conservative (Griffiths and Marquez, 2007; Nian *et al.* 2012, Ho 2014). In the application to engineering, solutions that are too conservative usually mean the cost will increase. Due to the progression of the computational program and tools, the 3-D FE analyses have demonstrated rigorous solutions and are not time consuming.

1.2 *Strength Reduction Method (SRM)*

In the conventional limit equilibrium method (LEM) for slope stability analysis, the critical slip surface has to be determined. The average shear strength along the slip surface is assumed to be the same along the entire slip surface. The factor of safety (FS) used in the LEM is the ratio of the average shear strength to the driving shear stress along the potential slip surface. However, in the FE analysis, the stress-strain relationship of the slope is considered and the potential slip surface need not be determined in advance. The analysis is based on how the FE analysis is unable to directly output a global factor of safety. In order to quantify an equivalent factor of safety from the limit equilibrium methods, the strength reduction method (Zienkiewicz *et al.* 1975; Ugai, 1989, Griffiths and Marquez 2007, Nian *et al.* 2012, Ho 2014, 2015) is adopted. The strength reduction factor (SRF) is applied to reduce the strength of the soil to the point of failure. The SRF can be regarded as the factor equivalent to the factor of safety in the limit equilibrium analyses. If the strength parameters of the soil are c and ϕ, the c_f and ϕ_f are the factors that will bring the slope to failure, which can be defined as:

$$c'_f = \frac{c'}{SRF} \quad (1)$$

$$\phi'_f = \arctan\left(\frac{\tan\phi'}{SRF}\right) \quad (2)$$

In the application of the strength reduction method in finite element analyses, successive applications of increasing the strength reduction factor are applied to reduce the soil strength in the model until the solution of the computation cannot converge any more. In the ABAQUS FE model, a field variable was created, which is the stress reduction factor supported in ABAQUS/CAE. The un-converged solution takes place whenever the critical strength reduction factor is reached. This ultimate strength reduction factor (SRF) is defined as equivalent to the factor of safety (FS). The FS concept is employed to illustrate the efficiency of stabilizing piles, and in order to illustrate the efficiency of the stabilizing pile to a slope, a stability improvement ratio (N_{pi}) is proposed to quantify the improved pile slopes. N_{pi} is defined as follows:

$$N_{pi} = \frac{F_p - F_s}{F_s} * 100\% \quad (3)$$

where F_p = the minimum factor of safety for the pile-slope system, and F_s = the minimum factor of safety for the slope stability problem without the pile.

In the slope stability analysis, several 2- or 3-D LE or FE solutions have been proposed. In certain cases, however, the 2-D solutions are still competitive if the limitations of the two-dimensional model can be identified. The 3-D model has been found to provide a more rigorous solution because the side effects and boundary conditions can be modeled closer to reality, particularly in the pile-stabilized slopes' problems.

This study presents the outcome of the 3-D FE analyses in slope stability in order to validate the accuracy of the model. The piles used to stabilize a slope containing a thin layer are explored. The solutions of the 2-D and 3-D analyses are compared, and their limitations are identified. In order to further verify the applications in pile-stabilized slopes, the piled-slope system was constructed using the 3-D FE technique, and the ABAQUS FE computation software was employed to compute the cases. The coupled finite element analysis is conducted, which considers both the slope stability and the pile response, simultaneously. Some influential factors, such as optimal pile location and the length of the pile verified in a 2-D model, are directly applied in the 3-D FE model. The improvement in the slope stability using piles, the influence of the pile head conditions, the failure mechanism resulted and the pile response are all addressed according to the results of the 3-D FE models. The solutions and findings are discussed, herein.

Table 1. Soil properties of slope.

c (kPa)	ϕ (°)	γ (kN/m^3)	H (m)	E (MPa)	ν (kPa)
40	20	20	40	100	0.3

2 OBJECTIVE AND SCOPE

The 3-D FE method that has been used is considered to be more rigorous and less conservative in the slope stability analysis, compared to 2-D FE technique. A 2-D FE analysis for slope stability is presented as Figure 1. In terms of equivalent plastic strain contour shown in the figure, the colored zone is where the potential slip surface will be formed according to the FE analysis. In many cases, particularly for un-stabilized slopes, the 2-D and 3-D slope stability analyses will not be given a significant difference in the results (Griffiths and Marquez 2007). However, it is improper to simplify the pile-slope model to a plane strain problem. In this type of problem, there are some limitations in the 2-D FE models that need to be overcome. The boundary conditions, the soil flow between piles and the arching effect should not be ignored. A 3-D FE technique is employed to analyze the stability of a homogeneous slope with foundation to improve the analysis result. In order to validate the results of slope stability analysis, this study adopts the benchmark example in the paper by Griffiths and Lane (1999). The parameters of soil and the height of the slope are summarized in Table 1. The geometry of the 3-D slope is shown as Figure 2.

Before the 3-D pile-slope problems are analyzed, the 3-D slope stability analysis was conducted using the powerful finite element package, ABAQUS 6.12 (2012). The Young's modulus (E) and Poisson's ratio (υ) of the soil are 100 MPa and 0.45, respectively. The 3-D stress and 8-node (C3D8) elements are employed in the 3-D model in ABAQUS/CAE. The soil material is modeled using the elastic-perfectly plastic constitutive model and the Mohr-Coulomb failure criterion is applied to the soil in the analyses. The pile members are meshed using the same elements, C3D8, and are assumed to be elastic. Therefore, only the soil failure has been considered in the pile-slope system without considering the plastic hinge formed in the pile members. The dilation angle in this model is assumed to

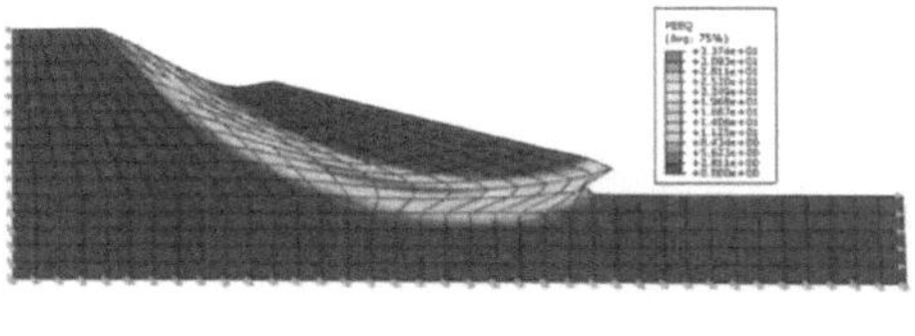

Figure 1. Plastic strain contour in 2-D finite element model in ABAQUS.

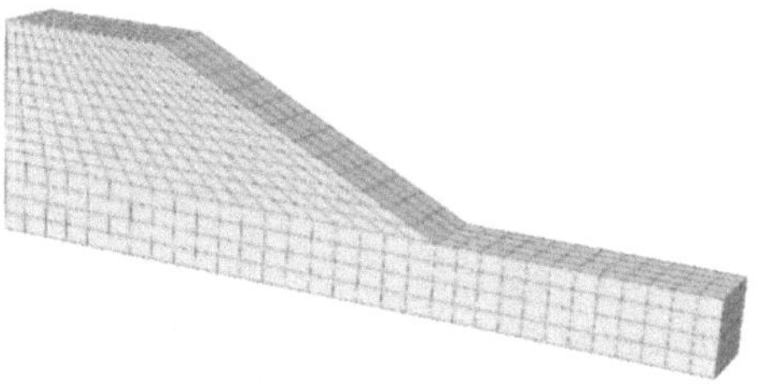

Figure 2. Meshed 3-D FE model of a homogeneous slope in ABAQUS.

be zero, hence, the plastic potential of the soil model is a non-associated flow. The boundary conditions on both sides in the third dimension (z-direction) are regarded as important influencing factors in the 3-D slope stability analysis.

According to Griffiths and Marquez (2007), the boundary can be assumed as three typical types: (1) smooth-smooth, (2) rough-smooth and (3) rough-rough, respectively. In this study, the two z-planes for the slopes are given roller-roller in the slope stability analysis. The movement of this two z-planes are not allowed, which is defined as the smooth-smooth boundaries in Griffiths and Marquez's study (2007). In addition, the pile model is incorporated into the numerical models to be analyzed.

2.1 *Pile-stabilized slope analyses*

It is improper to simplify the pile-slope problem into 2-D or plane strain problems. Besides complicated geometries, boundary conditions and soil variations, the soil flow between piles and the spacing effects are not easy to model using the 2-D technique. The 3-D finite element model for the pile-slope system is as shown in Figure 3, two piles are placed symmetrically on the slope. The notation of each symbol: H = height of the slope; X = the distance between toe and crest of the slope; S = distance from center to center of piles; X_p: the distance away from the toe.

In the pile-slope system, the piles are also modeled using the 3-D stress, 8-node element (C3D8). The interaction properties between the piles and the surrounding soil are defaulted as "penalty" and the friction coefficient is 0.51, which is dependent on the friction angle of the soil, ϕ and the material of the piles. The piles are assumed to be elastic without plastic deformation. The Young's modulus of the pile is 6000 MPa and the Poisson's ratio 0.28 is taken.

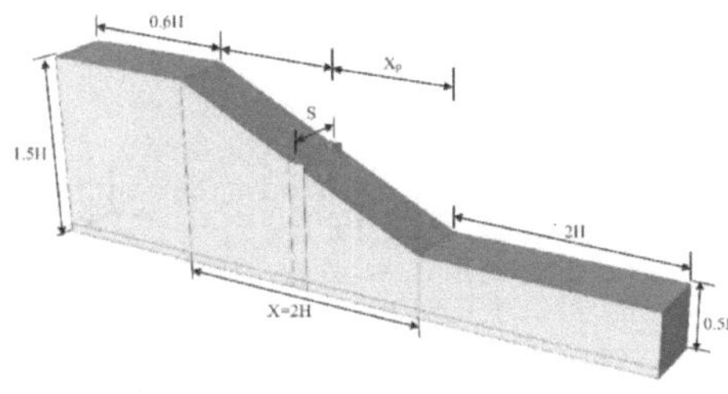

Figure 3. Pile-slope system model in ABAQUS ($X_p/X = 0.50$, L = 35 m).

Table 2. Summary of soil type, optimal location and analytical method for piled slopes.

Reference	Soil type	Recommended location	Analytical method	Comments
Itao *et. al* (1979)	Cohesive	Middle	Pressure-based	Infinite pile length
Poulos (1995)	Clay, Clay stone and Silt stone	Middle	Pressure-based	
Hassiots *et. al* (1997)	Cohesive	Upper to top	Pressure-based	Plane strain conditions
Lee *et. al* (1995)	Purely Cohesive	Toe and Crest	Displacement-Based	
Lee *et. al* (1995)	Upper soft Lower stiff	Between middle and crest		
Lee *et. al* (1995)	Upper stiff Lower soft	Toe and Crest		
Ausilio (2001)	Granular soil	Toe	Pressure-based	
Joeng et. al (2003)	c=10kPa ϕ=20°	Middle	Displacement method	Uncoupled analysis
Nian *et. al* (2008)	Cohesive soil	Toe	Pressure-based	Anisotropic and Non-homogeneous Soil

As mentioned earlier, the slope stability analysis for un-stabilized slope is 1.39 using the 3-D method.

2.2 *Optimal pile location*

As summarized in Table 2, some studies recommend the optimal pile location is the middle, some at the crest and some at the toe of the slope based on different analytical methods. To verify the best location of the piles to stabilize the slope, considering that the length of the piles is finite, and 3-D FE model is performed. The piles are placed in different locations along the slope, and the non-dimensional factor, X_p/X is employed. To make this factor consistent to the direction of the slope in this paper, the ratio, $X_p/X = 0$ is corresponding to the toe, and $X_p/X = 1.0$ is corresponding to the crest. The FS varies along with the location where the piles are placed, according to Figure 6. The results indicate that the highest FS for both free and fixed-head piles takes place in the middle of portion of a slope ($X_p/X = 0.5$). The $X_p/X = 0.25$, 0.50, 0.75 and 1.0 are examined. The lowest FS are found at the crest for both pile head conditions. The stabilizing piles are found to have little effect on the stability improvement when they are placed at the crest ($X_p/X = 1.0$), regardless of pile head conditions. The FS are both 1.39, the same for the un-stabilized slope. The piles placed at the position where the ratio, X_p/X equal to 0.75 and 0.25 lead to similar FS in both fixed and free pile head conditions, respectively. If the piles are placed in the middle part of the slope ($X_p/X = 0.5$), the factor of will increase to 1.72 for fixed-head piles and 1.46 for free-head piles. The optimal pile location in a slope can be concluded as the middle portion, where $X_p/X = 0.5$.

2.3 *Pile length optimized*

The literature addressed that the appropriate length of the pile based on the results of numerical analysis is very limited. However, a number of case studies presented can be observed regarding the pile length used in engineering applications. Most of the pile length used can be concluded that the potential slip surface to the entire length of pile is around 0.45 to 0.55 (Fukumoto 1975). A ratio of the actual pile length to the height of the slope, L/H, is used to investigate the effect of pile length. The actual length of the pile rages from 20 to 38 m and the height of the slope is 40 m. The range of the ratio, L/H, is from 0.5 to 0.95. Two results are presented in Figure 7 with respect to free and fixed pile heads. The FS jump from 1.56 to 1.73 when L/H is greater than 0.65, and then remain as 1.73 until L/H = 0.95. Thus, between L/H = 0.7 and 0.95, the FS will not increase along with the pile length any further. In terms of the stability improvement ratio, N_{pi}, the N_{pi} is between 8 to 24.5%. However, for the free-head piles, the FS increase from 1.41 to 1.47 when the L/H increases from 0.5 to 0.95. But the increase rate of the FS is very limited, with only a 4.3% difference. Based on the factor of safety, the length between 24 and 26 m in this case will result in higher FS for all cases. The results can be compared with the un-stabilized case, which presents the potential slip surface. The analyses can also support and validate that about half of the pile length has to extend below the potential slip surface.

The optimal length of the pile depends on the dimension of a slope. A non-dimensional ratio, L/H, is introduced in this paper to illustrate the length of pile compared to its height. In this paper, the L/H ratio for the fixed-head pile is about 0.7, and no particular optimal pile length ratio can be observed for free-head pile. The analytical results for the free-head pile indicate the FS increase along with the increase of the L/H ratio. However, the increase is very small, so in engineering application, the recommended L/H will not necessarily be the maximum value close to 1.0.

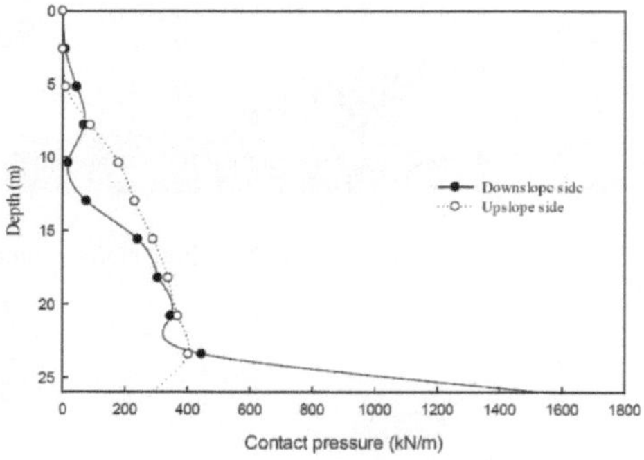

Figure 4. Deformation contour of pile-stabilized slope in ABAQUS/CAE.

The purpose of Figure 4 is to help realize the failure mechanism based on the soil pressure distributions on the front and back sides of the pile. The distribution of soil pressure along the pile is found to be non-uniform based on the 3-D FE analyses. The contact pressure distribution along the pile that is subjected to the slope movement is complex and unclear. Usually, either distribution of soil pressure or the soil movement has to be assumed when the analysis is conducted. However, in the 3-D FE analysis, no assumption has to be made.

3 ADVANTAGES OF THE 3-D FINITE ELEMENT ANALYSIS

The advantages of the FE analysis can be concluded as follows (Griffiths and Lane 1999): (1) No failure surface has to be presumed in advance, (2) Monitor the progressive failure, (3) Consider the stress-deformation relationship, (4) No need to assume the side forces which is not easy to be assumed in LE method, and (5) overcome the limitation in the 3-D LE method, which lacks a suitable method to determine the location of slip surface.

In addition, the 3-D FE is also considered to provide more rigorous solution analyses by reflecting the slope dimensions, boundary conditions and the realistic geometry compared to the 2-D FE analyses. The spacing ratio S/D is also considered to be one of the influencing factors that affect the slope stability improvement for the pile-slope system. The S/D ratio used in this study is 4.0 because it is a critical value to compare with the 2-D analyses. Moreover, S/D = 4.0 is also considered to be cost-effective in design. Based on results of 2-D and 3-D FE analyses, it can be inferred that the 2-D FE analyses will be less conservative if S/D greater than 4.0. In the 3-D model, the ratio S/D = 4.0 is found to obtain a factor of safety close to that of the 2-D model. If the spacing ratio keeps increasing, the FS will reduce to approach the un-stabilized cases. The 3-D FE analyses in this aspect can help consider the soil between piles, the group pile effect in the third direction, and arching effect.

4 FAILURE MECHANISMS IDENTIFICATION

The failure mechanism of the pile-slope system will change due to the presence of the stabilizing piles in different locations. In the 3-D FE analysis, the maximum lateral displacement appears at the toe of the slope and the soil flows through the space between the piles. The presence of the piles in different locations in the slope will alter the depth of the slip surface in this coupled analysis, which governs the global stability of a slope. In the computation using ABAQUS/CAE, the failure defined as the excess deformation occurred and the computation therefore stop due to unconverged solution. The corresponding SFR is equivalent to the FS using LE methods. Owing to the presence of piles, the failure mechanisms of pile-stabilized slopes also change. The failure mechanism changes depend on the position of piles, the length of piles and pile-head conditions.

5 ANALYSIS RESULTS

To investigate the pile-soil interaction mechanism in the pile-slope system, it is necessary to identify the contact pressure between the soil and piles. The contact soil pressure in the front side is different from the back side. Based on the pressure distribution along the piles, the stabilizing mechanism can be confirmed. The analysis result of the case where piles are placed in the middle portion of a slope. The corresponding SRF is 1.44. The SFR is 1.41 when piles are placed at $X_p/X = 0.25$. Two types of failure mechanisms can be observed in this paper due to the change of pile location. The failure type indicates that the soil movement at the crest can be successfully reduced, and the main deformation of soil is between the piles and the downslope side of piles. For the rest of pile locations, $X_p/X = 0.25$ to 0.75, the failure mechanism is governed by the circular planes, however, the depth of the potential slip surface will change subject to the pile locations.

As for the cases of the fixed-head piles used in the slopes. The pile heads are restricted as non-rotated and non-displaced. The piles are able to resist the soil movement upslope and the failure is governed by the soil deformation on the downslope side of piles. Hence, three different failure mechanisms can be observed if fixed-head piles are installed in different positions on the slope. The three failure mechanisms are governed by the downslope, upslope and both soil masses when the stabilizing piles are placed in different locations on the slope which are at crest, toe and middle, respectively.

6 SUMMARY AND CONCLUSIONS

The 3-D finite element method provides more realistic solutions compared to the 2-D finite element analysis. Moreover, the 3-D FE technique used to conduct the pile-slope problems is found to be useful and prevailed when compared to either pressure- or displacement-based analytical methods. Instead of pre-defining a sliding plane, a coupled analysis of a pile-stabilized slope is conducted, thus the pile response and the slope stability are considered, simultaneously. Through the numerical analyses, the different failure mechanisms can be identified, and the effectiveness of the piles can be evaluated. Several conclusions can be drawn as follows:

(1) The 3-D FE method used to analyze the slope stability of the pile slope system will consider the spacing effect, and the soil deformation between piles, which usually cannot be considered in 2-D model properly.
(2) The stabilizing piles placed in the middle portion of the slope are found to make the greatest contribution to the slope stability. The second option is to place the pile close to the toe of a slope. The least preferred location is the crest of a slope.
(3) The stabilizing piles will behave as a single pile if the S/D ratio $\approx$4.0. In the analysis, the pile deformed subjected to slope failure can be simplified to a single pile instead of considering a pile-row response.
(4) The piles in this paper that are assumed to be elastic are easier to realize that the maximum shear and moment will develop due to the slope failure. In design, if a target FS of a slope is determined after being reinforced, the pile capacity can therefore be reevaluated in order to provide the sufficient shear and moment capacity of a single pile.
(5) The length of pile used in a slope will change the depth of the potential slip surface, as well as the failure mechanism of s slope, L/H greater than 0.70 is the optimal pile length for slope reinforced with fixed-head piles.
(6) The fixed-head pile performs more efficiently to increase the stability when compared to the free-head pile and the pile head conditions will change the failure mechanism of the pile-slope system. Therefore, the FS, will change as well.
(7) The stabilizing mechanism of the piles mainly depends on the interaction between the piles and the surrounding soil. The change in the depth of the potential slip surface due to the presence of the piles needs to be carefully taken into account. Otherwise, the locations of the maximum shear force and moment will be misrepresented. The maximum moment is on the pile head for fixed-head piles and is close to the tip for free-head piles.
(8) The advantages of this proposed 3-D FE analyses for the pile-slope can conduct the coupled analyses, which considered the slope stability and the pile response, simultaneously. The computation will give a maximum SRF if the solution cannot converge any more. The maximum SRF is the FS of a piled slope. Moreover, the soil movement, the depth of the slip surfaces, and the soil pressure distribution along the pile are not necessarily to be pre-defined when the analysis is conducted.

REFERENCES

ABAQUS. 2012. ABAQUS Documentation, Version 6.12: Abaqus /CAE User's Manual. Simlulia, Providence, RI.

Ausilio, E., Conte, E., and Dente, G. 2001. "Stability analysis of slopes reinforced with piles." Computers and Geotechnics, 28(8): 591–611.

Cai, F., and Ugai, K. 2000. "Numerical analysis of the stability of a slope reinforced with piles." Soils and Foundation, 40(1):73–84.

Chen, R.H., and Chameau, J.L. 1985. Discussion: three-dimensional limit equilibrium analysis of slopes. Géotechnique, 35(2):215–216.

Chow, Y. K. 1996. Analysis of piles used for slope stabilization. International Journal for Numerical and Analytical Methods in Geomechanics, 20(9):635–646.

Fukumoto, Y. 1972. Study on the behavior of stabilizing piles for landslides. Soils and Foundations, 12 (2):61–73.

Galli, A., and Prisco, D. C. 2013. Displacement-based design procedure for slope-stabilising piles. Canadian Geotechnical Journal, 50(1):41–53.

Goh A. T. C., Teh, C. I., and Wong, K. S. 1997. Analysis of piles subjected to embankment induced lateral soil movements. Journal of Geotechnical and Geoenvironmental Engineering, ASCE, 123(9):792–801.

Griffiths, D. V., and Lane, P. A. 1999. Slope stability analysis by finite elements. Géotechnique, 49(3):387–403.

Griffiths, D.V. and Marquez, R.M. 2007. Three-dimensional slope stability analysis by elasto-plastic finite elements. Géotechnique , 57(6):537–546.

Hassiotis, S., Chameau, J. L, Gunaratne, M. 1997. Design method for stabilization of slopes with piles. Journal of Geotechnical and Geoenvironmental Engineering, ASCE, 123(4):314–323.

He, Y. *et al* . 2015. Estimation of lateral force acting on piles to stabilize landslides. Natural Hazards.

Ho, I. H. 2009. Optimization of Pile Reinforced Slopes Using Finite Element Analyses. Ph.D dissertation, Iowa State University, IA.

Ho, I. H. 2014. Parametric Studies of Slope Stability Analyses Using Three-Dimensional Finite Element Technique: Geometric effect. Journal of Geoengineering, 9(1):33–43.

Ho, I. H. 2015 Numerical Study of Slope Stabilizing Piles in Undrained Clayey Slopes with A Weak Thin Layer. International Journal of Geomechanics, 15(5):61048.

Hull, T.S., Lee, C.Y., and Poulos, H.G. 1992. Behavior of fixed and free head piles in a laterally sliding soil. Proc. 6th Aust.-N.Z. Conf. Geomechs, Christchurch, NZ, 151–156.

Ito, T., and Matsui, T. 1975. "Methods to estimate lateral force acting on stabilizing piles." Soils and Foundations, 15(4), 43–59.

Jenck, O., Dias, D., and Kastner, R. 2009. Three-Dimensional Numerical Modeling of a Piled Embankment. Int. J. Geomech., ASCE, 9(3):102–112.

Jeong, S., Kim, B., Won, J. and Lee, J. 2003. Uncoupled analysis of stabilizing piles in weathered slopes. Computers and Geotechnics, 30(8):671–682.

Kourkoulis, R., Gelagoti, F., Anastasopoulos, I., and Gazetas. 2011. Slope stabilizing piles and pile-groups: parametric study and design insight. Journal of Geotech and Geoenvironmental Eng., ASCE, 137(7): 663–678.

Lee, C. Y., Hull, T. S. and Poulos, H. G. 1995. Simplified pile-slope stability analysis. Computers and Geotechnics, 17, 1–16.

Muthukkumaran, K., 2013. Effect of slope and loading direction on laterally loaded piles in cohesionless soil. International Journal of Geomechanics, 14(1):1–7.

Nian, T. K., Chen, G. Q., Luan, M. T., and Zheng, D. F. 2008. Limit analysis of the stability of slopes reinforced with piles against landslide in non-homogeneous and anisotropic soils. *Can. Geotech. J.*, 45(8):1092–1103.

Nian, T. K., Huang, R. Q., Wan, S. S. and Chen, G. Q. 2012. Three-dimensional strength-reduction finite element analysis of slopes:geometric effects. Canadian Geotech. J., 49(5):574–588.

Poulos, H.G. 1995. Design of reinforcing piles to increase slope stability. Canadian Geotech. J., 32(5):808–818.

Reul, O. 2004. Numerical Study of the Bearing Behavior of Piled Rafts. Int. J. Geomech., ASCE, 4(2):59–68.

Won, J., You, K., Jeong. S., and Kim, S. 2005. Coupled effect in stability analysis of pile-slope systems. Computers and Geotechnics, 32(4):304–315.

Zienkiewicz, O. C., Humpheson, C., and Lewis,R. W. 1975. Associated and Non-Associated Visco-Plasticity and Plasticity in Soil Mechanics. Géotechnique, 25(4): 671–689.

Challenges and Innovations in Geotechnics – Zhussupbekov (Ed.)
 ISBN 978-1-138-03007-7

Numerical simulation for the estimation the jacking force in pipe jacking

J. Yen
CECI Engineering Consultants Inc., Taiwan

K. Shou
Department of Civil Engineering, National Chung-Hsing University, Taichung, Taiwan

ABSTRACT: For a pipe jacking construction, reducing the soil–pipe interface friction and providing enough jacking force are the most common approach to optimize the construction efficiency. In practice, jacking force is generally estimated by various empirical equations. However, the estimations using empirical equations frequently deviate from the reality. In this study, a model coupling finite element method and a displacement control method were applied to estimate the required jacking force in pipe jacking. Two cases were examined from Central Taiwan, where the primary geological foundation composed of gravel formations. Case A pertained to pipe jacking construction during which sewage pipes with a diameter of 2.4 m were utilized. The monitoring data from this case were used to establish the jacking force estimation model. The jacking force history observed in Case B, in which sewage pipes of 1.0 m diameter were used, was compared with those obtained by the developed model to demonstrate the applicability of the model. The results suggested the developed model can estimate the jacking force with a better accuracy towards the middle and the final stage of the pipe jacking process.

Keywords: Pipe jacking, Jacking force estimation, The displacement control method, FEM, interaction.

1 INTRODUCTION

In engineering design, numerical analysis is commonly applied to the simulation of engineering behavior. Through numerical simulation, the engineering behavior of soil–pipe interaction can be rapidly determined for use as the basis of a better engineering design. Most of the studies adopt the force control method, in which the force boundary conditions are given (Barla, 2006, 2013; Broere, 2007; Shou et al., 2010; Shou and Yen, 2010; Li,2012).

In the force control method, after the jacking force is obtained by using various empirical or theoretical equations, it is included as input data for the numerical model to simulate pipe jacking. If the jacking force is insufficient, the pipes can't be jacked to the appropriate position. In contrast, the jacking force may be excessive, and causing penetration through the excavation face, in which modifications will be required to obtain the suitable jacking force.

There have been numerous studies exploring and discussing the estimation of jacking force (Chapman and Ichioka, 1999; Staheli, 2006; Beckmann, 2007; Röhner, 2010). Numerical simulation can be conducted before the actual pipe jacking construction to estimate the required jacking force employed in various construction conditions and jacking distances, and to achieve the ultimate goal of improving pipe jacking efficiency.

In this study, the displacement control option in the numerical analysis software ABAQUS (Abaqus Inc., 2012) was used to designate the displacement at the end cross section of the pipe in the launch shaft. Accounting for the contact property and the contact range between the pipes and the soil during the jacking process, the stresses exerted on the pipes were used to back-calculate the jacking forces.

2 NUMERICAL ANALYSIS METHODS

2.1 *Numerical models for the cases*

The analyses focused on two construction cases in the Taichung Science Park. The inner diameters of the pipes used in Cases A and B were 2.4 m and 1.0 m, respectively. The details of Cases A and B are described and illustrated in Table 1 and Fig. 1.

The gravel layer is the dominant geological stratum of Central Taiwan as shown in Fig. 2. This formation is a composite geomaterial consisting of gravels and sandy soils. The grain sizes of the gravels, with a large volumetric percentage, are 5–20 cm on average and can be up to about 100 cm. The gravels in this area are originally from quartzite, which has uniaxial compressive strength of more than 1000 kg/cm^2. The material parameters of the soil and pipes are summarized in Table 2.

Table 1. The details of Cases A and B.

	Outer/inner diameter (m)	Jacking distance (m)	Pipe length (m)
Case A	2.85/2.4	116 m	2.50
Case B	1.28/1.0	63 m	2.35
	Overburden depth (m)	Pipe jacking method	Lubricants
Case A	13.0	Slurry	Bentonite of 120sec viscosity
Case B	10.8	Slurry	N/A

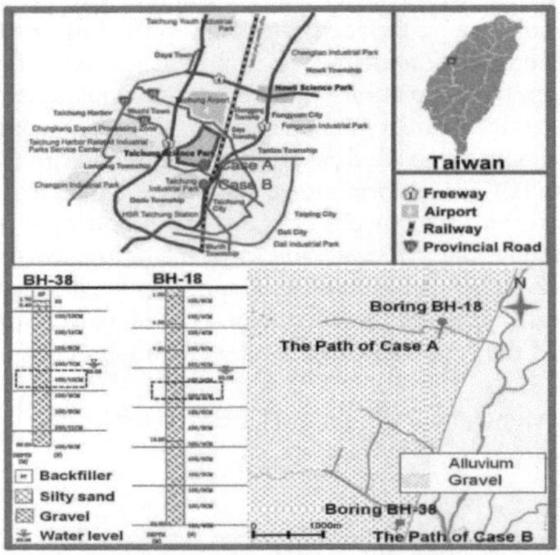

Figure 1. Locations and geology in Case A and Case B.

Figure 2. The gravel formation in Central Taiwan.

2.2 *Finite element meshes*

The objective of this study is to estimate the jacking force based on the stress exerted on the pipe by employing the displacement control method. Therefore, we focused on pipe elements rather than soil elements. The soil elements that account for the soil pressure exerting on the pipes are important, however, for the efficiency of computation, the size of the elements on the lateral sides of the pipe can be enlarged gradually to reduce the number of elements (as shown in Fig. 3). Fine elements were used only in the parts close to the pipe and the cutter head, as shown in Figs. 4 and 5. For Case A, 15,736 soil elements and 852 pipe element were used; however, 7616 soil elements and 720 pipe elements were used in case B.

When the displacement control method was used to designate pipe jacking movement, the selection of a specific group of elements or surfaces for the designated displacement could substantially influence the results. For example, when all pipe elements were selected, the jacking force obtained would be comparatively smaller. When only one element or node is selected, the jacking force obtained will be comparatively larger. In this study, the area of the red circle in Fig. 5 is adopted as the displacement designated location, as it is the end cross-section of the pipe that just crosses the mirror-face (the face where the pipe leaves the shaft to enter the soil mass) in the launch shaft.

Table 2. Material parameters for the gravel.

Parameter	Selection
γt (kN/m^3)	21.5
φ (°)	37
c (kN/m^2)	15
K0	0.398
N	0.3
E (kN/m^2)	330,000

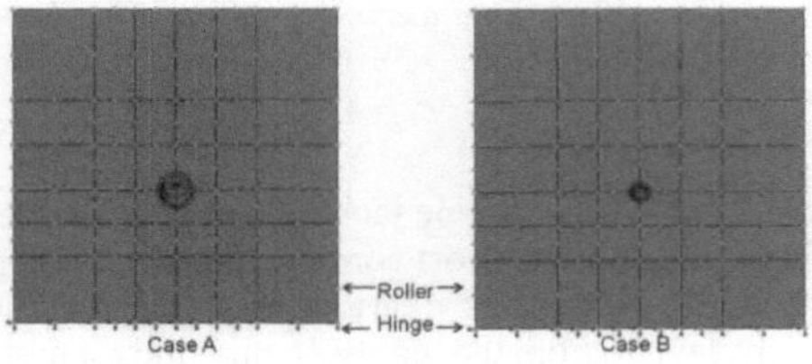

Figure 3. Thee finite element models of Case A and B.

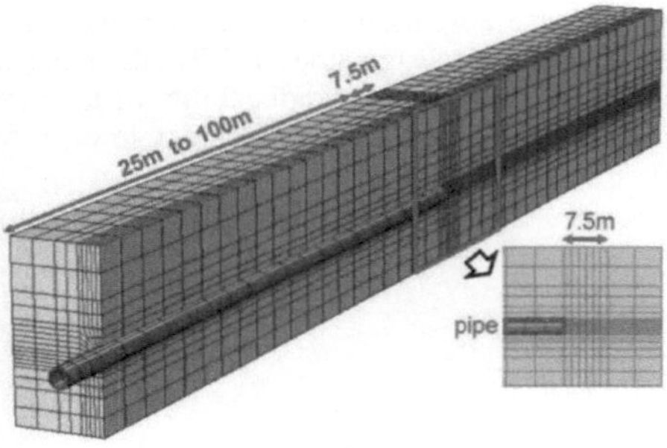

Figure 4. Cross-section of the model from 25 m to100 m.

2.3 *The contact properties*

To verify the jacking force obtained from the displacement control simulation, overcut and lubrication were included in the model by setting the contact range and

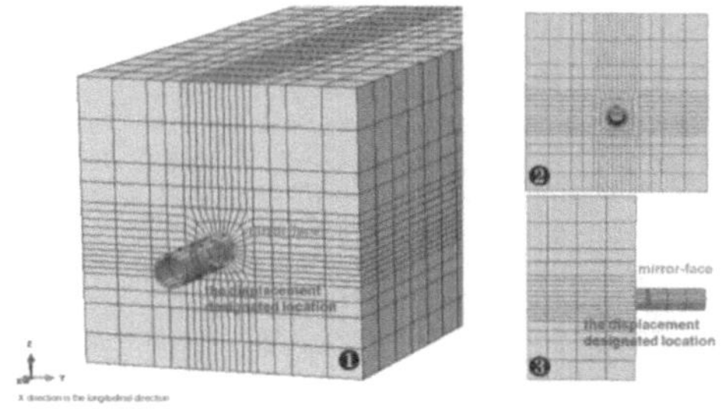

Figure 5. The pipe section used for the displacement control method. (1) Perspective view, (2) front view and (3) side view.

changing the frictional coefficient. To simulate overcut, the soil–pipe contact range was specified in the model. With the presence of the overcut, the soil and the pipe would not be in direct contact with each other, hence the contact property was set to be frictionless. In contrast, the other contact properties would be set according to the various frictional coefficients of the lubricants applied.

In Case A, the main jacks were used to push the first 116 m, and the length for each subsequent pipe jacking operation cycle was 2.5 m. Considering the pipes in Case A, seven numerical models were generated from 25 m to 100 m. In Case B, the length for each subsequent pipe jacking operation cycle was 2.35 m, five numerical models were generated from 20 m to 60 m.

For the contact property, the contact pair must be set first. This included the surrounding surface and the front face of pipe string. For the behavior in the tangential direction, a penalty function was adopted with frictional coefficients 0.19, 0.40, and 0.52.

In the Case A and Case B, the pipe jacking machines possess the copy cutters for the overcut excavation which helped to reduce the area of contact and the friction between the soil and pipes. The overcut can be simulated by changing the range of contact between the pipe and soil. Based on a preliminary sensitivity analysis, it suggests that the 1/2 and 1/3 contact conditions are more distinguishable and representative. Therefore, the contact range parameters were set to be full, 1/2, and 1/3 of the pipe surface in contact with the soil.

2.4 *Numerical simulation*

The numerical simulation of pipe jacking comprises the following steps: (1) obtain the initial stress in the soil mass by ally and balance the gravity; (2) before further simulation, apply the initial stress to deduct the initial settlement in the soil mass; (3) remove the elements representing the soil excavated by the pipe jacking machine to simulate the excavation; (4) define the contact pairs between the soil and the pipe according to the lubrication and overcut conditions; (5) enforce the initial stress to create the friction on the interface between soil and pipe; and (6) displace the pipes to the designated position (the displacement control assumption) to obtain the stress field and strain field in the soil and pipes.

By using the displacement control method, the pipe in the launch shaft was jacked forward one step (one operation cycle, was 2.5 m and 2.35 m in Cases A and B). The boundary conditions in the numerical analysis were employed to make the pipe in the launch shaft to be jacked forward by one step. In such a scenario, the pipes ahead of the jacks and the neighboring soil were influenced until the jacking distance was achieved. About the simulation of overcut, in order to avoid the possible simulation difficulties due to the thin gap between the pipe and soil, this study considers the interface be frictionless in the overcut ranges.

In the numerical simulation of Case A, three different friction lubricant coefficients were used. The coefficient was set to 0.52, 0.4, and 0.19, which were used to simulate the situation of no lubrication, applying bentonite with 120-s viscosity (commonly used in Taiwan, and was used in Case A), and applying bentonite together with polymer respectively. The three lubricant frictional coefficients in the numerical simulation were used for comparison and discussion on the three different lubrication conditions. However, as Case B involved small-diameter (1.0 m) and short-distance pipe jacking, no lubricant was used. And the frictional coefficient was set to be $\tan\phi$ ($\tan 37^o = 0.75$).

2.5 *The displacement control method*

In order to estimate the jacking force better, the displacement control method was proposed for this study. The displacement control option was used to designate pipe jacking movement to the end cross-section of the pipe just crossing the mirror-face in the launch shaft. Through the numerical analysis, we could obtain the stress distributions in the neighboring soil body and the pipes, including the stress distribution at the end of the pipe in the launch shaft. The jacking force could be obtained by integrating the stresses over the displacement designated area.

3 RESULTS OF NUMERICAL SIMULATION

3.1 *Stress distributions on the pipes*

The stress in the pipe was a function of the normal and lateral pressure from the soil in addition to the jacking. In order to simulate the pure component of stress caused by the jacks alone, only focused on the distribution of the longitudinal component (σ_{xx}) of stress on the pipe, without considering the other components caused by the overburden pressure. When the frictional coefficient was 0.4 and one-third of the pipe bottom surface was in contact with the soil, the distributions of normal stress in the longitudinal direction (σ_{xx}) at the end of the pipes with seven different jacking distances were shown in Fig. 6. The stress distributions were then used to obtain the jacking forces by the integration of the stress over the cross section areas.

Table 3. The jacking force estimations with different contact conditions.

Contact condition	all contact	1/2 contact	1/3 contact
Jacking force estimation for Case A*	5890 kN	5130 kN	4980 kN
Deviation (%)	69.74%	47.84%	43.52%
Jacking force estimation for Case B*	1270 kN	1250 kN	980 kN
Deviation (%)	15.45%	13.64%	10.91%

*numerical analysis results of Case A (jacking distance 50 m) and Case B (jacking distance 40 m).
**based on the monitoring results 3470 kN for case A and 1100 kN for case B.

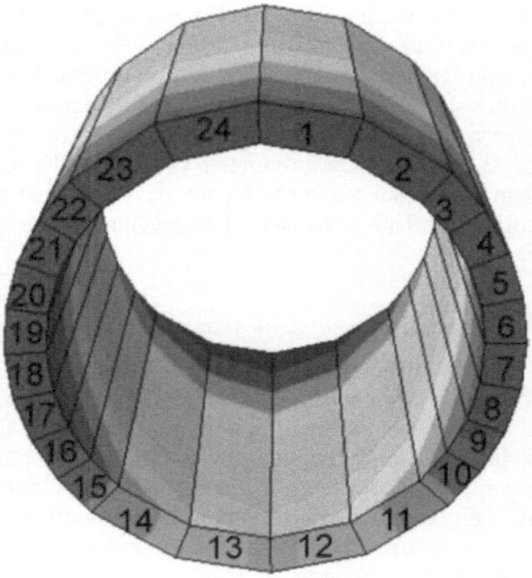

Figure 6. The indexes and locations of pipe elements.

In the simulation with the three frictional coefficients and seven jacking distances, the displacement designated pipe contained 24 elements, and the stress exerted on the pipe enables the back calculation of the jacking force.

For Case A of the full-contact model, the obtained jacking force was substantially greater than that obtained for the other lesser contact situations. Comparing the simulation results of the full-contact and half-contact situations, the data obtained from the simulation of the one-third contact condition is closer to the monitoring data (see Table 3). Thus, one-third of the pipe bottom surface is set to be in contact with the soil for the simulation model. In addition to the effects exerted by the copy cutter, the strong, highly self-sustainable gravel formations and the favorable lubricant enabled a satisfactory overcut condition to be maintained. Consequently, the area of contact between the pipe and soil could be reasonably reduced to one-third in contact. After the contact range was determined, the three different frictional coefficients, i.e., 0.52, 0.40, and 0.19, were used in successive simulations.

Table 4. The calculation of jacking force based on the numerical simulation results (for Case A,driving distance 100 m, 1/3 contact, frictional coefficient 0.4).

No.	σxx	No.	σxx	No.	σxx
1	3.70E+06	9	3.78E+06	17	3.72E+06
2	3.65E+06	10	4.00E+06	18	3.62E+06
3	3.59E+06	11	4.34E+06	19	3.59E+06
4	3.55E+06	12	4.75E+06	20	3.59E+06
5	3.52E+06	13	4.77E+06	21	3.61E+06
6	3.51E+06	14	4.40E+06	22	3.63E+06
7	3.52E+06	15	4.09E+06	23	3.67E+06
8	3.62E+06	16	3.88E+06	24	3.70E+06

No. 1–24 Element average of rxx = 3.82E+06 N/m^2.
The area of pipe section was 1.8545 m^2.
The jacking force = 3.82E+06/1.8545 = 7.084E+06 (N) = 7084 (kN).

3.2 *Calculation of jacking forces*

In Case A, the stress on the pipe elements varied with the location, the jacking force was calculated by integrating the stress with the cross-sectional area. The pipe elements coded with numbers, and the simulation results are illustrated in Fig. 6.

In a simple way, the jacking force can also be obtained by multiplying the mean stress with the cross-sectional area. For clarity, the jacking force calculation procedures for the Case A with driving distance 100 m, 1/3 contact, and frictional coefficient 0.4, is listed in Table 4.

3.3 *Comparison of jacking forces*

During pipe jacking construction, the jacking force was estimated according to geological conditions. Staheli (2006) proposed the following empirical equation to calculate the jacking force:

$$JF_{frict} = \mu_{int} \frac{\gamma \cdot r \cdot \cos(45 + \frac{\varphi_r}{2})}{\tan \varphi_r} \cdot \pi \cdot d \cdot l \text{.....................} (1)$$

where JF_{frict} is the frictional component of the jacking force, μ_{int} is the soil–pipe residual interface frictional coefficient, γ is the unit weight of soil, φ_r is the residual friction angle of soil, d is the outer diameter of pipe, r is the pipe radius, and l is the length of pipe.

The empirical equation proposed by the Japan Micro-Tunnelling Association (2000), which is commonly used in Taiwan, was also adopted to calculate the jacking force required. The jacking force F (kN) can be expressed as:

$$F = F_0 + \pi . B_c . \tau_a . L \text{...} (2)$$

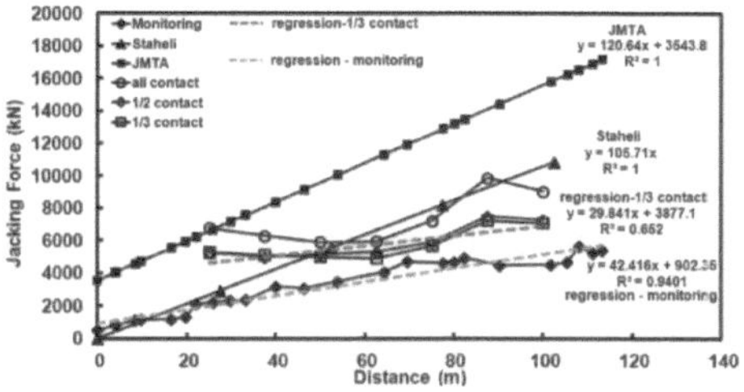

Figure 7. The comparison of jacking forces from monitoring, the empirical equations, and the displacement control method for Case A.

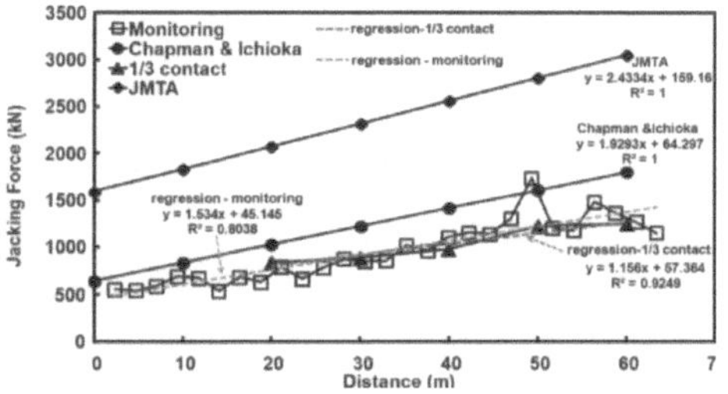

Figure 8. The comparison of the jacking force estimations for Case B.

where B_c is the outer diameter of the pipe, τ_a is the shear stress between the pipe and the soil, L is the length of jacking, and F_0 is the initial resistance force.

The jacking force estimated by the empirical equations of Staheli (2006) and the Japan Micro-Tunnelling Association (2000) were compared with those obtained by the displacement control modeling. The comparison in Fig. 7 reveals that the results obtained using the two empirical equations were different, because JMTA formula considers the face resistance, but the one of Staheli does not. And the numerical simulation with frictional coefficient 0.4 and the one-third contact condition can better predict the jacking force, especially at the middle and end stage of the pipe jacking work.

Table 5. The comparison of the regression lines with the empirical equations.

Origin of equation	Equation	Gradient Deviation (%)	Interception Deviation (%)
Case A			
Monitoring	Y = 42.416X + 902.35	N/A	N/A
1/3 cotact	Y = 29.841X + 3877.1	−29.65	329.67
Staheli	Y = 105.71X	149.22	N/A
JMTA	Y = 120.64X + 3543.8	184.42	292.73
Case B			
Monitoring	Y = 15.34X + 451.45	N/A	N/A
1/3 cotact	Y = 11.558X + 573.64	−24.65	27.07
Chapman	Y = 19.293X + 642.97	25.77	42.42
JMTA	Y = 24.334X + 1591.6	58.63	252.55

3.4 *The application to small-diameter pipe jacking*

Apart from the two aforementioned empirical equations that were applicable to large-diameter pipes, Chapman and Ichioka (1999) investigated pipes that had a diameter of less than 1.0 m and analyzed the monitoring data collected from such pipe jacking construction. They suggested calculating the jacking force required for the slurry pipe jacking method by the following equations:

$$F = f_0 + \pi DPL \qquad (3)$$

where f_0 is the primary resistance force in the excavation, P is the frictional resistance per unit area, D is the pipe outer diameter, and L indicates jacking length.

Based on the investigation of Case A, a contact range of one third of the pipe bottom surface was applied in Case B. The resultant calculated and monitored jacking forces were similar, indicating that the simulation outcome was quite accurate, as shown in Fig. 8 displays the jacking forces that were calculated by the empirical equation of Chapman and Ichioka (1999), JMTA, monitored during the actual construction, and generated by the displacement control simulation.

3.5 *The discussion of the results*

The deviation in the numerical results from the monitoring data is obvious for Case A, but the deviation is much smaller for Case B. The comparisons in Figs. 7 and 8 suggest that the displacement control numerical simulation can predict the jacking force better, especially at the middle and end stage of the pipe jacking work.

Based on the results numerical simulations and monitoring, the linear regression lines can be obtained as shown in Figs. 7 and 8. From those equations of regression lines, the intercepts (y values) can be further derived by setting x (distance) = 0, i.e., 3877.1 (kN) for Case A and 573.64 (kN) for Case B. These values stand for the face resistances, which are comparable to those suggested by other empirical formulas including the face resistance. In other words, this finding somehow confirms the face resistance was included, as the contact between the cutter head and the soil mass was considered in the numerical model.

For a better comparison, the equations of the regression lines and their characteristics are listed in Table 5. For Case A, the deviation of the 1/3 contact is −29.65%, in Case B, the deviation of 1/3 contact is −24.65%. Both of the numerical results were less within 24 ∼ 29% of the monitoring data. The extent indicates the contact range could be higher than the 1/3 contact. However, this modified estimation would be suitable only for the gravel formations of Central Taiwan. More numerical simulations are necessary to determine the soil–pipe contact range in other types of formations.

4 CONCLUSIONS AND SUGGESTIONS

Instead of time consuming trial-and error approach to estimate the jacking force, this study shows that the displacement control model can estimate the jacking force in a comparatively easier way. The displacement control model can obtain reasonable results for various contact conditions between pipe and soil. In cases A and B, the results suggest that the displacement control numerical analysis can produce reasonably good estimations of jacking force, especially at the middle and final stage of a pipe jacking work. The regression equations can be applied to pipe jacking construction involving various pipe diameters in geology similar to the gravel formations of Central Taiwan.

Because long-distance pipe jacking construction is becoming popular, it is worth testing and applying the displacement control method to simulate long-distance pipe jacking. Based on the jacking forces produced by various jacks, the locations of the main and intermediate jacking stations can be arranged and designed. In addition, the displacement control method also has the potential to analyze the curved pipe jacking. Although the mechanical behavior of curved pipe jacking differs from that of straight-line pipe jacking.

REFERENCES

Abaqus Inc, 2012. Abaqus/CAE User's Manual Version 6.12.

Barla, M., 2006. Analysis of jacking forces during microtunnelling in limestone.Tunn. Undergr. Space Technol. 21(6), 668–683.

Barla, M.,2013. A method to design microtunnelling installations in randomly cemented Torino alluvial soil. Tunn. Undergr. Space Technol. 33(1), 73–81.

Beckmann, D., 2007. CoJack–a new statics method of computing and controlling pipe jacking. Tunn. Undergr. Space Technol. 22(5), 587–599.

Broere, W., 2007. Modelling the boring of curves in (very) soft soils during microtunnelling. Tunn. Undergr. Space Technol. 22(5), 600–609.

Chapman, D.N., Ichioka, Y., 1999. Prediction of jacking forces for microtunnelling operations. Trench. Technol. Res., ISTT 14(1), 31–41.

Japan Micro Tunnelling Association, 2000. Pipe Jacking Application. JMTA, Tokyo.

Li, H., 2012. Analysis of Jacking Force for Rectangular Pipe Jacking Machine. Przeglad Elektrotechniczny, ISSN 0033–2097, R. 88 NR 9b/2012.

Marshall, M., 1998. Pipe-Jacked Tunnelling: Jacking Loads and Ground Movements, Ph.D. dissertation in University of Oxford.

Röhner, R., 2010. Calculation of jacking force by new ATV A-161. Tunn. Undergr. Space Technol. 25(6), 731–735.

Shou, K., Yen, J., 2010. On the behavior of a stuck curved pipe jacking. J. GeoEng. 5(3), 77–85.

Shou, K., Yen, J., Liu, M., 2010. On the frictional property of lubricants and its impact on jacking force and soil–pipe interaction of pipe jacking. Tunn. Undergr. Space Technol. 25(4), 469–477.

Staheli, K., 2006. Jacking Force Prediction: An Interface Friction Approach Based on Pipe Surface Roughness, Ph.D dissertation in Georgia Institute of Technology.

Challenges and Innovations in Geotechnics – Zhussupbekov (Ed.)
© 2016 Taylor & Francis Group, London, ISBN 978-1-138-03007-7

Application of 3-D static analysis program for residual deformation due to liquefaction

Yusuke Tadano, Hiroki Uno & Akira Tateishi
Taisei Corporation Technology Center, Yokohama, Kanagawa., Japan

ABSTRACT: A two dimensional static FEM analysis code, ALID, was developed to calculate residual deformation due to liquefaction. This is because most damages through liquefaction are induced by the residual deformation. Both parameter setting and evaluation of results are easy in ALID; therefore it has been applied in aseismic design of river dikes. However, three dimensional structures (*e.g.*, cylindrical tanks) must be approximately modeled in two dimensional analysis. For this reason, we introduced a methodology to a three-dimensional FDM analysis code FLAC3D. In this paper, the analysis methodology is outlined first. An analysis of the residential area is then conducted where liquefaction occurred in the 2011 off the Pacific coast of Tohoku Earthquake. We focus on the inclination and the settlement of the houses, and evaluate them from the viewpoint of the applicability of the analysis methodology in comparison to the investigation results by Tokimatsu and Katsumata (2014).

Keywords: Three dimensions, Liquefaction, Static analysis, Residual deformation

1 INTRODUCTION

The 2011 off the Pacific coast of Tohoku Earthquake occurred on March 11, 2011, and caused terrible ground deformation due to liquefaction in the Tokyo Bay area (Towhata et al. 2014).

It will require an enormous cost of measures if we want to completely prevent the ground deformation due to liquefaction. Instead allowing the liquefaction-induced deformation to some degree will enable us to apply economical measures. To do this, we have to precisely calculate the liquefaction-induced deformation.

As a prediction method of the ground deformation, two types of analysis method have been developed.

One is a dynamic effective stress analysis. This method has a merit that it can solve the deformation during and after an earthquake. While two-dimensional analyses are widely used, three-dimensional analysis are now ready for practical use. However, the method has a demerit that it requires immense skills in setting a lot of parameters and evaluating analysis results.

The other is a static analysis which solves the residual deformation due to liquefaction after an earthquake. This is because most damages due to liquefaction are induced by the residual deformation. Moreover, both parameter setting and evaluation of results are much easier than those of a dynamic effective stress analysis. Thus this method is applicable to a basic design stage in which information available on soil properties and seismic motions are limited.

The two dimensional analysis "ALID" (Analysis for Liquefaction-Induced Deformation) developed by Yasuda et al. (1999) is a leading program in the static analysis. This has been accordingly applied in aseismic design of river dikes in Japan. However, three dimensional structures (*e.g.*, cylindrical tanks) must be approximately modeled in two dimensional analyses such as ALID.

For this reason, we will introduce the methodology of ALID to a three dimensional FDM analysis code FLAC3D (ver 5.01) to evaluate a residual deformation due to liquefaction in three dimensions. FLAC3D is an FDM program which has been widely used for solving ground deformation, and it has variety of structural elements.

In this paper, we will first outline the analysis methodology in this three dimensional static analysis program. We will then conduct a simulation analysis to evaluate the applicability of the developed program. The analysis model used is the residential area in the waterfront of Urayasu, Japan where liquefaction occurred in the 2011 off the Pacific coast of Tohoku Earthquake.

The program has been applied to evaluate a liquefaction measures in existing road embankments (Kobayashi 2012), and to simulate centrifuge tests for groundwater level lowering method (Kobayashi et al. 2014).

2 METHOD OF ANALYSIS

2.1 *The methodology of ALID*

ALID focuses on residual deformation due to liquefaction after an earthquake. It is composed of the following two processes:

1) Liquefaction-induced flow due to reduction in shear stiffness of liquefied soil.
2) Consolidation settlement caused by dissipation of excess pore water pressure.

2.2 *Liquefaction induced flow*

Figure 1 schematically shows the stress-strain curves of liquefaction-induced flow in ALID proposed by Yasuda et al. (1999). Line 1 indicates an initial state before an earthquake, while Line 2 indicates a stress-strain relationship of liquefied soil using a bi-linear model. This curve has a concave shape, and the region from shear strain γ_A to γ_L with the gradient of G_1 is called the small resistant region (Figure 1). The value of G_1 is very small, and may be obtained from a safety factor of liquefaction F_L using the chart by Yasuda et al. (2005) shown in Figure 2. If a shear strain exceeds γ_L, the shear stiffness increases to G_2 through positive dilatancy. Both G_2 and γ_L can be calculated on G_1. (Yasuda et al. 1999)

Calculation process of liquefaction-induced flow is described below:

1) Deformation in the initial state is calculated by using the self-weight method with shear moduli G_i (point A in Figure 1).
2) The stress–strain relationship is changed in a liquefaction-induced flow state. The shear moduli G_i are changed into G_1 with condition of no volume change (point B in Figure 1).
3) Deformation in the liquefaction-induced flow state is calculated by using the unbalanced forces, shown as $\tau_A - \tau_B$ in Figure 1. If a shear strain exceeds γ_L, G_1 is changed into G_2(point C in Figure 1).

We call this constitutive model "ALID model" in this paper.

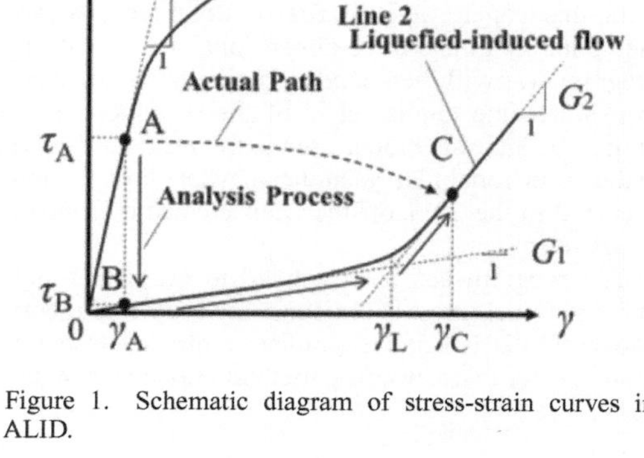

Figure 1. Schematic diagram of stress-strain curves in ALID.

2.3 *Consolidation settlement*

In order to predict a post-liquefaction volume change due to dissipation of excess pore water pressure, Ishihara and Yoshimine (1992) proposed a chart to determine the amount of post-liquefaction volumetric strain (Figure 3). The volumetric strain is determined using both safety factor of liquefaction F_L and relative density D_r of soil elements with this chart.

3 SIMULATION ANALYSIS

3.1 *Analytical model*

Figure 4 shows an analytical model which is a residential area in the waterfront of Urayasu, Japan where liquefaction occurred in the 2011 off the Pacific coast of Tohoku Earthquake. We assume eight houses on the ground in the analytical model. Each house has a clearance of 2.0 m with adjacent houses (LCAPTRC 2012). Displacements on the bottom boundary are perfectly

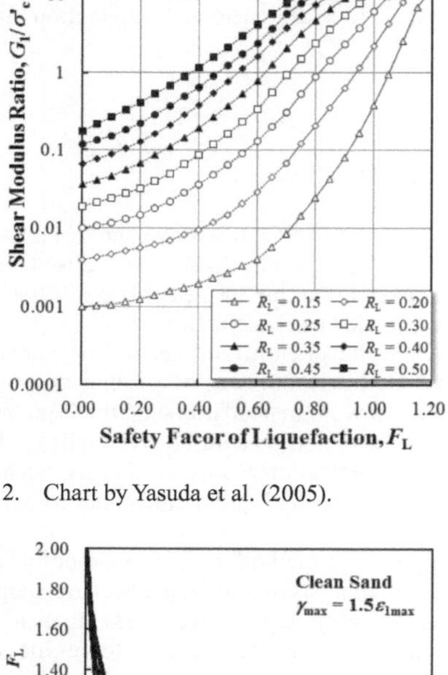

Figure 2. Chart by Yasuda et al. (2005).

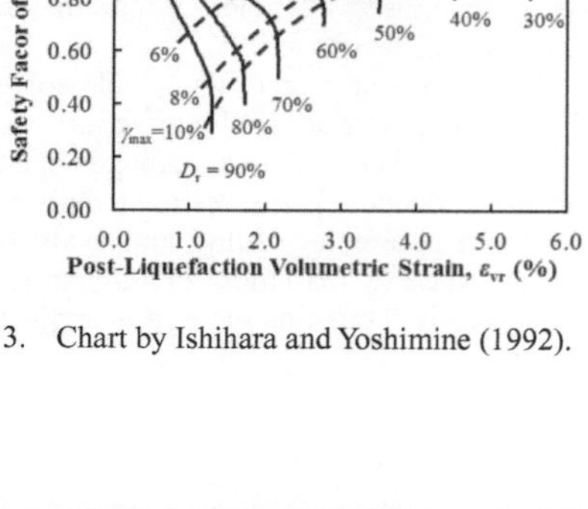

Figure 3. Chart by Ishihara and Yoshimine (1992).

Table 1. Soil properties used for the analysis

Layer	Thickness (m)	Fine content Fc (%)	Saturated density ρ_{sat} (g/cm^3)	Relative density Dr (%)	Cohesion c (kN/m^2)	Friction angle φ (deg.)	Poisson's ratio ν	Initial shear modulus G_0 (kN/m^2)	Shear modulus G_i (kN/m^2)	Liquefaction strength R_{L20}
▽ Bs (non sat.)	1	18.0	1.8	55.0	0	31	0.33	38,000	6,100	0.266
Bs (sat.)	1	18.0	1.8	55.0	0	31	0.33	38,000	7,500	0.247
Fs	6	22.0	1.8	76.3	0	30	0.33	29,000	6,800	0.188
As1	2	21.9	1.8	62.1	0	35	0.33	70,000	18,400	0.330
As2	2	31.0	1.7	72.9	0	30	0.33	40,000	10,900	0.228
Ac1	4	93.6	1.5	-	13	0	0.33	27,000	15,700	0.218

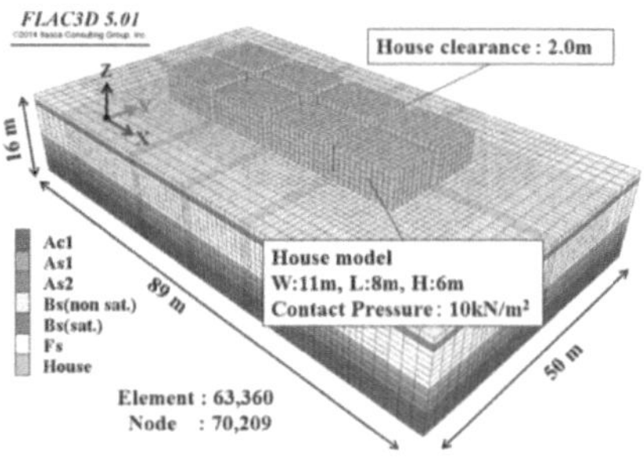

Figure 4. Analytical model.

fixed, whereas lateral boundary is restrained for horizontal displacement but is allowed to slide vertically.

3.2 *Properties of the model*

Table 1 shows the soil properties used in the analysis. Soil thickness and properties are average values determined for the liquefied area in Urayasu (LCP-TRC 2012). Shear moduli G_i at shear strain $\gamma = 0.1\%$ are obtained from the initial shear moduli G_0 using the strain-dependent characteristics of shear modulus (PHRI 1997). The water table is set at G.L. -1.0m.

3.3 *Seismic force*

Seismic force is also referred to the estimated value from the Urayasu committee report (LCAPTRC 2012). In the report, a peak ground acceleration of 1.61 m/sec^2 was obtained for this ground model in the seismic response analysis using the acceleration wave-form observed in Yumenoshima in the 2011 Earthquake. Accordingly, the committee determined the design value of acceleration α_{max} to be 1.70 m/sec^2. In this paper, we use a horizontal seismic coefficient $k_{hg} = 0.175 (= \alpha_{max}/g$, where g is gravitational acceleration) as an input seismic force.

3.4 *Constitutive models*

The non-saturated buried sand layer (Bs (non sat.)) and the alluvium clay layer (Ac1) are expressed in the elastic model as non-liquefied layers. The liquefied layers from the saturated buried sand (Bs (sat.)) to the second alluvium sand layer (As2) are expressed by the Mohr-Coulomb model in the initial state, by the ALID

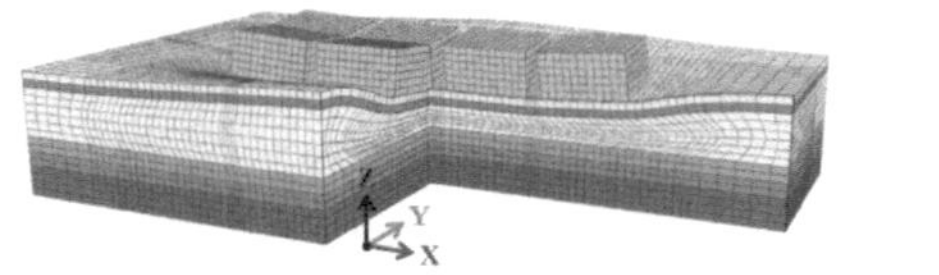

Figure 5. Residual deformation diagram (magnified 5 times).

model in the liquefaction-induced flow state, and by the elastic model in the consolidation settlement state, respectively.

4 RESULTS

Figure 5 shows the calculated residual ground deformation due to liquefaction. It may be seen that the houses settled and the liquefied soil (especially in the Fs layer) below the houses was pushed out laterally. In addition, the liquefied soil near the lateral boundary moved upward.

Figure 6 shows the contour of vertical displacement of ground surface as well as the inclination vector of each house. Note that the inclination vector expresses the inclination angle to the ground surface with colors, and the direction in which the maximum inclination angle occurred with an arrow. The settlement and inclination angle of a house are calculated from the following equations:

$$Sav = (s1 + s2 + s3 + s4)/4 \quad (1)$$

$$Sdif = S\max - S\min \quad (2)$$

$$\theta = Sdif/B \quad (3)$$

where S_{av} is the average settlement of a house; s_1 to s_4 are the settlements at each corner of a house; S_{dif} is the differential settlement of a house; S_{max}, S_{min} are the maximum and the minimum values among the settlements at the four corners of a house; θ is the maximum inclination angle of a house; and B is the distance between the corners where S_{max} and S_{min} occurred.

From Figures 5 and 6, it may be observed that the center of the model extremely subsided. The houses were very close to each other in this model, thus the vertical stresses due to self-weight were superposed in

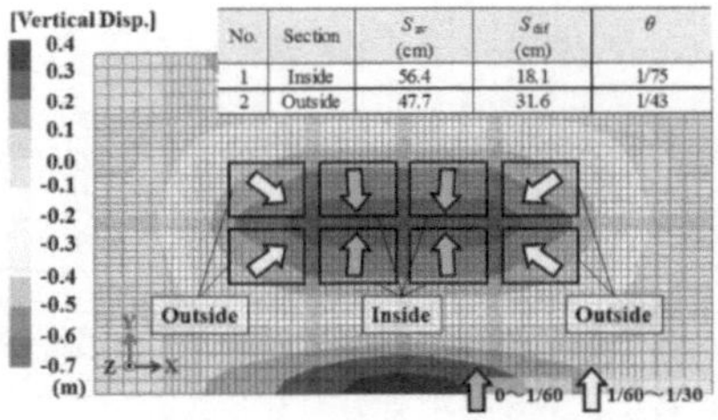

Figure 6. Contours of vertical displacement on ground surface and inclination vectors.

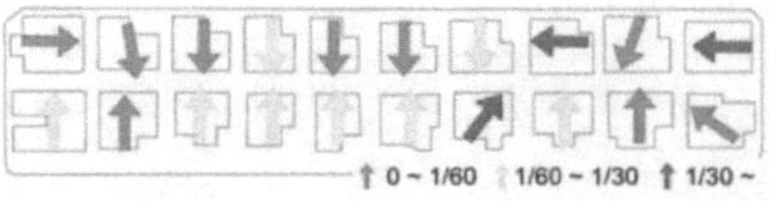

Figure 7. Inclination vectors in the block damaged due to liquefaction in Urayasu by Tokimatsu and Katsumata (2012).

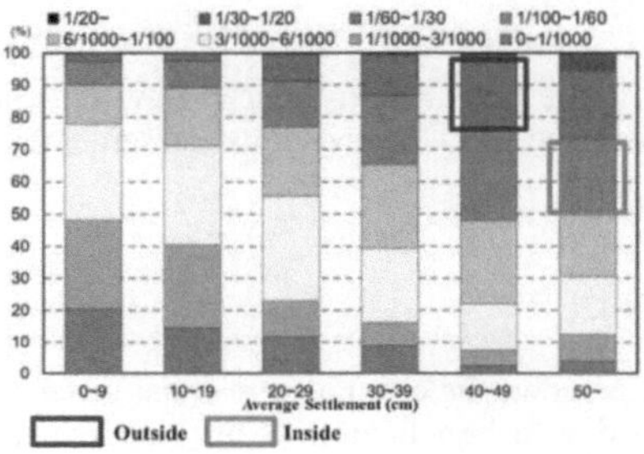

Figure 8. Liquefaction damage of houses in Urayasu by Tokimatsu and Katsumata (2012).

the ground. Meanwhile, shear moduli decreased from G_i to G_1 in the liquefaction-induced flow state, then the unbalance forces, $\tau_A - \tau_B$ in Figure 1, were applied to each node in the liquefied soil elements. The largest unbalanced forces occurred at the center of the analytical model, thereby the houses leaned over toward the center. It may also be seen in Figure 6 that the inside inclination vectors of houses were different from those outside.

Figure 7 shows the investigation result of inclination vectors due to liquefaction obtained by Tokimatsu & Katsumata (2012) in the block in Urayasu (expressed in the same manner as in Figure 6). In this figure too, the houses tended to incline toward the center of the block, and the directions of maximum inclination were generally consistent with the analysis result shown in Figure 6.

Figure 8 shows the investigation result in Urayasu by Tokimatsu & Katsumata (2012), in which the inclination angles of house were sorted with the average settlements. The maximum settlement was about 60 cm, and the inclination angle tended to increase with the increase in settlement. If we overlay the current analysis results shown in Figure 6, it is clear that the three dimensional static analysis program can simulate both average settlements and inclination angles.

5 CONCLUSIONS

We introduced the analysis methodology of ALID to three dimensional FDM analysis code FLAC3D. We carried out three dimensional static analysis of the damage of houses on liquefied ground in 2011 earthquake and compared it with the investigation result by Tokimatsu & Katsumata (2012).

As a result, it was found that the three dimensional static analysis program could estimate the damage of houses in terms of the amount of settlement and the inclination angle of houses, and that the planer inclination direction of houses could be evaluated in three dimensions.

REFERENCES

ALID Association 2007. *Two dimensional liquefaction-induced flow analysis program ALID/Win manual Ver4*: ALID Association. (in Japanese).

Japan Road Association 2002. *Part V, Seismic design, Specification for highway bridges*.JAPAN: JRA.

Kobayashi, M. 2012. A study on a reinforcement method for existing embankment on liquefiable ground. *Proc. of the 7th Asian Young Geotechnical Engineers Conference (7AYGEC)*: 117-122.

Kobayashi, M., Tateishi, A., Hiroshige, K., Hara, Y. 2014. Chikasuii teika kouhou wo tekiyou shita enshinmokei jikken no shimure-syon kaiseki sono 2 3 jigen seiteki zanryuu henkei kaiseki (Simulation analysis of centrifuge test for groundwater level lowering method, No.2 by using three dimensional static deformation analysis). *Proc. of the 69th Annual Conference of the Japanese Society of Civil Engineering*: 25-26. (in Japanese).

Liquefaction Countermeasure Achieved Possibility Technological Review Committee (LCAPTRC) 2012. *City liquefaction countermeasures achieved consideration investigation report.* Urayasu: Urayasu. (in Japanese).

Port and Horbour Research Institute (PHRI) 1997. *Handbook on liquefaction remediation of reclaimed land*: Port and Horbour Research Institute, Ministry of Transport. (in Japanese).

Tokimatsu, K. and Katsumata, K. 2012. Liquefaction damages to residual houses in Urayasu city caused by the 2011 Great East Japan Earthquake. *Proc. of the 47th Japan National Conference on Geotechnical Engineering*: 1489-1490. (in Japanese).

Towhata,I., Maruyama, S., Kasuda, K., Koseki, J., Wakamatsu, K., Kiku, H., Kiyota, T., Yasuda, S., Taguchi, Y., Aoyama, S. and Hayashida, T. 2014. Liquefaction in the Kanto region during the 2011 off the Pacific coast of Tohoku Earthquake. *Soils and Foundations* 54 (4): 859-873.

Yasuda, S., Inagaki, M., Nagao, K., Yamada, S. and Ishikawa, K. 2005. Stress-strain curves of cyclic softened soil including liquefied soils. Proc. of *the 40th Japan National Conference on Geotechnical Engineering*: 525-526. (in Japanese).

Yasuda, S., Yoshida, N., Adachi, K., Kiku, H., Gose, S. and Masuda, T. 1999. A simplified practical method for evaluating liquefaction-induced flow. *Journal of Geotechnical Engineering, Proc. JSCE* 638/III-49: 71-89 (in Japanese).

Challenges and Innovations in Geotechnics – Zhussupbekov (Ed.)
 ISBN 978-1-138-03007-7

Evaluation of damage morphology in three-hinge precast arch culvert based on shaking table tests and numerical analyses

Y. Sawamura, K. Kishida & M. Kimura
Kyoto University, Kyoto, Japan

H. Ishihara
Kajima Corporation, Hokkaido, Japan

ABSTRACT: The three-hinge precast arch culvert is a flexible structure, due to hinge functions in the main body, and is based on a different design concept than that of conventional culverts. Therefore, the conventional design method, which does not consider earthquake behavior, cannot be applied. Moreover, earthquake damage to several three-hinge precast arch culverts was reported after the 2011 Great East Japan Earthquake; and thus, it is necessary to investigate the dynamic behavior during strong earthquake motion. In this study, 1/5 scale shaking table tests, targeting a three-hinge precast arch culvert, were conducted to clarify the damage morphology of the culvert. In addition, two types of analyses, a dynamic analysis and a static analysis, were carried out for the shaking table tests to investigate the numerical analytical approach using an aseismic design for the culvert.

Keywords: Precast arch culvert, Damage morphology, Shaking table test, Numerical analytical approach

1 INTRODUCTION

The three-hinge precast arch culvert is a new type of culvert whose application can help save on labor at construction sites, shorten work periods and heighten quality control. The culvert consists of two arch members and three hinges in the body. Figure 1 shows the structure of the culvert. The shape is determined to minimize the tensile forces in the arch structure, thus creating an axially loaded structure. The hinge points are at the crown and both feet of the arch. The two arch members are set to incline toward each other to form the crown hinge, and the crown beam is cast in situ to support the arches against longitudinal loads. On the other hand, at each foot hinge, the arch member is supported by an independent concrete strip foundation or a concrete slab. The arch members are placed in simple keyways with joint filler which appends the hinge function at each foot.

As to the design of culverts in Japan, conventional culverts have been built over the past several years by applying methods which do not consider the seismic response. This is because such culverts have not suffered terrible damage in past earthquakes. However, the three-hinge precast arch culvert has hinge functions in the main body, and thus, is outside the range of conventional culverts and based on different design concepts. Furthermore, earthquake damage to several three-hinge precast arch culverts was reported after the 2011 Great East Japan Earthquake. Hence, it is necessary to investigate the dynamic behavior during strong earthquake motion and the critical state of the culverts.

Regarding the seismic performance of the three-hinge precast arch culvert, some experiments in the 1G gravitational field (Toyota and Takagai, 1999; Toyota and Itoh, 2000) and numerical analyses (Byrne et al., 1996; Wood and Jenkins, 2000) have been conducted. However, the damage morphology and the critical state of the culvert have not been clarified. In this study,

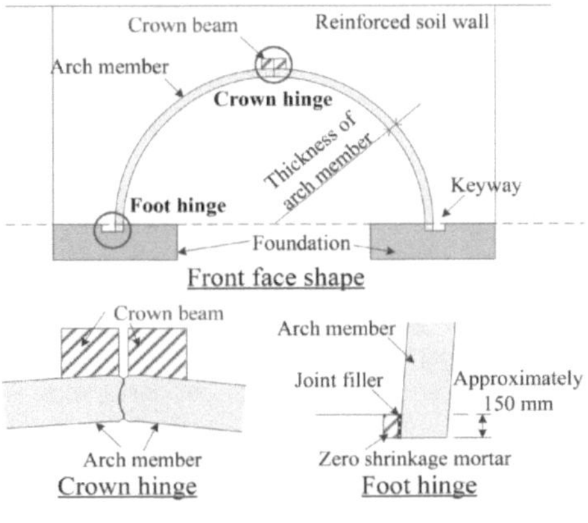

Figure 1. Structure of three-hinge precast arch culvert.

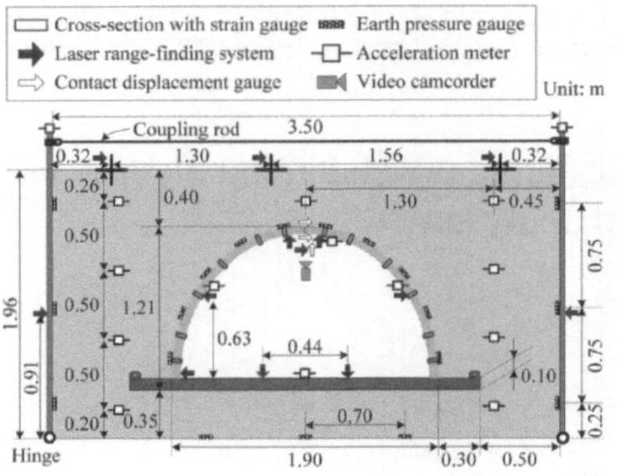

Figure 2. Set-up of culvert model and arrangement of sensors.

Table 1. Material constants of culvert model.

Concrete	Young's modulus E [kN/m^2]	3.40×10^7
	Compressive strength f_c [kN/m^2]	5.19×10^4
	Tensile strength f_t [kN/m^2]	2.86×10^3
	Poisson's ratio ν	0.2
Reinforcing bar	Young's modulus E [kN/m^2]	1.91×10^8
	Yield strength f_y [kN/m^2]	5.50×10^5
	Poisson's ratio ν	0.3

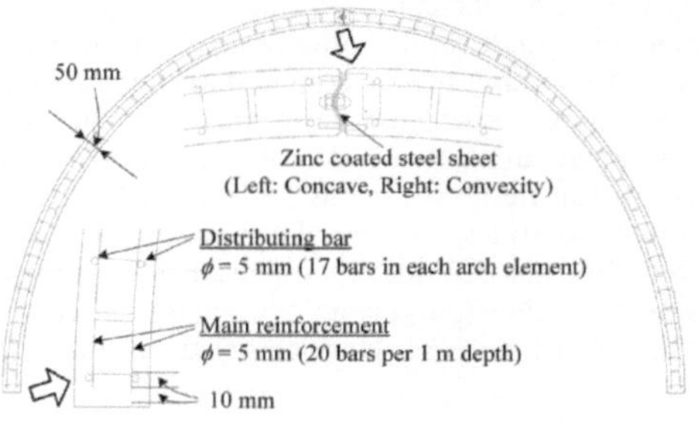

Figure 3. Bar arrangement drawing.

shaking table tests, targeting a three-hinge precast arch culvert of 1/5 scale, were compatibility of the deformation. Therefore, the axial-force dependency according to the variable axial force of the structure can be considered by using thisconducted to clarify the seismic behavior and the damage morphology of the culvert by using a strong earthquake response simulator. In addition, two types of analyses, a dynamic analysis and a static analysis, were carried out for the shaking table tests to investigate the numerical analytical approach using an aseismic design for the culvert.

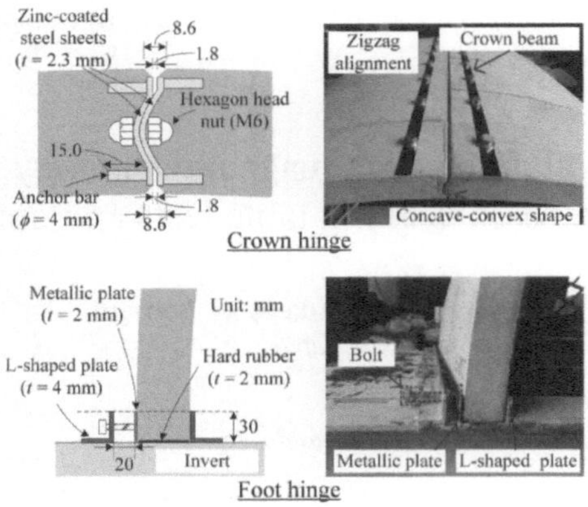

Figure 4. Modeling of hinges in experimental culvert.

Table 2. Material properties of Edosaki sand.

Specific gravity of soil particle G_s	2.73
Particle size distribution D_{50} [mm]	0.18
Internal friction angle ϕ [Deg]	38.3
Cohesion c [kPa]	14.0
Optimum moisture content w_{opt} [%]	20.8
Maximum dry density ρ_{dmax} [g/cm^3]	1.64

2 SHAKING TABLE TESTS

2.1 *Experimental conditions*

Experiments were conducted using the strong earthquake response simulator located at the Disaster Prevention Research Institute at Kyoto University.

Figure 2 shows the set-up of the culvert model and the arrangement of the sensors. A soil chamber, about 3.5 m long, 2.0 m deep and 1.0 m wide, was used. Since the lower part of the side wall and the bottom of the soil chamber are connected by a hinge, the side wall serves as a movable wall. Therefore, the soil chamber is a structure which permits simple shear deformation of the culvert and the model ground.

The culvert model was made from reinforced concrete. Table 1 shows the material constants of the model. Kagawa (1978) reported the similarity rule for model tests in the 1G gravitational field. If the similarity rule is applied, it is necessary to reduce the elastic coefficient of the culvert according to the model scale. However, it is difficult to control the elastic coefficient of concrete, and the plastic behavior does not satisfy the rule. Therefore, the 1/5 scale culvert model, whose material parameters for concrete and the reinforcing bar are not reduced, was used in these experiments. Figures 3 and 4 show a drawing of the bar arrangement and the modeling of the hinges in experimental culvert, respectively.

Both the foundation ground and the filling were made from Edosaki sand. Table 2 shows the material properties of Edosaki sand. The degree of compaction

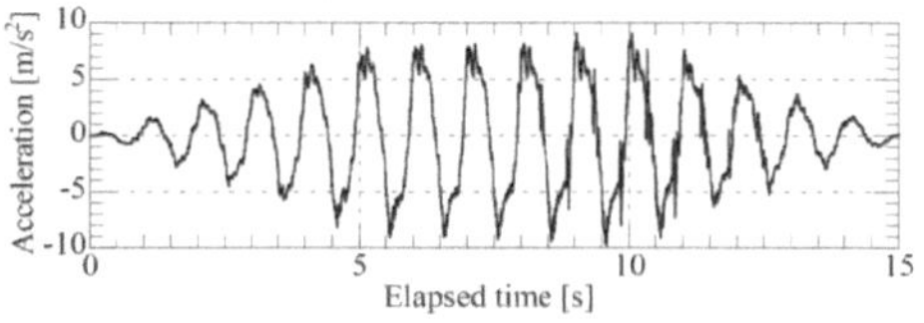

Figure 5. Input seismic motion.

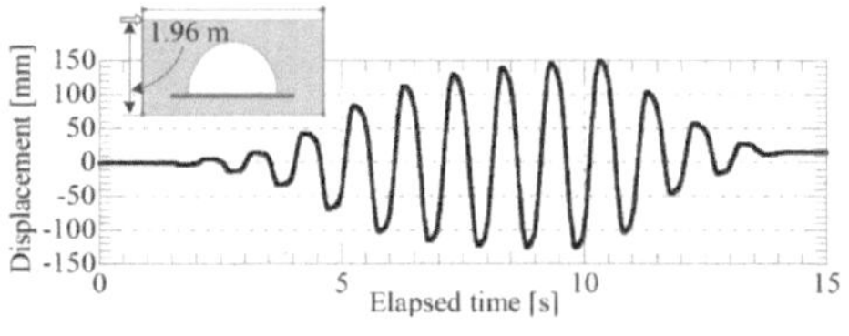

Figure 6. Time history of horizontal displacement at side wall.

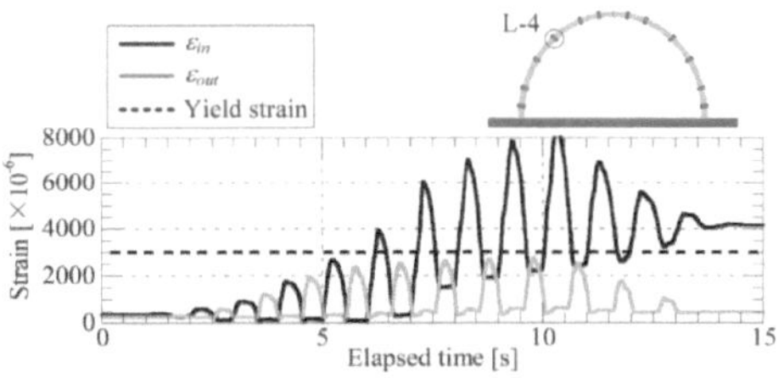

Figure 7. Time history of steel strain at L-4.

of Edosaki sand was set to 92%, which is the construction standard for backfill soil in a precast arch culvert. The sand was compacted with the prescribed water content ($w = 20.0\%$) in 39 layers for every 50 mm.

Figure 5 shows the input seismic motion. In this study, 1 Hz tapered sine waves were input to investigate the damage morphology of the culvert under huge seismic conditions. The sampling frequency was 1000 Hz and the total measurement time was 40 seconds.

2.2 *Experimental results*

Figure 6 shows the time history of the horizontal displacement at the side wall (at the height of ground level). The displacement means the relative displacement to the shaking table and the rightward displacement is defined as positive. The rightward displacement became maximum at t = 10.318 s and the maximum displacement at the wall was 149.69 mm. The shear strain, which is calculated by dividing the maximum displacement by the height of ground surface (1.96 m), is 7.64%. Hamada and Ohmachi (1996) reported that the amount of shear strain not accompanied by liquefaction was 1% in the 1995 Great Hanshin-Awaji Earthquake. In this experiment, therefore, it can be said that very huge shear strain occurred and that the model ground including the culvert greatly deformed.

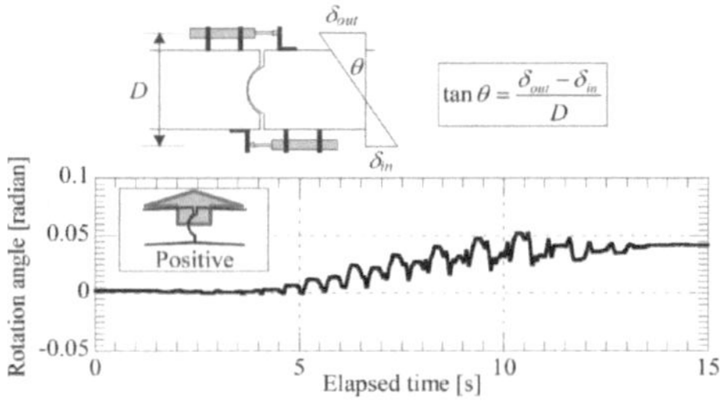

Figure 8. Time history of rotation angle of crown hinge.

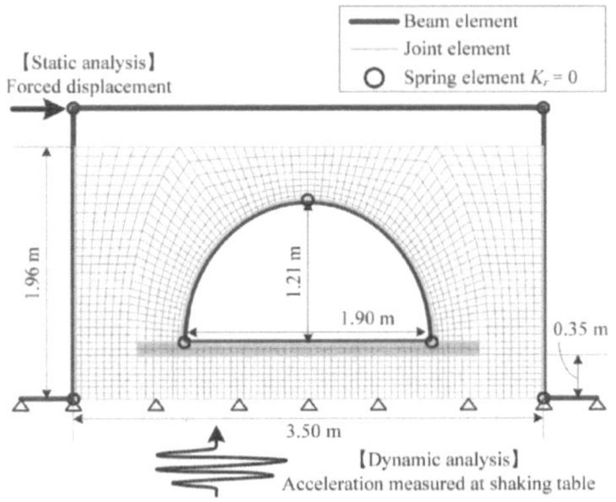

Figure 9. Analytical mesh and boundary conditions.

Figure 7 shows the time history of the steel strain at the left shoulder (L-4). The strain of the inside reinforcing bar gradually increased from the initial state and reached the yield strain at around 6 seconds. The plastic strain accumulates with the vibration, and finally, the strain reaches more than 8000 μ which is 2.8 times the yield strain.

Figure 8 shows the time history of the rotation angle of the crown hinge. The rotation angle accumulates as the whole arch transforms convexly. The residual value after the earthquake is 0.043 radian (2.45 degrees). Although it is feared that the crown hinge will drop out during an earthquake, it is confirmed that the possibility for the whole culvert to collapse due to the omission of a hinge part is low even when the shear strain of the ground is more than 7% and the strain of the reinforcing bar reaches 2.8 times the yield strain.

3 NUMERICAL ANALYSES

3.1 *Analytical conditions*

In this study, two-dimensional elastoplastic FEM analyses were performed to simulate the shaking table tests. The analytical mesh and the boundary conditions are shown in Figure 9. As the soil chamber is a structure which permits simple shear deformation, the soil chamber and the bottom of the side walls are modeled by sufficiently rigid elastic beam and rotation spring elements, respectively.

Table 3. Parameters of Edosaki sand.

Principal stress ration at critical state R_{cs}	4.0
Compression index λ	0.819
Swelling index κ	0.01
$N = e_{NC}$ at $p = 98$ kPa & $q = 0$ kPa	1.06
Poisson's ratio ν	0.276
Degradation parameter of overconsolidation m	0.02
Degradation parameter of structure a	0.65
Evolution parameter of anisotropy b_r	0.4
Wet Unit weight γ_t [kN/m^2]	17.74
Initial degree of structure R_0^*	0.158
Initial anisotropy ζ_0	0.5

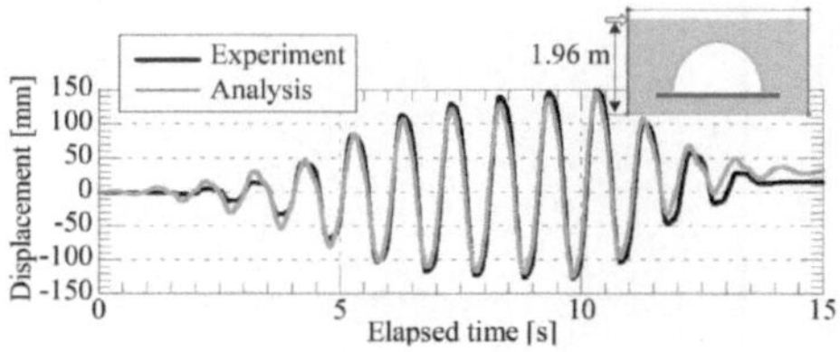

Figure 10. Time history of horizontal displacement at side wall (experiment and dynamic analysis).

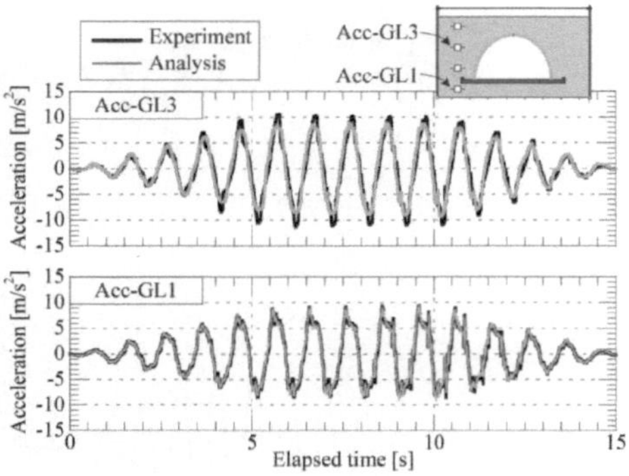

Figure 11. Time history of acceleration of Acc-GL1 and Acc-GL3 (experiment and dynamic analysis).

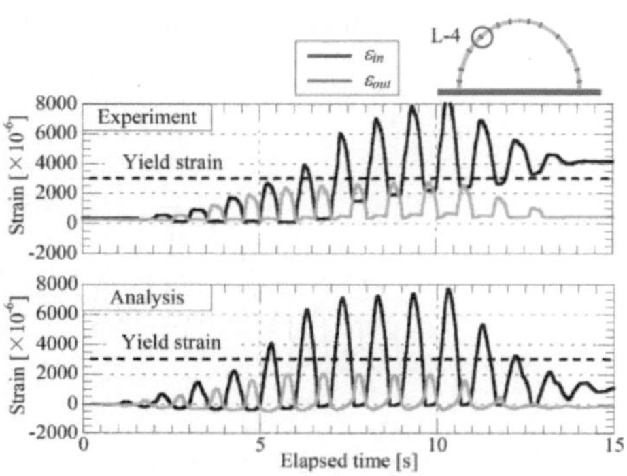

Figure 12. Time history of steel strain at L-4 (experiment and dynamic analysis).

The constitutive model for the Edosaki sand is the cyclic mobility model (Zhang et al., 2007). This model incorporates the concepts of subloading andsuperloading, as described by Hashiguchi and Ueno (1977) and Asaoka et al. (2002). The parameters of Edosaki sand are given in Table 3.

While modeling the culvert, the nonlinearity of the concrete and the reinforcing bar were also considered. For the culvert, the nonlinear moment-curvature relation was simulated using the AFD (Axial Force Dependent) model (Zhang and Kimura, 2002). This model introduces the concepts of the multi-spring model and the fiber model (Lai et al., 1984; Li and Kubo, 1999) into the finite element method by proposing a new weak form of the equilibrium equation for a beam, which satisfies the model. Both the crown hinge and the feet hinges are modeled by rotation spring elements whose rotation stiffness is zero.

In this study, two types of analyses, a dynamic analysis and a static analysis, were carried out for the shaking table tests. The input ground motion used in the dynamic analysis is the time history of the acceleration measured at the shaking table in the experiment, shown in Figure 5. The viscous damping and the direct integration method of Newmark-β ($\beta = 1/4$; $\gamma = 1/2$) are used, and the time interval of the calculation is 0.001 s. On the other hand, in the static analysis, the horizontal displacement measured at the left-side wall is applied as the forced displacement. The forced displacement was applied so that the shear strain of the soil chamber in each step would be 0.001%.

3.2 *Comparison of experiment and dynamic analysis*

Figure 10 shows the time history of the horizontal displacement at the side wall. The analytical results reproduce the experimental results comparatively well in both the amount of displacement and phase. Figure 11 shows the time history of the acceleration of Acc-GL1 and Acc-GL3. The acceleration obtained from the analysis tends to become small in the vicinity of the maximum and minimum values compared with the acceleration obtained from the experiment. However, the analysis precisely replicates the experiment as a whole.

Figures 12 and 13 show the time history of the steel strain at L-4 and the steel strain distribution at 10.318 s, respectively, when the rightward maximum displacement occurred. The analytical results reached yield strain earlier than the experimental results, and the resident value of the analysis was smaller than that of the experiment. However, the maximum strain and the strain distribution in the analysis sufficiently simulate those of the experiment.

Figure 14 shows the time history of the rotation angle of the crown hinge. Although the rotation angle in the analysis tends to become large earlier than that in the experiment, the rotation angle in both results

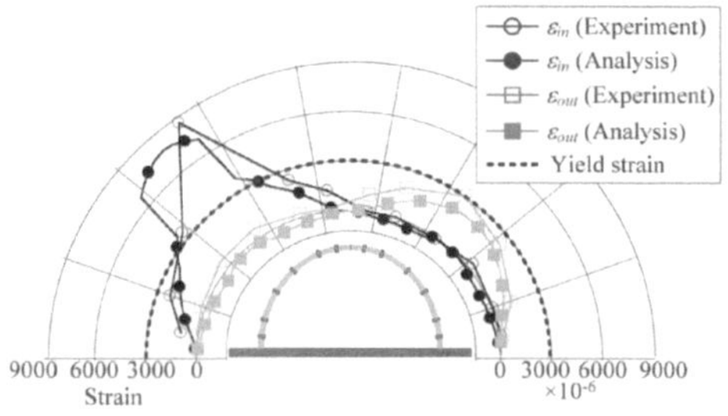

Figure 13. Steel strain distribution at 10.318 s (experiment and dynamic analysis).

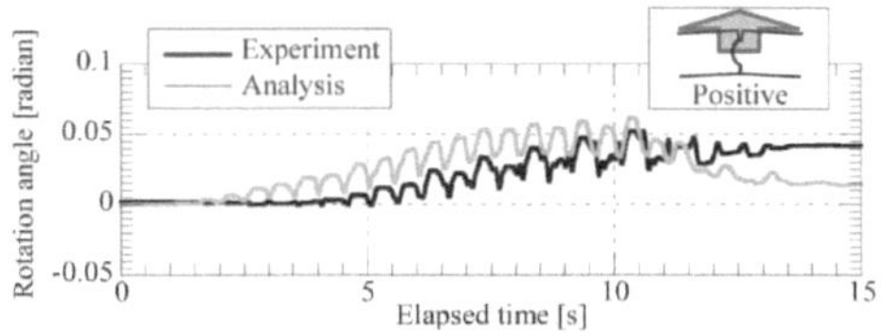

Figure 14. Time history of rotation angle of crown hinge (experiment and dynamic analysis).

become large at around 10.318 s when the shear strain of the ground becomes the maximum.

Based on the above results, the experimental study could be simulated accordingly in the present study by using a constitutive model for the ground and the arch culvert.

3.3 *Comparative results for damage morphology by experiment and numerical analyses*

Figures 15, 16 and 17 show the process of the reinforcing bars reaching the yield strain in the experiment, in the dynamic analysis and in the static analysis, respectively. In these figures, the time when the strain of the reinforcing bar reached the yield strain (3000 μ) for the first time at each measurement location is arranged. From Figure 15, it can be seen that the damage proceeds from the inside reinforcing bar. This is because the arch members are likely to deform inwardly during an earthquake. Firstly the inside reinforcing bar at the right shoulder (R-4in) reached the yield strain at 5.744 s, when the shear strain was 5.10%. Then, the damage expanded to the inside reinforcing bars. The shear strain was about from 4.5 to 5.5% when the inside reinforcing bars reached the yield strain. Finally, the outside reinforcing bar at the right shoulder (R-4out) reached the yield strain after the culvert had been heavily-damaged and the stiffness of the culvert had decreased. The damage process of the culvert (the inside reinforcing bars are damaged prior to the outside ones) is advisable from the perspective of maintenance. This is because visual confirmation is possible, although it is difficult to confirm the damage to culverts in comparison to bridges.

In the damage process of the dynamic analysis, the inside reinforcing bars around the left-side shoulder

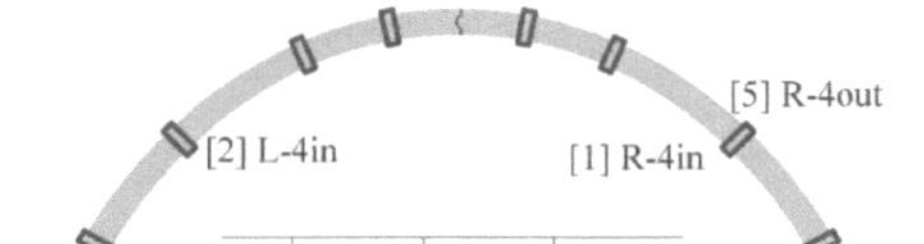

	t [s]	γ [%]	Member
[1]	5.744	5.10	R4-in
[2]	6.200	4.61	L4-in
[3]	6.236	5.42	L3-in
[4]	6.722	4.80	R3-in
[5]	9.268	6.92	R4-out

Figure 15. Process of reinforcing bars reaching yield strain in experiment.

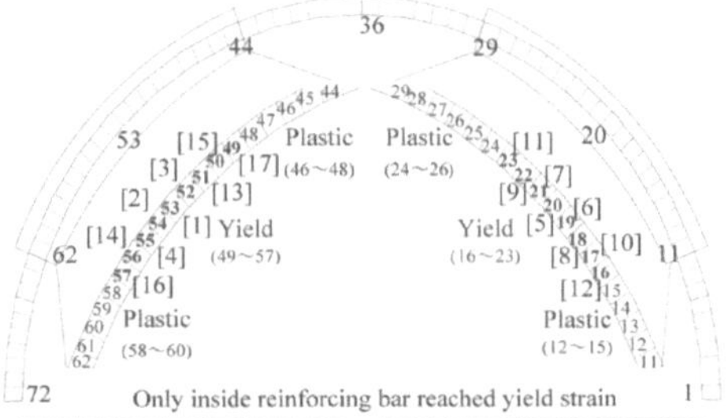

	t [s]	γ [%]	Member
[1]	5.257	4.09	53-in
[2]	5.259	4.11	54-in
[3]	5.266	4.17	52-in
[4]	5.273	4.22	55-in
[5]	5.708	4.05	20-in
[6]	5.710	4.10	19-in
[7]	5.714	4.18	21-in
[8]	5.718	4.26	18-in
[9]	5.725	4.39	22-in
[10]	5.732	4.51	17-in

	t [s]	γ [%]	Member
[11]	5.751	4.77	23-in
[12]	5.767	4.95	16-in
[13]	6.217	4.45	51-in
[14]	6.226	4.64	56-in
[15]	6.249	5.00	50-in
[16]	6.261	5.14	57-in
[17]	10.284	7.27	49-in

Figure 16. Process of reinforcing bars reaching yield strain in dynamic analysis.

reached the yield strain at about 5.2 seconds when the shear strain exceeded 4%. The area of damage expanded gradually by the repeating seismic ground motion, and thus, the inside reinforcing bars at both shoulders reached the yield strain. From the results, it is confirmed that the damage process of the culvert in the dynamic analysis reproduces that of the experiment well.

In the damage process of the static analysis, the inside reinforcing bar at the left shoulder was damaged first and then the area of damage expanded gradually. When the reinforcing bar first reached the yield strain, the shear strain of the soil chamber was 4.53%, which was almost the same as that of the experiment, although it was larger than that of the dynamic analysis. When the forced displacement is applied until the shear strain reaches 7.64%, which is the maximum shear strain in the shaking table tests, the area where the reinforcing bars reach the yield strain is same as for the dynamic analysis. From these results, it is confirmed that the static analysis would be an effective aseismic

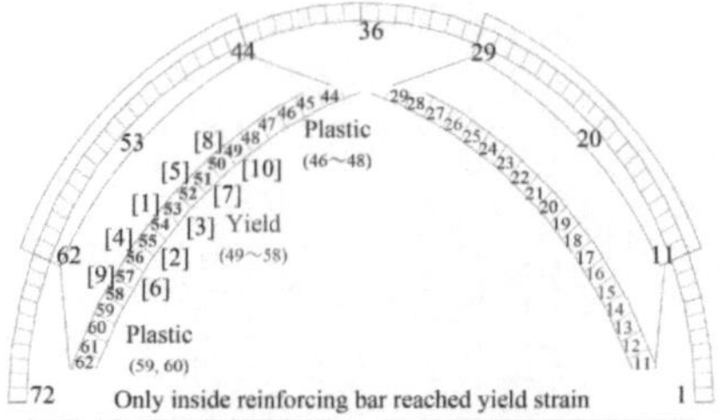

	t [s]	γ [%]	Member		t [s]	γ [%]	Member
[1]	-	4.53	54-in	[6]	-	5.34	57-in
[2]	-	4.55	55-in	[7]	-	5.51	51-in
[3]	-	4.61	53-in	[8]	-	6.33	50-in
[4]	-	4.73	56-in	[9]	-	6.44	58-in
[5]	-	4.92	52-in	[10]	-	7.40	49-in

Results are arranged up to shear strain of 7.64%

Figure 17. Process of reinforcing bars reaching yield strain in static analysis.

design method because the analysis reproduced the progression of damage to the culvert.

4 CONCLUSIONS

In this study, 1/5 scale shaking table tests targeting a three-hinge precast arch culvert were conducted to clarify the damage morphology of the culvert. In addition, two types of analyses, a dynamic analysis and a static analysis, were carried out for the shaking table tests to investigate the numerical analytical approach using an aseismic design for the culvert. The following conclusions can be drawn from the results of this study:

1) Although it is feared that the crown hinge will drop out during an earthquake, it has been confirmed that the possibility for the whole culvert to collapse due to the omission of a hinge part is low even when the shear strain of the ground is more than 7% and the strain of the reinforcing bar reaches 2.8 times the yield strain.
2) Damage to the three-hinge precast arch culvert was found to proceed from the inside reinforcing bar. It is advisable, therefore, that this damage process be viewed from the perspective of maintenance. This is because visual confirmation is possible, although it is difficult to confirm the damage to culverts compared to bridges.
3) The experimental study was simulated accordingly by using a constitutive model for the ground and the arch culverts.
4) It has been confirmed that the static analysis would be an effective aseismic design method because the analysis can reproduce the progression of damage to the culvert.

ACKNOWLEDGMENT

This research was supported by the National Institute for Land and Infrastructure Management, MLIT, Japan (a grant for the research and development of technologies for improving the quality of road policies, no. 24-4, 2012-2015).

REFERENCES

Asaoka, A., Noda, T., Yamada, E., Kaneda, K. and Nakano, M. 2002. An elasto-plastic description of two distinct volume change mechanics of soils, *Soils and Foundations*, 42(5), 47–57.

Byrne, P. M., Anderson, D. L., and Jitno, H. 1996. Seismic analysis of large buried culvert structures. Transportation Research Record 1541, *Transportation Research Board*, Washington, DC, 133–139.

Hamada, M. and Ohmachi, T. 1996. Evaluation of earthquake-induced displacement and strain of the surface ground in near-field, *Hanshin-Awaji Dai-shinsai ni kansuru Gakujyutsu-Kouenkai ronbunsyuu*, 69–80. (in Japanese)

Hashiguchi, K. and Ueno, M. 1977. Elastoplastic constitutive laws of granular material, Constitutive Equations of Soils, *Proc. 9th ICSMFE.*, Spec. Ses. 9 (eds. by Murayama, S. and Schofield, A. N.), Tokyo, JSSMFE, 73–82.

Kagawa, T. 1978. On the similitude in model vibration tests of earth-structures, *Proc. of Japan Society of Civil Engineers*, No.275, 69–77. (in Japanese)

Lai, S-S., Will, G. T. and Otani, S. 1984. Model for inelastic biaxial bending of concrete member. *Journal of Structural Engineering*, ASCE, 110 (11), 2563–2584.

Li, K. N. and Kubo, T. 1999. Analysis of circular RC member in MS/fiber model, *Summaries of Technical Papers of Annual Meeting*, Hiroshima St., II, AIJ.

Toyota, H., and Takagai, M. 1999. Dynamic behavior of 3-hinge arch in terre armee foundation, *Journal of Geotechnical Engineering*, No. 624/III-47, 255–266. (in Japanese)

Toyota, H. and Itoh, T. 2000. Effects of Shaking Conditions and Material Properties on Dynamic behavior of terre armee foundation and 3-hinge arch, *Proc. of Japan Society of Civil Engineers*, No.666/III-53, 279–289. (in Japanese)

Wood, J. H., and Jenkins, D. A. 2000. Seismic analysis of buried arch structures. *Proc. of 12th World Conf. on Earthquake Engineering*, New Zealand Society for Earthquake Engineering, Wellington, New Zealand.

Zhang F., Ye, B., Noda, T., Nakano, M. and Nakai, K. 2007. Explanation of cyclic mobility of soils: Approach by stress-induced anisotropy, *Soil and Foundations*, 47(4), 635–648.

Zhang, F. and Kimura, M. 2002. Numerical prediction of the dynamic behaviors of an RC group-pile foundation, *Soils and Foundations*, 42(3), 72–92.

Challenges and Innovations in Geotechnics – Zhussupbekov (Ed.)
© 2016 Taylor & Francis Group, London, ISBN 978-1-138-03007-7

Numerical study on electrical resistivity tomography for prediction of anomaly in mechanized tunnelling job-sites

Kang-Hyun Lee
Korea Expressway Corporation Research Institute, Hwaseong, Korea

In-Mo Lee
Korea University, Seoul, Korea

ABSTRACT: Anomalies, mixed grounds and fractured zones which are not detected by the geotechnical investigation performed during the design stage may cause difficulty when a tunnel is excavated. So, it is important to predict the ground condition ahead of a tunnel face during tunnelling to avoid accidents. Various methods to predict the ground condition ahead of the tunnel face have been studied and developed. The prediction methods such as tunnel seismic profiling, ground penetration radar, probe drilling frequently used at conventional tunnelling job sites. However, few studies applicable to mechanized tunnellingare tried because of the limitation of available spaces in Tunnel Boring Machine (TBM). So, this study proposes a prediction method applicable to mechanized tunnelling using the pre-drilled borehole. Numerical simulations were performed to adopt the most suitable electrode array for prediction of anomaly ahead of the tunnel face when electrical resistivity tomography is performed at mechanized tunneling job-sites. The pole-dipole and modified dipole-dipole array was found to be able to predict the location and condition of an anomaly reasonably well. A remote electrode is always needed to use the pole-dipole array, but it is very difficult to find an appropriate place to install a remote electrode. Accordingly, the modified dipole-dipole array does not need a remote electrode that is the most suitable one to be used for mechanized tunnelling job sites.

1 INTRODUCTION

Various kinds of geological surveys and geophysical surveys are generally conducted to interpret the ground condition of the proposed tunnel route to design the tunnel. The geotechnical investigations performed in the design stage chiefly focuses on predicting ground condition over a broad region. So, it is difficult to accurately predict the ground condition abutting on a tunnel route. As a result, there are many cases where unpredicted anomalies are found during tunnel excavation. These anomalies may decrease the stability of tunnel and generate temporal, economic losses.

For these reasons, various methods to predict the ground condition ahead of the tunnel face have been studied and developed. Prediction methods using various kinds of equipment include tunnel seismic prediction, horizontal drilling and/or boring, ground penetrating radar. However, most of the prediction methods are only applicable to conventional tunnelling job sites, which feature free access to the tunnel face.

In mechanized tunnelling job sites, it is not easy to handle adequately trouble due to sudden changes of the ground condition because the tunnel face is not visible with the naked eye. Various kinds of mechanical and electrical equipment are installed behind the cutter head, so there are spatial limitations on the survey equipment that should be installed to predict the ground conditions ahead of the tunnel face. So, few studies on the prediction technique, which applicable to mechanized tunnelling, have been conducted (Kneib et al., 2000; Dowden and Robinson, 2001; Yamamoto et al. 2003; Kaus and Boening, 2008; Richter, 2011). Kaus and Boening (2008) examined how to predict the ground condition ahead of a tunnel face from an electrical resistance and induced polarization by installing an array of electrodes in the cutter-head of TBM and the side of the main body. Richter (2011) conducted a study on predicting the ahead of tunnel face through an irregular borehole radar survey in a karst topography where stability is very low.

Generally, it is possible to drill a borehole by use of probe-drilling equipment in the TBM main body. Therefore, this study analyzes environment of the mechanized tunnelling job sites in order to apply the electrical resistivity tomography to mechanized tunnelling job site. We propose to install multiple electrodes to the pre-drilled hole to perform electrical resistivity tomography tests. Numerical simulations of multiple electrode arrays installed to the pre-drilled hole in the upper side of the TBM were performed to figure out the most suitable electrode array applicable to mechanized tunnelling job site.

2 BACKGROUND THEORY

2.1 *Characteristics of electrical resistivity*

Electrical resistivity is a constant representing the electrical property of a material regardless of its shape and size. The unit of electrical resistivity is the ohm-meter (Ωm). While electrical resistance represents the difficulty of passing an electric current through a material, the electrical resistivity represents an electrical resistance that a material of unit volume has. In an arbitrary conducting wire, the relationship between electrical resistance (R) and electrical resistivity (ρ) is as follows:

$$R = \rho F_s \tag{1}$$

Here, F_s is a shape factor that changes according to the shape of the material through which the current is flowing. For a cylindrical conducting wire, F_s is calculated by dividing the material's length by cross section.

Electrical resistivity in rocks changes by the degree of saturation, porosity, the electrical conductivity of groundwater and content of clay. Intact rock has low porosity and low clay content, so its electrical resistivity is high. On the contrary, the electrical resistivity of weathered rock with lots of joints is low due to low porosity and filling materials such as clay in voids. Fig. 1 shows the distribution of electrical resistivity classified by the type of soil and rocks. As shown in Fig. 1, the soil has comparatively lower electrical resistivity than rock, and rocks that have high uniaxial compressive strength tend to display higher electrical resistivity.

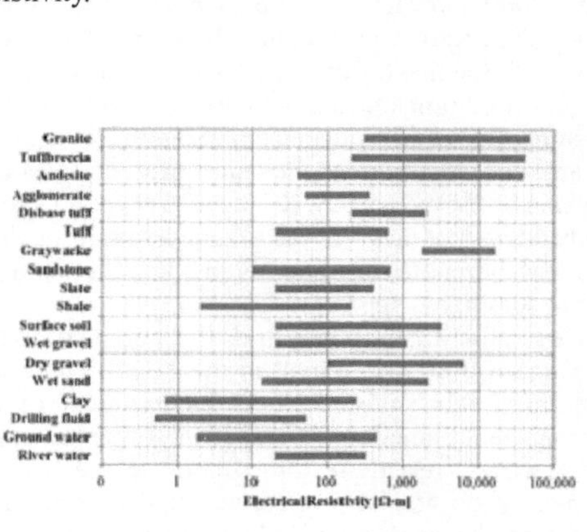

Figure 1. Distribution of electrical resistivity classified by the type of soil and rocks.

2.2 *Electrical resistivity survey and electrical resistivity tomography*

To predict the ground condition from electrical property of the ground, electrical resistivity and electrical resistivity tomography are performed. The electrical resistivity survey is trying to figure out the ground condition from distribution of electrical resistivity, which is an electrical property of the ground, and has widely been adopted in civil engineering. It is a method to examine the geological structure and condition by installing multiple electrodes on surface (Fig. 2). This method is economical and efficient to evaluate a wide range of ground conditions.

Electrical resistivity tomography performs measurements and analysis under the same principle with the electrical resistivity survey. In the electrical resistivity tomography, a borehole is drilled from the ground surface, multiple electrodes are installed the borehole letting the object for survey to be surrounded by the electrodes (Fig. 3). In short, electrical resistivity tomography to evaluate the ground condition between boreholes is performed. Compared with an electrical resistivity survey, electrical resistivity tomography can explore a narrower area and produce higher-resolution output.

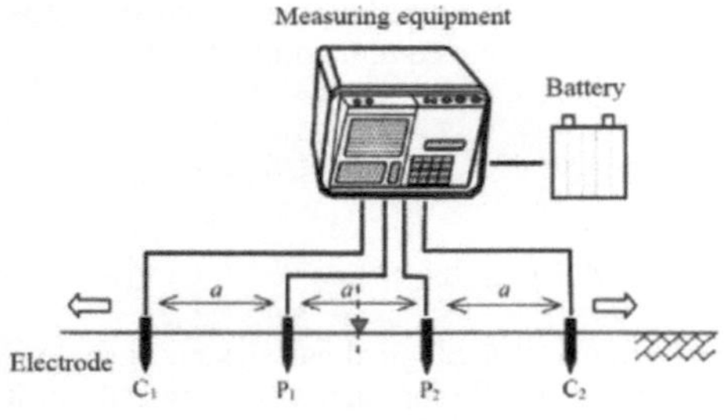

Figure 2. Schematic diagram of electrical resistivity survey.

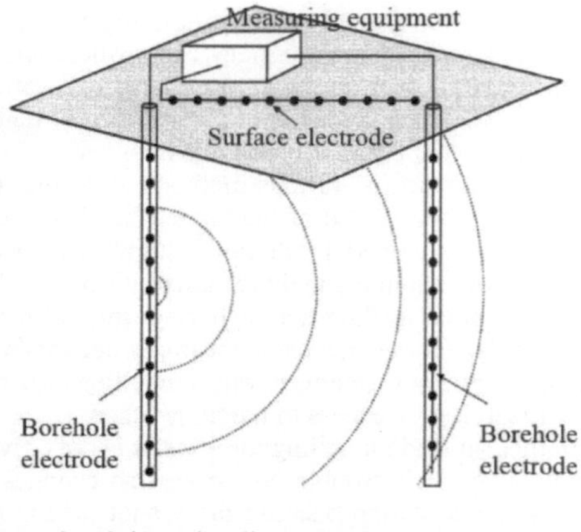

Figure 3. Schematic diagram of electrical resistivity tomography.

3 FEASIBILITY OF TOMOGRAPHY TESTING IN TBM

In mechanized tunnelling job sites, the ground ahead of the tunnel face contacts with the cutter head, and the rear of the cutter head has a lot of electrical and mechanical devices. So, if electrical resistivity is only measured around the cutter head, it is highly likely to have low reliability due to noise from metallic materials and various kinds of electric equipment.

In many cases of TBM, probe-drilling equipment is installed on the upper part of the machine. The ground condition of tunnelling route determines whether the probe-drilling equipment is necessary or not. Mostly the ground condition is poor, the equipment is installed. To conduct an electrical resistivity tomography, it is usually necessary to drill two full-length boreholes. However, it is hard to install the probe-drilling equipment on the TBM main body because various facilities and equipment are put in place at the back of the cutter-head.

Thus, in order to study the feasibility of using one long array along with another short array, we assumed that the probe-drilling equipment was installed in the upper part of the main body of the TBM. It means that a 20m borehole was drilled on the upper side. Also, a 1m borehole was assumed to be drilled on the lower side because of difficulties involved in drilling a borehole from the lower side of the main body of the TBM. We can perform electrical resistivity tomography to predict the ground condition ahead of the tunnel face using two boreholes. Figure. 4 represents the overview of the electrical resistivity tomography which is applicable to TBM proposed by this study.

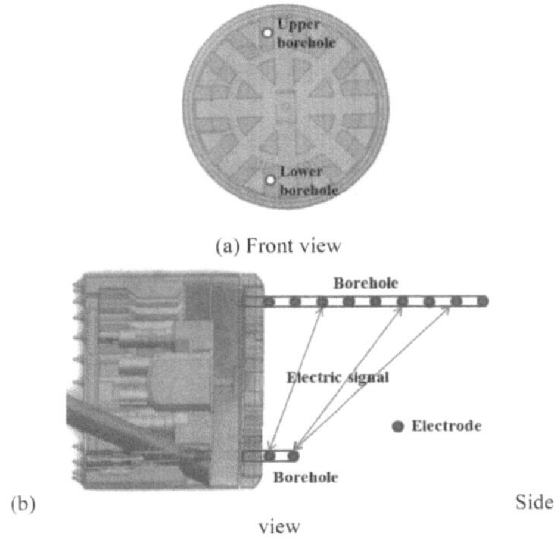

Figure 4. Schematic diagram of resistivity tomography applicable to TBM.

4 NUMERICAL SIMULATION OF THE GROUND WITH AN ANOMALY AHEAD OF THE TBM TUNNEL FACE

We used TomoDC ver. 1.2 to perform numerical modeling of the electrical resistivity tomography. The spacing between the upper borehole and the lower borehole was set to 8 m to reflect the TBM diameter mainly used for cross sections of road and subway tunnels.

It was assumed that an anomaly is location vertically as shown in Figure. 5. The vertical anomaly may cause problems when a tunnel is excavated with a TBM. With such an anomaly, the TBM equipment may have a chance to be jammed in the ground, and excessive groundwater can flow into the face through an anomaly and may create a mucking problem.

The electrical resistivity of bedrock was assumed to be 1,000 Ωm, and that of the anomaly (may be fracture rock), which had a lower strength than bedrock was assumed to be 100 Ωm, lower than that of bedrock. In modeling, the thickness of the anomaly was assumed to be 3 m and the distance between the cutter head and the center of the anomaly was assumed to be 10 m. the electrical resistivity of the cutter head was set to be 1 Ωm because the cutter head is metal (Fig. 5). The electrode spacing was set to be 1 m as shown in Fig. 5, and the electrode arrays used for numerical simulation were pole–pole, pole–dipole, dipole–dipole, modified pole–pole, and modified dipole–dipole. As properties of the modified dipole-dipole array change according to the sum of array constants, the sum of k and l array constants classified into seven cases: 4, 6, 8, 10, 12, 14 and 16.

Figure. 6 shows the numerical simulation results for various electrode arrays. Since the resolution of pole-pole array is relatively lower than other arrays, the anomaly and the cutter head appeared to be very wide (Fig. 6(a)). So, geological structure of the analysis result was dissimilar to that of modelling.

The dipole–dipole array has low signal to noise (S/N) ratio and a small measurement potential. Therefore, a large measurement potential is usually distributed only near the current dipole, and the measurement potential is very small in other regions. So, it can be said that good measurements will only be obtained from inline surveys in which the current dipole is close to the potential dipole. Because of this reason, the

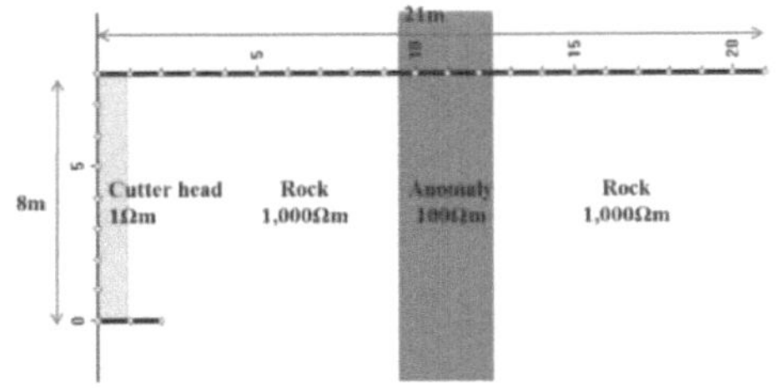

Figure 5. Numerical simulation of ground condition and overview of the boreholes.

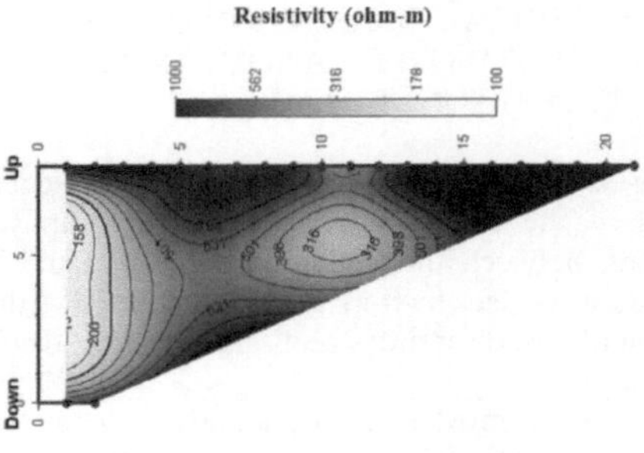

(a) Pole–pole array

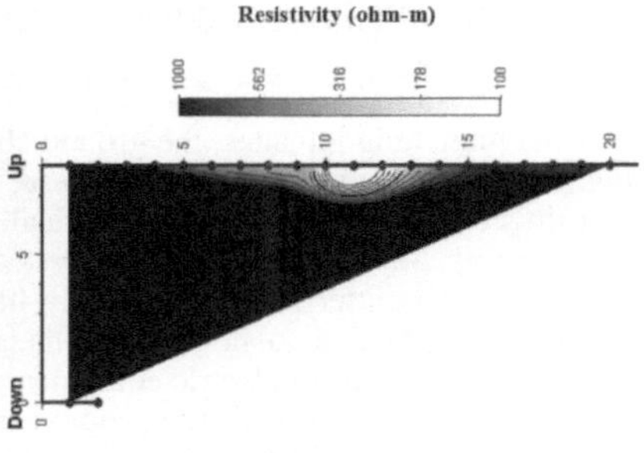

(b) Dipole–dipole array

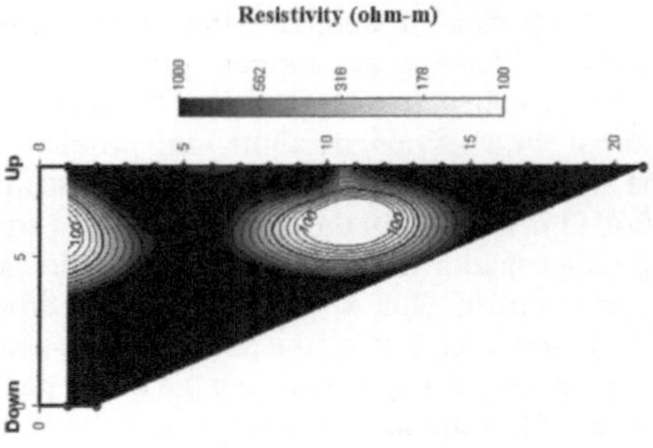

(c) Pole–dipole array

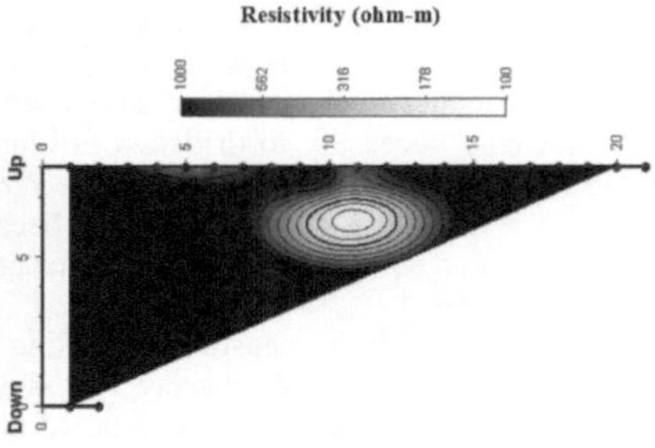

(d) Modified pole–pole array

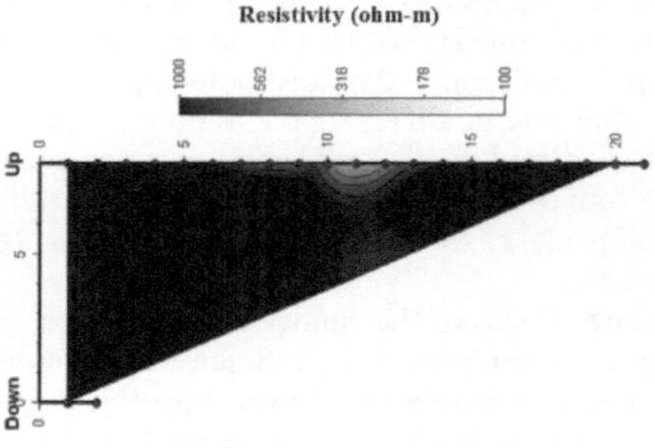

(e) Modified dipole–dipole array (k + l = 6)

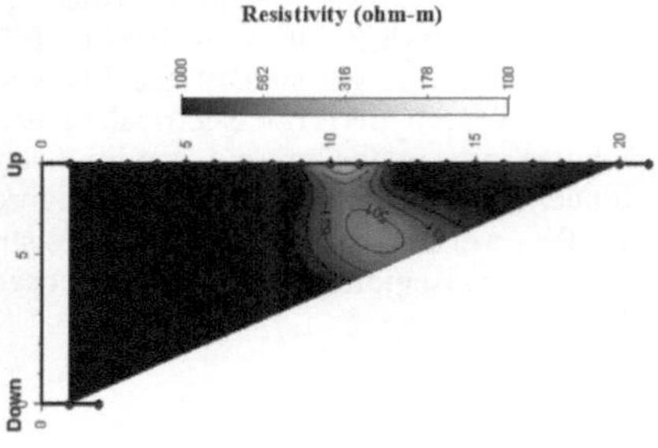

(f) Modified dipole–dipole array (k + l = 8)

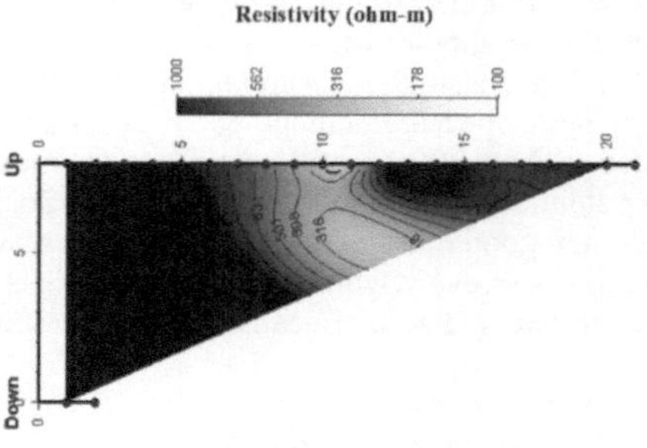

(g) Modified dipole–dipole array (k + l = 10)

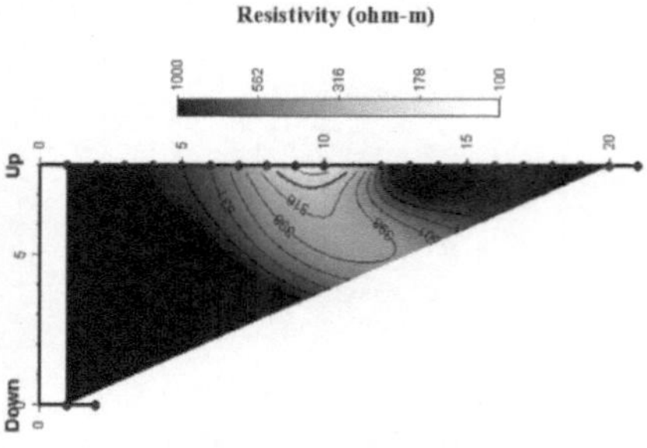

(h) Modified dipole–dipole array (k + l = 12)

Figure 6. Numerical simulation results with each electrode array.

anomaly only appeared in the zone close to the upper borehole, as shown in Figure. 6(b).

Although the pole–dipole array had higher resolution than the pole–pole array, it still has somewhat lower resolution than other electrode arrays.· The anomaly was distributed horizontally by about 6 m because of the relatively low resolution (Fig. 6(c)). However, the analyzed value of the electrical resistivity of the anomaly was quite close to that of the modelling value. The results with the modified pole–pole array were similar to those with the pole–dipole array (Fig. 6(d)).

In the case that the sum of array constants k and l was 6 or less, the modified dipole-dipole array is considered to predict the shape or location of anomaly just as dipole-dipole array (Fig. 6(e)). On the other hand, we were able to obtain the vertical anomaly which has the most similar shape to the modelling one (Fig. 6(f), (g)), with the modified dipole–dipole array, when the sum of array constants k and l were 8 or 10. The thickness of the anomaly was approximately 5m. The location of anomaly was more accurate than any other electrode arrays. But, the shape of anomaly appeared somewhat distorted because the borehole and the electrode location of modified dipole–dipole array were not symmetrical. Although the modified dipole–dipole array has better resolution than the pole–pole array, the electrical resistivity range of the anomaly was obtained to be 200~500 Ωm. The reason may be because the data in which the electrical resistivity dropped rapidly were removed while editing the data for the elimination of noise.

As a result of the numerical simulation using various electrode arrays, it has been concluded that the location and property of anomaly will be properly predicted if the electrical resistivity tomography survey is performed using a pole–dipole array at a mechanized tunnelling job site. A remote electrode is always needed to use the pole-dipole array. No adequate place to install a remote electrode far from electrical influence is found because a cutter head and a variety of electrical equipment in the rear of chamber installed at mechanized tunnelling job sites. If the remote electrode is installed in a wrong place, the measurement data may cause a serious doubt in terms of the reliability.

Among various electrode arrays, the modified dipole–dipole array having a sum of array constants k and l of 8 or 10 were found to predict the location and shape of an anomaly reasonably well. In addition, the modified dipole-dipole array does not require the remote electrode. Accordingly, we can conclude that the modified dipole-dipole array is the most suitable one to be used for mechanized tunnelling job sites.

5 CONCLUSIONS

This study analyzes characteristics of the mechanized tunnelling job sites in order to apply the electrical resistivity tomography to mechanized tunnelling job site. Also, numerical simulations were performed to choose the most appropriate electrode array for predicting anomaly ahead of the tunnel face using electrical resistivity tomography. The conclusion of our study can be summarized as follows.

(1) We expect that electrical noise will occur at mechanized tunnelling job sites because of the various kinds of electrical and mechanical equipment in TBM. In particular, it is highly unlikely to produce a reliable result in measuring the electrical resistivity on the ground because the metallic cutter head contacts the ground.
(2) When a pole–pole array was used for electrical resistivity tomography at mechanized tunnelling job sites, it is very hard to identify the ground condition ahead of the tunnel face due to the electrical influence from the cutter head and low resolution. Dipole-dipole array seems practically impossible to apply to electrical resistivity tomography due to its very low S/N ratio.
(3) The pole-dipole array turned out to predict well the location and property of anomaly. But the pole-diploe array is hard to use in mechanized tunnelling job site as it requires the remote electrode installation.
(4) The modified dipole–dipole array was found to predict the location and condition of an anomaly reasonably well than any other arrays when the sum of array constants k and l is 8 or 10. One advantage of utilizing this array is that it does not need a remote electrode.

The part of this article is based on a study first report in Ph. D. dissertation "Predictions of ground conditions ahead of tunnel face applicable to TBM" (Lee, 2014) and Proceedings of World Tunnel Congress "Numerical Study on the Resistivity Tomography Applicable to TBM" (Lee et al., 2015).

ACKNOWLEDGEMENTS

The author expresses their appreciation to my advisor, Dr. Nag-Young Kim. The support throughout the study was invaluable. This research was supported by a grant (Project number: 13SCIP-B066321-01 (Development of Key Subsea Tunnelling Technology)) from Infrastructure and Transportation Technology Promotion Research Program funded by Ministry of Land, Infrastructure and Transport of Korean government.

REFERENCES

Dowden, P.B., Robinson, R.A. 2001. Coping with boulders in soft ground TBM tunneling, *Rapid Excavation and Tunneling Conference, Littleton*, 961–977.

Kaus, A., Boening, W. 2008. BEAM-geoelectrical ahead monitoring for TBM-drives, Geomechanics and Tunnelling, Vol.1, No.5, 442–449.

Kneib, G., Kasselm, A., Lorenz K. 2000. Automatic seismic prediction ahead of the tunnel boring machine, *61st International EAGE Conference*, 295–392.

Richter, T. 2011. Innovative geophysical investigation technology in karstified and fractured rock formations, *1st Scientific Congress on Tunnels and Underground Structures in South-East Europe, Dubrovnik*, 20–21.

Yamamoto T., Shirasagi S., Yamamoto S., Mito Y., Aoki K. 2003. Evaluation of the geological condition ahead of the tunnel face by geostatistical techniques using TBM driving data, *Tunnelling and Underground Space Technology* 18: 213-221.

Challenges and Innovations in Geotechnics – Zhussupbekov (Ed.)
 ISBN 978-1-138-03007-7

Analysis of efficiency of fiber reinforced sand as a backfill of retaining walls

A.S. Grishina & A.B. Ponomaryov
Perm National Research Polytechnic University, Perm, Russia

ABSTRACT: For many years retaining walls remain popular type of structures in the transport and urban development. Sandy soils are mainly used to create ones. One of the main problems of such structures is large material cost to construct. The paper investigates the possibility of reduction of the facing material of retaining walls by applying the method of discrete reinforcement of sandy soils by polypropylene fibers. The studies found that the introduction of discrete fibers into sand has positive effect on shear parameters: specific cohesion and angle of internal friction. To evaluate the effectiveness of fiber reinforcement technology when constructing retaining walls numerical simulation using software "PLAXIS-2D" was made for test problems. The results show that the discrete soil reinforcement by polypropylene fibers can be an effective solution to reduce material consumption and increase the profitability retaining walls.

Keywords: soil strength properties, retaining wall, fiber reinforcement, numerical modeling

1 INTRODUCTION

1.1 *Review of existing researches*

Currently reinforced walls have become a popular type of retaining structures in terms of decrease the earth pressure and displacements of the walls. Metal mesh, various natural or synthetic additives can be used as the reinforcement. The main advantages of reinforced retaining walls are their fundamental simplicity, ease of construction, efficiency (Bartolomey et al. 1999). In addition to the linear reinforcement there is a technology there randomly distributed discrete fiber are introduced into the soil. This technique is called fiber reinforcement (Hejazi & Sheikhzadeh 2012, Kuznetsova et al. 2012). Fiber reinforced soil has improved strength characteristics, so one of its potential applications is to use as backfill of different retaining structures (Kuznetsova & Ponomaryov 2015).

Concept of fiber reinforcement was considered in some studies on earth-retaining problems. The first example of mass application of randomly oriented fibers for reinforcing retaining walls and protection of natural slopes was a polyester fiber named as "Texsol". "Texsol" is the technology of soil reinforcement, patented by the French government, which is the mix of soil particles and polymer yarns. The mixture is obtained by passing the soil through a pneumatic or a mechanical system simultaneously with the fibers (LeFlaive & Liausu 1986).

Literature review shows that fiber reinforcement can be used for reinforcement of slopes and retaining walls in combination with linear geosynthetic materials. Park and Tan concluded that the use of polypropylene fibers of 60 mm in combination with geogrid reinforced silty sand soil wall reduces the earth pressure and displacement of the wall and increases the stability of the wall (Park & Tan 2005).

Some researchers examined the use of fibers to improve properties of cement-stabilized soils. Nasr studied behavior of strip footing on fiber-reinforced cemented sand adjacent to sheet pile wall. Numerical and experimental results clearly showed that fiber inclusion into the cemented soil significantly reduces the lateral deflection of the sheet pile wall and increases the ultimate bearing capacity of footing (Nasr 2014).

Though concept of fiber reinforcing soil was originated in ancient times it is still a relatively new technique in geotechnical projects and requires father investigation. It seems that technology of soil discrete reinforcement can be an effective solution of earth-retaining problems (Spirin et al. 2016).

1.2 *The research aim, object and tasks*

When constructing structures for different purposes the main purpose is to ensure safety with lower costs. The aim of the paper is to investigate the possibility of reducing the facing material of retaining walls due to the decrease of the active earth pressure by applying the method of discrete reinforcement. The object of the research is a backfill of retaining wall. Authors propose the use of sand reinforced with polypropylene fiber as a backfill.

Research objectives include experimental laboratory studies to determine the physical and mechanical

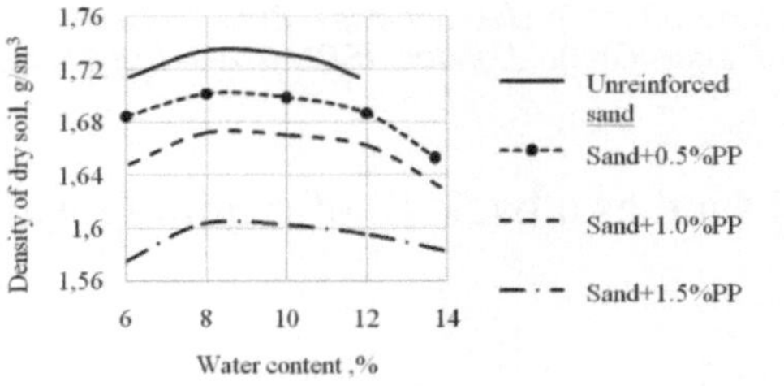

Figure 1. Dependences of dry soil density on soil water content for unreinforced and fiber reinforced sand.

properties of sand and fiber sand, analytical calculations and numerical modeling of retaining walls with the backfill of sand and fiber reinforced sand.

2 EXPERIMENTAL INVESTIGATION

2.1 *Material used for the testing*

Uniform fine sand of optimal water content and fixed particle size distribution was used in this study. Bulk volume density of sand was 1490 kg/m^3

Polypropylene (PP) fiber of 12 mm was used as reinforcement. Usually this fiber is used for the discrete reinforcement of concrete to improve its strength properties, reduce shrinkage for improving the crack resistance of concrete structures.

The content of reinforcement was 0.0, 0.5, 1.0 and 1.5% of polypropylene fibers by weight of dry soil. Fibers were randomly placed by mixing to form fiber reinforced sand (Kuznetsova & Ponomaryov 2014).

2.2 *Experimental program and results*

The first stage of the study was laboratory tests to determine physical and mechanical characteristics of the sand and fiber reinforced sand.

In accordance with the technology laying of backfill should be carried out when soil moisture content provides the achieving of required density which is achieved by the lowest labour inputs. To determine the optimal moisture content standard compaction test was carried out. The determination of maximum density showed that the introduction of polypropylene fibers reduces the soil compactibility and, therefore, reduces the maximum density of dry soil on average of 6–8% (see Figures 1-2). Optimal water content value does not depend on the presence of reinforcing fibers and is 8.5% (Fig. 1). Further tests to determine the strength parameters of sand and fiber sand composites were carried out with the optimum soil moisture.

Tests to determine the mechanical characteristics were performed in triaxial device. The tests were carried out according Russian standards. The specimens diameter was 38 mm and height was 76 mm. Consolidated drained tests were conducted at various levels of confining pressure (100, 200, 300 kPa). The experimental data were obtained using an automated software package for the testing system.

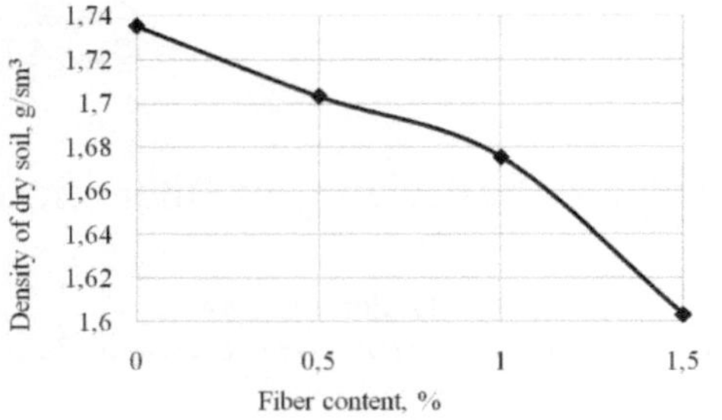

Figure 2. Dependences of dry soil density on fiber content.

Table 1. Soil strength parameters on the results of triaxial tests.

Material	Strength parameters		Relative parameters	
	ϕ,°	c, kPa	$\Delta\phi$	Δc
Sand	35.0	20	1	1
Sand+0.5%	39.3	67	1.12	3.35
Sand+1.0%	42.6	75	1.22	3.75
Sand+1.5%	49.4	37	1.41	1.85

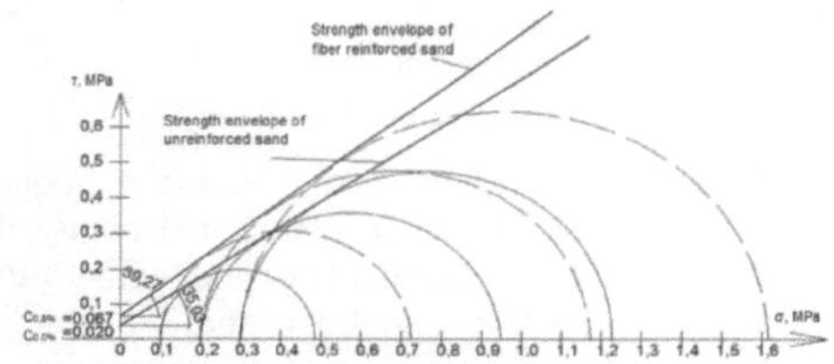

Figure 3. Mohr-Coulomb diagram for unreinforced sand and sand reinforced with 0.5% of polypropylene fiber after triaxial testing.

The experimental data were obtained using an automated software package for the testing system.

The test results are shown in Table 1. Studies found that the introduction of discrete fibers into sand has positive effect on shear parameters: specific cohesion and angle of internal friction.

To illustrate the tests results the Mohr-Coulomb criterion is given for unreinforced sand and sand reinforced with 0.5% of polypropylene fiber (Fig. 3).

When loading soils are working mainly on the shear plane where the load-bearing capacity is the lowest (Sipidin & Sidorov 1963). Therefore, the shear strength is a defining characteristic of the soil strength. The graphs clearly show that the reinforced specimen has higher strength characteristics as compared with unreinforced one. This is because that when polypropylene fibers are included into the soil internal adhesion increases due to contacting of sand and discrete fibers (Diambra & Ibraim 2010).

Test results show that due to the combined action of the sand and the fibers non-cohesive soil

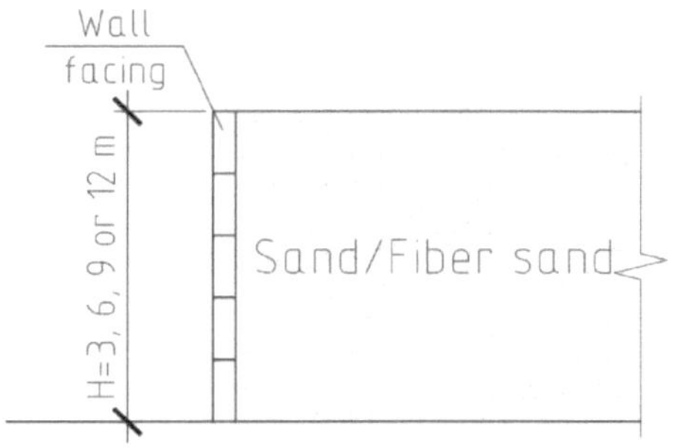

Figure 4. Design model of retaining wall.

acquires apparent specific cohesion and increased shear strength. The optimum percent of reinforcement is 0.5% by weight of dry soil. When reinforcement percent is 1% the material consumption increases in 2 times, and soil characteristics vary on average by 8–10% with respect to reinforcement of 0.5%. When introducing 1.5% of fiber the specific cohesion of soil varies slightly, since it is difficult to ensure uniform mixing with such content of polypropylene fiber.

3 NUMERICAL SIMULATION

In the next phase of the study the stability analysis of the retaining wall was carried out. Retaining walls work in difficult geological conditions and take significant horizontal forces of active earth pressure. To reduce the horizontal pressure is possible by using soil having specific cohesion as backfill. According to the standards backfill materials should have good drainage properties at the same time. Therefore, the use of fiber reinforced sand can be an effective solution to ensure safety and reduce costs when constructing retaining walls.

Design model of retaining wall is shown in Figure 4. The calculation was made for the pressure from the soil own weight. Unreinforced sand and fiber reinforced sand with optimum moisture content were simulated as wall backfill. The fiber content was 0.5 % by weight of dry soil as the optimum in concordance with laboratory tests results. Height of retaining wall was varied and equaled to 3, 6, 9, and 12 m.

Numerical simulation of retaining wall was performed by finite element analysis using the software package "PLAXIS-2D". Elastoplastic Mohr-Coulomb model was used when calculating. The model parameters were taken according laboratory tests results. Numerical simulation results are shown in Table 2.

The calculation for determining the active earth pressure of retaining wall backfill found that the use of soil reinforcement when constructing retaining walls is efficiently, since the value of the active earth pressure is greatly reduced. Retaining wall with a height of 3.0 m is exception. Fiber sand is most appropriate to apply for high retaining walls.

Table 2. The values of active earth pressure based on the results of numerical simulation.

Backfill material	Height of retaining wall, m			
	3.0	6.0	9.0	12.0
Sand	0.3	0.7	12.2	47.95
Sand+0.5%	0.3	0.52	0.67	0.82

4 RESEARCH RESULTS AND CONCLUSION

The possibility of reducing the active earth pressure on retaining wall was investigated by applying the method of discrete reinforcement. It was found that:

1. Inclusion of polypropylene fibers as reinforcing elements into the soil reduces the soil compactability of 6–8%. The optimum moisture content does not depend on presence of the reinforcing fibers and it equals to 8.5%.

2. Discrete soil reinforcement technology provides a significant increase in the soil strength characteristics. Non-cohesive soil acquires apparent specific cohesion and increased shear strength due to the combined action of the sand and the fibers. The analysis of laboratory testing results shows that the optimal percentage of reinforcement is 0.5% by weight of dry soil. When reinforcement percent is 1% the material consumption increases in 2 times, and soil characteristics vary on average by 8–10% with respect to reinforcement of 0.5%. When introducing 1.5% of fiber the specific cohesion of soil varies slightly, since it is difficult to ensure uniform mixing with such content of polypropylene fiber.

3. The calculation for determining the active earth pressure of retaining wall backfill found that the use of soil reinforcement when constructing retaining walls is efficiently, since the value of the active earth pressure is greatly reduced. Retaining wall with a height of 3.0 m is exception. Fiber sand is most appropriate to apply for high retaining walls (height of 6 meters or more).

Thus, the fiber reinforcement of soils can be an effective solution to reduce the thickness of the wall facing without loss of stability and safety. Technique proposed by the authors can be useful to decrease material consumption and increase the profitability of high retaining walls.

REFERENCES

Bartolomey, A. A. et al. 1999. The use of synthetic materials in the highway engineering in the Urals. *Geotechnical engineering for transportation infrastructure*; Proc. of the 12th European conference on soil mechanics and geotechnical engineering, Netherlands, Amsterdam, 1999: 1197–1202.

Diambra, A., Ibraim, E. 2010. Fibre reinforced sands: experiments and modeling. *Geotextiles and Geomembranes* 28: 238–250.

Hejazi, S.M., Sheikhzadeh, M. 2012. A simple review of soil reinforcement by using natural and synthetic fibers. *Construction and Building Materials* 30: 101–106.

Kuznetsova, A.S., Ofrikhter, V.G., Ponomaryov, A.B. 2012. Study of strength characteristics of sand reinforced with discrete polypropylene fibers. *PNRPU Construction and Architecture Bulletin* 1: 44–55.

Kuznetsova, A.S., Ponomaryov, A.B. 2014. Laboratory testing of the strength characteristics of fiber reinforced sand with various water saturations. *Bulletin of Civil Engineers* 6: 127–132.

Kuznetsova, A.S., Ponomaryov, A.B. 2015. Experimental design to investigate the stress-strain behavior of loaded fiber reinforced soil mass adjacent to retaining wall. *PNRPU Construction and Architecture Bulletin* 1: 135–148.

LeFlaive, E., Liausu, Ph. 1986. Texsol: Earth Threading Technology. *Geotechnical Fabrics Report* 4: 10–14.

Nasr, A. M. 2014. Behavior of strip footing on fiber-reinforced cemented sand adjacent to sheet pile wall. *Geotextiles and Geomembranes* 42: 599–610.

Park, T., Tan, S.A. 2005. Enhanced performance of reinforced soil walls by the inclusion of short fiber. *Geotextiles and geomembranes* 23: 348–361.

Sipidin, V.P., Sidorov, N.N. 1963. *The study of soils under conditions of triaxial compression*. Moscow: Gosstroyizdat.

Spirin, A.V., Grishina, A.S., Kleveko, V. I. 2016. Analysis of existing constructive decisions of bridge abutments and application design of fiber reinforced soil under their reconstruction. *PNRPU Construction and Architecture Bulletin* 1: 5–12.

Challenges and Innovations in Geotechnics – Zhussupbekov (Ed.)
© 2016 Taylor & Francis Group, London, ISBN 978-1-138-03007-7

Mathematical modeling of massif state around the mountain road slopes and highland pit

S.Zh. Kuvakov & K.Ch. Kozhogulov
Institute of Geomechanics and Development of Subsoil, National Academy of Sciences of the Kyrgyz Republic, Bishkek, Kyrgyzstan

G.Dz. Kabaeva
Kyrgyz-Russian Slavic University "B.N. Yeltsin", Bishkek, Kyrgyzstan

ABSTRACT: The paper considers the stress state of the slopes of mountain roads and highland pits. Research carried out by mathematical modeling with the use of the software implementing the finite element method. The computed results of stressed massif state around the slopes during the construction of the mountain road Bishkek-Osh and the open pit Makmal deposit presented.

1 INTRODUCTION

In recent years, intensive exploration of mountain regions of Kyrgyz Republic takes place mainly due to construction of mountain roads and upland quarries. So, in 2015 started the construction of an alternative road Bishkek-Balykchy-Kazarman-Osh and by open way are processed such deposits as the Kumtor Terekkan, Bozymchak and others. The slopes, on which are being constructed mountain roads and the quarries are characterized by a complex geological structure, the differences of the properties of the rocks, intensive tectonic disturbances. The ever-increasing volumes of works connected with the intensive construction of mountain roads require special attention to ensure long-term stability of slopes, which are largely determined by their stress-strain state. The full-scale experimental study of the stress state of rocks could provide information on the stress-strain state due to the influence of various factors. However, due to high laboriousness of the measurement, the results can be obtained only in limited quantities, and the experiment in the data points is not possible to be repeated. Therefore, the most complete information about the stress-strain state of the rock mass and its changes in the process of mining operations can be obtained only with a reasonable combination of field research with the research on the mathematical models. The data obtained in the process of mining studies are used as the boundary conditions for the simulation of the stress-strain state of rock mass. To study the geotechnical structures is an essential part of the survey of the object (geology, structure, formation, physical and mechanical characteristics, adding massive form, and so on), on which it built a mathematical model. As we know, that when making geotechnical problems are widely used analytical and numerical modeling. If the analytical method allows the functional dependence, using a numerical method, we have specific data, for example, when assessing a natural stress state of the massif, and after its development; assessing the road sustainability of slopes and so on. In this regard, in 80-90 s of the last century in Kyrgyzstan for solving various geomechanical problems, in particular, to determine the stress-strain state of untouched rock mass, as well as in the vicinity of the workings, the numerical methods started to be widely used, among which should be mentioned the finite elements method [1, 2]. Many software products implement finite element method [3]. In the early 90-ies of the past century, the software "Geomechanics" was first used in the assessment of the stress-strain state of the massif of bead of the career "Makmal" [4]. This computer programme allows getting elastic and elastic-plastic solution under plane strain with an arbitrary law of state in homogeneous and inhomogeneous-medium. In the study of stress-strain state of the Kumtor pit walls (Kyrgyzstan), Apatity (Russia), minerals Kostanay (Kazakhstan) and Terekkan (Kyrgyzstan), for modeling of the stress-strain state have been widely used "Stress" system [3, 5, 6]. Along with that the computer-based system Matlab and Plaxis software were widely used [8, 9].

2 COMPUTATIONAL DOMAIN

In this paper, we evaluated the stress state of the slopes at one of the Bishkek-Osh mountain road sections and highland pit of gold deposit Makmal (Kyrgyzstan) by the numerical finite element method.

Table 1. Geometric and mechanical parameters.

Height of the slope, H, m	The width of the selected part, L, m	b, m	α,°	β,°	a, m	The road width, b, m	c, m	γ, kN/m^3	E kN/m^2	ν
400	500	15	40	70	15	10	20	26.9	5.9 E+07	0.210

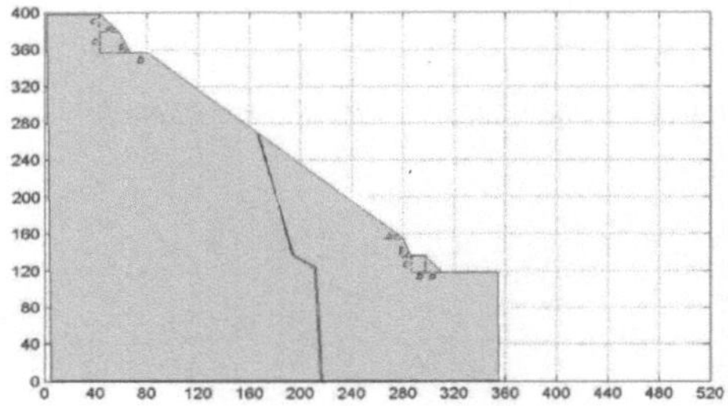

Figure 1. The calculated schema.

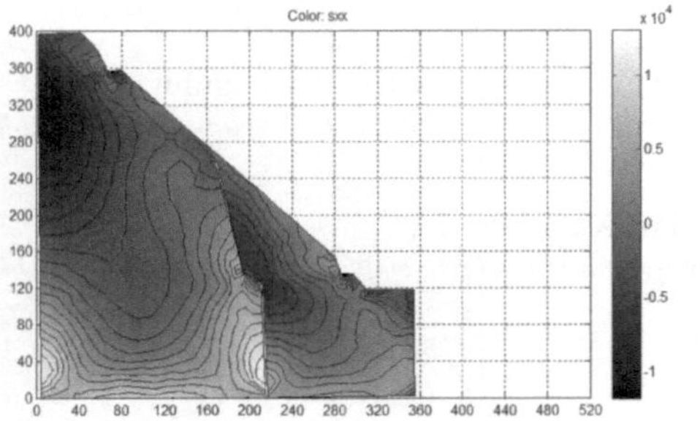

Figure 2. Contours of horizontal stress σ_{xx}.

Table 2. Physical and mechanical properties of the rock.

Name of the rock	σ_c, kN/m^2	σ_p, kN/m^2	γ, kN/m^2	E, kN/m^2	ν	φ,^{o}	C, kN/m^2
Marble limestone	1.16 E+05	5.8 E+03	26.5	6.200E+07	0.250	60	1.360E+04

2.1 *The stress state of the slopes of mountain road*

Have been considered two cases: 1) the road is laid at the top of the slope; 2) the road is laid at the foot of the slope. In addition, the massif has a vertical tectonic crack. The task was solved with use of Matlab. In this system for solving the tasks of elastic-plastic state of rocks can be used funds intended for the solution of partial differential equations by the finite element method. The solution to these equations in Matlab consists of the following steps: 1) construction area, which solves the equation; 2) entering of equations in partial derivatives; 3) definition of boundary and initial conditions; 4) triangulation of the region; 5) solution of equations; 6) visualization of results. Figures 1 and 2 demonstrated as example of a calculated schema and the result of the study of stress-strain state of the slopes at one of the Bishkek-Osh mountain road sections. The following geometric and mechanical parameters are used at computing (Table 1).

Calculations of stress-strain state of the slope showed that the distribution of horizontal stress σ_{xx} within the slope is uneven. Stress concentration is in all corners of the slope and the rocks near the contact, which is confirmed by the results of other researchers [10]. The calculation results also show the difference in slope of the stress state in the formation of slopes at its top and the foot.

2.2 *The stress state of the highland pit of gold deposit Makmal (Kyrgyzstan)*

In the simulation, we compared the stress in highland pit in cases of homogeneous and heterogeneous addition. Consider the stress distribution in the case of a homogeneous addition of highland pit, physical and mechanical properties are shown in Table 2.

To solve the task was used Plaxis software. Plaxis - a software system for the finite element analysis that is used to solve the dynamic and statistical problems of geotechnical engineering and designing, which is a package of computer programs for finite element calculation of mode of deformation of hydro- and geotechnical structures. In addition, there are different models of soil, depending on the structural state of the massif: a linear elastic model; Mohr-Coulomb model; model of soft soil creep; hardening soil model; model of the rock; Cam-Clay model. In addition, this software provides the ability to output the data tensions and deformation characteristics, to consider in detail and receive data diagrams of stresses on the cross sections in all directions, and export data. In the Figure 3 is shown stress distribution in highland pit in case of homogeneous addition.

On the distribution of the horizontal stresses revealed that on the basis of highland pit arise the compressive stress concentrations, which vary from 126 kN/m^2 to 405 kN/m^2. Now consider the heterogeneous addition of highland pit and the physical and mechanical properties which shown in Table 3.

The results of calculations for the heterogeneous addition of highland pit with isolines illustrated in Figure 4.

By the result of the evaluation of the stress state of the heterogeneous highland pit were received the following data – on the basis of highland pit evolved

Table 3. Physical and mechanical properties of the rock of the highland pit in the case of heterogeneous structure.

Name of the rock	σ_c, kN/m^2	σ_p, kN/m^2	γ, kN/m^2	E, kN/m^2	v	φ,^{o}	C, kN/m^2
Marble limestone	1.16E+05	5.8E+03	26.5	6.2E+07	0.250	60	1.360E+04
Granite	2.5E+05	1.56E+04	28	7.6E+07	0,220	61	1.8E+04

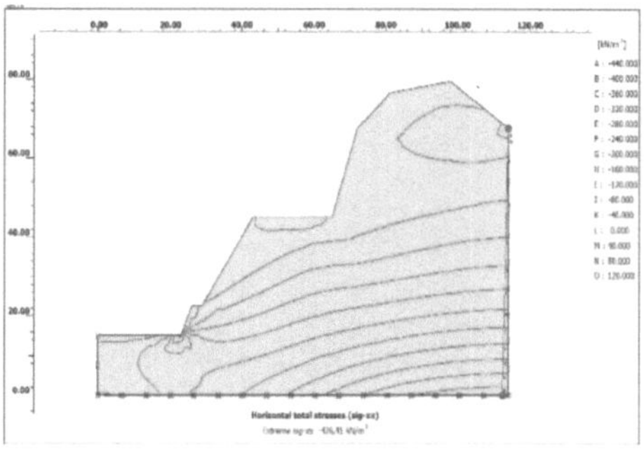

Figure 3. Horizontal stress distribution in the highland pit massif of the homogeneous addition.

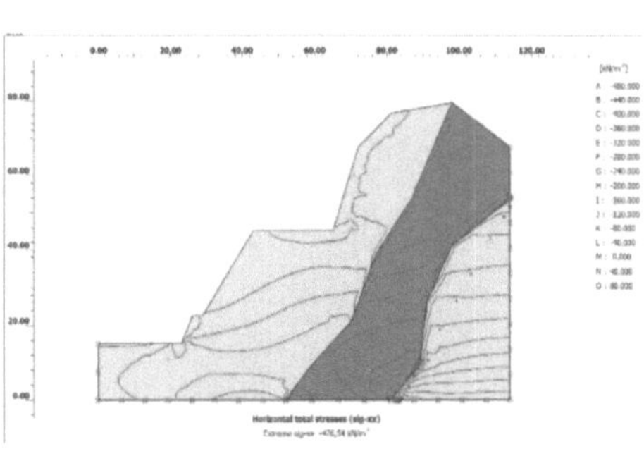

Figure 4. The horizontal stress distribution in the massif of heterogeneous addition of highland pit.

horizontal compressive stress concentration, which reached up to 288 kN/m^2.

As a result of numerical simulation state stress of highland pit came to the following conclusions: values of the horizontal compressive stress of heterogeneous addition of highland pit is 1.4 times less than the value of the horizontal compressive stresses in the case of homogeneous addition of highland pit. This is because the component of granite rock has large values of strength and density characteristics and form of its occurrence relieves stress state.

3 CONCLUSIONS

Calculations of stress state of the slope of the mountain road showed that the distribution of horizontal stress σ_{xx} within the slope has been uneven. Stress concentration is observed in all corners of the slope and the rocks near the contact. The calculation results also show the difference in slope of the stress state in the formation of slopes at its top and the foot. The stress values in the case of heterogeneous addition of highland pit were lower than the stress value in the case of homogeneous addition of highland pit due to the fact that the component of granite rock has large values of strength and density characteristics and form of its occurrence relieves stress state.

REFERENCES

Kovalenko, V.A. & Grigoriev, V.V. & Usmanov, S.F. 1987.Investigation of stress-strain state board of career finite element method. // Complex investigation of physical properties of rocks and processes-Moscow.

Aitmatov, I.T. & Abdyldaev, E.K. & Usenov, K.D. 1992. Calculation of the stress-strain state board to earn career. *Geomechanical substantiation methods stability calculation of exposures*. 3–9, Bishkek, Ilim.

Usmanov, S.F. 2009. *Prediction of stability of boards high-pits on the basis of modeling of stress-strain state*. 172 p. Bishkek KRSU.

Abdyldaev, E.K. *Stress-strain state of rock massif near the excavation*. 1990. Frunze: Ilim.

Kozhogulov, K.C. & Alibaev, A.P. & Mamatova, G.T. 2012. Evaluation of the stress-strain state of the slopes pit walls in finalizing with on-board ore reserves by underground mining stockpiled ore by underground methods. *Science and New Technologies*, 8, 9–12, Bishkek.

Osmonova, N.T. 2012. Evaluation of the stress state of the slopes of the pit walls during the mining of sub-pit reserves. Terekkan . P*roceedings of the universities*, 5, 41–44, Bishkek.

Tentiev, J.T. & Mekenbaev, B.T. & Kozhahmetov, K.H. 2000. *Calculation methods for evaluating slope stability.* – Bishkek, Kyrgyzstan, 163.

Ferreira, A.J.M. *MATLAB Codes for Finite. Solids and Structures*. 2009. Springer Science+Business Media B.V.

Plaxis version 8. *Material models manual* [Electron resource]/ Plaxis bv, AN Delft, The Netherlands. – Access mode: www.plaxis.nl.

The destruction of the earth's crust and the processes of self-organization in the areas of strong technogenic impact. 2012. Melnikov, N.N.(ed). Russian Academy of Sciences, CO, IGD – Novosibirsk, Publishing House of the SB RAS.

Challenges and Innovations in Geotechnics – Zhussupbekov (Ed.)
 ISBN 978-1-138-03007-7

Finite element analysis of deep excavation: A case study of underground parking underneath Russel Square, London

K. Bekish, A. Ulykbanov, T. Mukashev, Y. Serik, M. Nguyen & J. Kim
Nazarbayev University, Astana, Kazakhstan

ABSTRACT: This paper includes analysis of ground surface settlement and diaphragm wall displacement by using finite element analysis as part of underground car park design underneath Russel Square in central London, United Kingdom. Short-term (undrained), consolidation (time-dependent), and long-term (drained) analyses were performed by adopting Mohr-Coulomb soil model to ensure stability of diaphragm wall and predict the behavior of soil and wall. It was identified that the diaphragm wall is stable and both ground surface settlement and lateral wall displacement are within acceptable limits, thereby ensuring ultimate and serviceability limit states.

1 INTRODUCTION

Stability and deformation are the main contributing factors to the performance of deep excavation. The excavation can be called stable if collapse of the retaining wall and upheaval failure of base is eliminated, while deformation is associated with magnitude and pattern of ground movement around the excavation zone. One of the main factors affecting the excavation performance is soil condition. According to Ou (2006), excavation performed on clayey soil results in greater wall movement compared to the sandy and gravelly soil, 0.5% and 0.2% of the excavation depth, respectively. Excessive wall displacement and ground surface settlement induced by excavation can damage not only constructed structure itself, but also neighboring buildings. Therefore, wall displacement and surface settlement should be thoroughly controlled and prevented from reaching high values. This paper represents the analysis of soil settlement and retaining wall displacement by using PLAXIS software as part of deep underground car park design underneath Russel Square in central London, United Kingdom.

1.1 *Underground car park underneath Russel Square*

Deficiency of the parking spaces is becoming one of the main problems faced by large cities. Taking into account the growth in the number of vehicles, it can be predicted that this problem will remain acute in the future. As a result, car parks play a significant role in traffic management. Car parks can be incorporated into the design of multi-function buildings, or can be built separately as multi-story, underground or basement car parks. Currently, since land in cities is expensive and scarce, underground parking facilities are becoming prevalent. Therefore, the ground above underground car park can be used as parks and squares. In addition, underground car park enables decrease of fume and smoke amounts, noise levels, and car number in the urban environment.

By conducting preliminary research on car parks available in the vicinity and the requirements for provision of parking places, it was found out that the three-story car park with 750 spaces should be designed and constructed. By following the standards for space requirements, it was determined that area of the parking will be 76.1 m × 89.1 m.

Diaphragm wall was selected as the most appropriate vertical soil support system for underground car park design since it can be incorporated into the permanent structure and act as cut-off wall. By performing calculations, it was determined that the embedment depth should be 15 m, while the excavation depth is 12 m. Thus, the wall length is 27 m, and thickness of wall is 0.6 m.

1.2 *Ground conditions*

The soil profile in Russell Square was obtained from borehole scan provided by British Geological Survey. Soil profile under the Russel square is shown in Figure 1. It can be seen that at the depth of 2 m the soil profile is fill which is followed by weathered brown London Clay till 4 m. Below 4 m, the soil profile is sand and gravel, and starting from 6 m stiff silty clay soil occurs. Groundwater table was found at 4 m.

1.3 *Patterns of wall movement and ground settlement*

Based on the past experience, models for retaining wall displacement and surface settlements had been

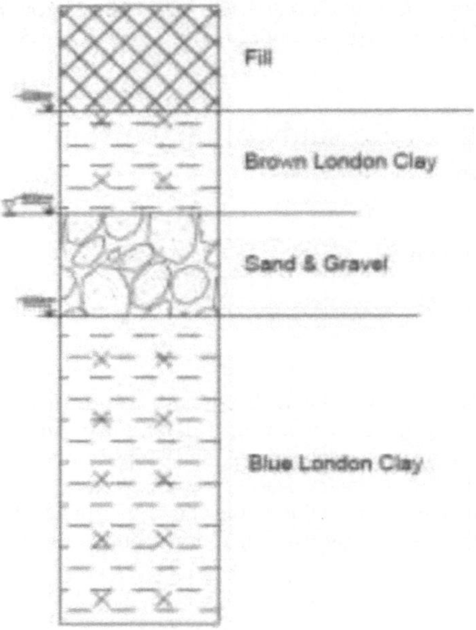

Figure 1. Borehole scan underneath Russel Square.

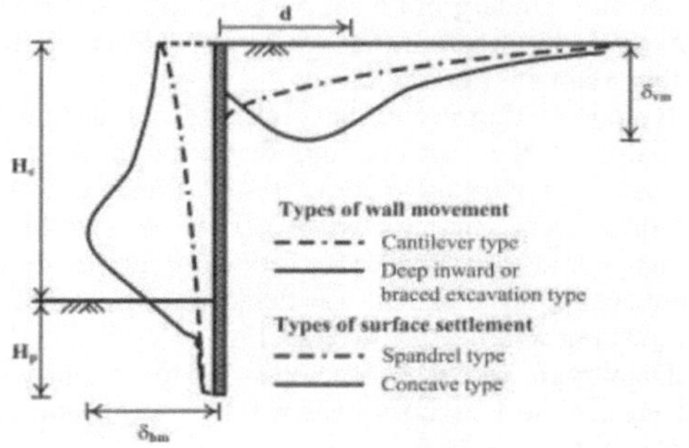

Figure 2. Types of wall movement and ground surface settlement. (Hsieh and Ou, 1998 and Ou, 2006, cited in Likitlersuang et al., 2013).

developed. Figure 2 shows the main shapes of the ground surface settlement which can be distinguished as concave type and spandrel type (Ou, 2006).

Magnitude and shape of wall displacement affects the surface settlement pattern. Spandrel type settlement occurs when the initial excavation stages are resulted in larger excavation values compared to the later stages. In this case, the maximum value of settlement is in the proximity to retaining wall. Concave type of settlement is characterized by deep inward movement of the vertical support structure, and the maximum settlement occurs at a distance behind the back of the wall. Thus, wall displacement can be differentiated as either cantilever or deep inward patterns. In deep inward displacement type, the maximum wall displacement occurs near the final excavation face. It was found out that excavation in soft clay tends to result in the concave type of settlement, while the spandrel type of settlement occurs during the excavation in sandy soil or stiff clay due to less deformation of the wall (Ou, 2006).

According to serviceability requirements, wall should displace less than 1% of its length (Chalmers University of Technology, 2014). Since the wall length is 27 m, in this case wall displacement should be less than 27 cm or 270 mm. Moreover, Ou (2006) states that lateral wall displacement for soft clay can be estimated by 0.5% of excavation depth. For 12 m excavation, wall displacement is 60 mm. Based on case histories, ground surface settlement can be predicted by taking 50%–100% of measured lateral wall displacement.

1.4 *Short-term, consolidation, and long-term analyses*

In order to ensure the stability of an earth structure, it could be necessary to consider both short-term and long-term stability. The former is associated with state during construction or immediately after construction, and it is assumed that no significant drainage due to excess pore water dissipation will occur in short-term condition; thus, undrained analysis should be adopted. According to the undrained analysis, excess pore water pressure is not dissipated ($u_e \neq 0$) (Ou, 2006). As a result, total stress method should be adopted. Total stress analysis is typically applied only to cohesive soils. In case cohesive soil is 100% saturated, undrained shear strength τ_u and $\varphi = 0$ parameters should be used (ibid.). Long-term analysis or drained analysis assumes that dissipation of all excess pore water pressure occurs ($u_e = 0$) (ibid.). As a consequence, volume change takes place. Drained analysis uses effective stress method and effective stress parameters such as c' and φ' (ibid.). For granular soil such as sand and gravel effective stress analysis is applied both in long-term and short-term conditions because excess pore water pressure in such soils dissipates quickly when shearing commences. However, long-term behavior of cohesive soils is also analyzed by using the effective stress method. It should be noted that consolidation has a significant impact on stability and deformation of retaining wall, especially in clayey soil (Bentler, 1998). Consolidation is associated with reduction of soil volume due to imposed stress, which packs soil particles closer (ibid.). The effects of consolidation are in particular substantial during the excavation process in clays, thereby causing changes in the shear strength of soils and time-dependent deformations. Thus, consolidation analysis should also be performed to ensure satisfactory behavior of earth retaining structure.

2 FINITE ELEMENT MODELLING

PLAXIS 2D is a finite element software that allows to perform two-dimensional analysis to determine deformation and stability of soil structures involving excavation, foundation, embankment and tunnel design and construction (PLAXIS, 2015). In order to obtain credible results from PLAXIS, input parameters should be carefully considered.

Diaphragm wall was modelled by using plate element, while floor slabs that act as props were modelled by fixed-end anchors. Input parameters of the plate and fixed-end anchor elements are presented in Tables 1 and 2. C40/50 concrete grade with elasticity modulus of 34 GPa was considered.

2.1 *Short-term analysis*

Plastic calculation type and staged construction loading type were assigned to all phases except Initial phase in which K_0 procedure was selected as a calculation type. In addition, phreatic pore pressure calculation was adopted for all phases. Drainage condition for fill and sand soil layers was set to drained, while clay layers were assigned Undrained C drainage type. Table 3 represents soil and interface properties for Undrained C analysis (PLAXIS, 2015), while construction stages are shown in Table 4.

Figure 3 illustrates vertical displacement of soil immediately after construction. It indicates that on the retained side soil moves downward which causes upward movement of soil in the excavation area. The vertical displacement of soil on the retained side can be seen from Figure 4, and concave type surface settlement can be distinguished. Concave surface settlement is characterized by increasing value of settlement to a certain point as distance from the wall increases. Maximum settlement of −14 mm occurs at 20–30 m away from wall. According to Puller (2003), vertical settlement due to wall installation is typically 0.05%

Table 1. Material properties of the plate (diaphragm wall).

Parameter	Name	Value	Unit
Material type	Type	Elastic/Isotropic	–
Normal stiffness	*EA*	1.8×10^7	kN/m
Flexural rigidity	*EI*	5.4×10^5	kNm²/m
Thickness	*d*	0.6	m
Weight	*w*	2.4	kN/m/m
Poisson's ratio	υ	0.2	–

Table 2. Material properties of the fixed-end anchor (floor slabs).

Parameter	Name	Value	Unit
Type of behavior	Material type	Elastic	
Normal stiffness	EA	1.99×10^{10}	kN
Spacing out of plane	$L_{spacing}$	7.7	m

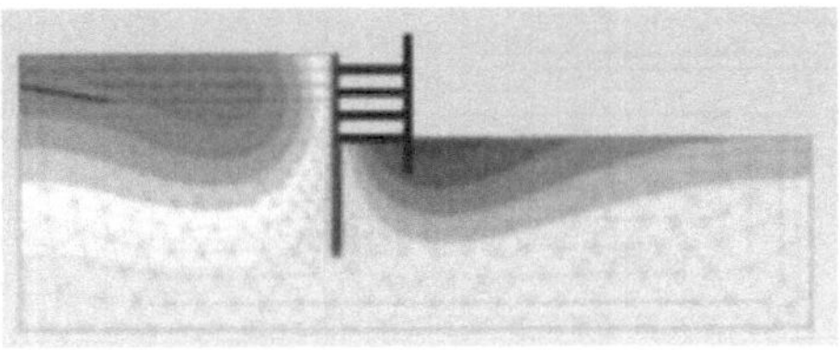

Figure 3. Vertical displacement of soil.
Legend: red= −14 mm and deep blue= +12 mm

Table 3. Soil and interface properties for Undrained C analysis.

Soil	Fill	Brown London Clay	Sand and gravel	Blue London Clay
General				
Material model	Mohr-Coulomb	Mohr-Coulomb	Mohr-Coulomb	Mohr-Coulomb
Drainage type	Drained	Undrained C	Drained	Undrained C
γ_{unsat} (kN/m³)	18	19.8	21	19.8
γ_{sat} (kN/m³)	21.5	19.8	23	19.8
E' or E_u (kN/m¹)	18×10^3	30×10^3	140×10^3	100×10^3
υ' or v_u	0.15	0.495	0.15	0.495
c′ or s_u (kN/m¹)	0.2#	60	0.2#	100
φ' or φ_u	33	0	39	0
ψ'	3	0	9	0
Flow parameters				
Data set	Standard	Standard	Standard	Standard
Soil type	Medium fine	Fine	Medium	Fine
k_x	1×10^{-7}	1×10^{-9}	0.5×10^{-3}	1×10^{-9}
k_y	1×10^{-7}	1×10^{-9}	0.5×10^{-3}	1×10^{-9}
Interfaces				
Strength	Manual	Manual	Manual	Manual
R_{inter}	0.8	0.7	0.8	0.7
Initial				
K_0	Automatic	Manual	Automatic	Manual
$K_{0,x}$	0.4554	1.5	0.3707	1.5

Table 4. Construction stages for short-term analysis.

Sequence	Construction activities
Phase 1	Diaphragm wall construction
Phase 2	Excavation –2 m
Phase 3	Roof slab construction
Phase 4	Excavation –4 m
Phase 5	Groundwater lowering till –5.2 m followed by excavation
Phase 6	Construction of the second slab
Phase 7	Groundwater lowering till –8.4 m followed by excavation
Phase 8	Construction of the third slab
Phase 9	Groundwater lowering till –11.6 m followed by excavation
Phase 10	Construction of base slab

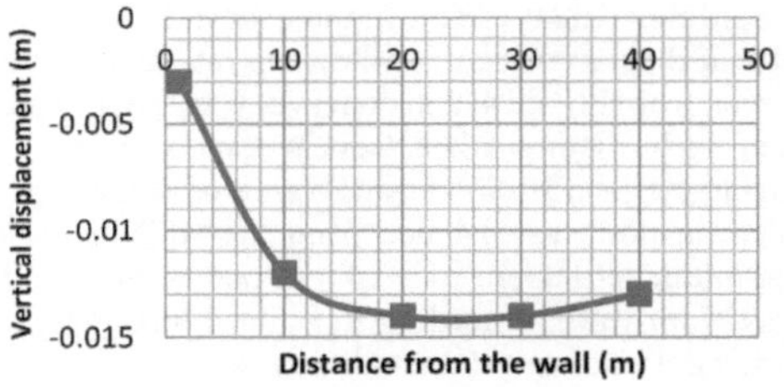

Figure 4. Vertical displacement of soil: max −14 mm.

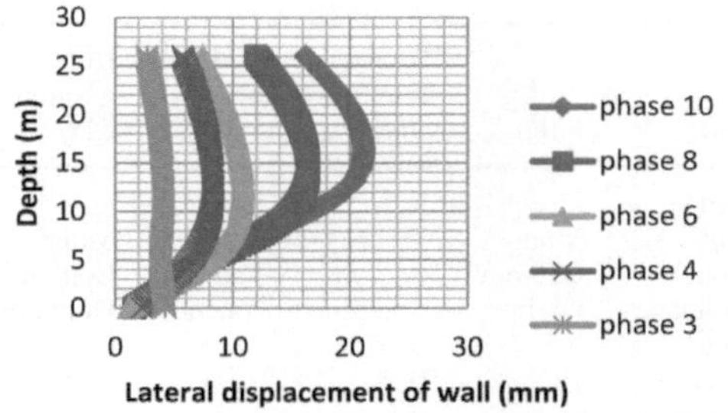

Figure 5. Lateral displacement of diaphragm wall: max 21 mm.

of wall depth. In our case, the wall depth is 27 m; thus, vertical displacement of soil at wall should be 13.5 mm. Since undrained analysis indicates the state right after construction, it can be seen that it is close to the value generated by PLAXIS. It can be viewed from the Figure 5 that as construction continues horizontal displacement of wall increases. It should be noted that during several construction phases, the lateral deformations of wall were identical, so phases with diverging wall displacements were plotted on the graph. It can be seen that in the beginning wall deforms as cantilever, but as construction advances the wall tends to follow the deep inward behavior. The maximum lateral displacement of diaphragm wall occurs during the last stage of construction, and the value for undrained analysis is approximately 21 mm.

2.2 *Long-term analysis*

For this analysis, Drained option was chosen for specifying drainage for all soil layers. Plastic calculation type and staged construction loading type were assigned to all phases except Initial phase in which K_0 procedure was selected as a calculation type. In addition, phreatic pore pressure calculation was adopted for calculation of pore pressure for soils above groundwater table. Since excavation proceeds below groundwater table level, water level was lowered and steady-state pore pressure calculation analysis was used to model long-term behavior. Construction stages of both short-term and long-term analyses were identical (Table 4). Table 5 represents soil and interface properties for Drained analysis.

Figure 6 shows that retained side soil moves downwards approximately by −10 mm. Figure 7 illustrates that maximum vertical soil displacement of soil on the retained side occurs at 30 m away from the wall, and its value is 6 mm. Figure 8 indicates that maximum horizontal displacement of wall occurs after the final

Table 5. Soil and interface properties for Drained analysis.

Soil	Fill	Brown London Clay	Sand and gravel	Blue London Clay
General				
Material model	Mohr-Coulomb	Mohr-Coulomb	Mohr-Coulomb	Mohr-Coulomb
Drainage type	Drained	Drained	Drained	Drained
γ_{unsat} (kN/m^3)	18	19.8	21	19.8
γ_{sat} (kN/m^3)	21.5	19.8	23	19.8
E′ or E_u (kN/m^2)	18×10^3	20×10^3	140×10^3	60×10^3
υ' or v_u	0.15	0.2	0.15	0.2
c′ or s_u (kN/m^2)	0.2#	12	0.2#	30
φ' or φ_u	33	20	39	20
ψ'	3	11	9	11
Flow parameters				
Data set	Standard	Standard	Standard	Standard
Soil type	Medium fine	Fine	Medium	Fine
k_x	1×10^{-7}	1×10^{-9}	0.5×10^{-3}	1×10^{-9}
k_y	1×10^{-7}	1×10^{-9}	0.5×10^{-3}	1×10^{-9}
Interfaces				
Strength	Manual	Manual	Manual	Manual
R_{inter}	0.8	0.7	0.8	0.7
Initial				
K_0	Automatic	Manual	Automatic	Manual
$K_{0,x}$	0.4554	1.5	0.3707	1.5

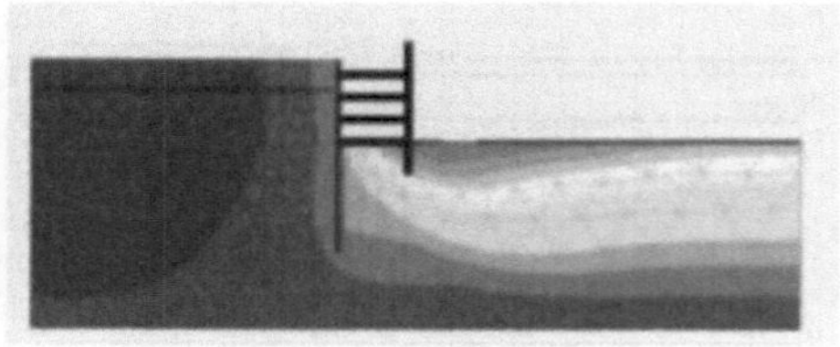

Figure 6. Vertical displacement of soil.
Legend: red= +120 mm and deep blue= −10 mm

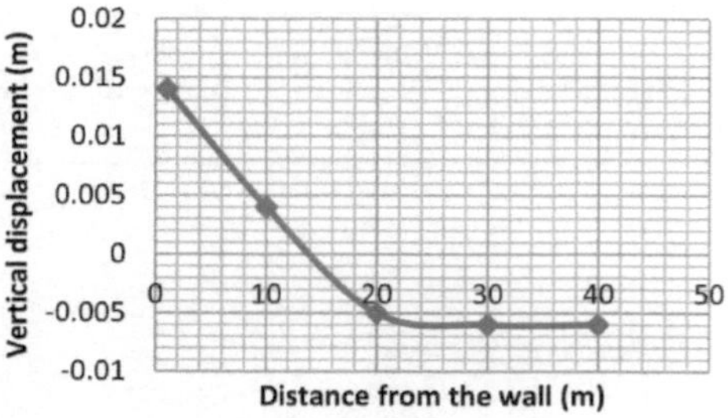

Figure 7. Vertical displacement of soil: max −6 mm.

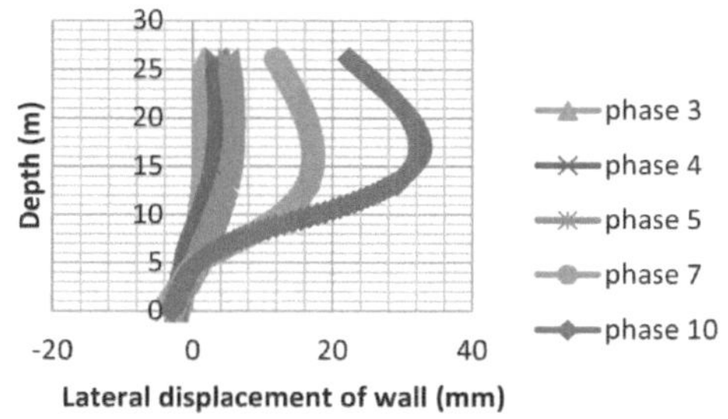

Figure 8. Lateral displacement of diaphragm wall: max 33 mm.

Table 6. Construction stages for consolidation analysis.

Sequence	Construction activities
Phase 1	Diaphragm wall construction
Phase 2	Excavation –2 m
Phase 3	Roof slab construction
Phase 4	Consolidation (20.3 days)
Phase 5	Excavation –4 m
Phase 6	Groundwater lowering till –5.2 m followed by excavation
Phase 7	Construction of the second slab
Phase 8	Consolidation (20.3 days)
Phase 9	Groundwater lowering till –8.4 m followed by excavation
Phase 10	Construction of the third slab
Phase 11	Consolidation (20.3 days)
Phase 12	Groundwater lowering till –11.6 m followed by excavation
Phase 13	Construction of base slab
Phase 14	Consolidation (20.3 days)
Phase 15	Consolidation (365 days)
Phase 16	90% consolidation

construction stage, and its value is 33 mm. It should be noted that the wall follows cantilever type of movement at the initial stages, and at the latter construction stages, deep inward movement pattern occurs.

2.3 *Consolidation analysis*

For this analysis, clay soils were set to Undrained A drainage type since it enables to generate excess pore water pressures which will be dissipated as a result of consolidation analysis. Time interval inserted in consolidation analysis corresponds to the time estimated to perform excavation and concrete hardening.

From Figure 9, it can be seen that the maximum vertical displacement on the retained side is –18.72 mm, while in the area in front of the wall is +14.69 mm. Figure 10 indicates that vertical displacement values of soil on the retained side is –19 mm. According to Figure 11, maximum lateral wall displacement is about 30 mm. Deep inward pattern of wall displacement can be differentiated. Figure 12 shows vertical displacement of soil for consolidation analysis with duration of 365 days. By comparing it with Figure 9, it can be

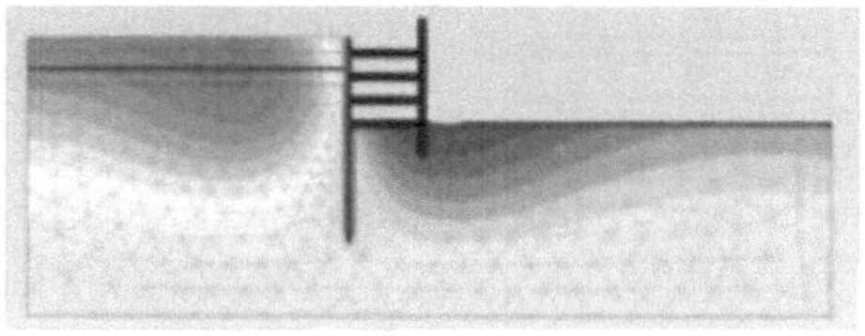

Figure 9. Vertical displacement of soil.
Legend: red= –20 mm and deep blue= +16 mm

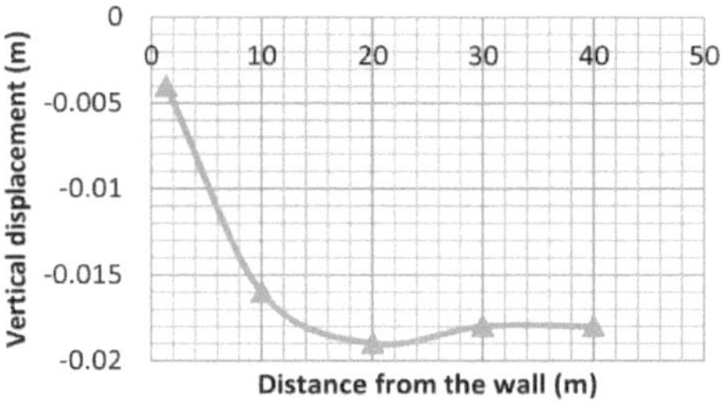

Figure 10. Vertical displacement of soil: max –19 mm.

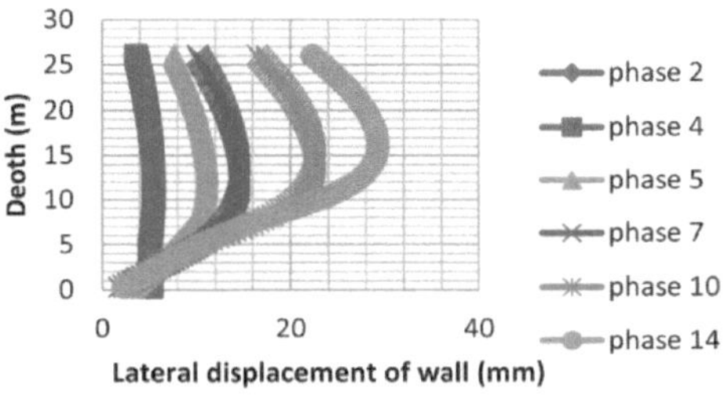

Figure 11. Lateral displacement of diaphragm wall: max 30 mm.

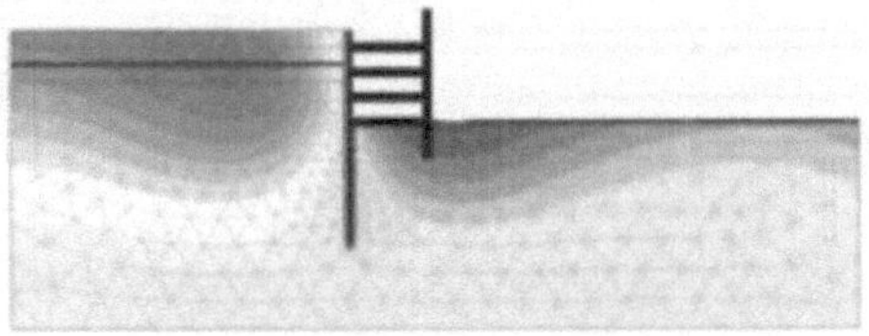

Figure 12. Vertical displacement of soil (365 days).
Legends: red= –20 mm and deep blue= +18 mm

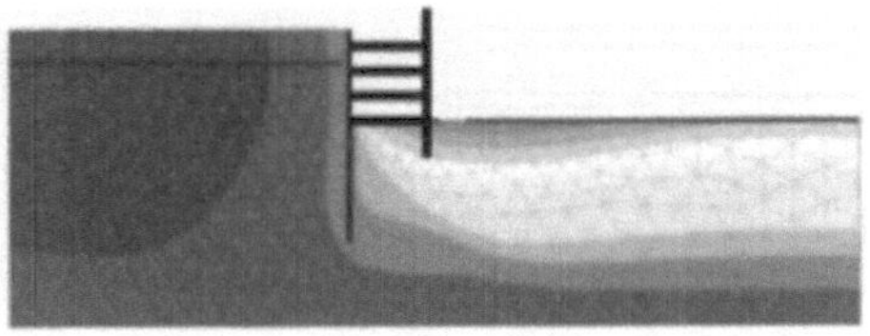

Figure 13. Vertical displacement of soil (90% consolidation).
Legends: deep orange= +110 mm and deep blue= –10 mm

seen that there is no significant difference in vertical displacement of soil. However, by comparing these two figures with vertical displacement of soil due to 90% consolidation on Figure 13, it can be viewed that maximum settlement value on the retained side reduces two times till 10 mm.

2.4 *Comparison of results*

By comparing analyses results, it can be said that regarding the vertical soil displacement on the retained side, consolidation analysis shows highest values with maximum of –19 mm at distance of 20 m from the wall. In addition, it has been found out that vertical soil displacement results of consolidation analysis immediately after construction and consolidation with duration of 365 days are similar. Nevertheless, settlement on the retained side decreases two times due to 90% consolidation analysis. Vertical displacement values from drained analysis seems to be the lowest one –6 mm, and undrained analysis output lies in between of these two analyses with –14 mm. All three analyses have shown that concave type of settlement occurs which is characterized by maximum settlement value at a distance from the wall. In all three cases, maximum settlement value occurred at a distance of 20 m from wall. Deep inward wall movement pattern was indicated, and maximum lateral wall displacement of 33 mm occurred during drained analysis, while consolidation and undrained analyses generated maximum lateral wall displacement of 30 mm and 21 mm, respectively. According to the obtained results, lateral deformation of wall is less than both 270 mm and 60 mm, thereby being within acceptable limit. By using empirical relationship, ground surface settlement can be calculated based on maximum lateral wall displacement of 33 mm, which gives range of 16.5–33 mm. It can be seen that maximum vertical displacement provided by PLAXIS lies on this range. In addition, ground surface settlement value is small, only 20 mm. According to the literature review, this value is within acceptable range.

3 CONCLUSION

This paper investigates the behavior of soil and retaining wall caused by excavation of 12 m depth. PLAXIS was adopted as the main tool for numerical analysis. Three types of analyses, namely undrained, consolidation, and drained, were performed by using Mohr-Coulomb soil model. Vertical soil displacement and lateral wall displacement were primarily examined. Results showed that maximum lateral wall displacement had occurred during long-term (drained) analysis, while time-dependent consolidation analysis had produced maximum ground surface settlement. Lateral wall displacement value is smaller than serviceability requirement, thereby ensuring that there is no excessive deformation. It has been indicated that empirical and numerical ground surface settlement values are close to each other. Thus, it can be stated that both ground surface settlement and lateral wall displacement are within acceptable limits.

ACKNOWLEDGMENTS

This paper is a part of the undergraduate capstone project by K. Bekish, A. Ulykbanov, T. Mukashev, and Y. Serik in the Department of Civil Engineering at Nazarbayev University. Dr. Nguyen and Prof. Kim are the supervisor and co-supervisor of this project. The authors thank Drs. Sudheesh Thiyya Kkandi and Dichuan Zhang for their valuable comments for this project.

REFERENCES

Bentler, D. (1998). *Finite Element Analysis of Deep Excavations*. Retrieved March 10, 2016 from http:/scholar.lib.vt.edu/theses/available/etd-9298-83534 unrestricted/chapter2.pd

Chalmers University of Technology. (2014). *Functional requirements for permanent diaphragm wall*. Retrieved April 20, 2016 from http://publications.lib.chalmers.se/records/fulltext/206907/206907.pdf

Likitlersuang, S., Surarat, C., Wanatowski, D., Oh, E., and Balasubramaniam, A. *Finite element analysis of deep excavation: A case study from Bangkok MRT*. Retrieved April 20, 2016 from http://www.sciencedirect.com /science/article/pii/S0038080613000954#gr4

Ou, C. (2006). *Deep Excavation: Theory and Practice*. The Netherlands: Taylor & Francis/Balcema

PLAXIS. (2015). *Material Models Manual*. Retrieved March 5, 2016 from http://www.plaxis.nl/files/files/2D-3-Material-Models.pdf

Puller, M. (2003). *Deep excavations: a practical manual*. London: Thomas Telford

Challenges and Innovations in Geotechnics – Zhussupbekov (Ed.)
© 2016 Taylor & Francis Group, London, ISBN 978-1-138-03007-7

Analysis of excavation-induced ground movements in urban areas

C. Hung
Department of Civil Engineering, National Cheng Kung University, Tainan, Taiwan

ABSTRACT: Applications of several enhanced bounding surface models for analyzing a deep excavation-induced ground movements in an urban area are presented in this paper. Such applications require knowledge of bounding surface theory within the concept of conventional constitutive modeling. Incorporations of the bounding surface models and a commercial finite element code provide a useful approach to evaluate the performance of deep excavations associated with soft soils. With respect to the evaluations, it was found that (i) agreement between finite element simulations and field observations in terms of lateral soil movements and nearby surface settlements was satisfactory; (ii) in most stages of the excavation the simulations obtained using the isotropic version of the enhanced bounding surface model were slightly improper than for the case of the anisotropic versions of the model; (iii) utilization of the advanced soil models (enhanced bounding surface models) can lead to a more realistic prediction of excavation-induced ground movements; however, careful consideration of detailed processes relating to constructions should be desired.

Keywords: enhanced bounding surface model; excavation-induced ground movements; finite element analyzes

1 INTRODUCTION

1.1 *Deep excavation*

The development of underground spaces often involves excavation. Excavation, particularly deep ones, are technically challenging undertakings that depend on site geological conditions, supporting systems, proximity of adjacent structures, and other factors. Although advanced excavation methodologies have been designed to mitigate ground movements, it is still difficult to predict and control deformations in deep excavations.

1.2 *Literature*

It is observed that there are no simple methods available for analyzing ground movements associated with deep excavations; studies of excavations have been either based on empirical observations (Peck, 1969) or simulated using numerical methods such as finite element of finite difference procedures. In the numerical modeling of deep excavations in clays, the soil has been idealized using constitutive models possessing varying levels of sophistication. These include quasi-linear elastic hyperbolic (e.g., Chang and Duncan, 1970; Ou and Lai, 1994), simple elastoplastic (e.g., Yong et al., 1989), and critical state (e.g., Borja, 1990) models. Instead of specialized finite element excavation computer programs, the use of commercial codes is gaining popularity in practice. However, the level of sophistication of constitutive models found in commercial codes is often restricted to nonlinear elastic, simple elastoplastic and critical state models.

1.3 *Incorporations of the advanced constitutive models and FEM*

In general, relatively simple constitutive models are unable to accurately simulate all aspects associated with an excavation. That is, if ground deformations are accurately simulated, wall bending moments, strut forces, etc. will likely not all be as accurate. For this reason, more advanced constitutive models have also been used to simulate deep excavation. Of particular interest to the present paper are bounding surface plasticity models (Finno et al., 1991; Whittle et al., 1993; Hashash and Whittle, 1996).

This paper describes the finite element simulation of a well-documented deep excavation case history in an urban area, namely the Garage at Post Office Square (POS). The soft soils at the site are characterized by an enhanced anisotropic bounding surface model with associative (Ling et al., 2002) and non-associative (Jiang and Ling, 2010; Jiang et al., 2012) flow rules. Along with the aforementioned versions of the models, the Sekiguchi-Ohta model (Sekiguchi and Ohta, 1977) is also used in the simulations.

2 SITE DESCRIPTION

2.1 *Garage at Post Office Square, Boston*

The garage at Post Office Square (POS) site was located in the Boston financial district, bounded by Milk, Congress, Pearl and Franklin Streets. The site occupied a plan area of 6,880 m^2 and was irregularly

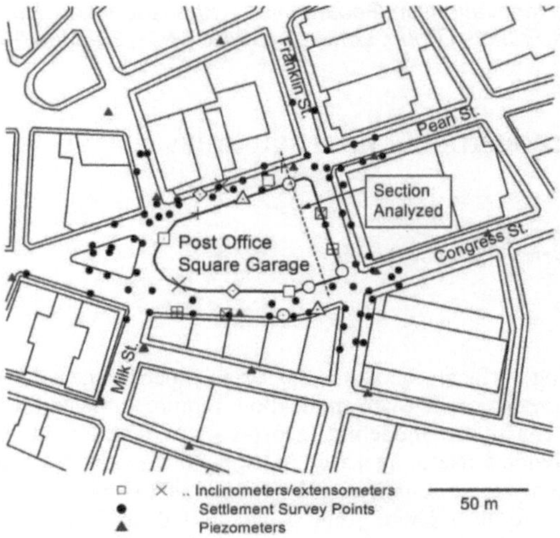

Figure 1. Plan view and instrumentation: POS site.

shaped. It was adjacent to several tall buildings up to 40 stories high and separated by as little as 18 m. A parking garage previously occupied the site, with two stories above, and one below ground. This structure was demolished in late 1988. The new structure consists of a 7-story underground parking garage that accommodates 1400 cars, with Norman B. Leventhal Park at the ground surface. It took approximately two years to complete this structure.

The soil at the site included a fine sand to sandy gravel fill layer at the surface. This was underlain by a thick deposit of Boston Blue clay (BBC), followed by silt or sand deposit, glacial till deposit, and bedrock. The thickness of the various soil layers varied across the site. The section used in the present study was identical to that used by Whittle et al. (1993, who approximated it as being 61 m wide.

The retaining system composed of a 90 cm tick reinforced concrete diaphragm wall embedded into the bedrock, 21 m below the ground surface. The final depth of excavation was 16.8 m below the ground surface.

Lateral wall movements at the site were monitored by a series of inclinometers cast within the diaphragm wall panels. Ground surface settlements were measured by optical survey. Observational wells and piezometers were installed inside and outside of the diaphragm wall. The maximum wall deflection was about 5.5 cm, which occurred at about 7 m below the ground surface. The maximum ground surface settlement was 5 cm, which occurred at about 90 cm behind the diaphragm wall.

Whittle et al. (1993) previously simulated the deep excavation at the POS site. Their simulation was based on the actual record of construction activities, and used the MIT-E3 form (Whittle, 1993) of the anisotropic bounding surface model.

3 FINITE ELEMENT ANALYSIS

3.1 *Implementation*

The BBC soil layers were characterized by the enhanced bounding surface models with both associative and non-associative flow rules, as well as by the Sekiguchi-Ohta model. The non-associative version of the model has the capability to simulate the strain-softening behavior of BBC. The formulation and predictive capabilities of the models, with both associative and non-associative flow rules, have been previously documented (Ling et al., 2002, Jiang et al., 2012). Hung et al. (2011; 2012) demonstrated the implementation procedures and calibrated the model parameters required to simulate BBC.

3.2 *Finite element modeling*

The site was modeled as two-dimensional boundary value problems and analyzed assuming plane strain conditions. The lack of sufficient spatial variations of soil profiles and construction details of POS site necessitated the two-dimensional analyzes, although the geometry was clearly three-dimensional.

The site was assumed to be geometrically symmetrical in the plane of analysis, thus, the solution domains consisted of only one-half of the entire excavation. Because of the aforementioned symmetry of the solution domains, horizontal displacements were prevented at nodes located along a side of the boundaries. In all simulations, the clay layers were assumed to be saturated. The displacements and stresses of the solid phase were assumed to be coupled to changes in excess pore pressure according to the Biot theory (Biot, 1941).

4 RESULTS AND DISCUSSIONS

Figures 2a–b present typical comparisons between field measurements as reported by Whittle et al. (1993), and the simulations obtained using the different versions of the enhanced bounding surface model. In spite of the scatter in the field measurements, it appears that the agreement between the simulated and measured lateral wall deflections is better at the later stages of excavation than for the earlier ones (i.e., stages 19 and 28 versus stage 10). The simulated results certainly fall within the range of measured values. Stage 10 represents a cantilever mode for wall deflections, which has been discussed by Whittle et al. (1993).

At the earlier stages, the simulated ground surface settlements were reasonably close to the measured values (note that Whittle et al. 1993 did not give specific locations of each measured settlement value). Once again, the measured data are rather scattered, likely due to the three-dimensional nature of the site geometry and slight differences in the excavation process at different locations at the site. In addition,

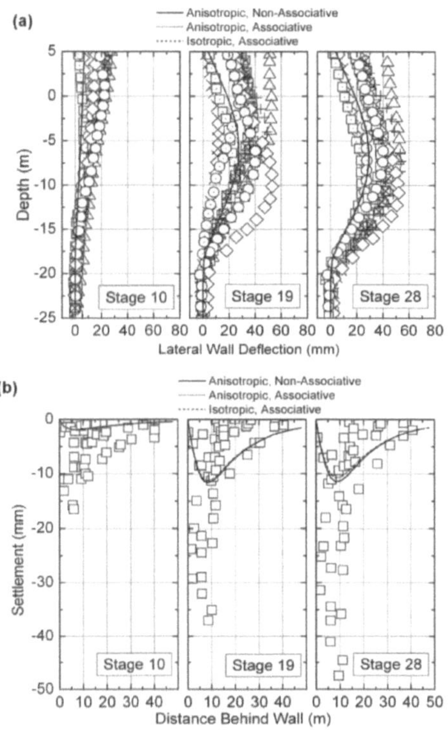

Figure 2. Comparisons between simulations using different versions of the enhanced bounding surface model and POS field measurements: wall lateral deflections and ground settlements.

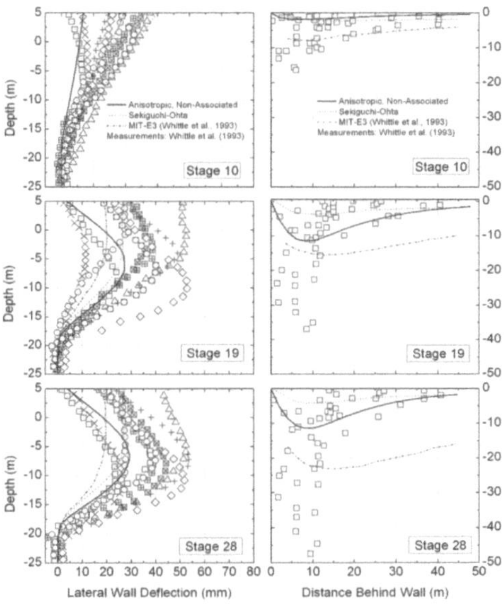

Figure 3. Comparisons between simulations using different soils models and POS field measurements: wall lateral deflections and ground settlements.

idealizations were made regarding the properties of cohesionless soil layers. As such, no attempts were made in the present study to improve the simulations by considering the roof displacement (shrinkage due to temperature change), as was done by Whittle et al. (1993).

The results obtained using the associative and non-associative versions of the enhanced bounding surface model were essentially identical. In the later stages of excavation, the simulations obtained using the isotropic version of the model tended to slightly underestimate the ground surface settlements.

Figure 3 compares the simulations obtained using the Sekiguchi-Ohta model with those obtained using the anisotropic bounding surface model with a non-associative flow rule, and with field measurements. Since Whittle et al. (1993) previously simulated the deep excavation at the POS site using the more advanced MIT-E3 model, these results are included for comparison. The simulations obtained using the MIT-E3 and Sekiguchi-Ohta models differed in magnitude from the simulations obtained using the enhanced bounding surface model (anisotropic version). The Sekiguchi-Ohta model seemed to underestimate the settlements, whereas the MIT-E3 model overestimated them.

5 CONCLUSIONS

The following conclusions are based on the simulations performed:

i) in light of the scatter in field measurements, the overall agreement between simulations obtained using the enhanced bounding surface model and field measurements was satisfactory. This included the lateral displacement of diaphragm walls and ground settlements.
ii) Since the aim of the non-associative model is to capture the post-failure softening, the difference between the non-associative and associative versions of the enhanced model in finite element excavation analyzes was rather negligible. This is explained by the fact that in the deep excavations simulated in this paper, the soil behavior is far from post-failure states.
iii) In most stages of the excavation, the simulations obtained using the isotropic version of the enhanced bounding surface model were less accurate than for the case of the other anisotropic versions of the model. This illustrates the importance of accounting for anisotropy in the deep excavations where the soil possesses inherent anisotropy. It is consistent with the findings of Clough and Hansen (1981) and Finno et al. (1991), who showed that the anisotropy of soils had a significant influence on the extent of ground movements associated with deep excavations.
iv) The overall agreement with field measurements of the simulations obtained using the advanced bounding surface models was better than that

obtained using the simpler Sekiguchi-Ohta model. Similar results would be expected for other classical elastoplastic constitutive models for cohesive soils.

v) the simulations of the POS site excavation obtained using the advanced bounding surface models were similar to, and sometimes more accurate than those obtained using the more advanced MIT-E3 model (Whittle, 1993).

ACKNOWLEDGEMENT

The author would like to thank the nomination of Kazakhstan Geotechnical Society (KGS) for this study. The author is grateful to Prof. Hoe I. Ling (Columbia U, U.S.A.), Prof. Askar Zhussupbekov (Eurasian National U, Kazakhstan), and Prof. Victor N. Kaliakin (U of Delaware, U.S.A.) for their suggestions and feedbacks in this study. This research was supported in part by (received funding from) the Headquarters of University Advancement at the National Cheng Kung University, which is sponsored by the Ministry of Education, Taiwan, ROC. The research is partly supported by the Ministry of Science and Technology, ROC (MOST104-2218-E-006-029).

REFERENCES

Biot, M.A. (1941). "General theory of three-dimensional consolidation." Journal of Applied Physics, 6 12, 155–164.

Borja, R.I. (1990). "Analysis of incremental excavation based on critical state theory." *Journal of Geotechnical Engineering*, ASCE, 116(6), 964–985.

Chang, C-Y. & Duncan, J.M. (1970). "Analysis of soil movements around a deep excavation." Journal of Soil Mechanics and Foundations Division, ASCE, 96(5), 1655–1681.

Clough, G.W. & Hansen, L.A. (1981). "Clay anisotropy and braced wall behavior." Journal of Geotechnical Engineering, ASCE, 107(7), 893–913.

Finno, R.J. et al. (1991). "Analysis of braced excavations with coupled finite element formulations." *Computers and Geotechnics*, 12(2), 91–114.

Hashash, Y.M.A. & Whittle, A.J. (1996). "Ground movement prediction for deep excavations in soft clay." *Journal of Geotechnical Engineering*, ASCE, 122(6), 474–486.

Hung, C. et al. (2011). "Calibration of Taipei silty clay using an enhanced anisotropic elastoplastic bounding surface model." Proceedings of the Fifth International Symposium on Deformation Characteristics of Geomaterials, 478–482.

Hung, C. (2012). "Implementation of an enhanced elastoplastic bounding surface models in plaxis." The 7AYGEC Conference, Tokushima, Japan.

Jiang, J.H. & Ling, H.I. (2010). "A framework of anisotropic elastoplastic model for clays." *Mechanics Research Communications*, 37, 394–398.

Jiang, J.H. et al. (2012). "An associative and non-associative anisotropic bounding surface model for clay." *Journal of Applied Mechanics*, 79(3), 031010–10.

Kung, G.T.C. (2010). "Finite element analysis of wall deflection and ground movements caused by braced excavations. Finite Element Analysis." David Moratal (Ed.), Sciyo.

Ling, H.I. et al. (2002). "An anisotropic elastoplastic bounding surface model for cohesive soils." *Journal of Engineering Mechanics*, ASCE, 128(7), 748–758.

Ou, C.Y. & Lai, C.H. (1994). "Finite-element analysis of deep excavation in layered sandy and clayey soil deposits." Canadian Geotechnical Journal, 31(2), 204–214.

Peck, R.B. (1969). "Deep excavation and tunneling in soft ground." Proceedings of Seventh International Conference on Soil Mechanics and Foundation Engineering, Mexico City, State-of-the-art Volume, 225–290.

Sekiguchi, H. & Ohta, H. (1977). "Induced anisotropy and time dependency in clays." *Proceedings*, 9th International Conference on Soil Mechanics and Foundation Engineering, Tokyo, 229–238.

Whittle, A.J. et al. (1993). "Analysis of deep excavation in Boston." *Journal of Geotechnical Engineering*, ASCE, 119(1), 69–90.

Yong, K.Y. et al. (1989). "Elasto-plastic consolidation analysis for strutted excavation in clay." *Computers and Geotechnics*, 8(4), 311–328.

Challenges and Innovations in Geotechnics – Zhussupbekov (Ed.)
© 2016 Taylor & Francis Group, London, ISBN 978-1-138-03007-7

Comparison analysis of features in Eurocode and Kazakhstan norms requirements

A.Zh. Zhussupbekov, A.S. Tulebekova, R. Lukpanov & I.T. Zhumadilov
Eurasian National University, Astana, Kazakhstan

ABSTRACT: As world practice shows construction development is directly reflects the economical position of state in whole world, has been one of the relevant fields for the future advance and development as a whole. In the part of the reform of technical regulation of the construction industry importance is harmonize of construction specifications and technical norms in the Republic of Kazakhstan with the requirements of international and European standards. Unfortunately, present Standards are confined application of modern technology of pile foundation installation, indicating incomplete usage of advanced technology. So necessities are occurred for scientific reasoning and using of traditional and international Norms give or take geotechnical situations. Eurocode 7 includes recommendations and requirements formodern advanced technologies and embraced many aspects of modern geoengineering design. This comparison will be basement for future modernization of Kazakhstan codes.

1 INTRODUCTION

Nowadays many international projects are realized in Kazakhstan. For example Italian and American companies which are work on western part of Kazakhstan met some complexity in non-correspondences of Kazakhstan codes to international. Kazakhstan engineers also feel some complexity during design modern projects where generally used advanced technologies.

Unfortunately, present Codes are confined application of modern technology of pile foundation installation, indicating incomplete usage of advanced technology such as CFA (continuous flight auger), DDS (drilling displacement system) or FDP (full displacement pile), Jet-grouting DMM (Deep mixing method) technologies and so on.

Aforementioned demands to using international Code, moreover, for realization unique project is required using leading foreign high-tech, economic, ecological and energy-efficient technology, including technology for pile installation, equipment for geological investigation, as well as laboratory testing.

Eurocode 7 – Geotechnical Engineering which was established in 2004 seems to be more reliable for adaptation for Kazakhstan construction condition. Many countries successfully accepted Eurocode 7 and during last years this Code becoming more international. Eurocode 7 is already show itself as very elaborate design code where given recommendations and requirements for most part of geoengineering process. It also allows using common international geoengineering terms and provides understanding between for designers, testing specialists, geotechnical engineers all over the World. Eurocode 7 include recommendations and requirements for modern advanced technologies and embraced many aspects of modern geoengineering design.

The results of research is directed to developing of recommendation for modernization of Kazakhstan Codes and oriented to adaptation of advanced geotechnologies. The modernization will allow to complete use of advanced technologies capabilities in existing construction condition of Kazakhstan. Is also allow greatly reduce the expenses of the null cycle, which forming significant part (20%) of the total construction project expense.

Development of recommendation for modernization of national geotechnical codes, oriented on adaption of advanced geotechnologies for piles installations in problematical soil conditions of Kazakhstan is very important for designers, testing specialists, geotechnical engineers as long as Codes of many countries has some differences due to of specific regional soil condition, and local specifications. Developing recommendation will be first stride for Kazakhstan construction to be of international part, to be understandable by world Geotechnics. We also believe that this recommendation will be basement for future modernization of Kazakhstan codes.

1.1 *Introduction to Eurocode 7*

The development of Eurocode 7 has been strongly linked to the development of En 1990: Eurocode: Basis of structural design (CEN, 2002) and the format for verifying ground-structure interaction problems is, of course common to both documents.

After giving the main contents of Eurocode 7, this contribution summarises the requirements relevant to pile design (without recalling the principles of LSD and of the partial factor method).

Design examples of piles under vertical compressive loadings can be found, for instance, in the proceedings of the workshop of ERTC 10 on the evolution of Eurocode 7

General rules is a rather general documents giving only the principles for geotechnical design inside the general framework of LSD. These principles are relevant to the calculation of the geotechnical actions on the structural elements in contact with the ground (footings, piles, basement walls), as well as to the deformations and resistances of the ground submitted to the actions from the structures. Some detailed design rules or calculation models, i.e. precise formulae or charts are only given in informative Annexes.

The Section on field tests in soil and rock includes cone and piezocone penetration tests CPT (U), Pressuremeter test PMT, flexible dilatometer test (rock and soil) FDT, code penetration test SPT, dynamic probing tests DP, weight sounding test WST, field vane test FVT, flat dilatometer test DMT and plate loading test PLT.

The Section on laboratory testing of soils and rocks deals with the preparation of soil and rock specimens for testing, tests for classification, identification and description of soil, chemical testing of soil and groundwater, strength index testing of soil, compressibility and deformation testing of soil, compaction testing of soil, permeability testing of soil, strength testing of soil, compressibility and deformation testing of soil, tests for classification of rocks, swelling testing of rock material and strength testing of rock material.

It also includes a number of informative Annexes with examples of correlations and derivations of values of geotechnical parameters from field test results. The informative Annexes D.6 & D.7 for CPT tests, and E.3, for PMT tests, are such examples for determining the compressive resistance of a single pile.

The core of Section 7 of EN 1997-1 is devoted to the behavior of pile foundations under axial (vertical) loads. The importance of static load tests is clearly recognized as the basis of pile design methods. An innovative concept introduced in this section, with regard to traditional pile design, is the use of correlation factors ξ for deriving the characteristic compressive and tensile resistances, of piles either from static pile load tests or from ground test results. In both cases, the correlation factors ξ depends mainly on the number of tests performed, whether pile load tests or profiles of ground tests.

2 COMPARISON OF EUROCODE WITH KAZAKHSTAN CODES

Nowadays is put into action recommendation of Eurocode 1–8 which are replacing National Codes. However some Countries arc transforming it's National Codes by taking into account Eurocode recommendation. For example Japan input some changes into ISO and as a result new published ISO 23469 is a copy of Eurocode 8.

The principal deference between Eurocode and Kazakhstan Code (SNiP RK) is absence of requirements for the geotechnical design in Kazakhstan Codes. In Eurocode the strategy of geotechnical design includes interaction of two researches – geological engineering and geotechnical. However today is difficult to design without qualitative geotechnical investigation. Geotechnical research include results of engineering and geological investigation which are had been used during definition of soil and foundation modeling. Recommendation of Eurocode promote to mutual researchers and designers work. Unfortunately in Kazakhstan practice engineering and geological investigation is one different part of the design, and frequently no interaction between researchers and designers. The program of the geological investigation rarely coordinate to designer and as a result there are absent common strategy of the design.

One another difference of Eurocode is design procedure. According to Kazakhstan Code the design of soil basement is recommended to carry out by three steps. During the first and second step of foundation design it is allowable to use preliminary strengthen and defomative properties of soil taken from table of SNiP RK, during the last third step it is required to perform both laboratory and field test to approve design project. According to Eurocode for all of this three engineering and geological investigation steps for definition strengthen and defomative properties of soil is required to use results of laboratory or field tests only. Moreover Eurocode use term ≪derive value≫ that mean value of geotechnical parameter of soil obtained by results of identical laboratory or field test of soil by using correlation relationship or using inverse calculation.

Design according to Eurocode uses over the limit state. For each geotechnical design situation is necessary to make sure that no possible limit state will not occur. The limit states can occur in the soil, in the construction or destruction as a combined diaphragm.

The possibility of limit state can be verified by calculation, models, field observations and tests, as well as exclude the adoption of prescriptive measures.

Checking the limit state by calculation method calculated model should describe the expected behaviour of the ground for the considerate limit states. The calculated model can be analytical, semi-empirical or numerical.

There are two groups of limit states: the absolute limit states (Ultimate limit states) and Serviceability limit state serviceability.

It demands checking of the proposed facility in order to make sure that it does not occur following the Ultimate limit state:

- loss of balance structures and soil as a rigid body (EQU);
- Internal failure or excessive deformation of the structure or its elements(STR);

Table 1. The values of reliability coefficients for limiting resistors.

Type of limiting state	symbol	EQU	STR/GEO			
			1.1	1.2	2	3
Capacity: (gF)	g_G	1,10	1,35	1,00	1,35	1,35
Constant, destabilizing	g_G	0,90	1,00	1,00	1,00	1,00
Stabilization	g_Q	1,50	1,50	1,30	1,50	1,50
Permanent, destabilizing	g_Q	0	0	0	0	0
Property (gM):						
tg j′	g_j	1,25	1,00	1,25	1,00	1,25
Cohesion	g_c'	1,25	1,00	1,25	1,00	1,25
undrained cohesion	g_{cu}	1,40	1,00	1,40	1,00	1,40
Specific gravity	g_s	1,00	1,00	1,00	1,00	1,00

- destruction or excessive deformation of the ground (GEO);
- Archimedean ascent structures or soil (UPL);
- removal of the soil caused by the hydraulic gradient (HYD).

When considering limiting states such as EQU, make sure that

$E_{dst,d}$ £ $E_{stb,d}$,
где $E_{dst,d} = E\{g_F F_{rep}, X_k/g_M, a_d\}_{dst}$;
$E_{stb,d} = E\{g_F F_{rep}, X_k/g_M, a_d\}_{stb}$;
$E_{stb,d}$ и $E_{dst,d}$ – design value of the stabilizing effect (holding) and the effect of destabilizing loads;
F_{rep} – a representative value of loads;
g_F – coefficient of load reliability;
X_k – the value of the properties of the soil;
g_M – safety factor on the ground;
a_d – design geometrical parameters.

The effect of stress – a free value determined by the load, for example, time shifting, or restraining forces.

The values of reliability coefficients for limiting states the EQU is type conditions which are given in Table 1. There are abbreviated reliability coefficient sets for the type of limit states STR and GEO.

The main types of limiting states that define the parameters of the bases are STR and GEO. In the analysis of such limit states should make sure that:

E_d £ R_d,
где$E_d = g_E\, E\{g_F F_{rep}, X_k/g_M, a_d\}$;
$R_d = R\{g_F F_{rep}, X_k/g_M, a_d\}/g_R$;
E_d и R_d design value of load effects and resistance to stress;
g_E и g_R – safety factors on the effect of stress and resistance;

EC7 offers three design approaches when considering the limit states STR and GEO. Each approach has its own set of safety factors values. The approaches differ in the stocks which are put in predominantly either load or resistance performance.

Project approach1. Make sure that the limiting state does not arise as in a set of safety factors 1.1 and 1.2 (see. Table 1). The package preferably 1.1 stock into the load in the set of 1.2 – in resistance performance.

Project approach 2. Make sure that the limiting condition does not occur in the recruitment of reliability coefficients 2. Here the stock is introduced into the load and resistance indicators.

Project approach 3. Make sure that the limiting state does not occur in the recruitment of reliability coefficients 2. Here the stock is introduced into the load or resistance indicators.

Quoted in the table EC7 reliability coefficients can be adjusted by national authorities.

The formulas and corresponding tables for absolute reliability coefficients of limiting state such as the UPL and the CEO.

Checking limit state deformations (serviceability buildings) performed by the condition of E_d £ C_d, where C_d – limiting design value of load effect (eg, sediment, banks).

Private reliability coefficients thus equal to 1.0.

For example deformation Modulus obtained by independence tests: laboratory tests, field tests by dilatometer, by correlation relationship with physical parameter, by results of well known settlement calculation (Boldyrev et al. 2010).

Laboratory tests which recommended by Eurocode are presented in Table 2. It is well known that the obtained by laboratory tests mechanical and physical properties of soil depends on quality of soil sample. Eurocode differ five categories of soil samples quality assuming that the properties of soil invariable due to sampling, packing and transportation.

According to Eurocode swelling of soil research for rock soil, whereas according to Kazakhstan Codes concern to unstable structure.

Comparison of national Codes with Eurocode (Boldyrev et al. 2010) shown that there are not yet developed recommendation for performing following field tests:

- Cone penetration test with analysis of pore water pressure (CPTU).
- Cone penetration lest by dynamic load (SPT).
- Dilatometer test.

However in Eurocode is absent some recommendation for field tests in condition of frozen soil (Boldyrev et al. 2010).

Modern megaprojects put forward modern requirements to engineers. This led to refuse from traditional out-of-dates technologies (traditional boring and driving diesel-hammer piles) and use new more economical and reliable technologies like CFA (continuous flight auger), DDS (drilling displacement system), steel "H" piles.

Another advantage of Eurocode is recommendation for design and calculation of modern pile technologies such as CFA, FDP and so on.

Table 2. Eurocode soil laboratory methods.

Parameter	Type of soil: Gravel	Sand	Sandy loam
Compression deformation Modulus (E_{oed}) Compression index (c_c)	OED, TX	OED, TX	OED, TX
Elastic Modulus (E) Shear Modulus (G)	TX	TX	TX
Consolidated Drained strength-parameters (φ, c)	TX, SB	TX, SB	TX, SB
Residual strength (φ_R, c_R)	RS, SB	RS, SB	RS, SB
Undrained strength, (c_u)	–	–	TX, DSS, SIT
Density of soil (p)	BDD	BDD	BDD
Consolidation (c_N)	–	–	OED, TX
Filtration	TXCH, PSA	TXCH, PSA	PTS, TXCH, PTF

Parameter	Type of soil: NC clay	OC clay	Silt
Compression deformation Modulus (E_{oed}) Compression index (c_c)	OED, TX	OED, TX	OED, TX
Elastic Modulus (E) Shear Modulus (G)	TX	TX	TX
Consolidated Drained strength parameters (φ, c)	TX, SB	TX, SB	TX, SB
Residual strength (φ_R, c_R)	RS, SB	RS, SB	RS, SB
Undrained strength, (c_u)	TX, DSS, SB, SIT	TX, DSS, SB, SIT	TX, DSS, SB, SIT
Density of soil (p)	BDD	BDD	BDD
Consolidation (c_N)	–	–	OED, TX
Filtration	TXCH, PTF, OED	TXCH, PTF, OED	TXCH, PTF, OED

Kazakhstan Code has not some recommendation for these piles technology, and so designers have to use recommendation for traditional out-of-dates pile technologies (traditional boring and driving diesel-hammer piles). As a result incomplete usage of modern technology has a place. Design of modem pile by Kazakhstan Code is not include many technologies factors such as high value of concrete pressure during CFA pile installation and soil displacement without excavation during DDS pile installation.

Comparison of safety factors recommended by Kazakhstan Code and Eurocode is presented in Table 3.

Table 3. Safety factor of pile tests comparison.

Code	Safety factor: Design	SLT	DLT	Number of test required
Eurocode		2.18		If number of tests equal or less than 2
		1.91		If number of tests equal or greater than 20
			2.23	If 2 DLTs is performed
			1.95	If number of DLTs greater than 20
Kazakhstan Code	1.5	1.4		SLTs on 0.5% of constructed piles on construction site (2 SLTs at least in a site)
			1.2	At least 6 DLTs in a site (or 1% of working piles on construction site)

In this table are listed design safety factor, safety factor for static load test, and safety factor for dynamic load test, together with the number of tests required or specified for a pile construction site.

3 CONCLUSION

Eurocode 7 – Geotechnical Engineering which was established in 2004 seems to be more reliable for adaptation for Kazakhstan construction condition. Many countries successfully accepted Eurocode 7. Eurocode 7 is already show itself as very elaborates design code where given recommendations and requirements for most part of geoengineering process. It also allows using common international geoengineering terms and provides understanding between for designers, testing specialists, geotechnical engineers all over the World. Eurocode 7 includes recommendations and requirements for modern advanced technologies and embraced many aspects of modern geoengineering design.

By result of Kazakhstan Code comparison with Eurocode it is become obvious one disadvantage of Kazakhstan Code is absence of recommendation for design, testing and calculation modern pile technologies such as CFA, DDS and so on. Eurocode is also presented by unified documentation for geoengineering comparing with Kazakhstan Code, where many Codes are and where someone may contradict to other.

Comparison also shown that in Kazakhstan Codes is not developed recommendation for performing following field tests: cone penetration test with analysis of pore water pressure, cone penetration test by dynamic load and dilatometer tests.

REFERENCES

EN 1997-1:2004. Eurocode 7. Geotechnical Design. Part I. General rules.

EN 1997-2:2007. Eurocode 7. Geotechnical Design. Part 2. Design assisted by laboratory and field testing.

SNiP RK 5.01-03-2002 – Pile foundation. Astana, KAZGOR, 2003.

Boldyrev G.G., Idrisov I.Kh. and Barvashov V.A. (2010). Comparison of Eurocode 7. Part 2 with Russian codes. Similarities and differences. Technical regulation. Engineering investigation, 22–26.

MSP 5.01-101-2003 – Design and installation of pile foundation.

Fadeev A.B., Ilichyev V.A. European code description geotechnical design: summary and coments.

Challenges and Innovations in Geotechnics – Zhussupbekov (Ed.)
 ISBN 978-1-138-03007-7

Stabilization of strain mound road to partially thawing permafrost base using geocells

N.V. Kushnareva, D.O. Grigoriev & S.A. Kudryavtsev
Far Eastern State Transport University, Khabarovsk, Russia

ABSTRACT: The scientific article is devoted to obtaining rational roadbed and array substrates subject to gravitational slope processes that are capable of as soon as possible to stabilize the deformation of the array with the structure of the existing highway. As a result, data analysis, decisions on the efficient operation of ground facilities with the use of geocell, which may contribute to the development of this direction.

1 INTRODUCTION

The high rate of motorization, particularly in the 90 years of the twentieth century, helped to raise attention to the condition of roads and transport facilities.

World practice of the use of synthetic materials in road construction to strengthen the foundations of roads, embankments and slopes recesses cones of bridges and overpasses has more than thirty years of experience.

Active use of plastic geocells in Russia had at the beginning of the 90s. three experimental test site highway, a length of 150 m at different variants bases were built.

Geocell, one of the types of geosynthetics, is a three-dimensional honeycomb structure, made of strips of polyester needle-punched fabric or plastic tape, held together by welds of high strength. When tension in the working plane forms a stable horizontal and vertical frame, which is intended for the reinforcement filler cell geogrid soil crushed stone, gravel, sand and other construction materials.

The great advantage of surround techniques have soil reinforcement, whereby the ground becomes virtually new material with higher mechanical properties. These methods include ways to strengthen the soil volume geocell with vertical walls.

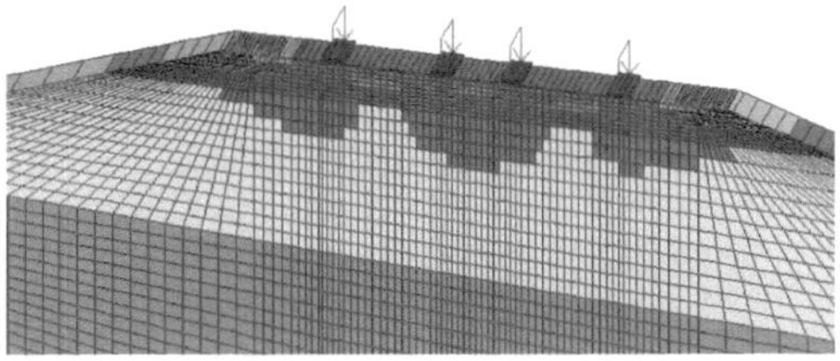

Figure 1. Zones of elastic and plastic deformations.

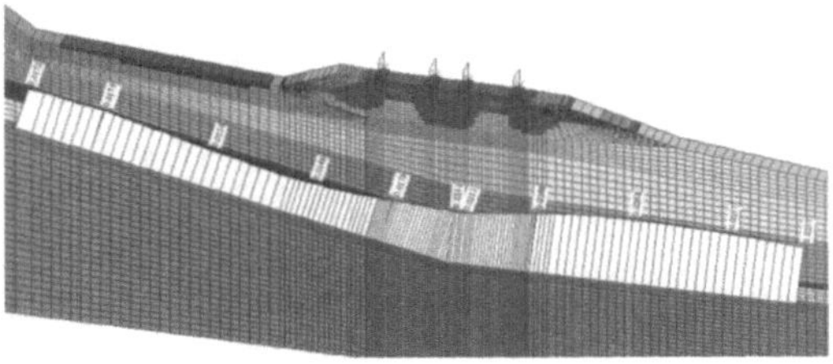

Figure 2. Plot of the horizontal displacement of the soil mass in the layer thawed frozen soil area, m.

2 ANALYSIS OF ENGINEERING GEOLOGICAL CONDITIONS OF CONSTRUCTION

To perform the calculations to determine the stress-strain state and stability of soil mass with structures can be adopted only real physical and mechanical properties of soils, actively participating in the process of shifing.

In an analysis of the original data and the performance of work established that in order to evaluate and compare the performance of strength, deformability is necessary to provide settlement and theoretical studies for existing road structures and substrate conditions, and for the developed designs and constructions. When analyzing the possible methods and means to achieve these goals, it is advisable to use several geotechnical software complexes "FEM models", "Termoground", capable in a complex simulated work facilities for its stress-strain state occurring thermodynamic processes in the annual cycle of freezing – thawing and assessing the sustainability of groundwater arrays and structures.

The analysis of the obtained data on the stress-strain state of elements of structures, nature and quantities of swelling heaving structures and foundation soil.

Figure 2 shows the elastic deformation zone, marked in blue, as well as the zone of plastic deformation in red.

The maximum horizontal displacement of the array in the interlayer zone thawed frozen ground reaches 2.7 cm.

Analysis of horizontal deformations indicative of progressive design on splyve interlayer thawed soil.

In addition, zones of plastic deformations occur in a layer of frozen soil thawed. This means that the entire array of soil prone to slipping on this interlayer.

To reduce or virtually complete stop plastic deformation necessary to take measures to increase road construction bearing capacity.

To reduce the risk of slipping on the ground thawed array interlayer is necessary to provide a set of measures to reduce soil thaw (control of surface runoff and insulation from moisture penetration into the base layer, the use of heat insulators).

Formulation of the problem. For a description of the construction work on the weak thawing grounds at this stage of the design applied elastoplastic model with extreme surface described by the Mohr-Coulomb criterion. The choice of this model due to the fact that its parameters can be taken from existing material standard of engineering and geological surveys. In this formulation, numerical calculations are in good agreement with the traditional engineering methods of calculation of sediment and can adequately describe the deformation structures on weak grounds thawing.

The processes of freezing and thawing in the «Termoground» program module described by the equation of heat conduction with phase transitions of ground water in the range of negative temperatures for transient thermal modes in three-dimensional space unpaved following equation:

$$\rho_d(C_{th(f)} + L_0 \frac{\partial W_w}{\partial T})\frac{\partial T}{\partial t} = \lambda_{th(f)}(\frac{\partial^2 T}{\partial x^2} + \frac{\partial^2 T}{\partial y^2} + \frac{\partial^2 T}{\partial z^2}) + q_v,$$

where $\boldsymbol{C}_{th(f)}$ – the specific heat of melting and frozen ground (Дж/кГ · ^{0}C); ρ_d – density dry soil (кГ/м^3); T – temperature (^{0}C); t – time (c); L_0 – the specific heat of the pre-rotation phase "water – ice" per unit weight – (Дж/кГ); $\lambda_{th(f)}$ – thermal conductivity or melting of frozen soil (Вт/м · ^{0}C); x, y, z – the coordinate (м); q_V – power internal heat sources (Вт/м^3); W_w – unfrozen water moisture in the soil.

The estimation of the parameters of thermodynamic processes of freezing – thawing, the definition of criteria changes temperature and humidity fields in the annual cycle of freezing and borders – thawing periods in the main seasons for recommended designs. The method of mathematical modeling of processes of freezing – thawing buildings and grounds determined absolute values of frost heave on the site for each decade.

Site selection is due to the need to simplify the work of the service of production technology of excavation in the summer to reduce the amount of work to replace the weak foundation soil thawing icy.

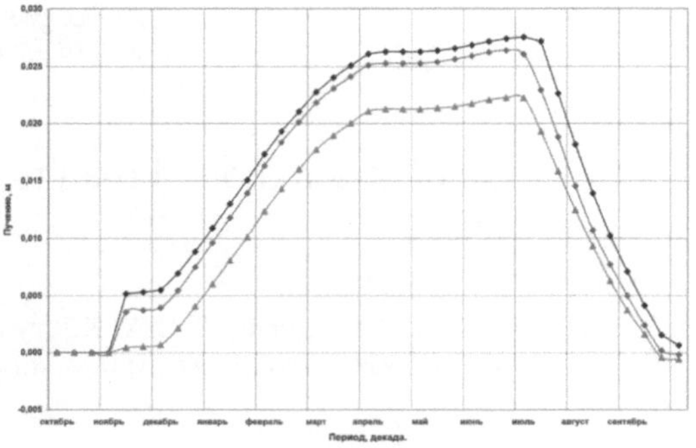

Figure 3. The values of swelling on the decades with a cumulative total in a year cycle design with increased geocell.

For the development of structures used in modern technology and geosynthetics.

On the basis of numerical modeling to establish the stress-strain state "building – the foundation" of the system comply with the decision of non-stationary thermal problems involving registration of phase transitions in the spectrum of negative temperatures. This enabled with sufficient accuracy to determine the parameters of the long-term strength and operational stability of the subgrade and base in cycles of freezing – thawing and under load.

Conclusion. Application in the construction of the reinforcing element in the form of high-strength cage geocheek geosynthetic material eliminates the replacement of the base soil, providing a much more rational indicators deformability and strength of the road construction.

REFERENCES

Kudryavtsev S.A. Numerical modeling of the freeing, frost heaving and thawing soils. Soil Mechanics and Foundation Engineering. 2004. Springer New York Consultants Bureauyu T. 41. No. 5. S. 177–184.

Kudryavtsev SA Settlement and theoretical basis of designing and building structures in a freezing soil heaving. The thesis for the degree of Doctor of Technical Sciences. PSTU. SPb., 2004. 344 p.

SA Kudryavtsev Geotechnical Modeling process of freezing and thawing soils Frost. The reconstruction of cities and geotechnical construction. Book No. 4. St. Petersburg; M .: Publishing House of the DIA, 2004. – 37 p.

SA Kudryavtsev Calculations of the process was frozen, thaw at "Termoground" program. The reconstruction of cities and geotechnical construction. Online Magazine. St. Petersburg, No. 8, 2004. S. 83–97.

SA Kudryavtsev The impact of migration of moisture in the process of seasonal freezing frost heaving soils. The reconstruction of cities and geotechnical construction. Online Magazine. St. Petersburg, No. 7, 2003–2004. S. 233–240.

Kudryavtsev S.A. Berestyanyy U.B., Valtseva T.U., Arshinskaya L.A., Zussupbekov A.Z. Developing design variants while strengthening roadbed with geomaterials and scrap tires on weak soil. Proceedings of the International Workshop on Scrap Tire Derived Geomaterials – Opportunities

and Challenges, IW-TDGM 2007. Yokosuka, 2008. pp. 171–178.
SA Kudryavtsev, VV Pupatenko, Danilyants ES Deformation test and operational capacity. Transport World. 2008. 6. T. number 2 (22). Pp 136–142.
SA Kudryavtsev, birchbark JB, TJ Valtseva Enhancing the way for the passage of heavy trains. Road and track facilities. 2008. No. 1. S. 27–29.
Kudryavtsev S.A., Berestyanyy Y.B., Valtseva T.Y., Goncharova E.D., Mikhailin R.G., Geosynthetical materials in designs of highways in cold regions of Far East. In the collection: Proceedings of the International Conference on Cold Regions Engineering Ser. "Cold Regions Engineering 2009: Cold Regions Impacts on Research, Design, and Construction" 2009. pp. 546–550.

Advanced soil mechanics (TS6)

Challenges and Innovations in Geotechnics – Zhussupbekov (Ed.)
© 2016 Taylor & Francis Group, London, ISBN 978-1-138-03007-7

Long-term durability of soil-bentonite cutoff walls used for containment of ground contamination

A. Takai, T. Katsumi & T. Inui
Kyoto University, Kyoto, Japan

M. Kamon
Research Institute for Environmental Geotechnics, Kyoto, Japan

S. Araki
Raito Kogyo Co., Ltd., Tokyo, Japan

ABSTRACT: Cutoff walls used for prevention of the migration of mobile contaminants in the aquifer must maintain its high barrier performance for a long period. Soil-bentonite (SB), which is a mixture of in situ soil and bentonite, has many advantages as a barrier material such as high softness, high homogeneity, and high material stability because the SB consists only of inorganic soils. To ensure long-term durability of the SB cutoff walls, we have studied factors affecting hydraulic conductivity of SB, self-recovery in the hydraulic conductivity against occurrence of hydraulic fractures, and the role of chemical diffusion in transport of mobile substances. These approaches confirmed that the hydraulic conductivity of the SB is strongly affected by chemicals in pore water and the content of bentonite powder. The effect of chemical diffusion on transport of mobile substances is not negligible because the relative concentration of a chemical substance attained 0.26 after 50 years only by the chemical diffusion when not considering the adsorption onto soil particles.

Keywords: Soil contamination, Hydraulic conductivity, Chemical diffusion, In situ containment

1 INTRODUCTION

The generation of waste materials, toxic chemicals, and other hazardous materials has become one of the most emergent problems to which society should find solutions not only in Japan but in the world. The disposal and dumping of such materials caused geo-environmental problems, typified by contamination in the subsurface ground. For the positive utilization of such contaminated land space especially in small countries such as Japan, various treatment techniques have been developed in these decades. Although excavation of the contaminated soil and subsequent disposal to landfills are one of the most conclusive methods in terms of complete removal of the contamination, with this method, the contaminants are just transferred to a different place. Thus, more reasonable and environmentally friendly countermeasures with minimum transference of the contaminants are required.

In view of this, in situ containment of contaminants using vertical cutoff walls with low-permeable steel or earthen materials is an effective method to prevent the migration of mobile contaminants in the aquifer, as shown in Figure 1 (Evans 1994, Yeo et al. 2009). To ensure strict isolation of the contaminants from surrounding environment, high barrier performance of the cutoff walls must be maintained for a long period. Soil-bentonite (SB) is one of widely used materials since it provides extremely low hydraulic conductivity (k) and high softness (e.g. Grube 1992). Besides, high

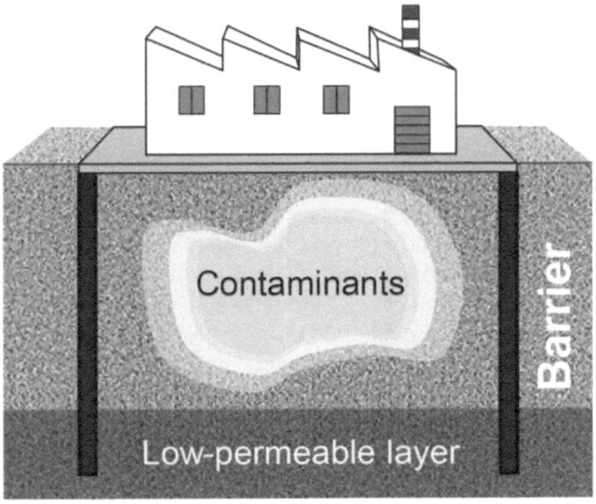

Figure 1. Schematic view of in situ containment of ground contaminants (Takai et al. 2016).

Table 1. Basic properties of soil used.

	Composite soil	Silica sand
Particle density (g/cm^3)	2.72	2.68
Grain size (%)		
Gravel [2 mm –]	5.6	0
Sand [75 μm – 2 mm]	70.8	99.6
Silt & clay [–75 μm]	23.6	0.4
Permeability (m/s)	1.5×10^{-7}	3.9×10^{-5}

homogeneity of the constructed SB cutoff walls can be expected, and material deterioration would not occur because the SB consists only of inorganic soils (Takai et al. 2016).

Low k of the SB attributes to the swelling characteristics of amended bentonite. Since the swelling characteristics of bentonite are strongly affected by chemicals and bentonite cannot swell sufficiently against solutions that have high concentration of inorganic chemicals or nonpolar liquids (Norrish and Quirk 1954), effect of chemicals in the groundwater at a side of concern must be clarified. Chemical diffusion which affects transport of mobile substances in the ground relatively in low-permeable materials such as SB is also important (Shackelford 2014).

This paper discusses the long-term durability of the SB cutoff walls from the viewpoints of factors affecting a k value, self-recovery capability against hydraulic fractures and the role of chemical diffusion in transport of mobile substances. We conducted hydraulic conductivity tests for the SB specimens made with various contents of bentonite powder and chemical concentrations in permeant and in soil pore water. The self-sealing capability of the SB was evaluated with specimens with an interface or a hole in the direction of permeation. The role of chemical diffusion in transport of mobile substances was numerically assessed after determination of some parameters that contribute to the chemical diffusion such as effective porosity and effective diffusion coefficient.

2 HYDRAULIC CONDUCTIVITY OF SB

2.1 *SB preparation*

We used two soils as base soil of SB as summarized in Table 1. Composite soil is a mixture of volcanic cohesive soil and sandy gravel obtained at a pilot scale test site and, and silica sand #7 is a commercially available geomaterial. The composite soil was prepared by mixing sandy gravel and volcanic cohesive soil with a ratio of 25:4 by dry weight according to a boring log at the site. The chemical concentration in pore water was adjusted to the target value by using distilled water, calcium chloride ($CaCl_2$) solutions, artificial seawater and heavy fuel oil A.

In construction procedures of SB cutoff walls by the trench cutting and re-mixing deep wall (TRD) method, first, the trench is cut with supplying bentonite slurry

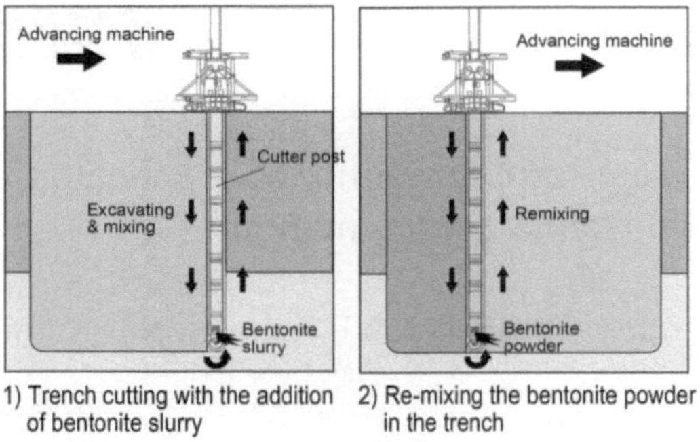

Figure 2. Installation of SB cutoff walls by TRD method (Katsumi et al. 2008).

to maintain the workability of the in situ soil as shown in Figure 2 (e.g. Katsumi et al. 2008). Second, bentonite powder is added to the soil-slurry mixture and re-mixed in the trench after the base machine goes back to the initial position.

To simulate these construction processes in the laboratory, first, water contents of the sandy gravel, volcanic cohesive soil and silica sand were adjusted to 27%, 70% and 26%, respectively. Second, 10%-concentration hydrated bentonite slurry was blended with the soil using a soil mixer for 2 min to achieve 150-mm flow value of the soil-slurry mixture according to JIS R 5201. Finally, bentonite powder was added with proportions of 25 to 150 kg/m^3 and re-mixed using the soil mixer for 2 min.

2.2 *Hydraulic conductivity test using a flexible-wall permeameter*

All testing conditions for the hydraulic conductivity test are summarized in Table 2. In all tests, we used a flexible-wall permeameter shown in Figure 3 according to ASTM D5084 which can minimize leakage along a side wall.

After the SB specimen was pre-consolidated at 39.2 kPa in a consolidation test apparatus and saturated using a vacuum deaerator for at least 2 days, the specimen with a height of approximately 2.0 cm and a diameter of 6.0 cm was placed between deaerated filter papers, geotextiles, and plastic caps with holes to connect the tubes. A confining pressure of 30 kPa and a hydraulic gradient of approximately 50 were applied during permeation.

In Table 2, P-series was designed to assess effect of chemicals in inflow groundwater. In this series, bentonite in SB was initially hydrated with pore water of the in situ ground containing no chemicals. $CaCl_2$ was selected because Ca^{2+} is commonly found in natural aqueous system as well as in water discharged from industrial processes. Seawater and heavy fuel oil A were used to evaluate the performance of the SB near seashore and at an oil contaminated site, respectively. In the N-series, the effect of the content of bentonite powder (C_{BP}) and the chemicals in original pore water

Table 2. Testing conditions and results of hydraulic conductivity test.

Test No.	Soil type	Chemical type in pore water *1	Type of permeant	Bentonite powder content	Hydraulic conductivity
				kg/m^3	m/s
P-1	Composite soil	Distilled water	Distilled water	100	5.0×10^{-11}
P-2			0.25 M $CaCl_2$		2.2×10^{-10}
P-3			1.0 M $CaCl_2$		1.4×10^{-10}
P-4			Seawater		1.2×10^{-10}
P-5			50% ethanol		4.9×10^{-11}
N-1	Composite soil	Distilled water	0.1 M $CaCl_2$	50	2.3×10^{-10}
N-2				100	1.2×10^{-10}
N-3		0.01 M $CaCl_2$	0.1 M $CaCl_2$	100	2.2×10^{-10}
N-4		0.025 M $CaCl_2$	0.1 M $CaCl_2$	50	$\mathbf{1.5 \times 10^{-9}}$
N-5				75	6.8×10^{-10}
N-6				100	5.6×10^{-10}
N-7		0.05 M $CaCl_2$	0.1 M $CaCl_2$	75	$\mathbf{1.1 \times 10^{-9}}$
N-8				100	1.0×10^{-9}
N-9				125	7.4×10^{-10}
N-10		0.1 M $CaCl_2$	0.1 M $CaCl_2$	100	$\mathbf{1.3 \times 10^{-9}}$
N-11				125	$\mathbf{1.1 \times 10^{-9}}$
N-12				150	7.0×10^{-10}
N-13		Artificial seawater	Artificial seawater	100	9.8×10^{-10}
N-14		5,000 mg/kg Heavy fuel oil A	0.1 M $CaCl_2$	100	1.0×10^{-10}
N-15		10.000 mg/kg Heavy fuel oil A	0.1 M $CaCl_2$	100	8.3×10^{-11}
S-1	Silica sand	Distilled water	Distilled water	100	1.1×10^{-10}

*1 $CaCl_2$: concentration in pore water of base soil, Heavy fuel oil A: concentration in parent material (dry weight basis), Seawater: pore water of the parent material was replaced by the seawater.

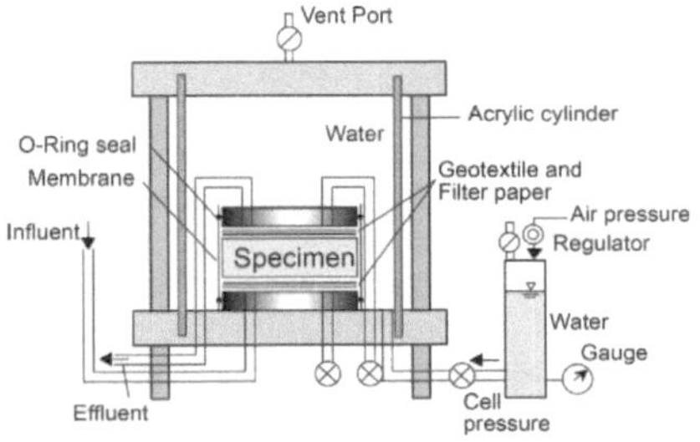

Figure 3. Diagram of a flexible-wall permeameter.

on the k was verified. In this series, swelling of bentonite is impeded due to the chemicals in the soil pore water on site. SB made with silica sand was tested by S-1 to evaluate the effects of the type of base soil.

2.3 *Chemical compatibility of SB*

Figure 4 illustrates the k values for $C_{BP} = 100\,kg/m^3$, affected by the chemicals in the permeant (Kamon et al. 2006). In the case when SB is permeated with 0.1 M (= mol/L) $CaCl_2$ solution, the k value is 3.5 times higher than the case of distilled water. Another finding is that the difference in k is negligible for higher $CaCl_2$ concentration than 0.1 M. SB can achieve its low k of 1.2×10^{-10} m/s even against the sea water. Hydraulic conductivity value against 50%-ethanol permeation is almost equal with that against distilled water permeation. Intrinsic permeability values, which take into account the viscosity of the permeant, are 5.0×10^{-20} m^2 against 50%-ethanol permeation and 3.0×10^{-20} m^2 against distilled water permeation. Therefore, the effect of organic solvent on hydraulic barrier performance is small. These observations confirm that the k of SB does not increase significantly with any permeants if the bentonite in SB is initially well hydrated with soil pore water.

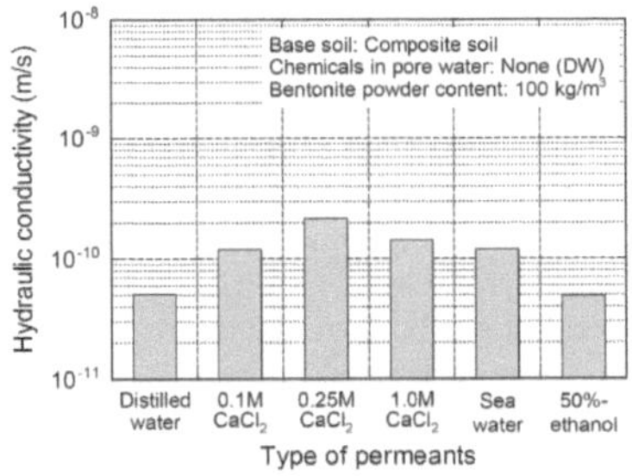

Figure 4. Effects of chemicals in permeant (edited from Takai et al. 2013).

Figure 5 shows the k values of SB ($C_{BP} = 100\,kg/m^3$) for different chemicals in pore water of the original soil. The higher k values were obtained with the higher $CaCl_2$ concentrations. Although the pore water containing calcium ions is diluted by the water fraction of the bentonite slurry, there was a significant increase in k. Hydraulic conductivity values increased linearly

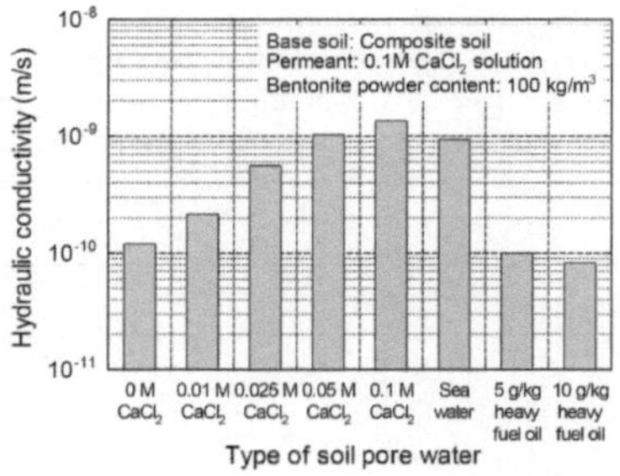

Figure 5. Effects of chemicals in soil pore water (edited from Takai et al. 2013).

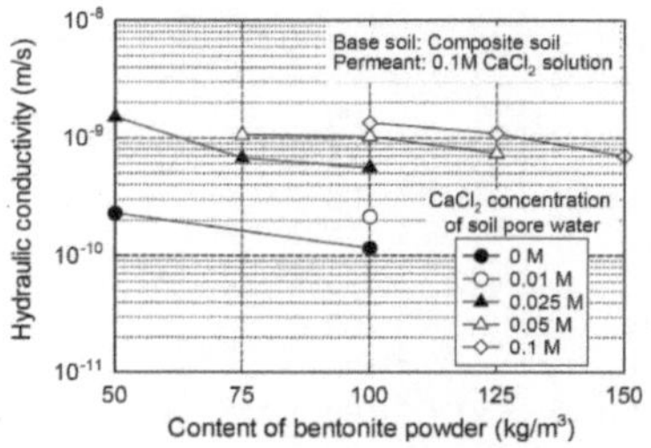

Figure 6. Effects of content of bentonite powder and $CaCl_2$ concentrations in pore water. (edited from Takai et al. 2013).

in association with $CaCl_2$ concentration in the range of 0–0.05 M. However, for concentrations of $CaCl_2$ higher than 0.05 M, it still increased but at a smaller rate. For 0.1 M, the k reached a value higher than 1.0×10^{-9} m/s. When the pore water was replaced by seawater, the k of SB reached the same values as in the case of 0.05 M $CaCl_2$. In the cases that heavy fuel oil is contained in the parent material, SB displays very low k values due to both of the hydrophobicity of oil and the swelling of bentonite.

2.4 *Effect of bentonite powder content*

Figure 6 illustrates the k values in N-series for various C_{BP}. The higher C_{BP} results in obtaining the lower k, even when the $CaCl_2$ concentrations in the pore water were relatively high (0.05 or 0.1 M). Even for the highest $CaCl_2$ concentration of 0.1 M, the k was lower than 1.0×10^{-9} m/s in the case of $C_{BP} = 150\,kg/m^3$. This can be attributed to the shorter distance of montmorillonite layer by adding the larger amount of bentonite powder. Montmorillonite is a mineral and is main component of bentonite. From these observations, when the SB cutoff walls are installed at a site where the inorganic chemical concentration of the groundwater is high, its hydraulic barrier performance can be enhanced by adding a larger amount of bentonite. In the case that the SB was prehydrated ($CaCl_2$ concentration in pore water = 0 M), a k lower than 1.0×10^{-9} m/s was achieved with $50\,kg/m^3$ of powder bentonite for the composite soil used in this study.

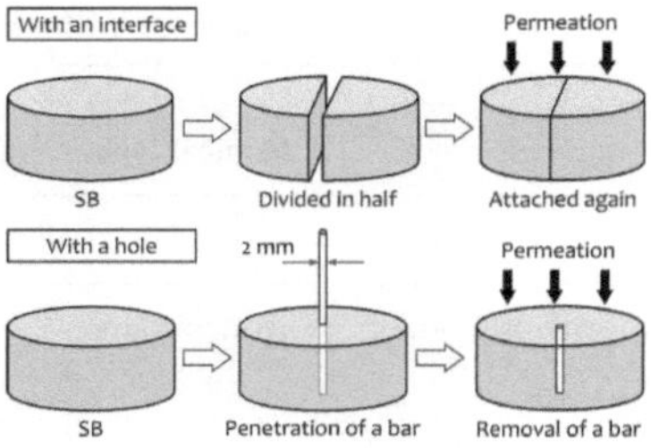

Figure 7. Preparation of specimens with an interface and a hole (edited from Takai et al. 2014).

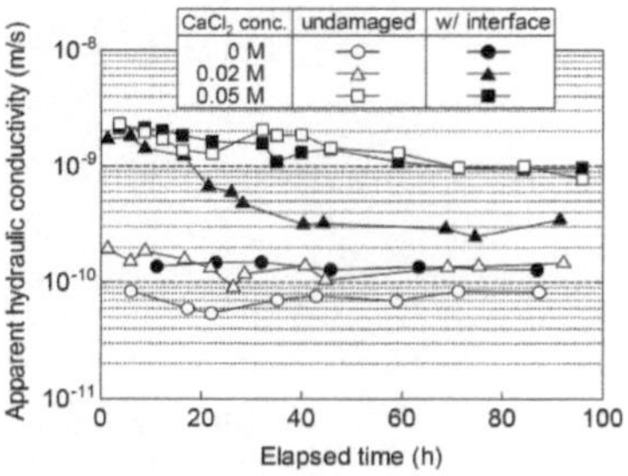

Figure 8. Change in apparent k with and without an interface (edited from Takai et al. 2014).

2.5 *Self-recovery in hydraulic conductivity*

Since SB maintains high softness unlike solidified materials, SB cutoff walls can deform as conforming to adjacent ground. This is an important function of SB as a barrier material because the SB cutoff walls may maintain its soundness even when earthquake occurs. Besides, even if hydraulic fractures occur in the direction of permeation, self-sealing capability of SB attributed to the softness and the swelling of bentonite can be expected. To assess this function, SB specimens with a hydraulic fracture—an interface or a hole—were also subjected to permeation. For the preparation of specimens having an interface, the cylindrical specimen demolded after pre-consolidation was diametrically halved. After the two semicircular specimens were attached again, the specimen was subjected to permeation. To prepare specimens with a hole, a stainless bar with a diameter of 2.0 mm was penetrated at the center of specimen in the direction of the permeation. These procedures for preparation of specimens with a fracture are shown in Figure 7. The k of SBs with a hydraulic fracture was calculated with the section area of specimen ($\approx 28.27\,cm^2$) and the volume of effluent as the apparent k on the assumption of Darcian flow in the specimen.

Figure 8 illustrates temporal changes in apparent k of the specimens with and without an interface. In the case that the $CaCl_2$ concentration is 0.02 M, significant increase in apparent k value was observed at the beginning of permeation when the specimen had an interface. However, even in such case, the apparent k decreased with time and converged within several times to the k of undamaged SB after 40 hours or more.

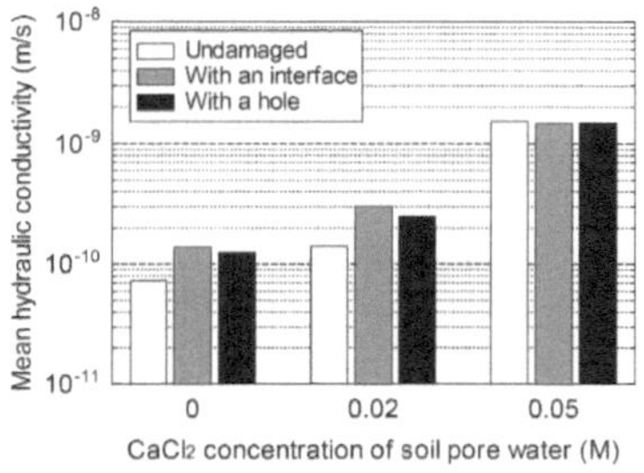

Figure 9. Comparison of mean k with and without a fracture.

Since there was no significant difference between the cases with and without an interface when the $CaCl_2$ concentration of soil pore water was 0.05 M, it can be assumed that the permeated volume through the interface is not much compared to total flux through the specimen when the k of undamaged SB is relatively high. Figure 9 shows mean k after 40 hours for the specimens without a fracture, with an interface and with a hole. As seen in this figure, when $CaCl_2$ concentration in pore water was 0 or 0.02 M, mean k of specimen with an interface and with a hole increased several times higher than those of undamaged SBs. Although this indicates that flow volume through the fractures was relatively larger than that through sound planes, the presence of fractures cause no significant change in flux through the whole domain. These observations confirm that bentonite particles in SB were dispersed into the permeant due to high dispersivity and re-swelled to the permeant within the fractures. Furthermore, the interface was firmly attached and the hole shrank by confining pressure during permeation because of the softness of SB. Thus, even if fractures appear in SB cutoff walls after an earthquake, the SB can recover its k under a certain level of overburden pressure regardless of $CaCl_2$ concentration in pore water at the site of concern.

3 THE ROLE OF CHEMICAL DIFFUSION

3.1 *Measurement of diffusion parameters*

Silica sand-based SB with C_{BP} of 50 kg/m^3 was filled into an acrylic cell with a diameter of 10.0 cm and a height of 1.5 cm to achieve a porosity of 0.47 between porous stones and filter papers. After the specimen was saturated by injecting distilled water from a bottom port, 0.05 M NaCl solution was circulated in a top porous stone with a peristaltic pump to keep the concentration gradient constant. A bottle connected to bottom ports was lowered to apply a water level difference of 90 cm. The mass and the Cl^- concentration of effluent were measured once at 24 h intervals. Cl^- concentration, which is not adsorbed onto soil particles, was measured, and the test was continued for 10 to 15 days. In this study, the mass transport parameters were determined by fitting analytical results to experimental ones regarding the mass flux of Cl^- leaching out from the bottom porous stone. Analytical results were obtained using Dtransu-2D, which is a two-dimensional advection-dispersion analysis program. We used a one-dimensional model which represents the SB specimen and the bottom porous stone in the diffusion test system. Each mesh has 1-mm thickness. Input k was an average one obtained during the test with a constant hydraulic gradient of 20 in consideration of an upward osmotic pressure caused by the concentration difference. Dispersion coefficient was set to 0 because the model was so thin and advective velocity in the SB can be negligible because of low k. A retardation factor of 1 and an attenuation coefficient of 0 were applied because Cl^- is not adsorbed.

A breakthrough curve obtained from the numerical analysis was shifted so that the initial rise in the Cl^- concentration coincides with the experimental one. Based on results of sensitivity analysis, effective diffusion coefficient, D_e, and effective porosity, n_e, of the SBM were determined with 2 significant digits so that the residual sum of squares at the initial three days becomes minimum. As a result, k, n_e and D_e of the silica sand-based SB with C_{BP} of 50 kg/m^3 were calculated to be 5.8×10^{-10} m/s, 0.22, and 3.0×10^{-10} m^2/s, respectively.

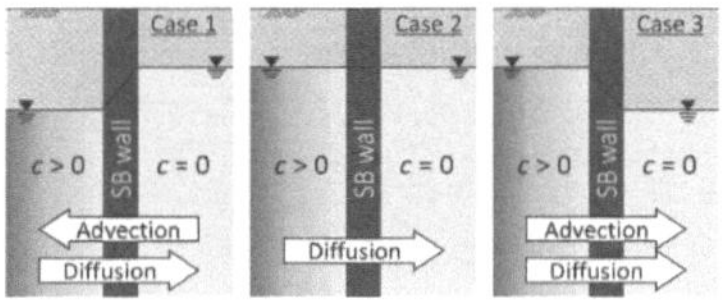

Figure 10. Possible scenarios of groundwater condition.

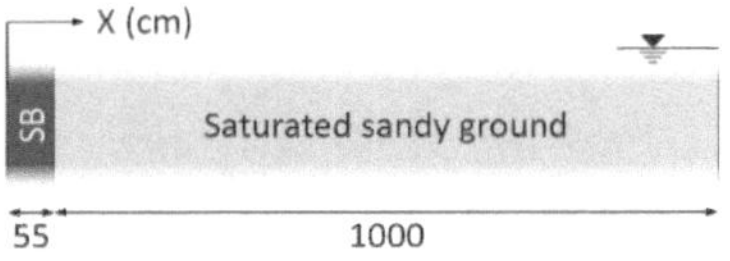

Figure 11. Analytical model used.

3.2 *On site transport of mobile substances*

Figure 10 illustrates three possible scenarios of hydraulic conditions on site which are; 1) advection and diffusion in the opposite direction, 2) only diffusion, and 3) advection and diffusion in the same direction. The chemical diffusion drives mobile substances only toward outside the cutoff walls because the concentration of toxic elements should be relatively high inside the cutoff walls. Horizontal mass transport of substances was numerically evaluated under these conditions by using an analytical model shown in Figure 11, which simulates the SB cutoff wall and an adjacent saturated sandy ground with a horizontal length of 10 m outward from the outer border of the cutoff wall. Thickness of the cutoff wall was 55 cm which corresponds to a typical thickness on site. Sorption of the substances onto soil particles was not considered in this analysis. In the model, the SB

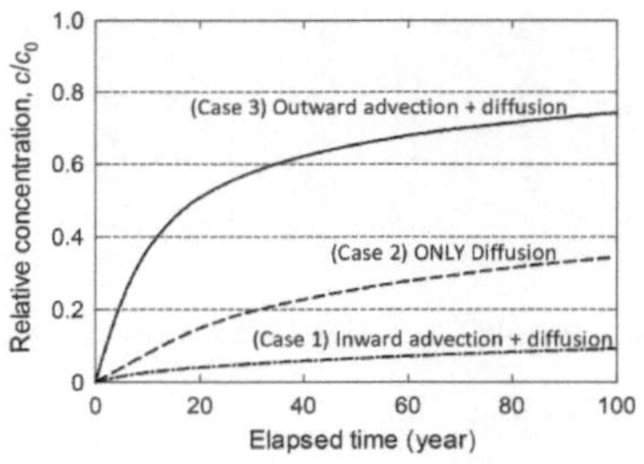

Figure 12. Transport of mobile substances.

cutoff walls and the sandy ground were divided into 1-cm and 10-cm thickness, respectively. A difference in the groundwater level was set to 55 cm, which corresponds to a hydraulic gradient of 1. Input k, D_e and n_e of the SB cutoff wall were 5.8×10^{-10} m/s, 3.0×10^{-10} m^2/s and 0.2, respectively, based on experimental results. The diffusion coefficient of Cl^- ion in water of 2.0×10^{-9} m^2/s was input as the D_e of the sandy ground. n_e of 0.2 and k of 1.0×10^{-5} m/s were used.

Figure 12 shows temporal change in relative concentration, which is a concentration normalized by the inflow concentration, at the outer border of the SB wall. The relative concentration after 50 years was 0.26, 0.66, and 0.07 in case 1, 2, and 3, respectively. Since influence of the advection is relatively large, the transport of contaminants can be restrained by preventing the advection with maintaining the inside groundwater level lower than the outside. Since the relative concentration in 50 years was 0.26 even in the case 1, the influence of diffusion is not negligible on the mass transport. However, since the relative concentration can be lowered to 0.07 when maintaining the groundwater level inside the cutoff wall lower than the one of the outside, control of groundwater level is effective to prevent the migration of mobile contaminants for a long period.

4 CONCLUSIONS

In this paper, long-term durability of SB cutoff walls used for ground contamination was discussed from the viewpoints of factors affecting a k value, self-sealing capability, and the chemical diffusion on transport of mobile substances. The conclusions are summarized as follows:

1) Sufficient hydration of bentonite in SB can maintain a low k against permeation of any chemical solutions, whereas calcium ions in soil pore water have more significant influence. Even when cations exist in pore water with a relatively high concentration, the k can be enhanced by mixing a higher content of bentonite powder.
2) Even if hydraulic fractures appear in the SB cutoff walls after an earthquake, self-sealing hydraulic barrier performance would not be significantly affected under a certain level of overburden pressure regardless of $CaCl_2$ concentration in pore water at the site of concern.
3) Although influence of chemical diffusion is not negligible in transport of mobile substances in the ground, the concentration can be lowered when maintaining the inside groundwater level lower than the outside one.

ACKNOWLEDGMENT

The authors are particularly grateful for the contributions given by former and current students at Kyoto University including Mr. K. Sano, Mr. Y. Nakazawa and Mr. K. Tatsumi.

REFERENCES

Evans, J.C. 1994. Hydraulic conductivity of vertical cutoff walls. In: D.E. Daniel & S.J. Trautwein (eds), *Hydraulic Conductivity and Waste Contaminant Transport in Soils*, ASTM STP 1142, ASTM, Philadelphia: 79–94.

Grube, W.E. 1992. Slurry trench cut-off walls for environmental pollution control. In: D.B. Paul, R.R. Davidson & N.J. Cavalli (eds), *Slurry Walls: Design, Construction, and Quality Control*, ASTM STP 1129, ASTM, Philadelphia: 69–77.

Katsumi, T., Kamon, M., Inui, T. & Araki, S. 2008. Hydraulic barrier performance of SBM cut-off wall constructed by the trench cutting and re-mixing deep wall method. In: M.V. Khire, A.N. Alshawabkeh, & K.R. Reddy (eds), *GeoCongress 2008: Geotechnics of Waste Management and Remediation*, GSP 177. ASCE, New Orleans: 628–635.

Norrish, K. & Quirk, J. 1954. Crystalline swelling of montmorillonite. Use of electrolytes to control swelling. *Nature*, 173, 255–257.

Shackelford, C.D. 2014. The ISSMGE Kerry Rowe Lecture: The role of diffusion in environmental geotechnics, *Canadian Geotechnical Journal*, 51, 1219–1242.

Takai, A., Inui, T., Katsumi, T., Kamon, M., & Araki, S. 2013. Hydraulic barrier performance of soil bentonite mixture cutoff wall, In: M. Manassero, A. Dominijanni, S. Foti, & G. Musso (eds), *Coupled Phenomena in Environmental Geotechnics – From Theoreical and Experimental Research to Practical Applications*, CRC Press, Torino: 707–714.

Takai, A., Inui, T., Katsumi, T., Kamon, M., & Araki, S. 2014. Experimental study on the self-sealing capability of soil-bentonite mixture cutoff walls, In: A. Bouazza, S.T.S. Yuen, & B. Brown (eds), *Proc. of the 7th International Congress on Environmental Geotechnics – Lessons, Learnings & Challenges–*, Engineers Australia, Melbourne: 411–416.

Takai, A., Inui, T. & Katsumi, T. 2016. Evaluating the hydraulic barrier performance of soil-bentonite cutoff walls using the piezocone penetration test. *Soils and Foundations*, 56(2), 277–290.

Yeo, S.S., Shackelford, C.D. & Evans, J.C. 2009. Hydraulic conductivity and compressibility of soil-bentonite backfill amended with activated carbon. *Journal of Geotechnical and Geoenvironmental Engineering*, 135(5): 664–672.

Challenges and Innovations in Geotechnics – Zhussupbekov (Ed.)
 ISBN 978-1-138-03007-7

Strength and deformation characteristics of sand-geosynthetic interfaces through digital image analysis

Prashanth Vangla & Madhavi Latha Gali
Civil Engineering Department, Indian Institute of Science, Bangalore, India

ABSTRACT: Effect of morphological characteristics and size of sand grains and surface features of geosynthetics on the strength and deformation characteristics of sand-geosynthetic interfaces were investigated in this study through digital image analysis. A special symmetric loading interface direct shear test-setup which minimizes the boundary effects of interface testing and allows High Definition (HD) image capturing of interface shearing was developed. Systematic series of interface direct shear tests were performed with sands of different sizes and morphological properties interfacing with different geosynthetics, namely geomembrane, geotextile and wire meshes. The interface shear stress- shear strain response of sands with similar morphological characteristics and different grain sizes revealed that with increase in roughness of geosynthetic materials, friction and dilation angles increase substantially. Further, the effect of aperture ratio of the interfacing materials and the surface roughness of the geosynthetic material were found to be governing the interfacial shear strength to a larger extent.

Keywords: Geosynthetics, Interface, Image analysis, Particle size

1 INTRODUCTION

Strength and deformation characteristics of sand-geosynthetic interfaces hugely depend on the morphological characteristics of sand grains, their size and the surface features of the geosynthetic material in contact. Several researchers demonstrated the importance of understanding morphological characteristics and grain size of sand on its shear behaviour (Rouse *et al* 2008; Göktepe and Sezer 2010). Few studies are also available on the effect of surface features of geosynthetic material on the interface shear behaviour of sand-geosynthetic interfaces (Subaida *et al* 2008; Fuggle and Frost 2010). Though the interfacial strength and deformation are very important for the design of reinforced soil structures, accurate estimation of these parameters through laboratory studies has not been given enough importance in earlier studies. Many existing designs and numerical models follow approximate and empirical correlations for the estimation of these parameters, compromising on the accuracy (Moradi 2014). Laboratory testing of interfaces coupled with digital image techniques and laser based measurement techniques hold great promise for the accurate estimation of strength and deformation characteristics of interfaces. Morphological and grain size characteristics of sand and the interaction of sand and geosynthetics during shearing can be closely captured through digital images and non-contact optical profiling, which can be successfully used for the estimation of interfacial strength and deformation. Limited studies are available on these topics (Yesiller and Cekic 2005; Frost *et al* 2012) and a huge gap exists between the laboratory characterizations and field practice. This paper studies the combined effect of particle morphology and size of sand grains and surface features of geosynthetics on the interfacial shear strength and deformations through most advanced imaging and profiling techniques.

2 MATERIALS USED IN EXPERIMENTS

2.1 *Sand*

Four different sands were used in the experiments. Among these, three sands, namely coarse sand (CS), medium sand (MS) and fine sand (FS) were obtained from scalping specific size fractions from a river sand and the fourth sand, namely angular coarse sand (ACS) was obtained from a local quarry. All sands are classified as poorly graded sands (SP) as per Unified Soil Classification System. Grain size distributions of these sands are presented in Figure 1 and properties of these sands are given in Table 1.

2.2 *Geosynthetics*

Five different types of interfacing materials are used in this study, a geomembrane (GM), a geotextile (GT)

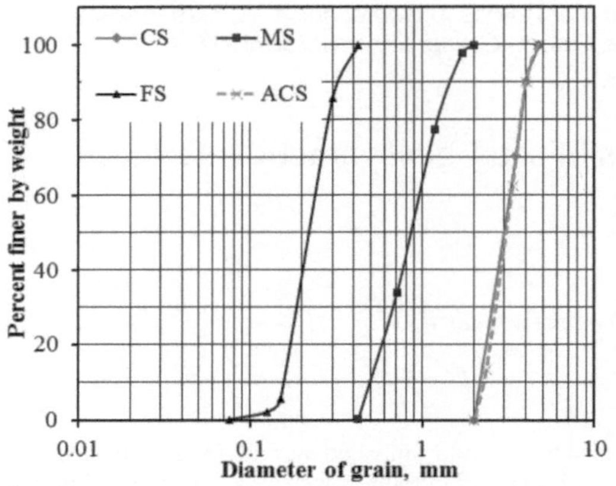

Figure 1. Grain size distributions of sands used in experiments.

Table 1. Properties of the sands used in the study.

Property	CS	MS	FS	ACS
D_{10}	2.18	0.5	0.16	2.29
D_{30}	2.57	0.68	0.19	2.67
D_{50} (mean size)	3	0.87	0.22	3.08
Coefficient of uniformity (C_u)	1.49	1.96	1.51	1.44
Coefficient of curvature (C_c)	0.93	0.97	0.93	0.95
Maximum unit weight γmax, kN/m^3	15.88	16.09	16.05	14.23
Minimum unit weight γmin, kN/m^3	13.96	13.59	13.1	11.73
Maximum void ratio (e_{max})	0.82	0.87	0.95	1.2
Minimum void ratio (e_{min})	0.6	0.58	0.6	0.81

and three wire meshes, namely, WM1, WM2 and WM3 with different asperity spacing (AS) and asperity height (AH). AS is defined as the clear space between the two consecutive strands of the square shaped wire mesh. AH is defined as the height of the topmost point on the surface ridge of the wire mesh from the smooth surface. Properties of these materials determined from laboratory tests are presented in the Table 2. Tensile strength of geotextile and wire meshes were determined as per ASTM D 4595 and ASTM D 4964 respectively and the tensile strength of geomembrane was taken as per the manufacturers report.

3 TESTING METHODOLOGY

3.1 *Morphological characteristics of sand grains through image analysis*

Morphological characteristics of sand grains represent the overall geometry and shape of the sand grains. Some of these parameters considered in this study

Table 2. Properties of geosynthetics used in the study.

Property	GM	GT	WM1	WM2	WM3
Thickness (t), mm	1.5	1	1.2	0.8	0.4
Mass per unit area, g/m^2	1326	118	2400	2152	1254
Tensile strength, kN/m	45	51	39	34	18
Failure strain, %	700	31	12	11	9
Asperity spacing (AS)	–	–	3	0.9	0.2
Asperity height (AH)	–	–	1.2	0.8	0.4

Table 3. Morphological characteristics of sands.

Morphological Descriptor	**Formula**	**Reference**	**ACS**	**CS**	**MS**	**FS**
Roundness (***R***)	$\frac{4\pi A}{P^2}$	Janoo (1998)	0.64	0.74	0.73	0.74
Sphericity (***S***)	$\frac{2\sqrt{\pi A}}{P}$	Sympatec (2008)	0.80	0.86	0.85	0.86
Regularity (***ρ***)	$\frac{(R+S)}{2}$	Cho et al. (2006)	0.72	0.80	0.80	0.80
Pictorial representation of notation and description	***P*** is the perimeter of any horizontal projected section of the particle at rest; ***A*** is the area of the profile of the particle projection					

include roundness, sphericity and regularity. Physical representation of these parameters along with the formulae used for their calculation are presented in Table 3. Recent developments in digital vision along with the availability of powerful image processing software has replaced the ocular based morphological classification and quantitative judgment analysis. In this study, microscopic images of 50 particles of each type of sand at same magnification were taken and used for the determination of their morphological characteristics. This process involves converting the images to binary images and using the pixel information to obtain the geometrical parameters of the grains to calculate the required morphological characteristics as per the formulae given in Table 3. An algorithm was written in MATLAB for the digital image analysis of grains to obtain their morphological characteristics. The morphological characteristics thus determined for all four types of sand are given in Table 3. According to the standard particle shape reference charts compiled by Janoo (1998), sands CS, MS and FS are classified as sub-rounded and sand ACS is classified as angular.

3.2 *Symmetric direct shear test setup*

A symmetric interface direct shear test (SDST) setup which can hold large samples (300 mm length × 300 mm breadth × 160 mm height) was developed for this study. This test setup minimizes the boundary effects of conventional interface testing and allows high definition (HD) image capturing of interface shearing through transparent side wall. The schematic diagram of this test set up is shown in Figure 2. The important feature of this setup is that the

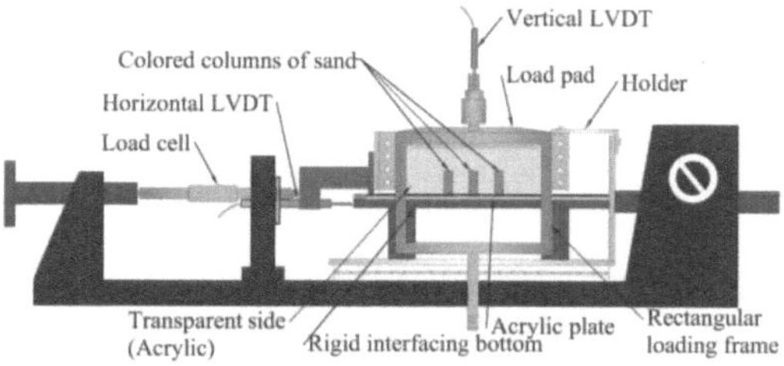

Figure 2. Schematic diagram of modified symmetric interface direct shear test set-up.

soil sample in the shear box is subjected to more symmetric stresses and strains during the test. To avoid the obstruction while capturing images and videos through the transparent side of the shear box during the test, a rectangular supporting frame connected to a vertical rod is provided (Fig. 2). The interfacing materials are fixed to the bottom rigid platform by specially designed clamps to avoid sagging and wrinkling of interfacing planar material during the test. This test setup has a facility to shear the sample up to a larger displacement (75 mm) to understand the post peak behaviour. The set-up is connected to an automatic data acquisition system which obtains the load and deformation data from a load cell and two LVDTs connected to the sample.

3.3 *Sample preparation and testing*

The bottom half of the shear box has flat and smooth surface to fix the planar interfacing materials. While the geomembrane (GM) was directly fixed to the platform with the help of clamps, geotextile (GT) was glued to a thin acrylic sheet prior to fixing it with clamps to the platform of bottom half. Wire meshes (WM) with different aperture sizes were attached to a 1 mm thin sheet of aluminum sheet prior to fixing it with clamps to the platform of bottom half. These arrangements ensure that the interfacing materials do not sag or wrinkle during the test. All test samples were prepared to a relative density of 70%. To study the shear band formations at the interface of the sand and geosynthetics, three equally spaced vertical colored sand columns of 2 mm thickness and 20 mm width and relative density of 70% were created along the transparent side of the shear box with the help of plastic channels (Fig. 2). All tests were carried out at a constant displacement rate of 1 mm/min under three normal stresses (23 kPa, 37 kPa and 53 kPa).

Initially, a series of sand alone direct shear tests were conducted to understand the shearing response of different sands. Systematic series of interface tests were then performed with four different types of sands and five different types of geosynthetics to understand the influence of morphology of the sand and surface features of the geosynthetics on the strength and deformation response of their interfaces. To understand the effect of grain size, first series of interface tests were performed on sands of similar morphological characteristics and different grain sizes (D_{50} varying from 3 mm to 0.22 mm). GT and GM were used as interface materials in these tests. The next series of tests focused on understanding the effect of relative size of sand grains with respect to the aperture size or asperities of interfacing materials by conducting tests with coarse sand and three wire meshes of different aperture sizes, 3 mm, 0.9 mm and 0.2 mm. In the next series of tests, effect of morphological characteristics of grains was investigated by conducting interface tests with two coarse sands (CS, ACS) of similar grain sizes but different morphological characteristics. GM was used as interfacing material in these tests to obtain better representation of grain impressions on the surface while shearing.

4 TEST RESULTS AND DISCUSSION

Friction angles obtained from sand alone tests were 40.76° for coarse sand (CS), 40.62° for medium sand (MS), 40.23° for fine sand (FS) and 35.84° for angular coarse sand (ACS). Though the grain sizes of CS, MS and FS are different, there is not much variation in their peak friction angles. One of the reasons for this is that the void ratios for these samples are almost same (about 0.7). Literature suggests that particle size does not have any effect on the peak friction angle if the void ratio is same (Holtz and Kovacs, 1981). Difference in morphological characteristics can alter the friction angles of sands of same grain size and void ratios. When the morphological characteristics of these sands are analyzed, it was found that these three sands have same roundness, sphericity and regularity (Table 3), supporting the similar friction angles obtained for these sands. The comparatively lower friction angle obtained for ACS is due to its relatively higher void ratio (0.928) at 70% relative density.

4.1 *Effect of grain size on interface shear behaviour*

Three sands (CS, MS and FS) of different grain sizes and similar morphological characteristics as described earlier were selected for understanding the effect of grain size on the strength and deformation response of these sands interfacing with GT and GM. Results of sands alone and interface shear tests for all three sands (CS, MS, FS) with GT and GM under a normal stress of 53 kPa is presented in Figure 3. Figures 3a–c show the variation of friction and dilation angles during shearing for CS, MS and FS respectively. Interface shear tests with GM did not show any dilation because the surface was smooth without any asperities. Among these three sands, coarse sand with GT and GM showed lowest interfacial strength (Fig. 3d), because coarse sand could not establish sufficient interlocking with asperities of GT because the grains are bigger than the asperities present in GT. Further tests were conducted to understand this interlocking effect using wire meshes of different aperture sizes, which are explained in latter sections. When deformations were compared in terms of dilation, coarse sand showed lower dilation angle in sand alone as well as interface shear tests with GT compared to medium and fine sands. This observation is against the general perception that the coarser the sand, higher the friction and dilation angles will be. Analysis of geomembrane after the test through microscopic images revealed that the effective contacts (contacts that cause grooves of considerable depth) per unit area in coarse sand were less compared to those in medium and fine sands because of bigger size particles in CS. MS and FS showed highest interfacial friction (δ) because of increased number of contacts per unit area and effective contacts, though the groove depths are less in case of FS.

Figure 3 also brings out the effect of surface roughness of the geosynthetic material on the interfacial shear behaviour of sands of different grain sizes. Root mean square roughness (S_q), which is defined as square root of the sum of the squares of the individual heights and depths of the profile from the mean line was determined as 0.6 μm for geotextile and 0.06 μm for geomembrane using non-contact laser based 3D optical profilometer. Comparison of interfacial shear response with GT and GM for any specific sand showed that sand-GT interfaces showed higher friction and dilation angles for all cases because of higher roughness of GT (Fig. 3).

In direct shear tests, failure and strain localization are often related to the formation of shear bands. Several earlier researchers correlated the shear band thickness to the failure strain (Cerato and Lutenegger, 2006). Thickness of shear bands formed in the interface shear tests was determined through image segmentation technique, which is based on the principle of pixel classification, as explained by Vangla and Latha (2014). Videos of deforming colored sand columns were captured and the images at regular displacement intervals (about 0.1 mm) were extracted and used to find the shear band thickness. Shear bands were not formed in case of geomembrane because of its smooth surface and non-dilative behaviour but tests with sand-GT interfaces showed clear shear band formation. Figure 4 shows the deformed (20 mm) and initial (0 mm) profiles of colored sand for CS-GT, MS-GT and FS-GT traced and overlapped one over the other so as to see the net deformation and shear band thickness (t). As observed from Figure 4 that the thickness of shear bands is proportional to the grain size of sand. Normalized thickness of shear bands (t/D_{50}) for CS, MS and FS is 3.31, 8.17 and 14.50 respectively, indicating that t/D_{50} values decrease with the increase in D_{50}, which is in agreement with the observations of earlier researchers.

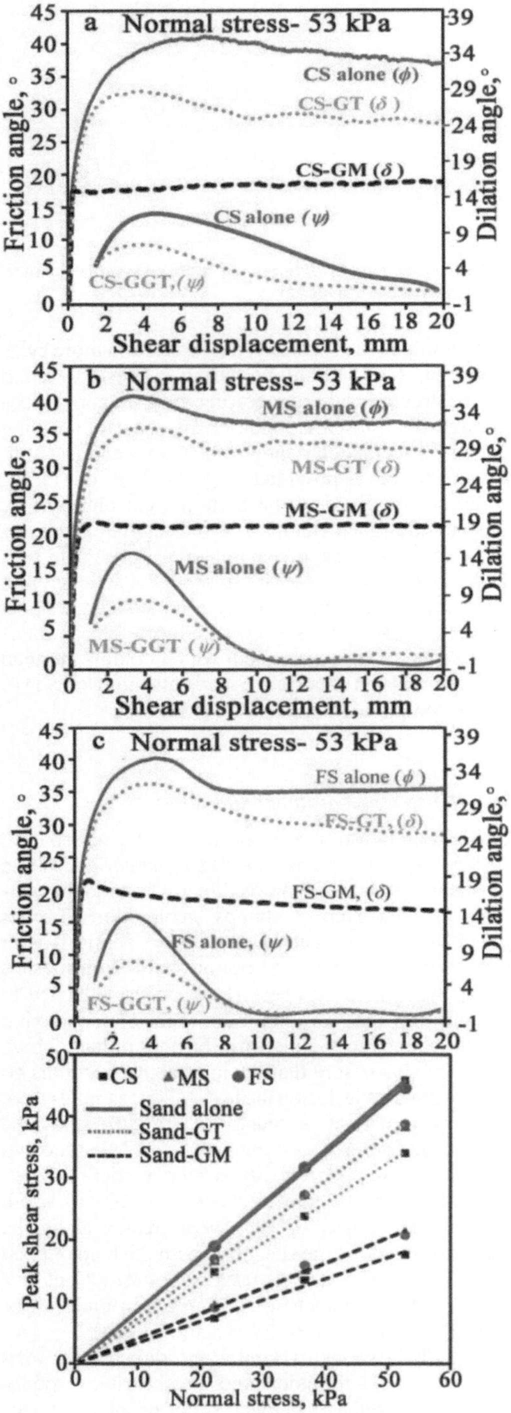

Figure 3. Results of direct shear tests with sands of different grain sizes interfacing with GT and GM. a), b) and c) Variation of friction and dilation angles with displacement for CS, MS and FS d) Failure lines for different tests.

4.2 *Effect of aperture ratio on the interface shear behaviour*

Athanasopoulos (1993) and Lopes et al. (2001) clearly explained the effect of relative size of geosynthetic asperities/aperture and the grain size of sands on their

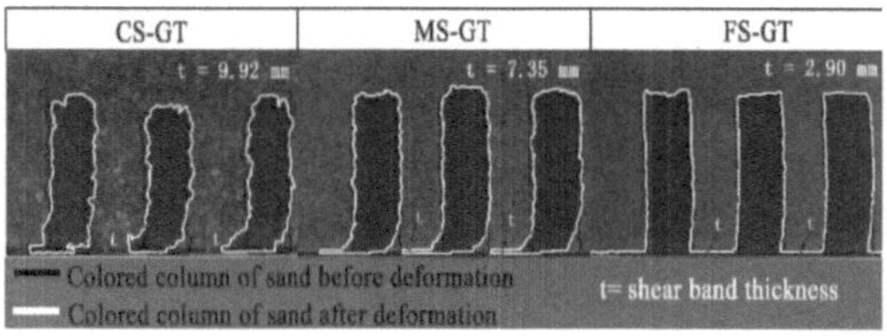

Figure 4. Estimation of shear band thickness from initial and final profiles of colored sand columns

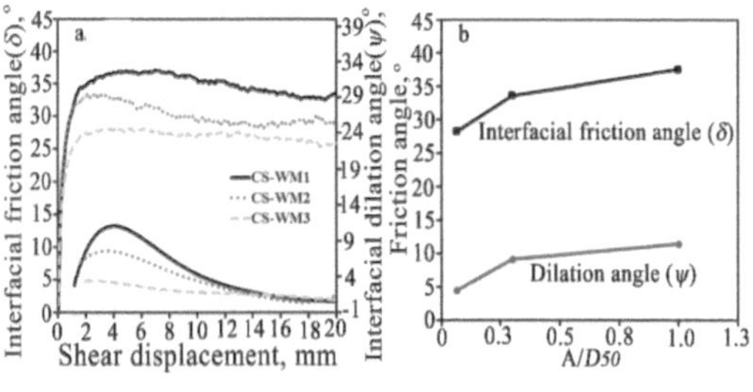

Figure 5. a) Variation of friction angle and dilation angle with shear displacement for sand-WM interfaces b) Variation of interfacial friction angle and dilation angle with aperture ratio.

interfacial shear behaviour. From the interface shear tests on sand-GT interfaces, it was clear that coarse sand could not yield higher friction angles as it could not establish enough interlocks with the asperities present in GT because of the bigger grain size compared to the size of asperities. To understand the effect of aperture ratio (A/D_{50}), which is defined as the ratio of aperture size (A) of the interfacing material to the median grain size (D_{50}), interface tests were carried with wire meshes (WM1, WM2 and WM3) of different aperture sizes. For coarse sand, the aperture ratios were calculated as 0.07, 0.3 and 1. Variation of friction and dilation angles with displacement for sand-WM interfaces of different wire meshes are presented in Figure 5. Figure 5b shows the variation of friction and dilation angles with aperture ratios. With the aperture size getting closer to the D_{50} of the grains, friction and dilation angles increased due to effective interlocking, highest values measured for an aperture ratio of 1.

Results of tests with varying aperture ratio also include the effect of roughness of the wire meshes, which is different for different aperture sizes. An analytical expression was developed to determine the surface roughness of the wire meshes as it is difficult to obtain the roughness of wire meshes through 3D optical profilometer like in case of GT and GM due to the less reflective index of WM and difficulties in preparation of representative sample due to its structure. Figure 6 shows the schematic representation of wire mesh and the parameters used to calculate the surface roughness. In the figure, l represents the clear spacing between the strands, which is same in both x and y directions, r is the radius of the strand, which is uniform for a mesh, θ is the apex angle of the

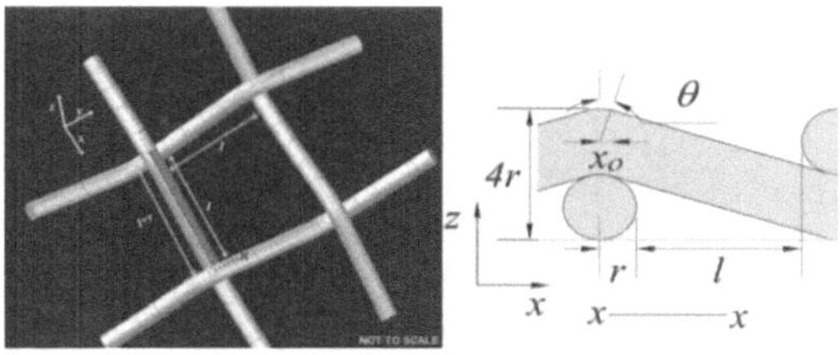

Figure 6. Schematic diagram of wire mesh showing a symmetric segment used for deriving the analytical expression.

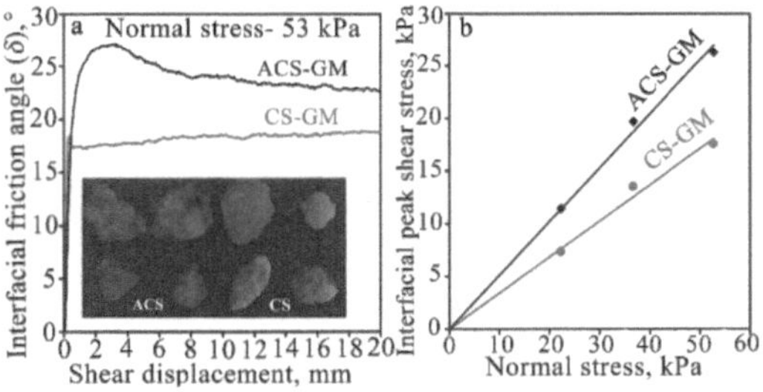

Figure 7. a) Variation of friction angle with displacement for CS-GM and ACS-GM interfaces b) Failure lines for CS-GM and ACS-GM interfaces.

mesh as shown in Figure 6 and x_o is the length of the curved portion of the strand. Root mean square roughness (S_q) for the wire mesh, which is same as the S_q of the element shown in the figure, can be obtained as follows. The S_q values calculated for wire meshes of clear spacing between the strands (l) as 0.2 mm, 0.9 mm and 3 mm with radius of strands (r) as 0.1 mm, 0.2 mm and 0.3 mm respectively are 141 μm, 305 μm and 408 μm respectively. From Figure 5, it is evident that increase in roughness of the mesh resulted in higher friction and dilation angles.

$$S_q = \sqrt{\frac{1}{Ae}\left(4 \times \int_0^r \int_0^{l+r} (Z(x,y) - \bar{Z})^2\, dxdy + \int_0^l \int_0^l (-\bar{Z})^2 dxdy\right)} \quad (1)$$

where Ae is the area of the element, $Z(x,y)$ is the profile height from the base and $Z(x,y)$ is the average height of the area profile.

$$Z(x,y) = Z_1(x,y) + Z_2(x,y) \quad (2)$$

Where, $Z_1(x,y) = 2r + \sqrt{r^2 - x^2} + \sqrt{r^2 - y^2}, x\epsilon(0, x_0)$ (3)

$$Z_2(x,y) = 2r + tan\theta\left(\frac{l}{2} + r\right) + \sqrt{r^2 - y^2} - xtan\theta, \quad x\epsilon(x_0, l + r), y\epsilon(0, r) \quad (4)$$

$$\bar{Z} = \frac{1}{\mathrm{Ae}}\left(4 \times \int_0^{\mathrm{r}} \int_0^{\mathrm{l+r}} Z(x,y)\, \mathrm{dxdy}\right) \quad (5)$$

Also, $sin\theta = \frac{2r}{l+2r}$; $x_0 = rsin\theta$; $A_e = (l + 2r)^2$ (6)

4.3 *Effect of morphology of sand grains on the interface shear behaviour*

To study the effect of morphology of sands on the interfacial strength, tests were carried out with coarse and

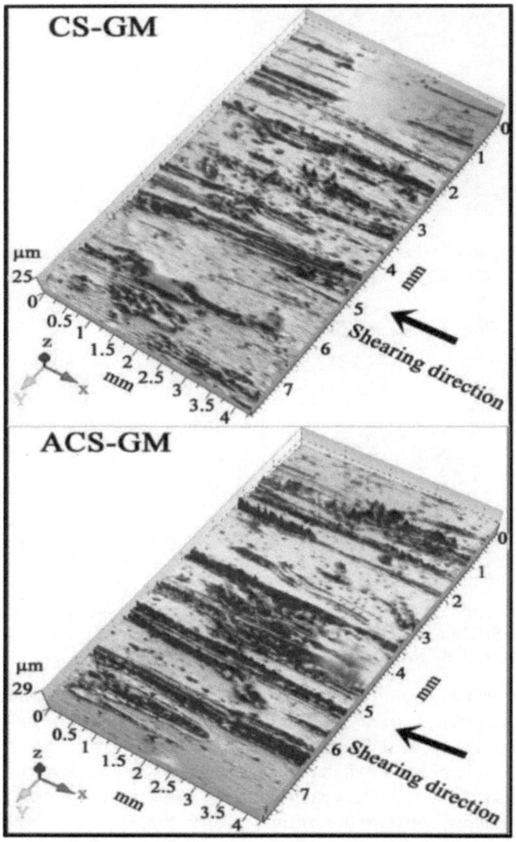

Figure 8. 3D view of surface topography of exhumed specimens of geomembrane from CS-GM and ACS-GM tests.

angular coarse sands, whose grain sizes are same and the morphological characteristics that include roundness, sphericity and regularity are different. Table 3 presented the morphological characteristics of these two sands. Figure 7 presents the interfacial shear behaviour of these two sands with the geomembrane at a normal stress of 53 kPa. With the increase in the roundness from 0.64–0.74 for ACS and CS, which represents reduction in angularity, the peak interfacial shear stresses decreased, as observed from the failure lines plotted in Figure 7b. Further, damage analysis of exhumed geomembranes from tests at 53 kPa with ACS and CS was carried out using laser based 3D optical profilometer on representative strips of size 7.5 mm × 4.5 mm. The surface topographies of geomembranes indented by angular sand (ACS) and sub-rounded sand (CS) are shown in Figure 8.

The average S_q values for the surfaces of geomembranes indented by ACS and CS obtained from the area profiling of surface topographies were 2.11 μm and 1.41 μm respectively. It can be seen that the angular sand particles could penetrate deeper into the geomembrane, causing deeper grooves formed on the surface, representing more effective contacts per unit area compared to sub-rounded coarse sand, which is the reason for angular sands showing higher friction angle.

5 SUMMARY AND CONCLUSIONS

A large symmetric interface direct shear test set-up that enables video capturing was developed for the study. Systematic series of sand alone and interface shear tests were carried out on sands of different grain sizes and morphological characteristics and geosynthetics of different surface roughness. Advanced imaging and profiling techniques were used for material characterization and for post processing of results. Effects of grain size and morphological characteristics of sand, aperture ratio and surface roughness of geosynthetic material on the strength and deformation response of various sand-geosynthetic interfaces were studied. Friction angles were not affected by the grain size for sands of similar morphology tested at same void ratio. However, interfacial friction angles of these sands with geotextile and geomembrane were quite different and depended on the effective contacts per unit area and effective interlocking governed by asperities of geosynthetics. Interface friction angles increased with the surface roughness of the geosynthetic material and decrease in the roundness of the sand grains. Analysis of shear bands and 3D topographical profiles of exhumed geosynthetics provided important insights into the strength and deformation response of sand-geosynthetic interfaces.

REFERENCES

Athanasopoulos, G. A. 1993. Effect of particle size on the mechanical behaviour of sand-geotextile composites. *Geotextiles and Geomembranes* 12(3): 255–273.

Cerato, A. B. & Lutenegger, A. L. 2006. Specimen Size and Scale Effects of Direct Shear Box Tests of Sands. *ASTM geotechnical testing journal* 29(6): 507–516.

Cho, G. C., Dodds, J., & Santamarina, J. C. 2006. Particle shape effects on packing density, stiffness, and strength: natural and crushed sands. *Journal of Geotechnical and Geoenvironmental Engineering* 132(5): 591–602.

Frost, J. D., Kim, D., & Lee, S. W. 2012. Microscale geomembrane-granular material interactions. *KSCE Journal of Civil Engineering* 16(1), 79–92.

Fuggle, A. R., & Frost, J. D. 2010, April. Particle size effects in interface shear behavior and geomembrane wear. In *Proceedings of International Symposium on Characterization and Behavior of Interfaces* (pp. 51–57), Atlanta, IOS.

Göktepe, A. B., & Sezer, A. 2010. Effect of particle shape on density and permeability of sands. *Proceedings of the Institution of Civil Engineers-Geotechnical Engineering* 163(6), 307–320.

Holtz, R. D., & Kovacs, W. D. 1981. *An introduction to geotechnical engineering*. Prentice-Hall, Inc Englewood cliffs, New Jersey.

Janoo, V. 1998. Quantification of shape, angularity and surface texture of base coarse materials. Cold Regions Research and Engineering Laboratory. *US Army Corps*

of Engineers, Vermont Agency of Transportation, Special Report, 98–101.
Lopes, P. C., Lopes, M. L., & Lopes, M. P. 2001. Shear behaviour of geosynthetics in the inclined plane test–influence of soil particle size and geosynthetic structure. *Geosynthetics International* 8(4): 327–342.
Moradi, G. 2014. Seismic Response Analysis of Geosynthetic Reinforced Soil Retaining Wall. *Electronic Journal of Geotechnical Engineering.*, 19: 3819–3835.
Rousé, P. C., Fannin, R. J., & Shuttle, D. A. 2008. Influence of roundness on the void ratio and strength of uniform sand. *Géotechnique* 58(3): 227–231.
Subaida, E. A., Chandrakaran, S., & Sankar, N. 2008. Experimental investigations on tensile and pullout behaviour of woven coir geotextiles. *Geotextiles and Geomembranes* 26(5): 384–392.
Vangla, P., & Latha, G. M. 2014. Image segmentation technique to analyze deformation profiles in different direct shear tests. *ASTM geotechnical testing journal* 37(5): 828–839.
Yesiller, N., & Cekic, A. 2005. Determination of surface and thickness characteristics of textured geomembranes using Image analysis. *Geotechnical Testing Journal*, 28(3): 275–287.

Challenges and Innovations in Geotechnics – Zhussupbekov (Ed.)
 ISBN 978-1-138-03007-7

Analysis of stress-strain state of anisotropic soil basement

L.V. Nuzhdin & K.V. Pavlyuk
Novosibirsk State University of Architecture and Civil Engineering (Sibstrin), Novosibirsk, Russia

ABSTRACT: The article analyses the results of the numerical experiment of the stress-strain state of the soil basement folded by anisotropic soils using the ANSYS software system. Since one of the objectives of the study is to evaluate the possibility of using this software in the designing of the foundations and basements and improve the forecast accuracy of the foundation settlement according to the Rules and Regulations SP 22.13330.2011 "Foundations of buildings and structures" (Russia), we considered different calculation schemes "foundation – basement".

It should be noted that we testedin advance the effect of Poisson's ratio and modulus of soil deformation on the stress-deformed state of isotropic and anisotropic soil basement. By analyzing the results of thenumerical experiments, we derived correction coefficients for calculation of foundation settlement by layer-by-layer summation according to the Rules and Regulations SP 22.13330.2011, which consider the anisotropic properties of soil and allow improving the forecast accuracy.

Keywords: stress-strain state, soil basement, ANSYS, stress, deformations, anisotropic soil, isotropic soil.

1 INTRODUCTION

An advance of calculation methods of soil basement and foundations plays an important part in indicators increase of construction efficiency, as their processing may constitute up to 20% of total architecture building cost. Considering that today's construction trends show growth of building's storey numbers, erection of unique and vast buildings that are made sometimes in complicated geological engineering conditions, it is necessary to take a full account of soil basement actual properties. Thanks to many researches done by specialists in this field, it is known that almost all non-rocky basements have more or less anisotropic properties. Despite this, following actual construction engineering regulations for the territory of Russia SP 22.13330.2011 "Foundations of buildings and structures" when designing a soil basement for second category of limit states, deformational anisotropy is not taken into account. As far as current methods of forecasting stress-strain of soil basements and foundation settlements cannot result in a proper evaluation of ground mass behavior, it is necessary to improve the methods and increase the accuracy of the calculations.

2 ANALYSIS OF STRESS-STRAIN STATE OF SOIL BASEMENT

To evaluate a stress-strain state of soil basement, ANSYS software system was used. Calculations were carried out using a model ofelastic linear material strain. Strain properties of environment are determined by strain module E and Poisson's ratio μ. As under side foundation pressure p was assumed as equal to reference resistance of soil R, when determining the strain, occurrence and development of plastic strain was not taken into account. Two types of foundations were concerned: pier foundation with plain dimensions 3×3 m and continuous foundation with underside width $b = 3$ m. Assumed as homogeneous: clayey soil with module $E = 10$ MPa. Dimensions of ground mass – $20 \times 20 \times 40$ m. Load is assumed as evenly distributed – 233.7 kPa.

2.1 *Influence of various interaction schemes of foundation-basement system*

As one of the research purposes is an evaluation of possible use of the software when designing basements and foundations, several tasks of various interaction schemes basement-foundation were concerned. It is known that if foundation hardness is concerned, there is stress redistribution on foundation underside with endlessly big contact stress concentrated on foundation circuit. It is worth noting that calculated in this way, strain values of soil basement are close tosettlement values, determined using layer-by-layer summing method.Results of foundation settlement calculations, determined using different methods are given in Table 1.

Data received indicate a rather good coincidence of settlement values calculated using layer-by-layer

Table 1. Results of foundation settlement calculations.

	Settlement S, sm	
Calculation method	Pier foundation	Tape foundation
Layer-by-layer summing method	4.54	7.12
ANSYS (hard)	4.02	7.41
ANSYS (flexible)	5.7	9.73

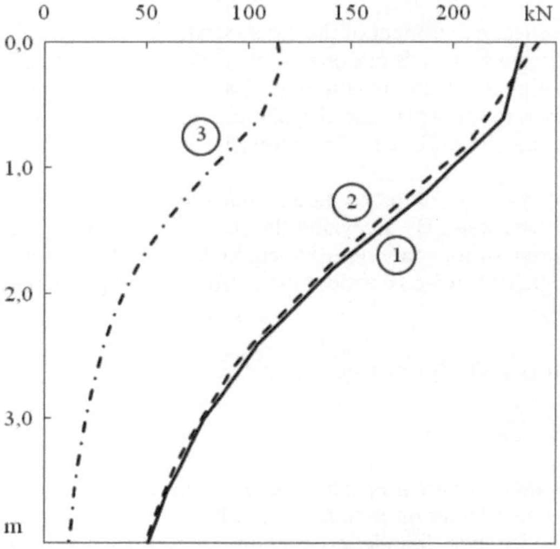

Figure 1. Scheme of stress distribution underside of externally loaded pier foundation, dimensions $A = 3 \times 3$ m: 1 – according to layer-by-layer summing method (SP); 2 – results calculated in ANSYS (flexible foundation); 3 – results calculated in ANSYS (hard foundation).

summing (SP) and ANSYS software system, including foundation hardness. Settlement difference of pier foundation is 11.5%, of continuous foundation – 3.9%. Determination of settlementof flexible foundation is incorrect.

For flexible foundation, results are evenly distributed stress to depth, most corresponding to scheme of linear strained semi-space, suggested in SP 22.13330.2011 "Foundations of buildings and structures". The greatest divergence of stress of soil basement with hard and flexible foundation may be observed in area close to foundation underside. Results of strain to depth distribution are given in Figure 1.

It should be noted that for continuous foundation with isotropic soil in its basement analysis of stress-strain state beside a 3D solution there is a plain 2D setting. Calculation results are in accordance with the data received and published by famous researchers earlier, see Figure 2.

Analysis of the data received indicates that flat task solution give sin creased values of soil basement stress compared to volume task solution (up to 43.7%). A good coincidence of results was received

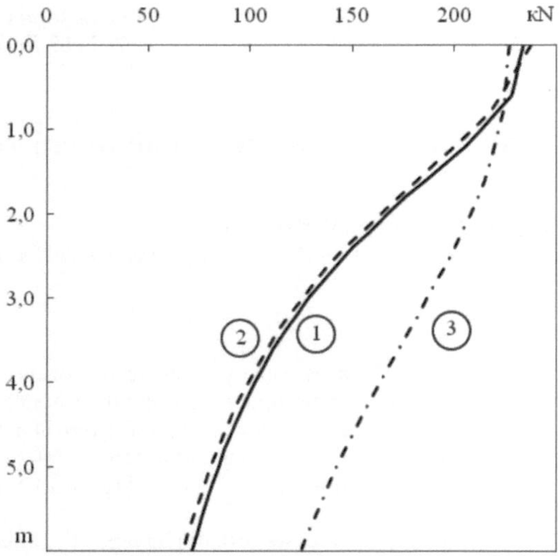

Figure 2. Results of stress calculations of continuous foundation soil basement $b = 3$ m: 1 – according to layer-by-layer summing method (SP); 2 – results calculated in ANSYS (volume task); 3 – results calculated in ANSYS (flat strain).

for volume task solution with data of layer-by-layer summing method, recommended SP for calculations of foundations basements of buildings and structures.

2.2 *Effect of Poisson's ratio and deformation modulus*

It is worth noting that besides the above said, an effect of Poisson's ratio and deformation modulus were tested for stress-strain state of isotropic and anisotropic soil basement. Analysis of data received indicates that Poisson's ratio has a sufficient influence on calculation results, while regulatory methods of the settlement forecasting and evaluation do not take it into account. Given Poisson's ratio $\mu = 0.42$ stress values were received corresponding mostly to reference scheme of linear-stressed semi-space, stated in designing regulations. In percentage, difference of results received makes up to 30.8% for pier and up to 45.3% for continuous foundations. It should be noted that Poisson's ratio has influence on deformation of soil basement. When $\mu = 0.42$ there are minimal values of settlement both for pier, as well as for continuous foundation. Difference of settlements makes up to 1.13 times for pier foundation and up to 1.2 for continuous foundation.

Deformation modulus, as expected, does not influence stress distribution in ground mass.

2.3 *Stress-strain anisotropy influence of stress distribution*

For evaluation of stress-strain state of anisotropic soil basement transverse-isotropic environment with given degree of anisotropy is concerned. Degree of stress-strain anisotropy was evaluated based on ratio

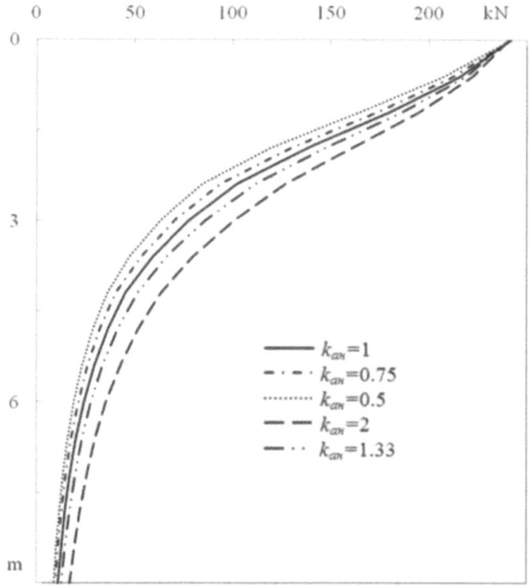

Figure 3. Stress distribution under centre of pier foundation A = 3 × 3 m including stress-strain anisotropy of soil basement.

of deformation modules in vertical E_z and horizontal E_x directions. Taking into account results of many years of experimental researches, conducted by department of geological engineering, soil basements and foundations of the Novosibirsk state architecture and civil engineering university (Sibstrin) (Nuzhdin et al. 2014) a range of most probable indicator of anisotropy $k_a = E_z/E_x$ is assumed from 0.5 to 2.

Values of Poisson's ratio μ assumed as equal in horizontal and vertical directions.

Values of shift modulus G_{xz} may be determined by formula 1 (Ambartsumyan, 1982):

$$G_{xz} = \frac{E_x E_z}{E_x(1+\mu_z) + E_z(1+\mu_x)} \tag{1}$$

where E_zиE_x = deformation modulus of vertical and horizontal directions; μ_zиμ_x = Poisson's ratio correspondingly.

Calculation results of stress-strain state of anisotropicsoil basement are given in Figure 3.

Resulting from calculation experiments, transverse-isotropic environments were concerned with coefficients of stress-strain anisotropy $k_a = 0.5$; 0.75; 1.33 and 2. Here stress to depth distribution were concerned for continuous foundations, square, rectangular and round in plain view. Given below are stress distributions for isotropic and anisotropic soil basement of pier foundations (see Figure 4,5).

Based of processed results correction coefficients were received α_a for determination of normal stress from external loading anisotropic soil basement including different deformability of soil in vertical and horizontal dimensions. Stress calculation should be

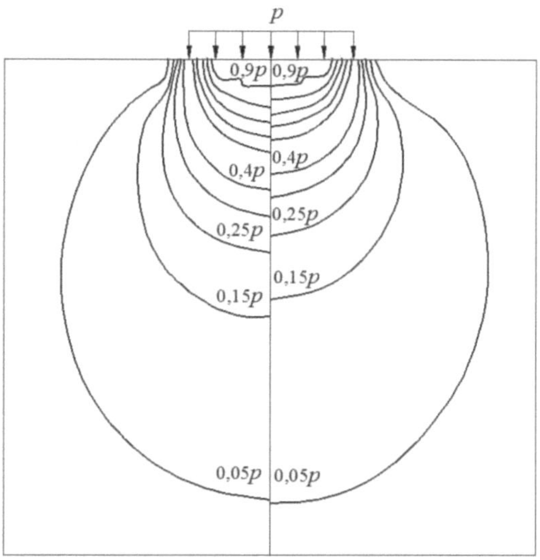

Figure 4. Stress distribution under side of foundation with isotropic (left) and anisotropic $k_{an} = 0.5$ (right).

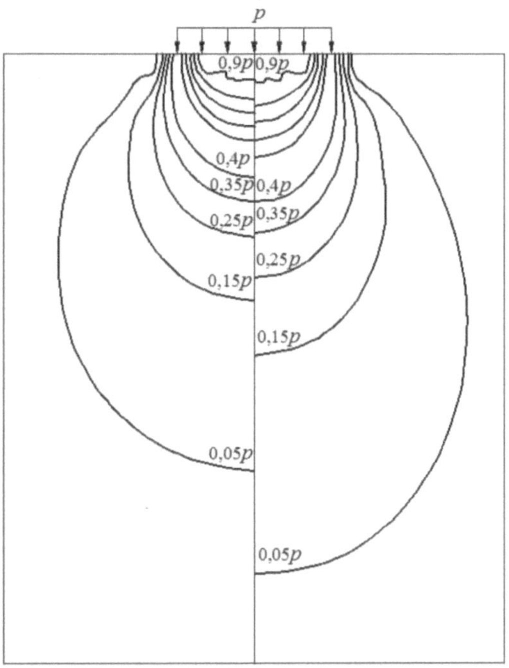

Figure 5. Stress distribution under side of foundation in isotropic (left) and anisotropic $k_{an} = 2$ (right) soils.

determined according to scheme of linear-deformed semi-space with additional coefficient in the formula:

$$\sigma_{zp,a} = \alpha \times \alpha_a \times p \tag{2}$$

where σ_{zp} = vertical stress from additional load in anisotropic basement; α = coefficient as summed

Table 2. Correction coefficients values α_a for rectangular foundations with sides proportion $\eta = l/b = 1$.

	Correction coefficient α_a with anisotropy coefficient k_a, value			
$\xi = 2z/b$	0.5	0.75	1.33	2
0	1.000	1.000	1.000	1.000
0.4	0.926	0.948	0.975	0.992
0.8	0.881	0.928	0.996	1.043
1.2	0.843	0.917	1.043	1.139
1.6	0.808	0.895	1.062	1.209
2.0	0.804	0.899	1.098	1.286
2.4	0.790	0.887	1.109	1.331
2.8	0.766	0.866	1.100	1.353
3.2	0.769	0.869	1.119	1.388
3.6	0.763	0.863	1.115	1.405
4.0	0.759	0.852	1.120	1.426
4.4	0.758	0.846	1.110	1.429
4.8	0.766	0.844	1.117	1.442
5.2	0.746	0.836	1.104	1.433
5.6	0.759	0.828	1.103	1.448
6.0	0.765	0.824	1.098	1.451
6.4	0.778	0.822	1.111	1.467
6.8	0.775	0.825	1.100	1.475
7.2	0.778	0.833	1.083	1.472
7.6	0.813	0.844	1.125	1.500
8.0	0.828	0.862	1.103	1.517
8.4	0.846	0.846	1.115	1.538
8.8	0.833	0.875	1.125	1.542
9.2	0.864	0.864	1.136	1.545
9.6	0.850	0.900	1.150	1.600
10.0	0.842	0.842	1.105	1.579
10.4	0.882	0.882	1.176	1.647
10.8	0.875	0.875	1.188	1.625
11.2	0.867	0.867	1.133	1.667
11.6	0.929	0.857	1.143	1.643
12	0.923	0.923	1.154	1.692

Table 3. Correction coefficients values α_a for continuous foundation ($\eta \geq 10$).

	Correction coefficient α_a with anisotropy coefficient k_a, value			
$\xi = 2z/b$	0.5	0.75	1.33	2
0	1.000	1.000	1.000	1.000
0.4	0.962	0.969	0.983	0.992
0.8	0.947	0.959	0.988	1.010
1.2	0.942	0.959	1.005	1.046
1.6	0.933	0.950	1.008	1.064
2.0	0.935	0.951	1.016	1.080
2.4	0.931	0.945	1.015	1.088
2.8	0.919	0.933	1.007	1.086
3.2	0.914	0.928	1.005	1.088
3.6	0.905	0.920	1.000	1.089
4.0	0.895	0.908	0.990	1.085
4.4	0.886	0.896	0.982	1.082
4.8	0.876	0.888	0.977	1.081
5.2	0.866	0.879	0.971	1.079
5.6	0.857	0.865	0.960	1.072
6.0	0.846	0.856	0.952	1.077
6.4	0.837	0.842	0.944	1.071
6.8	0.822	0.832	0.935	1.065
7.2	0.811	0.817	0.926	1.069
7.6	0.807	0.807	0.922	1.066
8.0	0.791	0.797	0.911	1.063
8.4	0.787	0.787	0.907	1.073
8.8	0.776	0.783	0.902	1.077
9.2	0.766	0.774	0.898	1.073
9.6	0.758	0.758	0.886	1.076
10.0	0.754	0.746	0.889	1.079
10.4	0.738	0.738	0.877	1.082
10.8	0.735	0.726	0.872	1.085
11.2	0.726	0.717	0.867	1.088
11.6	0.716	0.706	0.862	1.092
12	0.698	0.689	0.849	1.094

according to table 5.8 SP; α_a = correction coefficient, regarding stress-strain anisotropy of soil basement; p = average pressure of foundation underside.

Correction coefficients α_a were determined depending on depth ξ resulting $2z/b$, including division of stressed layers of basement thickness into reference layers not more than $0.4b$. Correction coefficients values, recommended for determination of stress, are given in Table 2.

2.4 *Determination of foundation settlements including stress-strain anisotropy*

Settlement of anisotropic foundation basement using scheme of linear deformed semi-spaces should be determined with improved layer-by-layer summing method including different deformability of soil in vertical and horizontal direction under vertical stress $\sigma_{zp,a}$, calculated including stress-strain anisotropy:

$$S = \beta \sum_{i=1}^{n} \frac{(\sigma_{zp,a,i} - \sigma_{z\gamma,i}) h_i}{E_i} + \beta \sum_{i=1}^{n} \frac{\sigma_{z\gamma,i} h_i}{E_{e,i}} \quad (3)$$

Stress of groundmass's own weight $\sigma_{z\gamma,i}$ assumed according to calculation method of SP 22.13330.2011 "Foundations of building sand structures". Values of vertical stresses from external load $\sigma_{zp,a}$ should be determined with formula 2.

Earlier, an approach of foundation basements stress-strain calculation was suggested by the authors based on direct data of field research of soils, carried out with a wedge dilatometer (Nuzhdinetal.2013a). This method assumes thickness of reference layers 0.2 m, corresponding to intervals of standard determination of stress-strain modulus with dilatometer RD-100. Using of machine calculation simplifies process of calculations sufficiently, for which a pack of specialized software was developed and adapted to materials of processed results of soils research withwedge dilatometer. Use of actual values of the stress-strain modulus E_{WD} in calculation facilitates to include actual summing of soil basement, which increases accuracy of calculation. Though, for foundations with relatively small plain dimension sit is made due to regard of uneven change of soils deformability in relation to actual zone depth of basement, for plate

Table 4. Correction coefficients values α_a for round foundations.

	Correction coefficient α_a with anisotropy coefficient k_a, value			
$\xi = 2z/b$	0.5	0.75	1.33	2
0	1.000	1.000	1.000	1.000
0.4	0.894	0.918	0.948	0.964
0.8	0.873	0.923	1.000	1.054
1.2	0.834	0.912	1.055	1.172
1.6	0.800	0.890	1.074	1.246
2.0	0.796	0.895	1.109	1.323
2.4	0.785	0.883	1.117	1.364
2.8	0.764	0.861	1.109	1.382
3.2	0.769	0.087	1.123	1.415
3.6	0.764	0.858	1.123	1.425
4.0	0.759	0.851	1.115	1.437
4.4	0.753	0.849	1.110	1.438
4.8	0.758	0.839	1.113	1.452
5.2	0.755	0.849	1.113	1.453
5.6	0.761	0.826	1.109	1.457
6.0	0.775	0.850	1.125	1.475
6.4	0.750	0.833	1.083	1.444
6.8	0.806	0.839	1.129	1.516
7.2	0.786	0.857	1.107	1.500
7.6	0.833	0.875	1.167	1.583
8.0	0.864	0.864	1.182	1.591
8.4	0.810	0.857	1.095	1.524
8.8	0.842	0.842	1.105	1.526
9.2	0.882	0.882	1.176	1.588
9.6	0.875	0.875	1.125	1.563
10.0	0.867	0.867	1.133	1.533
10.4	0.857	0.857	1.143	1.571
10.8	0.846	0.846	1.154	1.615
11.2	0.917	0.833	1.167	1.583
11.6	0.909	0.909	1.182	1.636
12	0.900	9.000	1.200	1.700

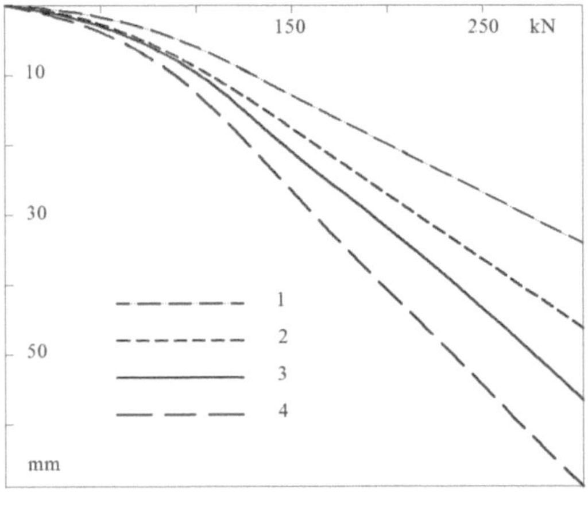

Figure 6. Reference settlements of foundations based on data of field tests of soils with thickness of reference layers $h_i = 0.2$ m: 1 – pier foundation $A = 3 \times 3$ m (E_{WD}); 2 – pier foundation $A = 3 \times 3$ m (E_{cal}); 3 – continuous foundation $b = 3$ m (E_{WD}); 4 – continuous foundation $b = 3$ m (E_{cal}).

Figure 7. Erected 26-storeyed residential building on Zalessky Street.

foundations, when calculating settlements it is recommended to evaluate their possible unevenness in plain view (Nuzhdinetal.2013b).

Comparative analysis of results from different reference calculation schemes of basement is made when designing a number of residence and public buildings in various soil conditions of Novosibirsk. Including sites surveyed by homogeneous and non-homogeneous basements, consisting of weak, as well as relatively low-compressible soils. Comparative calculations according to the SP 22.13330.2011 with different thickness of reference layers and stress-strain modules of soils, taken directly from results of dilatometer researches and after their statistical processing with in geological engineering elements were made for plate foundations, rather big in dimensions, as well as for pier and continuous foundations (see Figure 6).

Given above examples of foundation design calculation solutions for residence and public buildings of Novosibirsk city are carried out presently at construction sites (Figure 7,8). Foundation settlement of all buildings in construction stage are under geodesic observation.

Comparison of calculation and observation results indicates that suggested approach of foundation settlement calculations in most cases leads to increase of forecasting accuracy due to decrease of reference

Figure 8. 28-storeyed residential building in the course of construction on Vybornaya Street.

values. An especially good coincidence of reference settlements with stress-strain modules E_{WD} actual values was observed on initial stages of plate foundations loading (where traditional calculations with E_{cal} would lead to increased reference settlements up to 3 times). More close to actual values of settlements were also received for small foundations, based on low-deformable soils. Calculation accuracy when using E_{WD} increases with decrease of foundation load.

3 CONCLUSIONS

Results of there search carried out indicate that to increase forecasting accuracy of stress-strain soil basements and foundations settlements, not only calculation methods should be improved, but actual stress-strain characteristics of soils lying in foundation basements should be regarded more thoroughly.

Consideration of anisotropic soil properties provides a more thorough approach to assigning foundation dimensions and their settlement forecasting.

Results' analysis of stress-strain soil basement calculations according to the method suggested, reveals settlements increase with anisotropy indicator $k_a < 1$ and their decrease correspondingly with $k_a > 1$.

REFERENCES

Ambartsumyan S.A. 1982. Multimodulus theory of elasticity. Moscow: Nauka Publishers. Main editorial board of physics and mathematics literature.

Nuzhdin, L.V. 2013a. Improvement of forecast accuracy of foundations based on field research deformability of the soil wedge dilatometer [Text] / L.V. Nuzhdin, M.L. Nuzhdin, K.V. Kozminykh // Scientific works: edition: Industrial Engineering & Construction. Issue 3 (38).V. 2. – Poltava: PoltNTU, 2013. – pp. 258–273.

Nuzhdin, L. 2013b. Evaluation of deformation properties of soils using a wedge dilatometer and improvement of forecast accuracy of buildings settlement [Text] / L. Nuzhdin, M. Nuzhdin, X. Kozminykh, et al. // Challenges and innovations in Geotechnics: Proceedings. of ATC7 Workshop, 18ICSMGE. – Paris: Presses des Ponts, 2013. – P.68-75

Nuzhdin, L.V. 2014.The practical calculation method of foundation settlements with regard strain anisotropy of the soil basement [Text] / L.V Nuzhdin, O.A. Korobova, M.L Nuzhdin // Deep foundations and problems of underground space development: collected works of All-Russian conference with international participation Part 4. – Perm: PNRPU, 2014. – P.154-162.

Challenges and Innovations in Geotechnics – Zhussupbekov (Ed.)
 ISBN 978-1-138-03007-7

The use of express method for determining the modulus of deformation of fill soil

A.B. Ponomaryov & S.A. Sazonova
Perm National Research Polytechnic University, Perm, Russia

ABSTRACT: Engineers often encounter with errors either at the design stage or at the stage of the construction despite the great experience on construction of fill soils and construction of embankments. According to the authors, it is necessary proper assessment of possible technologies, taking into account a large experience in order to avoid errors at the design stage. At the stage of construction of embankments quality control is very important, starting with the input soil control in accordance with the project. What's more important is the operational control and a close interrelation between designers and contractors. This will control the quality of work quickly and efficiently, as well as to predict the settlement of embankments of soil, backfill, etc. The authors conducted a series of experiments to obtain the dependence of the modulus of deformation from compaction ratio, which includes three stages. The results are reported in this article.

Keywords: fill soil, dynamic density meter, compacting factor, compression modulus, dynamic modulus.

1 INTRODUCTION

Fill soils have a number of specific features and differ from the natural soils. There are several problems with the construction of such soil conditions: the variability of the composition of the addition, the variable density, increased and uneven compressibility, unstable structures, and others. Physical and mechanical properties of soil can vary within wide limits. Research of U. Abelian, V. Krutov, Y. Zaretsky, V. Didukh, P. Konovalov, N. Tanatarov and many other scientists devoted to research of fill soil, the design and construction of soil conditions (Krutov 1988, Krutov & Kogaj 2008).

Despite the great experience on construction of fill soils and construction of embankments (Ponomaryov & Zolotoubov 2014, Rücknagel et al. 2013, Ter-Martirosjan & Mirnyj 2013), engineers often encounter with errors either at the design stage or at the stage of the construction. Errors at the design stage are related to the underestimation of the properties of fill soil, or an incorrectly selected compaction technology. Errors at the stage of production work are a departure from the project and a lack of quality control of the works.

Special attention should be given to the production of works on creation of embankments (Rozhkov & Solovyov 2006):

- the quality of fill soil (lack of organic matter and frozen clods, the required moisture content, etc.);
- the thickness of the layers, that are dependent on the chosen compression technology;
- evenly compaction over the entire area backfilled layer;
- quality control of compaction of each layer of soil, and, if necessary, after the construction of the embankment as a whole throughout its thickness.

Quality control is an essential and indispensable in the construction process, but often it does not pay enough attention. Primarily this is due to unscrupulous contractors. It should be noted that the costs for additional studies of fill soil can be several times higher than the value of the cost when there is a properly operating quality control. As a method of operating the quality control it is recommended to use direct methods for determining the density of the soil. Recently, express methods, that using different densitometers, are gaining popularity (Sulewska 2004).

According to the authors, it is necessary proper assessment of possible technologies, taking into account a large experience in order to avoid errors at the design stage. Moreover correctly chosen technology is based on technical and economic comparison of options can lead to a significant reduction in cash costs. At the stage of construction of embankments quality control is very important, starting with the input soil control in accordance with the project. What's more important is the operational control and a close interrelation between designers and contractors. Operational control of the quality of the construction of embankments will not only improve the quality of work, but also gives the opportunity to make the necessary changes to the project. The rate of flow of information is important for close work. Also selected terminology plays a major role. The main controlled variable in the construction of embankments is the compaction ratio, however, deformation characteristics, primarily deformation module, are important for the design of foundations. Summarizing the above, we can conclude that the interrelation of the monitored parameters on a construction site with characteristics that are used in the design of the soils should be determined. This will control the quality of work quickly and efficiently, as well as to predict the sediment of embankments of soil, backfill, etc. (Sazonova & Ponomaryov 2015).

2 METHODOLOGY OF INVESTIGATION

The authors conducted a series of experiments to obtain the dependence of the modulus of deformation from compaction ratio, which includes three stages:

Stage 1 – the definition of the density of the samples in accordance with GOST 5180-84 and the determination of the maximum density and optimum moisture content in accordance with GOST 22733-2002;

Stage 2 – in parallel with the determination of the compaction factor assess the density of the soil mass express method using a dynamic density meter of soil DPG-1.2;

Stage 3 – compression tests with subsequent determination of the compression module according to GOST 12248-2010.

The object of the study were fill soils of sandy pillow under foundation slabs and reverse fillings sinus foundations on one of the objects of the Perm city. Size distribution of the soil was determined by sieve method in accordance with GOST 12536-79 and according to GOST 25100-2011 soil was gravelly sand gravel with an average content of 30%.

Determination of the maximum density was carried out in accordance with GOST 22733-2002 with the help of the device of standard compaction GT 1.4.1. The maximum density of dry soil was determined by the results of tests $\rho_{d\ max}$, the mean value was 1.9 g/cm^3 with a range of values up to 8% in both directions for different areas. The average value of the optimum humidity was $\omega_{\text{opt}} = 9{,}2\%$ with a range of values up to 10% in both directions for a variety of platforms. Next compression ratio is defined as the ratio of the dry soil density to the maximum density of dry soil. The value of the compaction factor is over 0.92 in all tests, which corresponds to the standard value for the type of the fill soils.

At the second stage of the study the soil mass density assessment conducted by express method with using dynamic densitometer DPG-1.2 (fig. 1).

According to the instruction manual DPG-1.2 is designed for the determination of the dynamic elastic modulus E_d of the soil and the bases of roads by simulating a vehicle travel on the road surface, according to the main provisions of the document ST SEV 5497-86. The principle of operation of the device is the measured

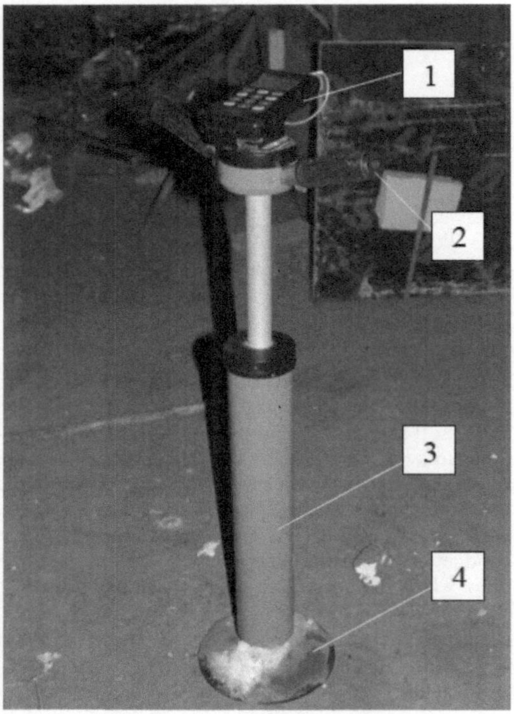

Figure 1. Dynamic density meter DPG-1.2: 1 – electronic unit; 2 – reset knob; 3 – mechanical percussion device; 4 – round stamp.

settlement S under the soil round stamp (platform, stove), when subjected to a shock load (force) F. During the strike electronic unit automatically records the signals from the force sensors and platforms settlement. Simultaneously, the microprocessor produces a double integration of the acceleration signal, and calculates a soil settlement. The dynamic modulus of elasticity that characterizes the deformability of soil, calculated as follows:

$$E_d = \frac{\pi \cdot D \cdot \sigma}{4 \cdot S} \cdot (1 - \mu^2) \tag{1}$$

where S = amplitude of the total soil settlement under the stamp, mm; D = diameter of the stamp, mm; μ = Poisson's ratio, for sand and sandy loam is 0.3 according to GOST 20276-85; σ = contact stress under the stamp, is calculated using the formula, MPa:

$$\sigma = \frac{4 \cdot F}{\pi \cdot D^2} \tag{2}$$

where F = impact force (load), H.

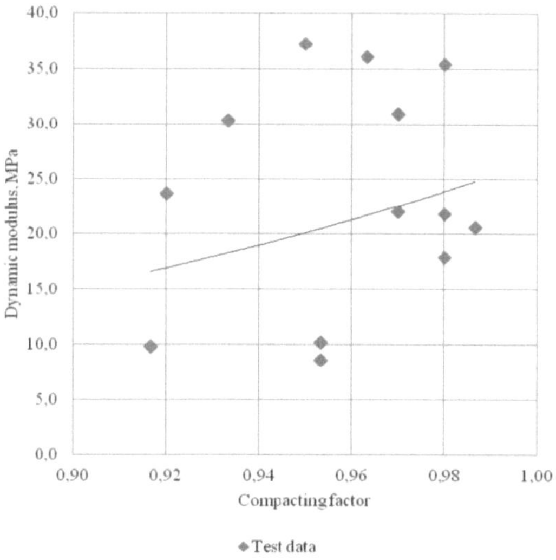

Figure 2. The dependence of dynamic modulus on the compacting factor.

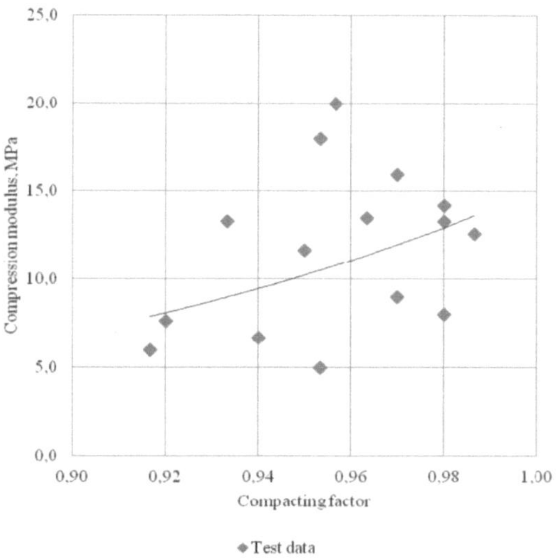

Figure 3. The dependence of compression modulus on the compacting factor.

The magnitude of impact force and the duration of its effect is determined by the mass of a free falling weight and stiffness of the spring damper.

In previous studies the authors strongly influenced by humidity readings DPG-1.2 was observed. With humidity over 12% of the value of the dynamic modulus of elasticity E_d grossly underestimated (about 10 MPa) at a relatively high density of the soil (K_{com} = 0,95). This can be explained by the fact that the wave velocity increase significantly with increasing humidity, and as a consequence of the elastic modulus is significantly reduced (Ponomaryov & Shutova 2013).

At the last stage of research determining compressive modulus of deformation produced in accordance with GOST 12248-2010. The first stage pressure is 0,025 MPa. Subsequent pressure levels were taken to be twice the value of the previous step (0.025; 0.05; 0.1; 0.2; 0.3 MPa), a total of five stages. Compression module calculated in the pressure range of 0.1–0.2 MPa, which is due to operational loads on the fill soil.

3 RESEARCH RESULTS

Figure 2 and 3 shows the dependence of the dynamic modulus and compression modulus of compression deformation ratio.

Analyzing the graphs, we can conclude about the lack of convergence of the results (the magnitude of reliability of approximation of about 0.2). As for depending E_d on K_{com}, according to the authors, the spread of values can be attributed to the strong dependence of the recorded data DPG-1.2 on the experimental conditions and soil humidity. To find the required dependence authors plan to conduct additional experiments in the laboratory with constant control of humidity and temperature. The authors also plan to conduct additional studies in view of the contents of large gravel particles with a size greater than 2 mm with different percentages in the mixture with a parallel determination of strength characteristics to determine the dependence of E_k on K_{com}.

4 CONCLUSIONS

To date, increase the speed and quality of works is the main trends in construction. Therefore, the role of express methods increases. Scientists from the road industry trying to find dependence of dynamic and static modulus. This dependence can be of interest for constructors, as dynamic methods are often used in soil compaction. The dependence of the static modulus of deformation from compacting factor makes it possible to efficiently control the quality of the embankment, which is very important for geotechnical monitoring.

REFERENCES

Krutov, V.I. 1988. *The base and foundation on fill soil.* Moscow: Stroyizdat.

Krutov, V.I., Kogaj, V.K. 2008. Device planning embankments in preparation areas for development. *Mechanization of construction* 8: 21–24.

Ponomaryov, A. B., Shutova, O. A. 2013. Study of vibration impact of vehicles on the foundations of buildings in dense urban areas. *Internet-vestnik VolgGASU* 2(27).

Ponomaryov, A., Zolotoubov, D. 2014. Several approaches for the design of reinforced bases on karst areas. *Geotextiles and Geomembranes* 42: 48–51.

Rozhkov, S.V., Solovyov, I.A. 2006. Investigation of the characteristics of fill soil at the base of construction. *Zhilishhnoe stroitel'stvo* 10: 9–12.
Rücknagel, J., Götze P., Hofmann, B., Christen, O., Marschall, K. 2013. The influence of soil gravel content on compaction behaviour and pre-compression stress. *Geoderma*: 209–210.
Sazonova S.A., Ponomarev A.B. 2015. Experimental design to investigate the dependence between the coefficient of compaction and the modulus of deformation. *Herald of Perm National Research Polytechnic University, Construction and architecture* 3: 67–80.
Sulewska, M. J. 2004. The application of the modern method of embankment compaction control. *Journal of civil engineering and management* 10(1): 45–50.
Ter-Martirosjan, Z.G., Mirnyj, A.Ju. 2013. Influence of inhomogeneities on the mechanical properties of soils. *Bases, foundations and soil mechanics* 6: 27–32.

Challenges and Innovations in Geotechnics – Zhussupbekov (Ed.)
© 2016 Taylor & Francis Group, London, ISBN 978-1-138-03007-7

Experimental research of deformed state of uniform soil massif depending on the step of barretts

K.Y. Stepanischev & P.V. Buryak
Moscow State Building University, Moscow, Russian Federation

D.A. Vnukov & F.F. Zekhniev
NIIOSP "N.M. Gersevanov"–Research Centre of Construction JSC, Moscow, Russian Federation

ABSTRACT: Deep foundations are used in the design of high-rise buildings, because it connects with the transfer of large pressure on the soil base. As the foundations of the building sometimes use barretts, which are able to perceive significant vertical and horizontal pressures due to improved lateral surface. Barretts have increased load bearing capacity as compared with large diameter piles. In modern practice, as a rule, the interaction of barretts is investigated by analytical and numerical methods and has no sufficient experimental confirmation. In this article the review of the planned experimental methods of research of the intense deformed state depending on a step of elements in the form of barretts at interaction of model of the soil massif with deep foundation will be made. Experimental research are planned with using of the laboratory stand on the basis of laboratory No 2 of "The bases and foundation on soft soil" for the purpose of obtaining qualitative data on interaction barretts with an assessment of adequacy of settlement model, and also at research of the intense deformed state by numerical methods.

1 INTRODUCTION

Large number of the experimental and theoretical works is devoted to questions of interaction of piles with an environmental soil and nature of their work under loading .The main results were saved up when carrying out experimental and field works which then were generalized in the form of theoretical provisions.

The researches conducted by V. V. Znamensky, N. M. Doroshkevich, N.Z. Gotman, E.E. Devaltovsky, Whitaker, J. Hanisch, etc., have shown distinction in transfer of load of soil between a single pile and a bush of piles. At a step of piles in a bush more than 5 … 6d there is no influence of a tension of the massif of soil around a pile on next and imposings of zones of mutual shift of soil, that they work as single. In turn at decrease in a step between axes of piles in a bush to 3 … 4d there is a decrease in friction forces on a lateral surface and there is a phenomenon of blocking of interpile soil (the pile and the soil massif work as a unit), and also there is an uneven distribution of loading between piles (Strunin, 2013).

In modern practice these researches are applied to barretts and used at design of the barrett's foundation which pay off as boring support or boring piles according to provisions of SP 24.13330.2011 "Svayniye fundamenti".

At the same time distinctive features barretts in comparison with piles (their geometrical form, the size of cross section) aren't considered. (Sidorov, 2013).

At interaction research barretts with the soil massif as a rule numerical and analytical methods, and also natural experiments are used. However, carrying out laboratory experiments allows to obtain with the smallest expenses of time and resources data on change of fields of displacement of the soil massif of physical model at interaction with him both single, and groups of barretts, and also to estimate adequacy of the results received by a numerical method.

2 EXPERIMENTAL EQUIPMENT

The barretts on change of a strained state of the ground massif is planned to execute research of influence of a step with use of the automated exploring complex (AIK "ASIS") consisting of the laboratory test stand (a flat tray) and a measuring system of "ASIS", production of NPP GEOTEK, Penza. Laboratory researches are conducted on the basis of laboratory No. 2 of "The bases and the bases on soft soils" of NIIOSP of N. M. Gersevanov for the purpose of studying of a strained state of the homogeneous ground massif depending on a step of the barretts within a master's thesis.

The stand is intended for carrying out laboratory and researches and allows to carry out, in the conditions of plane and axisymmetric deformation, test for various types of the bases or designs buried in a soil.

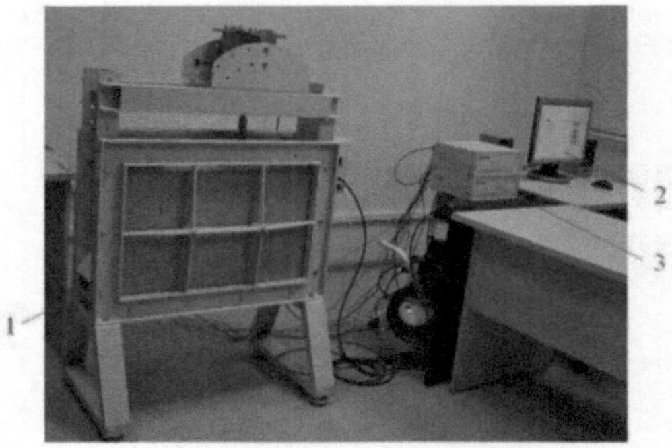

1 – a tray, 2 – the computer, 3 – the control unit and registration of signals.

Figure 1. General view of the laboratory stand and equipment.

The general view of inventory and measuring system is presented in figure 1.

Designly the stand is executed in the form of a tray with internal sizes: length – 722 mm; width – 156 mm; height – 536 mm. The stand represents a flat tray with the transparent the forward and back walls made of glass (duplex) 12 mm thick.

Two steel lattices in which clusters there are basic washers are applied to an exception of a deflection of the transparent walls. The external loading (a flat stamp) is created by means of a pneumatic loader. Vertical displacement of model of the base is measured by the displacement sensor, and loading – force sensor. Deformations (settling) of a surface of a soil are measured by sensors of displacement (VD), which are fixed on a traverse.

Control testing is performed by using the program "ASIS", designed for automated control of testing models of foundations in the laboratory.

Results processing and receiving results of experiment (the field of displacement sandy soil) is performed by digital processing of Particle Image Velocimetry (PIV). This method allows to perform non-contact measurement fields of displacement, has high accuracy and readability, does not require high security measures, expensive equipment or high labor input of processing of results and is widely used in the solution of research tasks in the field of geotechnics and construction in General. Experimental researches with using method of digital processing of images usually perform in conditions of plane strain. Images of particles are recorded on the electronic medium of the digital camera and further processing images allows to calculate the displacement of particles between a pair of processed images, for example, for the step load applied to the ends of barrett. The two-component field of displacement with further deformation analysis was made after measuring the displacement of particles (Melnikov, 2015).

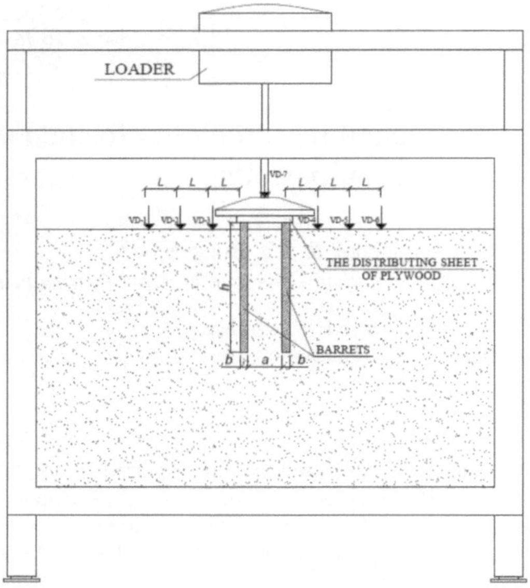

h - element height; b - element width; a - distance in the light between elements;
L - distance between sensors of vertical deformations; VD - sensor of measurements of vertical deformation

Figure 2. Scheme of the experimental tests.

3 METHOD OF CARRYING OUT EXPERIMENTS

Experiments are carried out by method of physical model operation for natural barrett of 10 m in depth and 0,6 m wide that taking into account the accepted scale factor 1:50 there corresponds to depth and width of barrett in model, equal 20 and 1,2 cm.

The physical analog is understood as the material system which properties and parameters are similar to actual object, and representing the simplified procreation of the processes happening in actual system. In the conducted researches the scale of model experiments for the linear dimensions and load of barrett's cap were accepted taking into account overall dimensions of the laboratory test stand.

As a physical analog of the massif of a soil sand in an air-dried state is used. Sand is characterized as average fineness on classification of GOST 25100-2011. As elements of a physical analog the barretts made of M50 brand concrete are accepted.

For carrying out tests in the laboratory stand a la carte the sandy soil of average fineness is fallen asleep with fiber consolidation. In process of dumping of sand according to the experimental scheme the made elements of a physical analog are established beforehand. Upon completion of the device of elements and a filling of a soil to the provided scheme of a mark

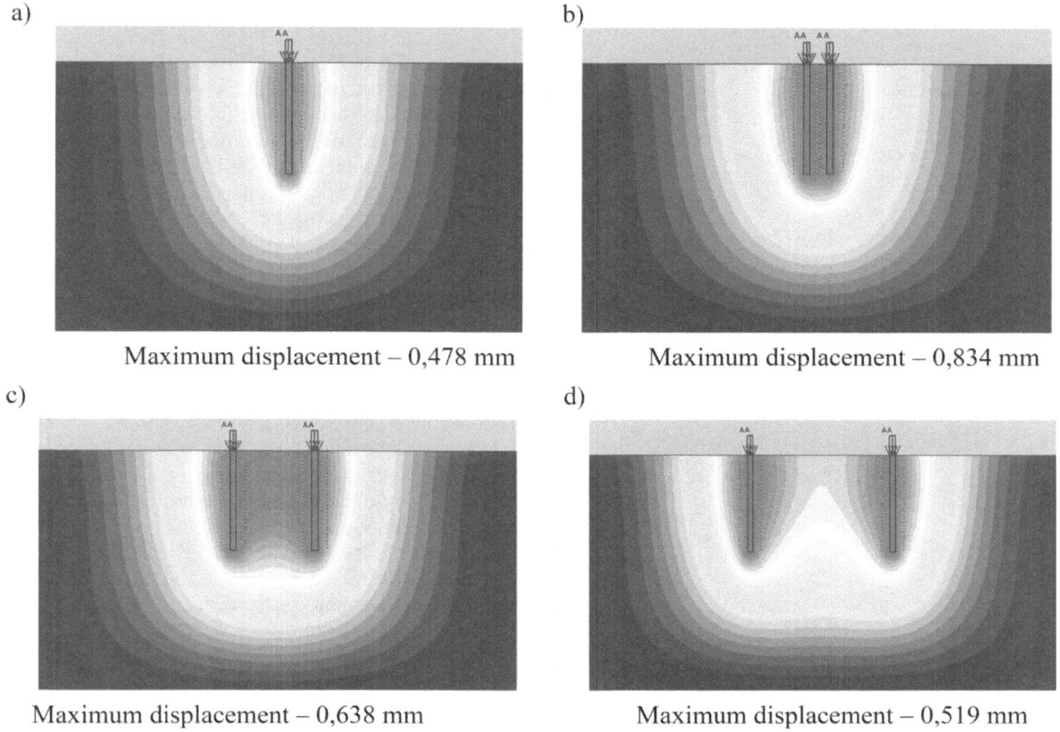

Figure 3. Isofields vertical displacements, obtained by calculation in PLAXIS 2D, depending on the step barretts; a – for single barrett; б – distance between the barretts 2 cm; В - distance between the barretts 12 cm; г - distance between the barretts 27 cm.

in a surface of the ground massif sensors of measurement of vertical deformations are installed. Just before carrying out experience installation, control and calibration of the photo equipment with illuminants, the displacements of a sandy soil used for the organization of non-contact measuring system is made. After preparation of a physical analog and setting up contact and non-contact count systems of displacements transition to immediate carrying out the experimental experience is carried out. The loading enclosed on the barrett's cap is accepted on the basis of beforehand the executed calculation of a carrying capacity natural the barrett on a soil according to provisions of SP 24.13330.2011 "Svayniye fundamenti". When carrying out calculation influence of face walls on a carrying capacity barretts was not considered.

The experimental scheme of research is submitted in figure 2.

By results of calculation for the first group of the limiting conditions the carrying capacity natural barretts on a soil made 1472 kN on 1 item of m. Considering a scale factor of a physical analog, we receive a carrying capacity barretts in the conditions of model of equal 29,4 kN. Proceeding from width of a tray, equal 155 mm, the loading enclosed on one barrett are defined as $N = 29{,}4 \times 0{,}155 = 4{,}5$ kN. Loading moves steps and corresponds to size 1/10 from the maximal value of loading. After giving of each of steps there is a keeping the barrett before the complete stabilization of settlings of elements of a physical analog. Automation of process of carrying out tests is carried out by means of the ASIS program allowing to fix vertical displacements of a stamp on the course of a loading of elements.

Before the beginning and after giving of each of steps of load of elements of the barrett's foundation photofixing for further processing of results and receiving fields of displacements of a sandy soil by method of digital processing of pictures of Particle Image Velocimetry (PIV) is carried out.

4 DEFINITION OF THE EXPERIMENTAL STEP BARRETTS BY METHOD OF NUMERICAL MODELLING

On the basis of the calculations executed by a numerical method with use of the program PLAXIS 2D complex before carrying out experiments the step of barretts for physical model was selected.

The minimum distance in light between elements was accepted not less than 1 m according to provisions of SP 24.13330.2011 "Svayniye fundamenti", as for a boring pile that taking into account large-scale coefficient corresponds to a step in physical model of equal 2 cm.

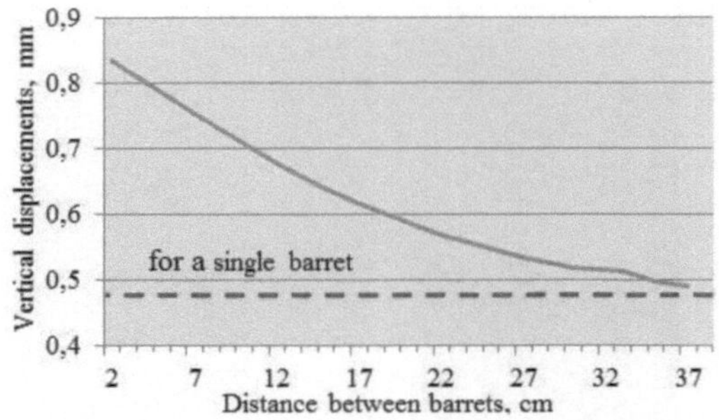

Figure 4. The graph of dependence of influence of vertical displacements on distance between barretts.

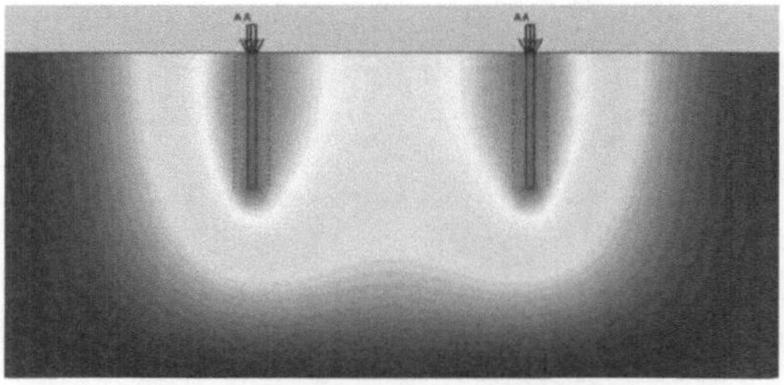

Figure 5. Isofield vertical displacements by the distance in light between barretts 37 cm. Maximum displacement 0,49 mm.

Calculations were carried out on the basis of use of elasto-plastic model of soil Mohr-Columbus which demands input of the following calculated parameters: G – the shift module, υ – Poisson's coefficient, φ – a corner of internal friction, c – coupling, ψ – a dilatancy corner. For interaction modeling barretts with surrounding soil the interface elements were used (contact between a design and surrounding soil). At the same time, the following calculated values of strength characteristics on contact "a design – the soil massif" are accepted: specific coupling $c_к = 0$, a corner of friction of soil on design material $\delta = \gamma_к \times \varphi$, where φ – a corner of internal friction of soil, $\gamma_к$ – the coefficient of working conditions (SP 22.13330.2011 "Osnovaniya zdaniy i sooruzheniy", 2011).

According to the results of calculation and analysis of the results were accepted in the distance between elements, which are used in the experiments. This research consists of 4 experiments with distances in light 2, 15, 30 cm, as well as for single barrett. Each experiment is repeated 3 times. Isofields of vertical displacements, obtained as a result of numerical modeling and conducted for full-scale physical model of natural barretts with distances between the elements, are shown in figure 3.

Compare isofields of vertical displacements and values, obtained by numerical methods and made for experiments, are shown various character of influence model elements on the strain state of the soil massif by their interaction, depending on the steps of elements. By results of calculations the graph of dependence of influence of size of draft on distance between barretts has been constructed, submitted in figure 4.

According to the graph it is visible that there is a reduction of size of draft in process of increase in distance between elements of model. In the range from 2 to 30 cm the linear dependence showing mutual influence barretts at each other is observed. At increase in distance between elements there is gradual stabilization of draft, values of vertical displacements decrease.

Comparison of results of numerical modeling for a single barrett and group barretts at distance of 37 cm in physical model that corresponds taking into account large-scale coefficient to natural distance in 18,5 m, is shown convergence of values of sizes drafts. From this we can conclude that the elements of barrett foundation when the distance between them in light 37 cm will work as a single and will not have mutual influence on the formation of a strain state of the soil massif. The results of the calculation, performed using the software package PLAXIS 2D, is given in the form of isofield vertical displacements by the distance in light between barretts 37 cm on figure 5.

5 CONCLUSION

Research of the deformed state at interaction of the soil massif with barretts is connected with using of numerical and analytical methods of research. In too time experimentally these questions aren't rather studied.

The description of the used equipment, structure and a technique of the experiments, based on preliminary numerical modeling by results of which the analysis of change of the deformed state depending on distance in light between barretts was made, is provided in this article. By results of calculations the preliminary distance between model elements, wherein mutual influence between barretts is excluded, has been defined and they work as single.

REFERENCES

P.V. Strunin (2013), the thesis on a subject "The intense deformed state the jet piles, interacting with the soil basis and interpile space".

V.V. Sidorov (2013), the thesis on a subject "The intense deformed condition of the non-uniform soil massif interacting with barretts of big length".

A.V. Melnikov (2015), the thesis on a subject "Experimental and theoretical researches of a method of static sounding of soils".

GOST 25100-2011, (2011), Grunti. Klassifikaciya, (Russia).

SP 24.13330.2011, (2011), "Svayniye fundamenti", (Russia).

SP 22.13330.2011, (2011), "Osnovaniya zdaniy i sooruzheniy", (Russia).

Challenges and Innovations in Geotechnics – Zhussupbekov (Ed.)
© 2016 Taylor & Francis Group, London, ISBN 978-1-138-03007-7

Estimation of railway ballast and subballast bearing capacity in terms of 300 kN axle load train operation

A.F. Kolos & A.A. Konon
Emperor Alexander I St. Petersburg State Transport University, St. Petersburg, Russia

ABSTRACT: The stated ballast and subballast bearing capacity calculation method allows to ensure track stable performance and optimize track lifecycle costs. The calculation method is based on limit equilibrium theory considering vibrodynamic impact and field test results of ballast stressed state and vertical and horizontal vibroaccelerations in ballast layer.

Railway track stability depends not only on the subgrade sustainability, ties, fastenings and rails, but also on the quality of all subballast elements and the ballast itself. The research of vibrodynamic load damping effectiveness of ballast have been conducted in USSR and Russia since 1980s for different car axle load and speed. In spite of that, the problem of ballast overhaul life is still very actual.

The main ballast characteristics determining its use in the track structure according to Russian Federation and other countries' national standards do not set values of subballast bearing capacity and deformability. The last are the criteria of track stable performance under the train dynamic load. On the basis of actual ballast bearing capacity and deformability one should plan ballast cleaning and renewal and track maintenance aiming to increase ballast bearing capacity and deformability.

Unfortunately, nowadays there are no methods allowing to predict ballast layer bearing capacity and deformability considering gross tonnage, climatic and other conditions. Modern approaches to this issue do not consider ballast performing in terms of advanced vibrodynamic loading. That's why strength and deformative properties must be defined in terms of accurately approximated vibrodynamic loading.

The main cause of ballast reliable performance is its bearing capacity. It depends from the level of vibrodynamic load applied to ballast layer and subballast characteristics. Ballast layer bearing capacity and deformability depend on ballast cohesion and friction angle, moduli of deformation and elasticity. These characteristics depend on vibrodynamic load level under the sleeper pad, ballast granulometric composition, fouling, density, roundness, subsleeper damper existence and its stiffness and so on. Values of these characteristics and their change according tonnage accumulation allow to predict ballast bearing capacity, deformability and overhaul life. Ballast strength and deformative properties and their change according tonnage accumulation in terms of vibrodynamic load haven't been studied before. Indirect methods of defining ballast strength and deformative properties are ungrounded, often uncertain and may give wrong calculation results.

1 THE STUDY OF BALLAST LAYER STRESSED STATE

To assess ballast overhaul life according its bearing capacity one should know the actual stress distribution in ballast layer for specified operation conditions.

To solve this issue Emperor Alexander I St. Petersburg State Transport University (PSTU) scientists held tests of ballast stressed state. Rolling stock on the site had axle loads from 230 to 300 kN and 70 kmph speed. Tests were held at Russian Railway Research Institute experimental track.

Track structure on the site was the following: 65 kg/m rails, concrete sleepers (2000 items/km), tension clamp fastenings ARS-4, thickness of granite ballast was 55 cm under the sleeper.

Vertical stress values in ballast layer were measured with soil pressure capsule set. Capsules were put under the sleeper and in the ballast layer. Along the sleeper capsules were set at the sleeper end, at underrail section, and near centre line of a track and up to 55 cm under the sleeper. Capsule placement in the ballast layer and subballast is shown in the Figure 1.

Test results showed that axle load increasing causes rize of stress values in the ballast layer and subballast. Vertical stress distribution charts at level of 10 and 55 cm under the sleeper pad are shown in the Figures 2 and 3.

Data analysis from Figures 2 and 3 shows that car axle load has the principle effect on ballast and subballast stressed state. Vertical stresses under the

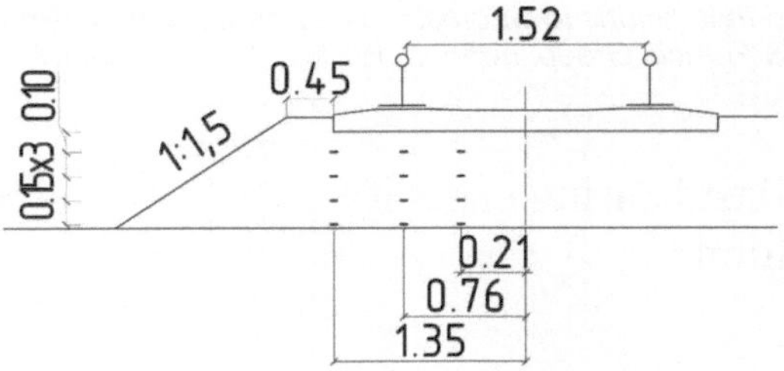

Figure 1. Capsule placement in the ballast layer and subballast.

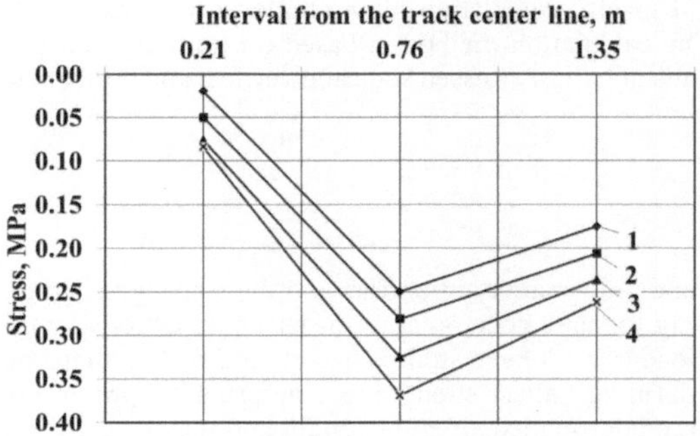

Figure 2. Vertical stress distribution at the level of 10 cm under the sleeper pad with axle load: 1 – 230, 2 – 250, 3 – 270, 4 – 300 kN/axle.

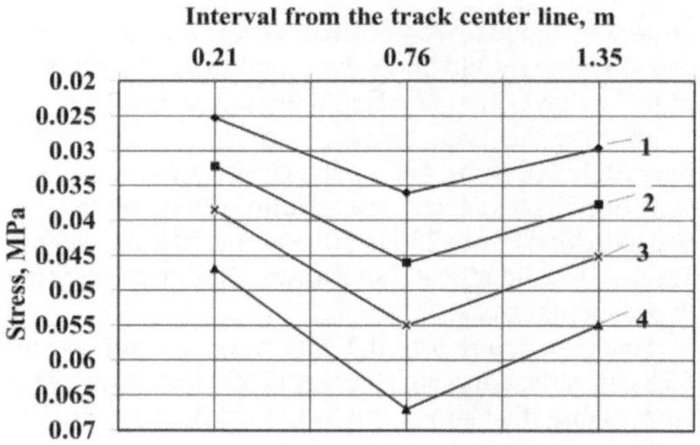

Figure 3. Vertical stress distribution at the level of 55 cm under the sleeper pad with axle load: 1 – 230, 2 – 250, 3 – 270, 4 – 300 kN/axle.

sleeper pad and under the ballast layer grow intensively as a result of axle load increasing. Thus, at the underrail section (point "0.76 m" in the Figures 2, 3) at the level of 10 cm under the sleeper pad stresses became 1.5 times as much and at the level of 55 cm under the sleeper pad they increased for 1.9 times. This stressed state change should be taken into account while designing ballast layer and subballast structure from the point of bearing capacity and deformability.

These results can be used for laboratory model tests of the crushed stone and for calculations of ballast and subballast bearing capacity of new or reconstructed railway lines.

2 THEORETICAL BASIS OF BALLAST AND SUBBALLAST BEARING CAPACITY CALCULATION

Ballast and subballast limiting equilibrium state is a state when a slight increase of static or dynamic load leads to an imbalance of the soil mass. At this state the limit equilibrium zones begin to develop, where the tangent stress reaches its limit. The applied force with the soil mass considered and the marginal load with the soil strength properties considered are equal and are in the limiting equilibrium state. Such a state of the structure is defined with relationships of stress tensor components and soil strength properties. So this ballast and subballast soil state defines bearing capacity which is one of the main criteria of stable track performance under the train load. Ballast and subballast strength is defined with the marginal stress under the sleeper pad.

Ballast layer performs in conditions of intensive vibrodynamic load so its strength characteristics should be taken into account with decrease considered.

Limit equilibrium theory (e.g. Sokolovsky (1960)) is a base of ballast and subballast bearing capacity calculation method developed by Prokudin (1982). It considers vibrodynamic impact and is built upon simultaneous solution of motion equations and limiting equilibration criterion with soil strength characteristics decrease due to vibrodynamic load and oscillation damping (1):

$$\begin{cases} \dfrac{\partial \sigma_z}{\partial z} + \dfrac{\partial \tau_{zy}}{\partial y} = Z + \rho \dfrac{\partial^2 U}{\partial t^2} \\ \dfrac{\partial \tau_{yz}}{\partial z} + \dfrac{\partial \sigma_y}{\partial y} = Y + \rho \dfrac{\partial^2 V}{\partial t^2} \\ \sigma_1 - \sigma_2 = \left(\sigma_1 + \sigma_2 + 2C_{dn} ctg\varphi_{dn}\right)\sin\varphi_{dn} \end{cases} \qquad (1)$$

where: σ_z, σ_y = normal stress components; τ_{zy}, τ_{yz} = tangent stress components; $\rho = \gamma/g$ = soil density; g = free fall acceleration, 9.81 m/sec^2; γ = bulk density of the soil; U, V = oscillation displacements in the direction of Z and Y axes; σ_1, σ_2 = maximum and minimum main stresses; C_{dn}, φ_{dn} = cohesion and the friction angle of the soil sustaining the vibrodynamic impact. C_{dn}, φ_{dn} strength characteristics depend on vibrodynamic impact level, position of the point and ability of the soil to sustain and damp vibrodynamic load.

PSTU scientists got the experimental relationships of vertical and horizontal vibroaccelerations damping

(Kolos & Morozova, 2014) in terms of train operation with axle load up to 300 kN (2), (3):

$$\frac{d^2U}{dt^2} = g_0^v \cdot \exp(-\delta_z^v \cdot z - \delta_y^v \cdot y) \quad (2)$$

$$\frac{d^2V}{dt^2} = g_0^h \cdot \exp\left(-\delta_z^h \cdot z + \delta_y^h \cdot y\right) \quad (3)$$

where: g_0^v = vertical vibroacceleration under the sleeper pad; δ_z^v = vertical vibroacceleration damping factor in the vertical plane; δ_y^v = vertical vibroacceleration damping factor in the horisontal plane; z, y = position of the point; g_0^h = horisontal vibroacceleration under the sleeper pad; δ_z^h = horisontal vibroacceleration damping factor in the vertical plane; δ_y^h = horisontal vibroacceleration damping factor in the horisontal plane.

Equations (2), (3) are put to the right side of the equation system (1). It is a first order system of hyperbolic type, and characteristic equations and characteristic relations can be derived according to higher mathematics methods.

As algebraic transformation result equations of characteristic lines and characteristic relations (4), (5) are received.

$$dz = dy \cdot tg\left(\delta \mp \frac{\pi}{4} \pm \frac{\varphi_{dn}}{2}\right) \quad (4)$$

$$d\sigma \mp 2\sigma \cdot tg\varphi_{dn} \cdot d\delta = (\gamma + B) \cdot (dz \mp dytg\varphi_{dn}) + D \cdot (dy \pm dztg\varphi_{dn}) \quad (5)$$

where:

$$B = \frac{\gamma}{g} \cdot g_0^v \cdot \exp(-\delta_z^v \cdot z - \delta_y^v \cdot y),$$

$$D = \frac{\gamma}{g} \cdot g_0^h \cdot \exp\left(-\delta_z^h \cdot z + \delta_y^h \cdot y\right),$$

δ = inclination angle of σ_1 stress to Y axis.

Simultaneous solving of equations (4), (5) lets the ballast and subballast bearing capacity be derived by tracing the slide lines (characteristic lines) net and calculating the δ angle and σ stress values in the net junction points. Equations of characteristic lines (4) and characteristic relations (5) are solved with the finite-difference method. For that approximate function values that define the characteristic lines shape and the stressed state are calculated sequentially in the net junction points. In equations (4), (5) the differentials are replaced by the differences of values in two points situated closely at the slide line.

To calculate ballast and subballast bearing capacity dimensions of ballast layer and subgrade cross-section and soil strength characteristics should be taken into account simultaneously. Railway track cross-section has complicated geometry with some points of contour line direction change. In these points stresses on the soil surface change stepwisely. Due to these peculiarities the authors suggested to calculate stresses in five zones, according to Figure 4.

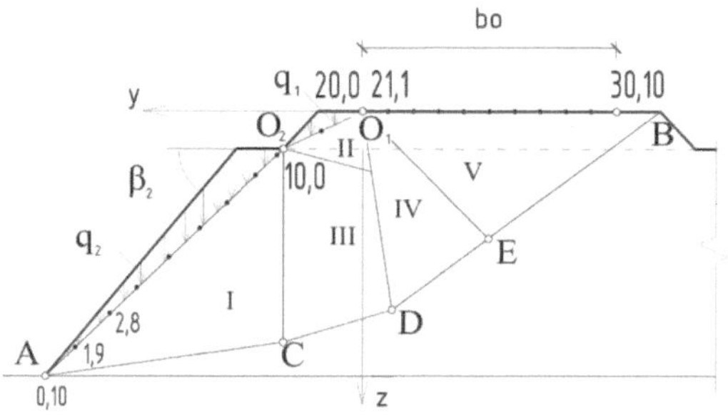

Figure 4. Ballast and subballast bearing capacity calculation zones.

Limit stress at the underrail section under the sleeper pad $[\sigma_z]$ is calculated with equation (6):

$$[\sigma_z] = \sigma_b(1 - \sin\varphi_b \cdot \cos 2\delta_b) - C_b \cdot ctg\varphi_b \quad (6)$$

where σ_b = ballast stress at the underrail section under the sleeper pad; φ_b, C_b = ballast strength characteristics; δ_b = inclination angle of σ_1 stress to Y axis in the ballast layer in the point where σ_b is calculated.

Operating stress values must not exceed the limit value according the safety factor. Ballast and subballast strength inequality (or condition) is (7):

$$\sigma_z \leq \frac{[\sigma_z]}{\gamma_n} \quad (7)$$

where: σ_z = operating stress value; γ_n = reliability factor.

3 PROBLEM SPOTS AND CONCLUSIONS

The stated method (Morozova, 2014) is to be improved with experimental data on effect of vibrodynamic impact on ballast strength characteristics. To solve this problem PSTU and Material and Constructions Test Center of Russian Railways (JSC 'RZD') planned the following test series:

- research of vibrodynamic load impact on ballast strength characteristics (cohesion, friction angle, moduli of deformation and elasticity);
- research of relationships between ballast strength and deformability characteristics and its density, humidity, fouling, accumulated tonnage;
- effect detection of mixing new and cleaned ballast on ballast mechanical characteristics;
- research of ballast deformability in terms of cyclic loading depending from density, humidity, fouling.

Using laboratory tests results and the stated mathematical model bearing capacity dependence on the operating parameters can be derived, such as accumulated tonnage, speed, axle load, undersleeper damping and so on.

These complex research will help solving scientific and practical problem of ballast and subballast bearing capacity and optimize track lifecycle costs.

REFERENCES

Kolos, A. & Morozova, A. 2014. Distribution survey for vibrational acceleration of ballast particles under condition of train traffic with increased axial loads. *Proceedings of Petersburg Transport University* 2(39): 29–35.

Morozova, A. 2014. *Ballast and subballast bearing capacity at railway lines with 300 kN car axle load. Candidate of Engineering Science thesis.* St.Petersburg: PSTU.

Prokudin, I. 1982. *Railway clay subgrade strength and deformability in terms of vibrodynamic load. Doctor of Engineering Science thesis.* Leningrad: LIIZT.

Sokolovsky, V. 1960. *Granular medium statics.* Moscow: State Publishing house of physico-mathematical literature.

Historical sites (TS7)

Challenges and Innovations in Geotechnics – Zhussupbekov (Ed.)
 ISBN 978-1-138-03007-7

The rehabilitation of historic buildings—case study Al-Thakana Al-Hamidieh in Syria

L. Awwad & N. Rahme
Damascus University, Syria

T. Awwad
Department of Geotechnical Engineering, Damascus University, Syria

ABSTRACT: There are many historic buildings that belong to different eras in Syria. Some of them dated back to thousands years ago. These buildings are different from each other, in the architecture, structure and building materials in accordance with the historical era they were built in. So, making a decision about treatment and/or rehabilitation, we should deal with each case separately. It is nearly impossible to make a general rehabilitation rule for all styles of historic buildings. This paper shows some case studies of historic buildings in Syria, the geotechnical and structural problems which cause some architectural flaws and it also goes further in a specific building in Damascus called Al-Thakana Al-Hamidieh or as it was known before Al-Keshleh, which means an accommodation for soldiers in Turkish language. This building was built in 1897. Its foundations and structural elements were made by big rocks which have been brought to the location from Damascus Citadel. The time gaps left an influence on the structure and architectural situation of the building. This building also suffered from a huge fire in 2004 which required rehabilitation. This paper points that supporting of foundations should not affect the style of this part of building, even though it is underground structure; and figures out the possibility of soil improvement by using the same materials which was used in the era of construction.

1 INTRODUCTION

Archaeologists consider Syria the center of one of the oldest civilizations on earth, wherein was the beginning of human settlement, the first urban planning and development of the alphabet. Syria has six World Heritage sites: the Ancient City of Aleppo, the Ancient City of Bosra, the Ancient City of Damascus, the Ancient Site of Palmyra, the Ancient Cities of Northern Syria (JEBEL SEMAN, JEBEL BARISHA, JEBEL AL A'LA, JEBEL WASTANI, AND JEBEL ZAWIYE), and Crac des Chevaliers and Qal'at Salah El-Din. Syria has seen some of the oldest and most important civilizations, which some of them dates back to over 8,000 years BC.

In Syria were the first cities in history, such as the ancient kingdom of Ebla, from 2500 to 2400 BC , the Kingdom of Mary about 2900 BC and Palmyra which flourished in the second half of the first cen-tury BC.

Figure 1. Overview of Syrian World Heritage sites.

2 CONSTRUCTION AND FOUNDATION TYPES OF HISTORICAL BUILDINGS

The heritage monuments include temples, theaters, baths, churches, monasteries, and Islamic monuments from the dawn of Islam to the present day. Most of these monuments are lightly-loaded structures composed of cupolas and stone walls based on strip foundations that seated directly on the soil. Therefore, expansive soils that existed in many parts of Syria can do considerable damage to these structures.

The main reason of soil movement is change in soil moisture which is occurred either artificially by construction works, or naturally as a result of change in rainfall between summer and winter [2].

Figure 2. Ebla Kingdom.

Figure 3. Mary Kingdom.

Figure 4. Cracks in Al-Sinaniya Mosque walls.

3 AL-SINANIYA MOSQUE

Al-Sinaniya Mosque is located in a historical popular area called Bab Al-Jabieh and it is considered one of the important archaeological palaces in Damascus. It was built by the ruler of Damascus Sinan Pacha in 1590.

3.1 *Construction and deformation*

Stone walls, stone arches, stone cupolas, and stone columns were used to make the palace spaces and transfer roofs loads to the foundations. The lime mortar was used to bond the stones. Stone walls and columns are based on strip stone foundations that seated directly on the soil.

It was observed in the recent few years the appearance of cracks in some walls. Engineering analysis and evaluation of this phenomenon stated that, there were two kinds of cracks: First kind: cracks resulted from change in temperatures, often they are vertical and occur in places with continues big diminutions. Second kind: cracks resulted from soil settlement, often they are inclined and not stable in width.

4 SOULIMAN TIKIE

It is a historical huge building in Damascus located in Al-marjeh area, and it was designed by order from the Sultan Sulaiman Alkanonee.

Construction works of Souliman Tikie started in 1554 and the building finished in 1559, after that it was started building a Souliman Tikie school and it was finished in 1566.

4.1 *Construction and deformation*

Souliman Tikie consist of two buildings eastern and western one, both of them were done according to ottoman architecture style.

Western part of Souliman Tikie consists of Tikie surrounded by fence with tree doors, in the middle there is a courtyard surrounded by open rooms covered by small cupolas, to the southern side there is a mosque with two minarets, in front of it there is fountain, to the northern side there are kitchen and storages, between the fence and the building there are gardens everywhere.

Eastern part of Souliman Tikie is a school consists of a big courtyard and a lot of rooms in addition to many open rooms covered by small cupolas, in the northern face of this part there are many shops. This part of Souliman Tikie is completely independent of the western part and the distance between them is about 15 m, this building is now a market of popular arts and traditions.

Repeated movement of the soil in the Souliman Tikie site leads to cracks in the in walls and cupolas in addition to waves (corrugation) in the soil surface under the slabs in the courtyard.

Because of the fact that, there is a phenomenon of reduction in the ground water level in the Barada valley, in the river itself and in Syria in general, therefore

Figure 5. Souliman Tikie part 2.

Figure 5a. Cracks in the cupolas of Souliman Tikie.

Figure 6. Waves (corrugation) in the soil surface under the slabs in Souliman Tikie.

the general direction of the building tends to be settlements as a result of shrinkage in the site soil, so we can see the waves (corrugation) in the soil surface under the slabs in the courtyard of the Souliman Tikie.

Figure 6a. Historical photos of the building and the site

5 AL-THAKANA AL-HAMIDIEH

Al-Thakana Al-Hamidieh is an architectural masterpiece in the heart of Damascus, surrounded by a lot of cypresses. Originally, it was built to be a university to join the ranks of other French universities in the region but it had been used as military barracks till it returned to its basic function.

It is located in an area which is nowadays considered one of the most important junctions of the city that connects many vital parts to each other. Many important buildings have been constructed in the area, for instance: Damascus University Head Quarter, The Opera House. In addition to many future projects are planned to be built nearby like Damascus Towers and Masar Rose.

After the Independence (The end of the French mandate of Syria) in 1946, the building started to be used for its original purpose. So it became the cornerstone of the Damascus University campus. That campus began

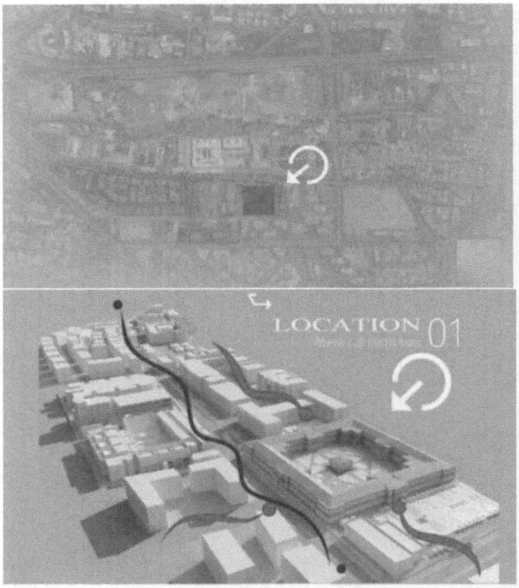

Figure 7. The current location of the building and its relationships to the nearby buildings.

- The blue arrow shows the main movement axis inside the university campus
- The red arrows show the main entrances of the building

to expand year after year till it covered the area illustrated in the pictures (fig.7). The blue shape represents the area of the university campus in 2016. The red shape represents the building area

5.1 *Construction and deformation*

This building was built in 1897. Its foundations and structural elements were made by big rocks which have been brought to the location from Damascus Citadel, while the roof is made by wood and bricks. The same load-bearing walls, that are used to bear the weight of the building, form the architectural spaces. Afterwards, wooden columns were added to the structure to carry the dropped ceiling which has been added to the building to hide the structural elements of the roof.

The time gaps left an influence on the structure and architectural situation of the building. This building also suffered from a huge fire in 2004 which required rehabilitation. Nowadays, there is a plan to turn this building into a university library that requires taking the maintaining of the main structure into consideration.

6 PREDICTED REASONS OF CRACKS AND RESTORATION WAYS

Nowadays most of the historic buildings have been exposed to many factors which had made them crack,

Figure 8. The construction elements of the wooden roof.

that is why they required the strengthening and rehabilitation. Some of the cracks are due to the time-effect and some of them are due to the human misuse. Because of the differences of the structures and the architectural materials, it is nearly impossible to make a general rehabilitation rule for all buildings, so we should deal with each case separately.

6.1 *Reasons of cracks*

Some of predicted reasons of cracks for above mention buildings are:

- Reduction in ground water level in the Valley of Barada River which is close to the Al-Azem Palace .this reduction in ground water level was occurred during the period of 1980s and 1990s.
- Waste water leakage from the near buildings which reached the foundation soil.
- Change in water content of the foundation soil as a result of water leakage from fountains and gardens irrigation, rainfall accumulation and water leakage from the roofs through walls to the foundations.

- Maintenance and rehabilitation measures done to the near building could affect constructions.
- Al-Thakana Al-Hamidieh also suffered from a huge fire.

6.2 *Restoration ways*

The historical character of the heritages and monuments should be kept, and any kind or method of restoration and maintenance must achieve construction safety, keep the nature of the building and not to change its historical characters.

Foundation restoration aims to transfer the building loads to more depth soil layers, which is less exposable to changes in moisture. There are a lot of methods for supporting and restoration of historical building foundation. Current methods of foundation restoration depend on soil improvement under the historical building foundations by cement mortar injection, or foundation supporting by micro piles, and using concert or metal beams under the foundation.

Both mentioned ways of soil improvement and foundation support have disadvantage of using modern material such as metal and concrete , which is not natural and different from the original foundation materials, Therefore we recommend soil improvement by lime , because it is natural material and doesn't detract the building's historical character. Deep soil improvement with lime can be divided into two main types, deep column mixing method and the method of drill hole lime pile or chemical lime pile. (Awwad, 2013).

7 CONCLUSION

Nowadays most of the historic buildings have been exposed to many factors which had made them crack, that is why they required the strengthening and rehabilitation. Some of the cracks are due to the time-effect and some of them are due to the human misuse.

Because of the differences of the structures and the architectural materials, it is nearly impossible to make a general rehabilitation rule for all buildings, so we should deal with each case separately.

Supporting of foundations should not affect the style of this part of building, even though it is underground structure. Instead, we should keep its influence for the next generations. Because of that we figured out the possibility of soil improvement by using the same materials which was used in the era of construction, and getting rid of some modern materials as they have negative effects on the historical value, even though they would be easier to use and have lower costs.

REFERENCES

AAAS, (2016) Ancient History, Modern Destruction: Assessing the Current Status of Syria's World Heritage Sites Using High-Resolution Satellite Imagery. http://www.aaas.org/.

Al-Rawas .A.A and Goosen. M.F.A (2006) Expansive soils recent advances in characterization and treatment .Johnson, H.L. 1965. Artistic development in autistic children. *Child Development* 65(1): 13–16.

Awwad, T. (2013). "The Role of Soil Improvement for Conservation of Syrian Historical Monuments". Proceedings of Workshop ATC19, 18th ICSMGE 18th ICSMGE World Congress, Paris, September, 2013. PP 89–94.

Awwad, T. (2014). "Analytical Soil-Structure Interaction Study Of a Historical Building Located Near A River Bed (Damascus Citadel Tower Near Banias River)", Journal of AL Baath University-Syria-Homs, Vol. (36). No. (1) 2014. P 153–168.

Dietz.K and Schürmann .A (2006). "Foundation Improvement of historic buildings by micro piles, Museum Island, Berlin and St. Kolumba, Cologne." 7th ISM workshop, Schrobenhausen 2006,

National Lime Association. (2001).Using lime for soil stabilization and modification. USA.

Study of Souliman Tikie restoration and maintenance project (2008) – Ministry religious issues – Syria.

Wong.P.K (2001). Ground improvement case studies chemical lime piles and dynamic replacement. Malaysia and Egypt.

Zhussupbekov,A. Zhunisov,T. Issina, and A. Awwad, T. (2013). "Geotechnical and structural investigations of historical monuments of Kazakhstan". Proceedings of Second International Symposium on Geotechnical Engineering for the Preservation of Monuments and Historic Sites, Naples, Italy, 2013. PP 779–784.

Challenges and Innovations in Geotechnics – Zhussupbekov (Ed.)
© 2016 Taylor & Francis Group, London, ISBN 978-1-138-03007-7

Engineering-geological conditions and analysis of deformation of historical monuments in Kazakhstan

A.Z. Issina, S.Sh. Sadykova, Zh.S. Bisenova & A.S. Temirova
Gumilyov Eurasian National University, Astana, Kazakhstan

ABSTRACT: This paper considered, climatic, geotechnical, hydrogeological characteristics of the studied objects, their constructive decision and the analysis of deformations of elevated and underground construction designs. Main reasons for deformations of monuments in Kazakhstan: uneven rainfall of soil of the basis in the most overloaded sites of designs of the building; seismic influences; violation of temperature moisture conditions and moisture; anthropogenous influence (especially notable in the historical cities of Samarkand and Bukhara where monuments of architecture are influenced by vibrations from traffic).

Keywords: Monument, Deformations, Ground waters, Heritage, Foundations

1 INTRODUCTION

The history of our ancient land leaves deep into the millennia. Holding an advantageous geographical position, the connecting North with the South, the East with the West, Central Asia was the important center on the road of a caravan which became history under the name of the Great Silk Way. On branches of this ancient transcontinental highway not only trade developed – there was an active process of mutual enrichment of ideas, cultures, traditions, religions, crafts and technologies.

Considering all importance of the huge cultural heritage, which got to us carrying out numerous researches and monitoring of technical condition of significant monuments of architecture of the Central Asian region for the purpose of its preservation is necessary for descendants.

One of the types of the works directly concerning cultural heritage, demanding greater financial influences, but which aren't receiving the due amount of financing is carrying out engineering-geological and geotechnical researches of historical monuments of architecture of Central Asia.

At the present stage the majority of monuments of architecture of the countries of Central Asia face a problem of destruction of their foundations under the influence of climatic factors. One of the most common causes of deformations is uneven rainfall which, in turn, cause deformations and destructions of the bearing designs – the foundations, walls, columns, overlappings, the arches, crossing points window and doorways.

2 GEOTECHNICAL, HYDROGEOLOGICAL AND CONSTRUCTIVE CHARACTERISTICS OF MONUMENTS OF ARCHITECTURE OF CENTRAL ASIA

2.1 *The mausoleum of Hodge Ahmed Yassavi (XIV-XV cen.) (Figure 1)*

Soils. The prolyuvial loams with a general power of 5–7 m quarternary allyuvial containing form the basis of a monument interlay also lenses of sandy loams. They are spread water containing gravel with the power from 4 to 24 m. Sandy loams are developed mainly in the top part of a section. Sandy loams of light brown color, macroporous, loessial, uniform. Brown color dense loam with inclusion of carbonate particles. Sand dusty, meets in the form of lenses, mainly on contact with a galechnik, color its gray and brown, uniform. Pebble soil is presented by sedimentary breeds, fragments well rounded, so-so thickened and extended, as filler sand serves.

Ground waters. Ground waters for June, 2011 are opened with developments at a depth of 7.6–7.8 m from an earth surface. In relation to situation level of ground water (LGW) for September, 1997, LGW in June, 2011 is recorded 1.4–1.5 m below.

Foundations. The main part of a construction is built on clay gulfs, in South side under portal part the rubble foundations up to 7 m in depth are executed.

In 1993 the technology of jet cementation of soil was applied to strengthening of the basis of a monument by the Turkish company. Strengthening and the device of piles was carried out to two stages. At the first

Figure 1. The mausoleum of Hodge Ahmed Yassavi.

stage of work were carried out for walls to 2 m. At the second stage strengthening of the basis of walls up to 4 m thick was made. Total length of a pile is about 15 m. Thus, the support of all construction through piles on thickness of solid soil is reached.

Deformations. Due to the lack of emergence of essential deformations on a monument to the first cycle it wasn't given due consideration though at this moment there was a deep soaking of soil and the basis of the foundations of monuments of the complex (Figure 2). The second cycle of rise in level of ground waters caused already external deformations of the building of a monument, in December, 1982 there was a destruction of the top restoration number of stalactites of a dome of the Main Hall – Kazanlyk.

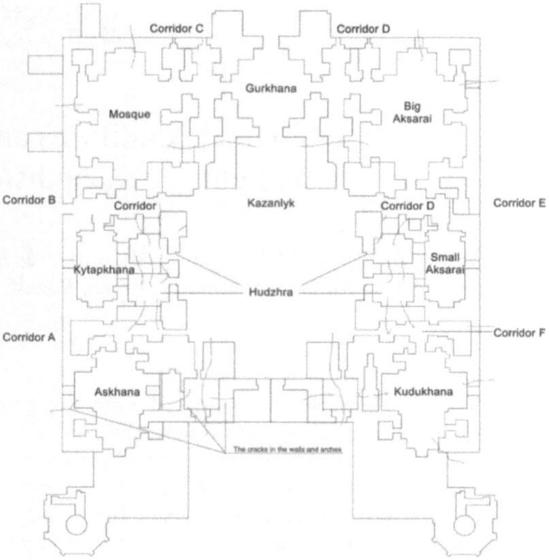

Figure 2. Scheme of deformations of the monument. Plan of the 1st floor.

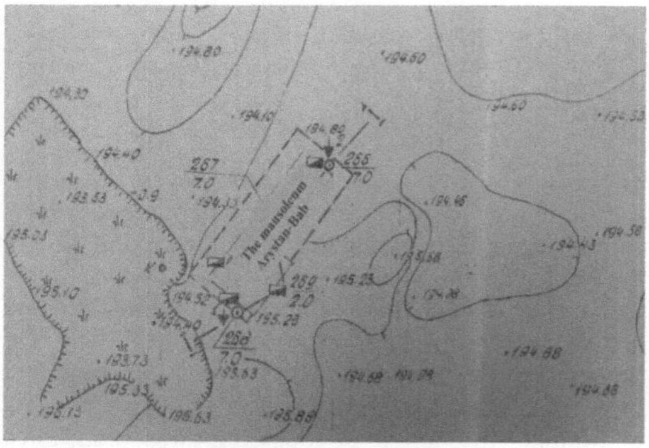

Figure 3. Plan of a platform of the mausoleum Arystan-Bab (1984).

2.2 *The mausoleum of Arystan-Bab (XIV-XV cen.)*

Soils. The analysis of average values physic mechanical property of soil on holes and an additional driving of holes allow directly under a sole without foundations walls and the bases with depth to 1.0 m to allocate 3 engineering-geological element (EGE) (Figure 3 and Figure 4):

1. EGE – loam of a firm-semi-firm consistence, not collapsible, uniform, with rare inclusions of fragments of a brick and the vegetable remains in the top part of a layer, 1.0–1.5 m (0.9–1.0 m – on wells), it is possible to assume that it is the loam layer, or the layer executed by method of "gulf" in a trench width exceeding thickness of walls which is artificially condensed in an open trench; it possesses quite certain and rather close indicators of physic mechanical properties 2 EGE;
2. EGE – loam dark brown a firm-semi-firm consistence, not collapsible, with roots of vegetation, 3.3–3.6 m (on wells);
3. EGE – sandy loam brown, a plastic and fluid consistence with pro-layers of sand dusty (to 0.2 m) with an opened power of 3.5–3.7 m (Researches, 1984).

Ground waters. LGW near a contour of the monument is opened at depths of 1.9–2.7 m (August, 1984), 2.75–2.85 m (May, 2004) and 3.14–3.37 m (August, 2004) (Report, 2004).

Foundations. The foundations are arranged under the most loaded parts of a construction. Under minarets the foundations are executed from a stone-plitnyak on clay solution. Materials of the bases are in a good shape.

On a hole No 10 up to 1.5 m in depth, in a place of a joint of a longitudinal wall on axis A and the left minaret, lack of the foundation under a wall which laying leans on three rows of preparation from a detrital brick is revealed (Figure 5). Their basis is the uniform layer of loam of the first EGE.

The hole No 11 from the outer side of a wall on axis 4 up to 0.7 m in depth, opened the concrete plate 17–20

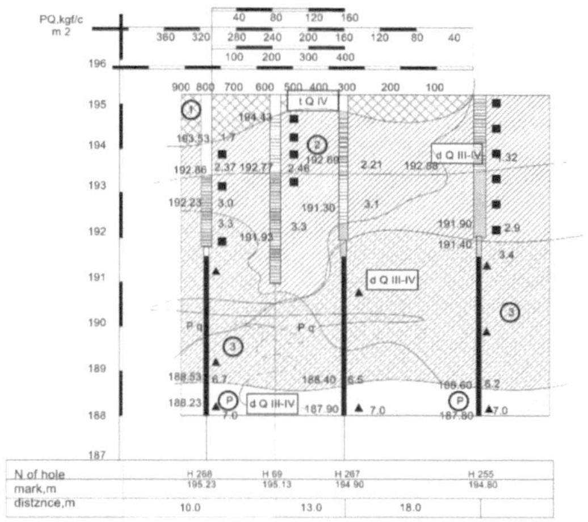

Figure 4. Engineering-geological section of a platform (1984).

Figure 5. Hole No 10. A joint without base wall (axis A) with the left minaret; the minaret base, the uniform basis under a wall.

Figure 6. Hole No 11. Concrete plate on soil; the old foundation under it isn't found.

Figure 7. The modern view of the mausoleum Arystan-Bab.

thick see above it the layer of dry, fragile roofing material, and over it one more concrete layer about 9 cm thick, revetted outside with a thin facing tile is found. Over this layer there is a wall bricklaying (Figure 6).

Deformations. In a laying of walls cracks on all facades are noted. During observation it is established that width of disclosure of cracks makes from 0.1 to 2 mm. In a place of interface of a minaret to a wall (a northeast facade) the crack reaches 2 cm and goes mainly on laying seams. All cracks in a laying come to an end, without reaching a socle (Figure 7).

Cracks, generally in places of reduction of thickness of walls settle down. Indoors tombs No 1 are noted the deformations in a laying menacing to safety of a construction. These are cracks and smashing of a brick in a laying of basic part of arches, split of a brick in places of support of angular arches in northeast and northwest walls. In a northeast wall of a crack in a laying of basic part of arches have extent to 2 m (Issina & Zhussupbekov, 2015).

3 CONCLUSIONS

1. The bases of historical monuments of Central Asia are presented in lump by collapsible loams, in rare instances not collapsible. However and in this case the increase in porosity and coefficient of porosity, reduction of number of plasticity is a consequence of washout of loam the water stream and carrying out filtering through it thus of the smallest clay particles and fluctuation of level of underground waters.
2. In the course of complex research of historical monuments of architecture of Central Asia the main reasons for an unsatisfactory condition of some constructive elements of a monument are established:

   1) uneven rainfall of soil of the basis in the most overloaded sites of designs of the building;
   2) seismic influences;
   3) violation of temperature moisture conditions and moisture.
   4) anthropogenous influence.

3. Engineering-geological researches have to be directed on obtaining such information which could promote development of measures for elimination of the reasons leading to development of deformation of monuments.
4. Necessary materials are among data on results of purposeful engineering-geological researches; these surveys of design features of a monument, types of the foundations and about distribution of loads of soil of the bases; data on nature of deformations and speed of their development; materials of supervision over shrinkage of a monument or its elements.

REFERENCES

The hydrogeological researches and the project of removal of ground waters from the mausoleum Arystan-Bab in the village Kogam of the Kzyl-Kum area of the Shimkent Region. 1984. Kzyl-Orda.

Report on engineering-geological researches on object "Mausoleum Arystan-Bab, YuKO". 2004. Shimkent.

Issina A.Z., Zhussupbekov A.Zh. Analysis of geotechnical properties of soils and ground water of Mausoleum Arystan-Bab in South Kazakhstan. *Proceedings of 6th International symposium on Disaster Mitigation in Special Geoenvironmental Conditions. – Chennai, India, 2015:* 277–280.

Challenges and Innovations in Geotechnics – Zhussupbekov (Ed.)
© 2016 Taylor & Francis Group, London, ISBN 978-1-138-03007-7

Strengthening of hydraulic structures on undermined territories

A.K. Aldungarova, M.K. Kuderin & V.A. Kozionov
Pavlodar State University "S. Toraighyrov", Pavlodar, Kazakhstan

ABSTRACT: A series of tests with models of the dam (reinforced and unreinforced) to compare the stability of structures with horizontal and vertical deformations. The results of the experiment are shown in the form of photos and graphics. The objects of study in this article are the deformation model of the dam, located on the subgrade. Purpose of this research a comparison of admissible deformations models earth dams, reinforced and non-reinforced, with the partial collapse of the base and at the same time the horizontal and vertical strains of dam.

1 INTRODUCTION

Relevance of hydraulic engineering structures reliability problem is quite important nowadays. The reason for this is a number of regular negative accidents currently happening throughout the world. These are various catastrophes related to dam collapses and bursts, which have a large percentage of wear at the moment due to their long-term service lives. The issue is interesting not only for specialists, but also for Kazakhstani and foreign researchers. Sufficient amount of hydraulic engineering constructions reinforcement variations have been proposed for the time being. But unfortunately the problems of soil dams protection from hazardous impact of different factors such as seismic or dynamic actions, foundation undermining, floods, underground mining activities, etc., in Kazakhstan still remain reasonably complex and poorly explored. In theory, there is a lot of suggestion on hydraulic engineering constructions reinforcement, but in reality there is not enough implementations about this critical problem. As a proof, disasters like soil weirs undermining with human deaths still take place despite the fact that we live in the 21st century. This article examines behavior of soil dam model in different variations of crack resistance deformations. Several model trials were carried out. One of them is an experimental investigation of reinforcement effects on stability of dam model impacted by horizontal deformations of soil undermining foundation.

2 EXPERIMENTS ON THE STAND

Experiments were carried out with the aid of three-dimensional test stand (Fig. 1). The test stand (three-dimensional) for soil dam prototype deformations modeling is made in the form of separate U-shaped cross-sections (1). Elastic rubber pads of thickness = 10 mm are installed between the sections. Side ledges of U-shaped sections (1) are equipped with bolted joints (3) in the upper and lower levels horizontally. There are end face walls (4) in a cradle. The lower part of U-shaped sections (1) isequipped with adjustable footings (5), made in the form of roller supports, installed on a bed frame (6).

The test stand for deformations modeling [Zhussupbekov & Bazarov (1991)] operates in the following way:

U-shaped sections (1) compression or tension is conducted with the aid of bolted joint (3) together with material deformation in the cradle. Horizontal deformations of soil tension occur due to compressed elastic (rubber) pads (2) flexible strain forces by loosening bolted joints (3). Horizontal deformations of soil compression occur due to elastic (rubber) pads (2) compression by bolted joints (3) pulling U-shaped sections (1) closer to each other. Vertical deformations occur due to a step-by-step lowering of U-beams (7) installed before the experiment start in accordance with junctions A and B (Fig. 1).

As a material for dam and soil foundation model a mixture consisting of 97% fine silica sand and 3% straw oil by weight was chosen. The mixture has strong cohesion which enables to make prototypes of cohesive soils [Zhussupbekov (1994)].

In order to determine mechanical strength and deformability properties of actual soils and equivalent materials under the vertical loadings a compression kind of stabilometer was applied for horizontal deformations taking place in undermining conditions.

Sample put into compression device is matured till full consolidation under the given loading equal to 0.3 MPa.

To applied weightings of vertical loadings in 0.05–0.1 MPa limits.

Vertical deformations of soil sample were measured by clock-face type indicators with scale interval equal to 0.001 mm. Transmission of vertical loadings to the sample was conducted by weighting mechanisms

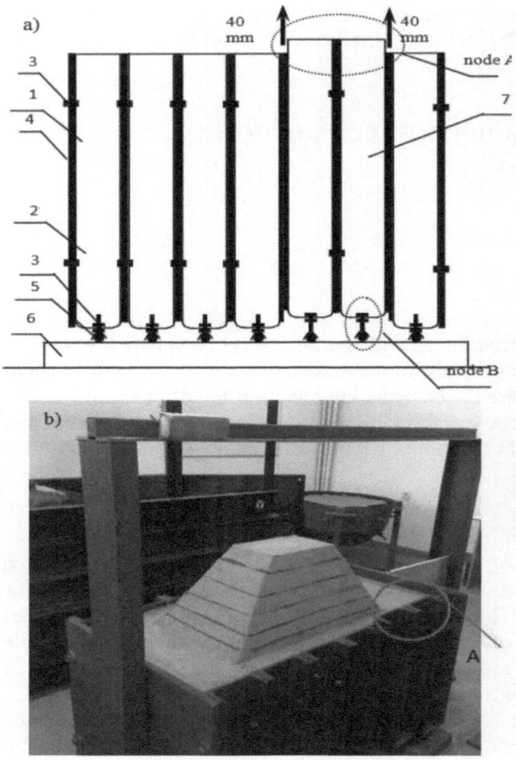

Figure 1. Three-dimensional test stand for deformations modeling of soil foundation: a – 3D test stand with soil dam model image, b – 3D test stand diagram (plane view).

Table 1. Physical parameters of full-scale and modeled dams.

Type of soil	γ (kN/m^3)	c, (kPa)	$\varphi(°)$, (deg)	E (MPa)	υ
Full-scale soil					
1. Loam	20,5	40	22	20	0,3
Model of the dam (model soil)					
2. Sand – 97% + 3% – spindle oil	17,7	0,90	39	0,27	0,25

through DOCM-3-5 dynamometer. The pressure was measured by pressure-gauge.

Required parameters are obtained by the results of the testing trials E, c, φ, γ (Table1).

Substitute the corresponding values for modeled and full-scale soil into the equation (1) and obtain linear scale of modeling.

$$m_c = c_m/c_n \times \gamma_n/\gamma_m = 0{,}9/40 \times 20{,}05/17{,}7 = 1/40 \quad (1)$$

Hence, linear scale of model and full-scale object (buildings, foundations, structures) is calculated as a proportion of strength properties (cohesion) of clay and equivalent material and equals 1:40.

Figure 2. Dam model level-by level placement without reinforcement.

Figure 3. Reinforced dam model level-by-level placement.

As a soil dam model an embankment with the corresponding dimensions was chosen (Fig. 1a):

- 700 mm*350 mm (dam model foundation);
- 430 mm (dam model height).

a) Foundations placement

Before laying a soil foundation, test stand should be installed in such a way so that in the future 1/3rd part of dam model footing was placed to the foundation, lifted to a certain distance with the aid of U-shaped sections (Fig. 1, nodes A, B). The beams (Fig. 1–7) are uplifted by bolts to 40 mm.

After preparation of equivalent material foundation can be placed to the 3D test stand. Equivalent material was arranged in layers of 7 cm and was compacted by a rolling press (7 full compaction cycles).

During the foundation preparation process mechanical strength of material should be checked carefully.

b) Dam model placement without reinforcement (Fig. 2), level-by-level placement in 6 layers of 7 cm plus compaction. Colored sand of thickness equal to ≈2 mm was placed between each layer. A soil sample of each layer should be taken to determine soil density.

c) Reinforced embankment model preparation (Fig. 3) is carried out in a way of level-by-level placement in 6 layers of 7 cm plus compaction. An embankment was placed with the aid of special shape. Colored sand of thickness equal to ≈2 mm was placed between each layer.After arrangement of each layer plus coloured sand, a reinforcing net of area equal to dam model's piling layer area was installed.

A soil sample of each layer should be taken to determine soil density.

d) The average density of foundation soil and soil of dam model with/without reinforcement without consideration of aluminum cup weight ($\rho_m = 13$), (g/cm^3) is presented in Table 2.

Table 2. Average density of foundation soil and soil of dam model with/without reinforcement.

	Average density (ρ (g/cm^3))					
Layer	1	2	3	4	5	6
Foundations w/o reinforcement	118	119	119	120	119	
Foundations with reinforcement	119	119	118	119	118	
Dam model w/o reinforcement	119	119	118	119	119	118
Dam model with reinforcement	119	120	119	119	118	119

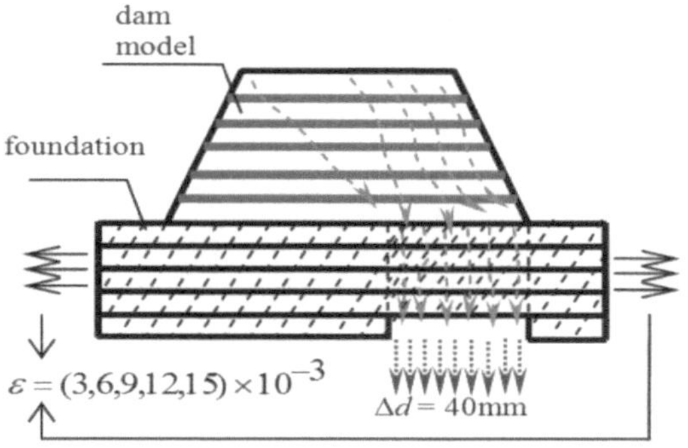

Figure 4. Horizontal and vertical deformations effect on dam model functioning diagram.

The process of dam model cracks, deformations and failure development in condition of both horizontal tension and vertical deformation happening in a soil foundation at the same time can be observed and fixed with the usage of a photo camera.

An invention and development of digital photography allowed to scheme out a contactless photogrammetric method of prototype systems cracks and other deformations lifecycle monitoring.

Vertical and horizontal deformations of embankment foundation and model during experiments conduction process were obtained using photogrammetric approach. This method helps to determine deformations which occur in plane and are useful to examine flat objects. The method assumes that several images of prototype system can be obtained from one fixed point, e.g. first image obtained before deformations, second – during deformations development and the third – after deformations. Thus,camera should be installed in such a way that plane of applied frame was parallel to plane of an object where image orientation elements should be preserved. In this case a periodic shoot by equipment with high matrix resolving capacity (2000 pixels per 1 cm^2) should be implemented.

In the given article Canon EOS RebelT3/DS126291 camera with matrix resolution equal to 12,2 MP was used. The shooting data was recorded for documentation of mechanical measurements at dam model slopes and crest.

The task was to examine model stability in 5 stages of horizontal deformations $\varepsilon = (3, 6, 9, 12, 15) \times 10^{-3}$ and simultaneous vertical failure (Fig. 4), using bolted joints to assemble soil foundation part with dam model in variations with and without reinforcement in order to determine conditions of embankment's critical state [Tanaka, Zhussupbekov & Aldungarova 2014].

A 3D test stand allows to create independent tension and vertical uplift lowering deformations in a significant range.

The following trial series were carried out:

(a) Testing of dam model at different conditions of soil foundation part's simultaneous lowering and horizontal tension of foundation without preliminary reinforcement.

Figure 5. Comparison of stability modeling key stages for dam model (a) without reinforcement; b) with reinforcement).

(b) Testing of dam model at different conditions of soil foundation part's simultaneous lowering and horizontal tension of foundation with reinforcement.

In accordance with accepted model investigations programmer the following trial series were conducted:

(I) foundation deformation in horizontal direction without dam model reinforcement up to $\varepsilon = 3 \times 10^{-3}$ and vertical lowering of soil foundation to $\Delta d = 8$ mm;

(II) foundation deformation in horizontal direction without dam model reinforcement up to $\varepsilon = 6 \times 10^{-3}$ and vertical lowering of soil foundation to $\Delta d = 16$ mm;

(III) foundation deformation in horizontal direction without dam model reinforcement up to $\varepsilon = 9 \times 10^{-3}$ and vertical lowering of soil foundation to $\Delta d = 24$ mm;

(IV) foundation deformation in horizontal direction without dam model reinforcement up to $\varepsilon = 12 \times 10^{-3}$ and vertical lowering of soil foundation to $\Delta d = 32$ mm;

(V) foundation deformation in horizontal direction without dam model reinforcement up to $\varepsilon = 15 \times 10^{-3}$ and vertical lowering of soil foundation to $\Delta d = 40$ mm;

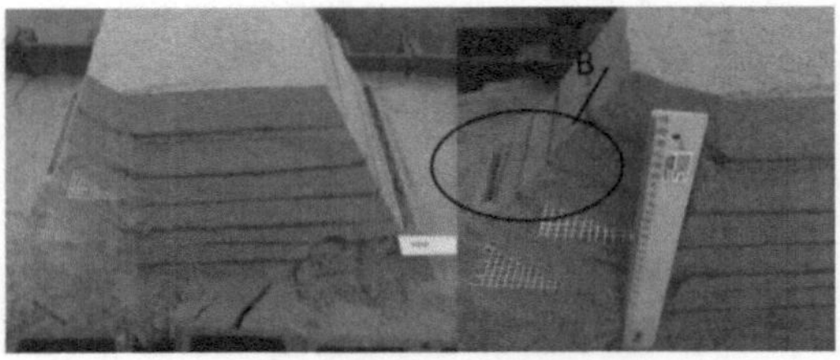

Figure 6. Reinforced dam model shift.

(VI) foundation deformation in horizontal direction with dam model reinforcement up to $\varepsilon = 3 \times 10^{-3}$ and vertical lowering of soil foundation to $\Delta d = 8$ mm;

(VII) foundation deformation in horizontal direction with dam model reinforcement up to $\varepsilon = 6 \times 10^{-3}$ and vertical lowering of soil foundation to $\Delta d = 16$ mm;

(VIII) foundation deformation in horizontal direction with dam model reinforcement up to $\varepsilon = 9 \times 10^{-3}$ and vertical lowering of soil foundation to $\Delta d = 24$ mm;

(IX) foundation deformation in horizontal direction with dam model reinforcement up to $\varepsilon = 12 \times 10^{-3}$ and vertical lowering of soil foundation to $\Delta d = 32$ mm;

(X) foundation deformation in horizontal direction with dam model reinforcement up to $\varepsilon = 15 \times 10^{-3}$ and vertical lowering of soil foundation to $\Delta d = 40$ mm;

After each trial soil was extracted from the tray and a new foundation was prepared for the following trial series.

Figure 5 shows comparison of dam model stability modeling key stages, as demonstrated on a 3D test stand with the usage of equivalent material.

3 CONCLUSIONS

Following the experiment results, conclusions can be made.

As it can be observed, the crest of a reinforced dam has remained in original state, without cracks which is different to dam model without reinforcement.

Cracks on an unreinforced dam model begin spreading parallel all over the shape with the very first seconds of deformations whether development of cracks on a reinforced model occurs under the reinforcement netting bed (Fig. 5).

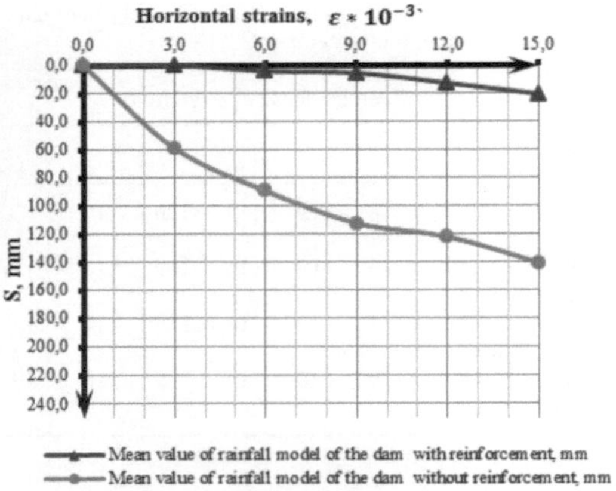

Figure 7. Reinforced/unreinforced dam model stability dependence on horizontal and vertical deformations plot.

Model of a reinforced dam is more prone to shear than to crack formation and collapse; the upper part of model placed above the reinforcement net remained in initial state, without cracks in fact, under the condition of dam model shift to 2 cm (Fig. 6, node B).

It can be concluded from the plot (Fig. 7) that strengthening of dam model with the use of reinforcement net significantly affected its stability at horizontal and vertical deformations. Given choice of strengthening can be applied for hydraulic engineering structures as one of the methods to increase stability and safety.

3D test stand allows to examine behavior of dam model with/without reinforcement impacted by different combinations of soil foundation deformations.

REFERENCES

Patent 1250808. 1991. *Test stand for undermining buildings' foundations deformations modeling*/Invented by Zhussupbekov A.Zh., Bazarov B.A., published in B.E. No 48.

Tanaka T., Zhussupbekov A.Zh., Aldungarova A.K.. 2014. *The influence of the stress-strain state of the soil on the stability of the dam model*. "Perspective trends of theory and practice development in soil rheology and mechanics", XIV Soil rheology International Symposium materials, KazGASU–Kazan.

Zhussupbekov A.Z. 1994. *Structural properties of buildings' foundations in undermined areas*. - Almaty: Gylym, – p. 162.

Challenges and Innovations in Geotechnics – Zhussupbekov (Ed.)
© 2016 Taylor & Francis Group, London, ISBN 978-1-138-03007-7

Author index